Clinical Prognostic and Predictive Biomarkers

Clinical Prognostic and Predictive Biomarkers

Editors

Yuli Huang
Yong Yuan
Peisong Chen

MDPI

Basel • Beijing • Wuhan • Barcelona • Belgrade • Novi Sad • Cluj • Manchester

Editors

Yuli Huang
Shunde Hospital, Southern Medical University
Foshan
China

Yong Yuan
Zhongshan City People's Hospital
Zhongshan
China

Peisong Chen
The First Affiliated Hospital, Sun Yat-sen University
Guangzhou
China

Editorial Office
MDPI
St. Alban-Anlage 66
4052 Basel, Switzerland

This is a reprint of articles from the Special Issue published online in the open access journal *Diagnostics* (ISSN 2075-4418) (available at: https://www.mdpi.com/journal/diagnostics/special_issues/Clinical_Biomarkers).

For citation purposes, cite each article independently as indicated on the article page online and as indicated below:

Lastname, A.A.; Lastname, B.B. Article Title. *Journal Name* **Year**, *Volume Number*, Page Range.

ISBN 978-3-7258-0587-7 (Hbk)
ISBN 978-3-7258-0588-4 (PDF)
doi.org/10.3390/books978-3-7258-0588-4

Contents

About the Editors

Yuli Huang

Dr. Yuli Huang is a professor of cardiology at Shunde Hospital, Southern Medical University. He is also a senior honorary research fellow at the George Institute for Global Health, Australia. He majored in pathogenesy and therapeutics for cardiometabolic disease; risk factors of cardiovascular disease; epidemiology of cardiovascular disease; and metabolic therapy for heart failure. As the leading author, he has published more than 100 publications including in *The BMJ* (3 papers), *Neurology* (2 papers), and *BMC Medicine*. The total citations of Dr Huang's publications are up to 6000 with an H index of 33 as of Feb 2024. Fifteen of his publications have been cited by multiple international clinical guidelines, including the 2017ACC/AHA hypertension guideline, 2018 ESC hypertension guideline, and ADA diabetes guideline (from 2019 to 2026, six editions). Dr Huang is a Member of the Committee of the Chinese Cardiovascular Association; Editorial Board Member of the *Journal of Geriatric Cardiology*; Guest-Associated Chief Editor of *Frontiers in Cardiovascular Medicine, Frontiers in Bioscience-Landmark; Reviewer for The Lancet Diabetes & Endocrinology, EHJ, BMJ, International Journal of Epidemiology*, and *American Journal of Kidney Disease* (100+ SCI journals on Publons record).

Yong Yuan

Professor Yong Yuan is a chief physician of cardiology at Zhongshan People's Hospital, Sun Yat-sen University. He serves as the vice president of Guangdong Clinical Medical Association, vice president of Guangdong Hospital Association, and as a member of the emergency and critical care group of the fifth committee of the Cardiovascular Physician Branch of Chinese Medical Doctor Association. He has been engaged in cardiovascular medicine for more than 20 years and has rich clinical experience in the prevention, diagnosis and treatment of cardiovascular diseases, especially in the rescue and treatment of cardiovascular emergencies. He was the first to apply extracorporeal membrane oxygenation (ECMO) to the rescue of cardiac emergencies, the first to put forward the standards for the use and withdrawal of ECMO, and the first to actively promote the popularization and application of ECMO technology in the province and the country.

Peisong Chen

Peisong Chen is a PhD in clinical laboratory diagnostics from Sun Yat-sen University and is a chief technician and master tutor. He has been working at the Clinical Laboratory of the First Affiliated Hospital of Sun Yat-sen University since 2010, accumulating extensive experience in clinical biomarkers, laboratory diagnostic technology, infectious disease and big data research. With a track record of over 50 SCI articles published, he has successfully led and contributed to numerous national and provincial projects.

Preface

Clinical biomarkers are various measurable biological variables that can be detected in organ tissues, blood or other body fluids. Currently, biomarkers can be divided into two main types: prognostic and predictive biomarkers. Prognostic biomarkers are associated with the clinical outcomes of the disease of interest (e.g., disease progression and relapse, death) and are mainly used to identify more aggressive disease states. Predictive biomarkers are used to identify individuals who respond better to a particular treatment, thereby better identifying those patients who are more likely to benefit from the treatment regimen. Usually, a clinical biomarker can be divided into prognostic or predictive markers, but, in some cases, some biomarkers can be used to predict both prognosis and treatment response. With great progress in proteomics, metabolomics, functional genomics, and bioinformatics, increasing numbers of new biomarkers are being discovered. They are of great significance in multiple fields of clinical medicine, including cardiovascular disease, diabetes, and cancer, as they play an important role in identifying high-risk individuals, diagnosing disease states, and predicting treatment response and prognosis. In addition, the discovery of new biomarkers has enabled us to better understand the mechanisms and molecular pathways of disease occurrence and development, and the in-depth exploration of biomarkers has also provided more perspectives for the development of new precision and personalized treatments.

Yuli Huang, Yong Yuan, and Peisong Chen
Editors

diagnostics

MDPI

Article

Second-Line Treatment of Metastatic Renal Cell Carcinoma in the Era of Predictive Biomarkers

Andreea Ioana Parosanu [1,2], Catalin Baston [3,4,*], Ioana Miruna Stanciu [1,2], Cristina Florina Parlog [1,2] and Cornelia Nitipir [1,2]

1 Department of Medical Oncology, Elias Emergency University Hospital, 011461 Bucharest, Romania; andreea-ioana.parosanu@drd.umfcd.ro (A.I.P.); ioana-miruna.stanciu@drd.umfcd.ro (I.M.S.); cristina-florina.pirlog@drd.umfcd.ro (C.F.P.); cornelia.nitipir@umfcd.ro (C.N.)
2 Department of Oncology, Faculty of Medicine, "Carol Davila" University of Medicine and Pharmacy, 050474 Bucharest, Romania
3 Department of Urology, Fundeni Clinical Institute, 022328 Bucharest, Romania
4 Department of Urology, Faculty of Medicine, "Carol Davila" University of Medicine and Pharmacy, 050474 Bucharest, Romania
* Correspondence: catalin.baston@umfcd.ro; Tel.: +40-765210001

Abstract: Background: Over the past few years, significant advancements have been achieved in the front-line treatment of metastatic renal cell carcinomas (mRCCs). However, most patients will eventually encounter disease progression during this front-line treatment and require further therapeutic options. While treatment choices for mRCCs patients are determined by established risk classification models, knowledge of prognostic factors in subsequent line therapy is essential in patient care. Methods: In this retrospective, single-center study, patients diagnosed with mRCCs who experienced progression after first-line therapy were enrolled. Fifteen factors were analyzed for their prognostic impact on survival using the Kaplan–Meier method and the Cox proportional hazards model. Results: Poor International Metastatic RCCs Database Consortium (IMDC) and Memorial Sloan-Kettering Cancer Center (MSKCC) risk scores, NLR value > 3, clinical benefit < 3 months from a therapeutic line, and the presence of sarcomatoid differentiation were found to be poor independent prognostic factors for shortened overall survival. Conclusions: This study provided new insights into the identification of potential prognostic parameters for late-line treatment in mRCCs. The results indicated that good IMDC and MSKCC prognostic scores are effective in second-line therapy. Moreover, patients with NLR < 3, no sarcomatoid differentiation, and clinical benefit > 3 months experienced significantly longer overall survival.

Keywords: mRCCs; late-line treatment; prognostic factors; risk classification

Citation: Parosanu, A.I.; Baston, C.; Stanciu, I.M.; Parlog, C.F.; Nitipir, C. Second-Line Treatment of Metastatic Renal Cell Carcinoma in the Era of Predictive Biomarkers. *Diagnostics* **2023**, *13*, 2430. https://doi.org/10.3390/diagnostics13142430

Academic Editors: Yuli Huang, Yong Yuan and Peisong Chen

Received: 25 June 2023
Revised: 13 July 2023
Accepted: 14 July 2023
Published: 20 July 2023

1. Introduction

Kidney cancer is not just one disease but one with great complexity. Renal cell carcinomas (RCCs) are a heterogeneous group of disorders with unique distinct morphological and biological profiles. The 5th edition WHO Classification of Urinary and Male Genital Tumors now recognizes more than seventeen RCC subtypes. Clear cell renal cell carcinoma (CCRCC) is the most common histological subtype of kidney cancer, which comprises 65–70% of all RCCs [1].

RCCs rank as the 14th most frequently diagnosed cancers worldwide with approximately 430,000 new cases and nearly 180,000 associated deaths estimated in 2020. Its incidence and mortality trends have changed over time [2]. Over the past two decades, the incidence of RCCs increased by 1–2% annually. In contrast to the rising incidence rates, the overall mortality rate from RCCs has decreased from 69.71% in the early 1990s to 38.96% in the late 2010s. This improvement can be attributed to significant advancements in treatment [3–5].

Currently, approximately one-third of RCCs cases are diagnosed with metastatic disease and a poor five-year survival rate of 17% [6,7]. Unfortunately, patients diagnosed with advanced unresectable or metastatic RCCs often experience disease progression during front-line treatment, and only 60% of them survive long enough to receive second-line therapy. As a result, there has been a recent introduction of numerous second-line treatment regimens for mRCC with promising results (Figure 1) [8].

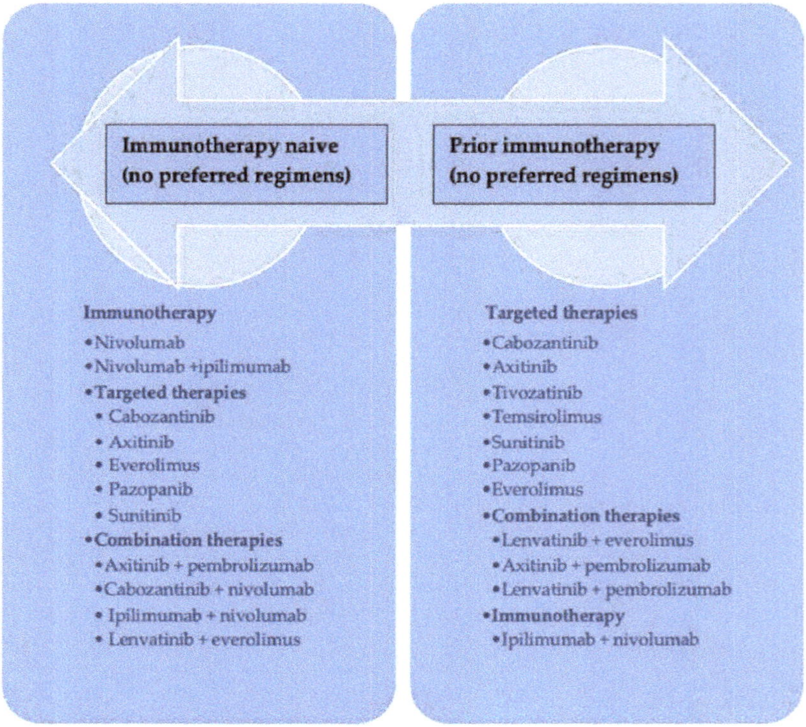

Figure 1. Algorithm for the second-line treatment selection for mRCCs.

The treatment landscape is rapidly evolving, and there are currently several therapeutic options available for second-line therapy, including dual immune checkpoint inhibition (ICI) or a combination of immunotherapy ICIs and antiangiogenic tyrosine kinase inhibitors (TKI) [9].

The second-line therapeutic approaches are determined by the mechanisms of acquired resistance to first-line regimens and require alternative treatment options.

For patients who experience progression on immunotherapy as their first-line treatment, a TKI inhibitor should be considered a second-line therapy option. Conversely, if an ICI plus TKI combination was utilized as the initial therapy, patients should receive a regimen that includes a different TKI either in monotherapy or in combination with immunotherapeutic agents [10–12].

Despite recent advancements in mRCC, there is limited knowledge regarding prognostic parameters for second-line therapy. Several prognostic models have been refined over time [13–17]. One of the most widely used models is the Motzer score or Memorial Sloan Kettering Cancer Center (MSKCC) risk model, initially developed in the cytokine era and validated in molecular-targeted therapy [18,19]. Similarly, the International Metastatic Renal Cell Carcinoma Database Consortium (IMDC) model, proposed by Heng et al., was developed and validated for patients receiving first-line antivascular endothelial growth factor therapy. Most recently, the IMDC model demonstrated prognostic value in patients receiving immune checkpoint inhibitors [20–22].

The IMDC and MSKCC models for prognosis in mRCCs and the front-line therapy administered have been identified as the only prognostic factors in patients receiving second-line therapy [23–25]. Previous reports indicated that patients with mRCC and early progression on the front-line regimen had a significantly higher probability of not receiving subsequent therapy. Moreover, there is good evidence that the timing of treatment benefits may serve as a reliable surrogate endpoint for progression-free survival [26,27].

There are several reasons why inflammation is a recognized hallmark of cancer (Figure 2) [28]. First, increased neutrophils can lead to excessive inflammation and a pro-inflammatory condition. Second, neutrophils can reduce lymphocyte T proliferation and inhibit antitumor T-cell responses. Lastly, lymphocyte depletion reflects a defective host immune system that fails to control tumor occurrence and growth.

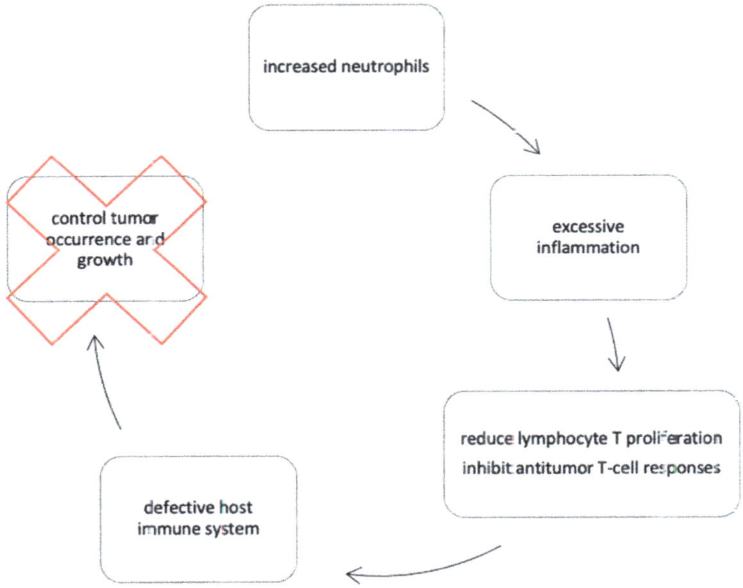

Figure 2. Inflammation as a hallmark and cause of cancer.

In the context of cancer, a neutrophil-to-lymphocyte ratio (NLR) indicates an optimal balance between pro- and antitumor immunity [29,30]. NLR is a well-known prognostic marker in several malignancies, including RCCs. Elevated pre-treatment NLR was associated with poor survival and an unfavorable response to first or subsequent-line therapy treatment [31,32].

A particular feature of RCCs is sarcomatoid differentiation, which indicates aggressive behavior and poor prognosis. Even a small amount of sarcomatoid differentiation might independently predict poor overall survival and limited therapeutic options. Also, when adjusted for therapeutic response, patients with sarcomatoid RCCs have worse survival [33,34].

However, additional research warrants validation of these potential markers' prognostic accuracy and consistency in mRCCs, especially when treated with late-line therapy. To address this knowledge gap, we conducted the following retrospective analysis to evaluate the prognostic significance of clinical and biological factors in patients undergoing second-line therapy.

2. Materials and Methods

This is a single-center retrospective observational study of mRCCs patients treated in our oncology department. Clinical data were extracted from patients' medical records in concordance with the recommendations of the ethics committee and the Declaration of Helsinki.

The principal inclusion criteria for the study were documented metastatic clear cell renal cell carcinomas (mCCRCCs), progression on first-line therapy, available clinical and imaging data before initiation of each treatment line, willingness to provide written informed consent, and age $\geq$ 18 years.

Exclusion criteria include short-term follow-up (<6 months), active autoimmune disease, evidence of active infection before initiating any systemic therapy, second primary cancers, brain metastases, and histotypes other than clear cell renal cell carcinoma.

Between January 2020 to October 2022, 74 patients diagnosed with mCCRCCs initiated first-line therapy. The majority of patients (89.4%) had received first-line treatment with TKI (57.3% sunitinib, 41.1% pazopanib, 1.4% sorafenib), while only 10.5% had received immunotherapy with nivolumab and ipilimumab. During the follow-up period, out of 74 patients treated in our department, 51.3% (38 patients) required second-line treatment. Nivolumab (3 mg per kilogram) plus ipilimumab (1 mg per kilogram) every 3 weeks for four cycles, followed by nivolumab monotherapy, was the most frequently (39.4%) administered second-line regimen, followed by cabozantinib (26.3%) at a daily dose of 60 mg orally once daily, axitinib (21%) at a dose of 5 mg orally twice daily, and pazopanib (13.1%) at a dose of 800 mg orally twice daily.

Patients were followed from the date of first-line administration to the date of last follow-up or death. The median follow-up was 15.3 (interquartile range: 8.3–22.6) months. The median time to progression on first-line therapy was 7.5 months (HR = 0.49; 95% CI: 0.215–1.65). As of the last follow-up in October 2022, thirty-three patients receiving second-line therapy were alive, while five died. The median OS was 10.5 (range 5–22.5) months. Treatment and therapeutic monitoring were conducted based on computed tomographic scans performed every 3 months.

The data were analyzed using SPSS 23.0. The chi-square test and the t-test were used to compare the categorical and continuous variables, respectively. We identified clinical characteristics with p-values $\leq$ 0.1 in the univariate analysis and further included them in the multivariate Cox regression analysis ($p < 0.05$). We further calculated the curves of overall survival (OS) using Kaplan–Meier survival curves and the log-rank test. The optimal cut-off point of the neutrophil-to-lymphocyte ratio (NLR) was determined by the receiver operating characteristic (ROC) curve analysis. OS was defined from the time of first-line treatment initiation until death or the last follow-up.

3. Results

3.1. Baseline Patient Characteristics and First-Line Therapy

A total of 74 mCCRCCs patients were included in the study. The median age was 62.8 years (interquartile ranger, 43–88 years), and 70.3% were males. Most patients (77%) had good performance status, defined as a Karnofsky performance status equal to or greater than 80%.

Before the start of first-line treatment, 62 (85%) patients with clinically diagnosed mCCRCCs underwent surgical intervention (radical or partial nephrectomy), and only 12 (15%) patients were diagnosed based on the results of a tissue biopsy of the primary tumor. The most frequently observed metastatic sites were the lymph nodes (13.5%), lung (23%), liver (39.2%), and bones (28.4%). The most common WHO/ISUP (WHO/International Society of Urological Pathology (ISUP) grading system) grades were II (43.2%) and III (44.6%). Only 12.2% of tumors were WHO/ISUP grade IV. At Histopathological analysis, sarcomatoid differentiation was evident in seven (9.5%) cases.

The baseline characteristics are presented in Table 1.

Table 1. Demographic and clinical characteristics of patients with mCCRCCs starting first-line therapy.

Patient and Disease Characteristics	Count	Count%
Patients treated with first line therapy	74	
Age median, range (years)	62.8 (range 43–88)	
Gender		
Male	52	
Female	22	
Surgical treatment		
Radical nephrectomy	48	64.8%
Tumor biopsy	11	15%
Partial nephrectomy	15	20.2%
First line therapy		
TKIs	66	89.4%
Immunotherapy	8	10.6%
The main sites of metastasis		
Lung	17	23%
Distant lymph nodes	10	13.5%
Liver	29	39.2%
Bones	21	28.4%
WHO/ISUP grades		
2	32	43.2%
3	33	44.6%
4	9	12.2%
Sarcomatoid differentiation		
Present	67	90.5%
Absent	7	9.5%
Karnofsky Performance Status		
<80%	17	23%
≥80%	57	77%
Time interval from diagnosis to treatment		
<12 months	52	70.3%
≥12 months	22	29.7%
Hemoglobin		
<lower limit of normal	41	55.4%
≥lower limit of normal	43	44.6%
LHD		
≥1.5× upper limit of normal	12	6.2%
<1.5× upper limit of normal	62	83.8%
Serum-corrected calcium		
≥upper limit of normal	16	21.6%
<upper limit of normal	58	78.4%
Platelets		
≥upper limit of normal	11	14.9%
<upper limit of normal	62	83.8%
Neutrophils		
≥upper limit of normal	24	32.4%
<upper limit of normal	50	67.6%
IMDC score		
Favorable	5	6.3%
Intermediate	38	51.4%
Poor	31	41.9%
MSKCC score		
Favorable	8	10.8%
Intermediate	49	66.2%
Poor	17	23%
Poor		

We estimated the prognostic scores using baseline clinical and laboratory characteristics. Over 50% of patients with mCCRCCs at first-line treatment were classified as intermediate-risk, according to the IMDC (51.4%) and MSKCC (66.2%) prognostic models.

In the first-line therapy, 89.4% of patients received tyrosine kinase inhibitors (57.3% sunitinib, 41.1% pazopanib, 1.4% sorafenib), and only 10.6% of patients received immunotherapy with nivolumab plus ipilimumab. The median PFS was 7.5 months in patients treated with first-line therapy (Figure 3).

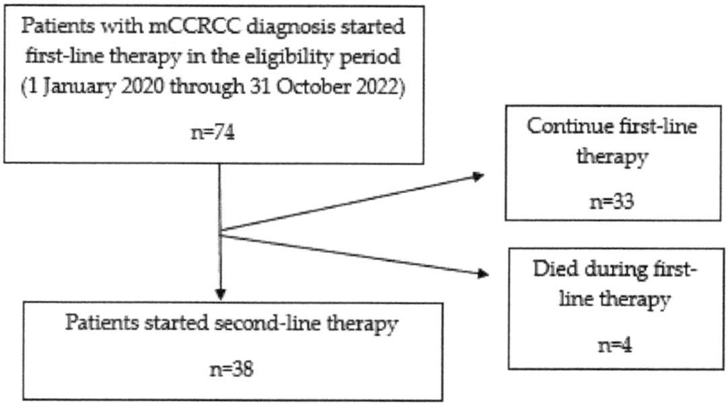

Figure 3. Identification and attrition of the first-line treatment and second-line treatment patient.

3.2. Patient Characteristics before the Start of Second-Line Systemic Treatment

A total of 38 (51.3%) patients were considered resistant to first-line treatment. The mean age was 63.9 years; 71.1% of the patients were males, and 48.7% had a good performance status prior to second-line initiation. Radical (65.7%) and partial nephrectomy (26.3%) were the most common surgical treatments, and only 7.8% of patients underwent tumor biopsy. Two (7.8%) patients had sarcomatoid features in their tumors.

Table 2 summarizes patients' demographics and characteristics.

Table 2. Demographic and clinical characteristics of patients with mCCRCCs starting second-line therapy.

Patient and Disease Characteristics	**Count**	**Count%**
Patients treated with second line therapy	38	51.3%
Gender		
Male	27	71.1%
Female	11	28.9%
Age median, range	63.92 ± 8.03 (48–78)	
Karnofsky status		
>80%	18	48.7%
<80%	19	51.3%
Tumor location		
Left side	20	52.6%
Right side	18	47.4%
Sarcomatoid differentiation		
No	35	92.1%
Yes	2	7.8%
Surgical treatment		
Radical nefrectomy	25	65.7%
Partial nefrectomy	10	26.3%
Biopsy	3	7.8%
Second line therapy		
Nivolumab + Ipilimumab	15	39.5%
Cabozantinib	10	26.3%
Axitinib	5	13.1%
Pazopanib	8	21.1%

Table 2. *Cont.*

Patient and Disease Characteristics	Count	Count%
IMDC score		
Favorable	1	2.6%
Intermediate	26	68.4%
Poor	11	28.9%
MSKCC score		
Favorable	1	10.8%
Intermediate	31	81.6%
Poor	6	15.8%

The prognostic risk scores were calculated at the start of the second-line therapy with most patients classified as having intermediate risk according to both the IMDC (68.4%) and MSKCC (81.6%) risk classifications. In second-line therapy, 15 patients (39.5%) received immunotherapy with Nivolumab plus Ipilimumab, 10 (26.3%) with Cabozantinib, 5 (13.1%) with Axitinib, and 8 (21.2%) with Pazopanib.

We calculated the neutrophil-to-lymphocyte ratio in all patients receiving second-line therapy. The median NLR value was 2.85 ± 2.05 with a range between 0.8–10. We identified the cut-off point using ROC curves in logistic regression (Figure 4). We obtained a value of three as the optimal cut-off point, and we further divided patients into high (>3) and low (>3) NLR.

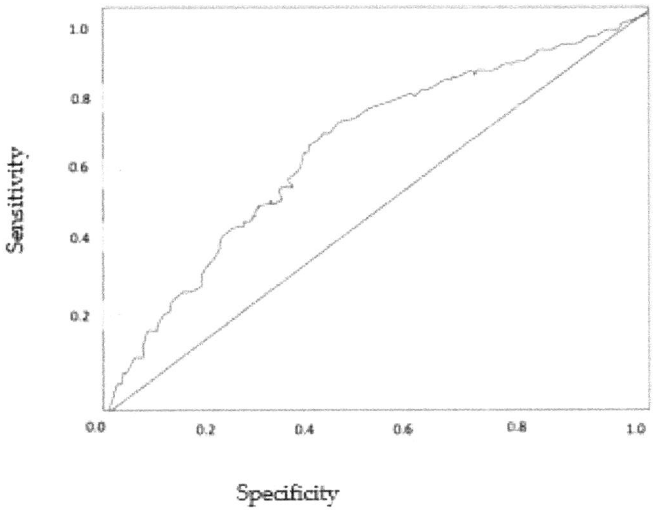

Figure 4. ROC curve for NLR.

3.3. *Statistical Analysis*

We carried out exploratory investigations of the parameters associated with better outcomes in mCCRCCs patients treated with second-line therapy. In particular, it was queried whether different clinical, biological, or histological factors might be associated with higher OS.

Univariate analysis (Table 3) demonstrated that the high IMDC (p = 0.028) and MSKCC (p = 0.003) scores, as well as elevated NLR (p = 0.002), were significant prognostic factors for poor OS.

Table 3. Univariate analyses for overall survival.

Baseline Parameters	N (%)	Univariate HR (95%CI)	*p* Value
Patients treated with second line	38 (51.3%)		
Gender			
Male	27 (71.1%)	0.314 (0.068–1.458)	0.139
Female	11 (28.9%)		
Age	63.92 ± 8.03	1.048 (0.957–1.148)	0.309
(median, range)	(48–78)		
Tumor location			
Left side	20 (52.6%)	0.512 (0.157–1.669)	0.267
Right side	18 (47.4%)		
Time from diagnosis to initial systemic treatment			
<12 months	12 (31.6%)	2.272 (0.488–10.569)	0.295
>12 months	26 (68.4%)		
Karnofsky status			
>80%	18 (48.7%)	1.766 (0.554–5.633)	0.336
<80%	19 (51.3%)		
Hemoglobin			
<12 g/dl	18 (47.4%)	1.444 (0.454–4.587)	0.534
>12 g/dl	20 (52.6%)		
Platelets			
<upper limit of normal	35 (92.1%)	2.030 (0.248–16.615)	0.509
>upper limit of normal	3 (7.9%)		
LDH			
>1.5× upper limit of normal	16 (33.3%)	1.443 (0.421–5.224)	0.301
<1.5× upper limit of normal	22 (66.6%)		
Corrected calcium			
>upper limit of normal	10 (26.3%)	0.421 (0.171–1.655)	0.254
<upper limit of normal	28 (73.7%)		
Neutrophils			
>upper limit of normal	21 (55.2%)	1.261 (0.415–5.503)	0.313
<upper limit of normal	17 (44.7%)		
NLR (median range)	2.85 ± 2.05 (0.8–10)	1.324 (1.078–1.625)	0.007
NLR			
<3	25 (65.8%)	9.599 (2.299–40.072)	0.002
>3	13 (34.2%)		
IMDC score			
favorable	1 (2.6%)		
intermediate	26 (68.4%)	1.826 (1.068–3.122)	0.028
poor	11 (28.9%)		
MSKCC score			
favorable	1 (10.8%)		
intermediate	31 (81.6%)	8.907 (2.148–36.935)	0.003
poor	6 (15.8%)		

There were no statistical differences in terms of gender (p = 0.13), tumor location (p = 0.26), time from diagnosis to initial systemic treatment (p = 0.29), performance status (p = 0.33), LDH (p = 0.30) hemoglobin (p = 0.53), calcium (p = 0.25), platelets (p = 0.50) or neutrophils (p = 0.31).

The multivariate analysis offered a more complete examination of these data. NLR was the only independent predictor of OS with an HR = 8.672 and p = 0.010 (Table 4; Figure 5).

Table 4. Multivariate analyses for overall survival.

Variables	*p* Value	HR 95.0% CI
NLR	0.010	8.672 (1.658–45.349)
IMDC score	0.320	3.092 (0.334–28.661)
MSKCC score	0.961	506.651 (0.000–8.078)

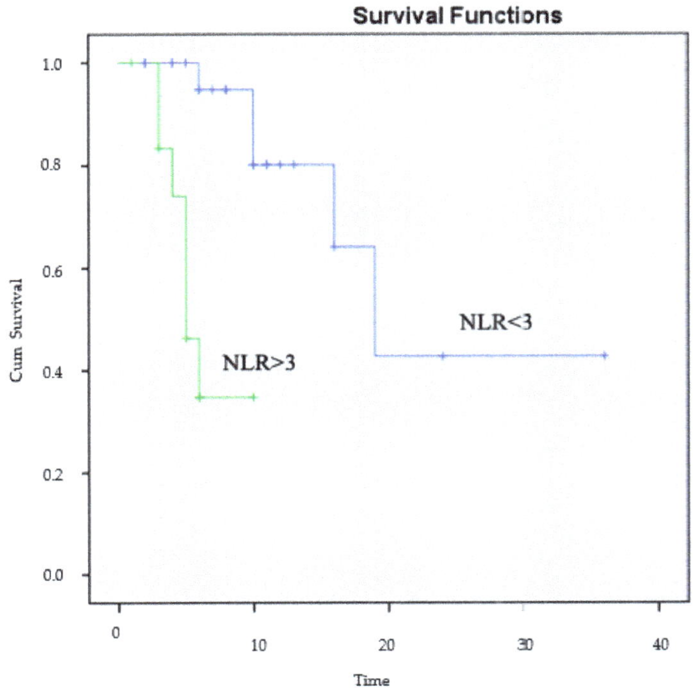

Figure 5. Kaplan–Meier estimates of OS according to the NLR.

The Kaplan–Meier survival curves of NLR showed significant differences in OS between patients with low (<3) and high (>3) NLR.

Time without disease progression was essential for all patients. The duration of treatment and its impact on prognosis was also examined (Table 5). Patients who experienced a clinical benefit for more than three months on a therapeutic line had better overall survival in the first line (22.25 months versus 7.62 months) and the second line (25.22 months versus 12.33 months).

Table 5. The therapeutic response and clinical outcome in first and second-line therapy.

Therapeutic Response		**Overall Survival**				
		Mean	**Standard Deviation**	**Median**	**Min.**	**Max.**
First line therapy	<3 months	7.62	4.84	8.00	1	19
	>3 months	22.25	18.07	15.00	4	84
Second line therapy	<3 months	12.33	9.58	9.00	6	31
	>3 months	25.22	16.95	19.00	8	72

Most patients had a multimodal treatment consisting of surgery and systemic therapy.

Almost 90% of patients had undergone initial surgical resection, including radical nephrectomy with lymph node dissection (24.3%) or without lymph node dissection (55.4%) and partial nephrectomy (9.5%). Minimally invasive biopsy procedures were performed in 10% of cases.

The overall survival did not differ significantly between patients treated with radical nephrectomy (21.73 ± 18.24 months) and partial nephrectomy (22.43 ± 23.25 months) but was reduced in patients with a diagnostic biopsy (8.88 ± 5.98 months), $p = 0.150$ (Table 6, Figure 6).

Table 6. The surgical approach and clinical outcome.

Surgical Approach	N (%)	OS (Months)
Radical nephrectomy	41 (55.4%)	21.73 ± 18.24
Radical nephrectomy with lymph node dissection	18 (24.3%)	18.72 ± 15.68
Partial nephrectomy	7 (9.5%)	22.43 ± 23.25
Tumor biopsy	8 (10.8%)	8.88 ± 5.98

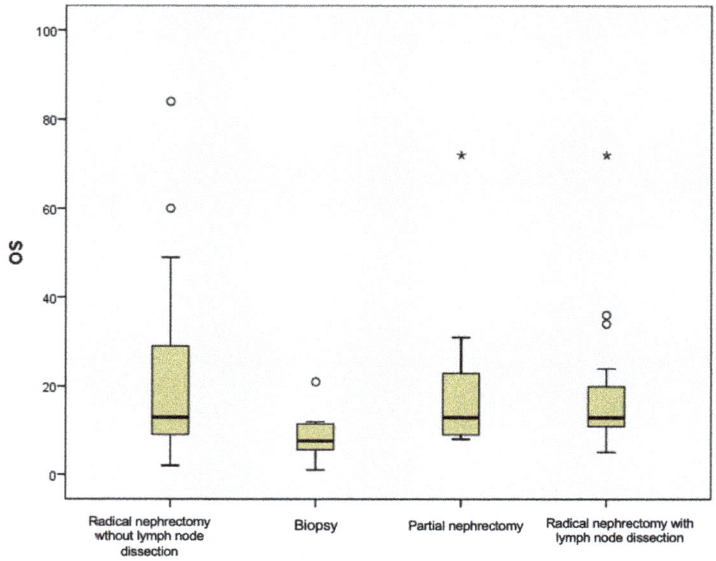

Figure 6. Box plot demonstrating survival difference in mCCRCCs patients according to surgery approach. The outliers (°), the extreme outliers (*)

Unfortunately, we found no statistically significant correlation between the surgical approach and clinical benefit in first and second-line therapy (Table 7).

Table 7. The therapeutic response, surgical approach and clinical outcome in first and second-line therapy.

Therapeutic Response		Radical Nephrectomy	Tumor Biopsy	Partial Nephrectomy	Radical Nephrectomy with Lymph Node Dissection	p Value
				N (%)		
First line therapy	<3 months	9 (21.95%)	1 (12.55)	2 (28.55)	1 (5.55)	0.382
	>3 months	32 (78.05)	7 (87.55)	5 (71.45)	17 (94.45)	
Second line therapy	<3 months	3 (13.65)	1 (25.05)	1 (16.65)	1 (16.65)	0.953
	>3 months	19 (86.35)	3 (75.05)	5 (83.35)	5 (83.35)	

Another interesting aspect was the sarcomatoid differentiation, which was observed in two (7.8%) patients with metastatic mCCRCCs receiving second-line therapy (Table 8, Figure 7). The results suggested that sarcomatoid features were associated with worse clinical outcomes (14.86 months OS versus 19.97 months OS, $p = 0.033$).

Table 8. The presence of sarcomatoid differentiation and clinical outcome.

Tumor Caracteristics	N (%)	Overall Survival (Months)
No sarcomatoid differentiation	35 (92.1%)	19.97
With sarcomatoid differentiation	2 (7.8%)	14.86

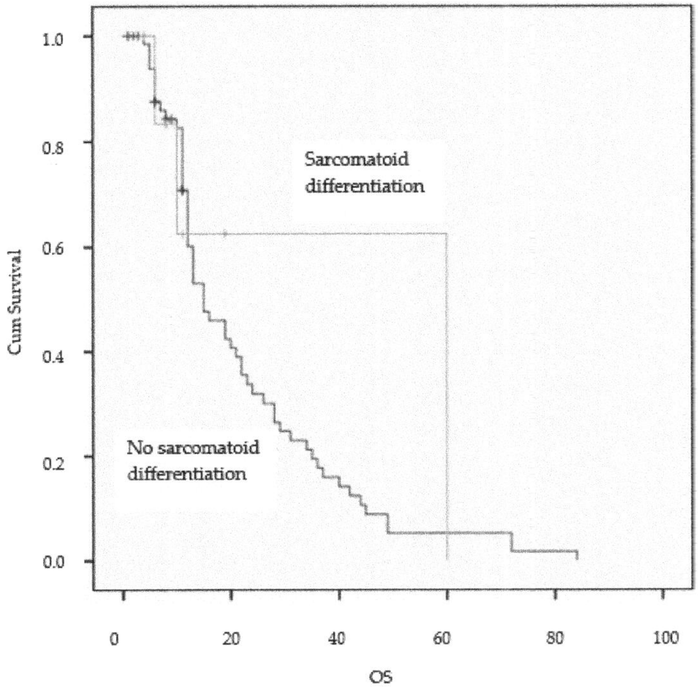

Figure 7. Kaplan–Meier estimates of OS according to the sarcomatoid differentiation.

Moreover, the clinical benefit was significantly correlated with the sarcomatoid differentiation in first therapy ($p = 0.004$) and second-line therapy ($p = 0.029$) (Table 9).

Table 9. Clinical benefit for patients with sarcomatoid differentiation.

		Sarcomatoid Differentiation		
	Therapeutic Response	**No**	**Yes**	***p* Value**
First-line therapy	<3 months	9 (13.43%)	4 (57.14%)	0.004
	>3 months	58 (86.57%)	3 (42.86%)	
Second-line therapy	<3 months	6 (18.18%)	4 (80%)	0.029
	>3 months	27 (81.82%)	1 (20%)	

4. Discussion

RCCs are highly vascular tumors with strong immunogenicity and an unpredictable natural history. The treatment of advanced and metastatic RCCs has revolutionized in the past three decades. Interferon and interleukin two immunotherapies have been a treatment options for over 30 years. Recently, immune checkpoint inhibitors and vascular endothelial growth factor receptors have become promising therapeutic strategies [35].

The TNM classification is the most important prognostic factor in mCCRCCs patients treated with first-line therapy. Other independent predictors of long-term survival include

MSKCC and IMDC risk models. Few sources of data on potential prognostic factors in mCCRCCs patients receiving second-line therapy exist. Despite the significant progress in the treatment landscape of metastatic RCCs, only a minority of patients have access to second-line therapies [21]. Therefore, understanding the baseline parameters of patients who are eligible for second-line therapy is crucial. This study provides real-world evidence on the characteristics and outcomes of these particular patients.

In this study, 51.3% of mCCRCCs patients received second-line therapy, which is consistent with previous reports [21,36,37]. We evaluated the baseline characteristics of the patients before initiating second-line therapy. Most of the patients had intermediate-risk levels based on the prognostic models.

The IMDC and MSKCC models were originally developed and validated for metastatic mCCRCCs patients who received front-line treatment [38–40]. Recently, these models have been shown to profile risk in late-line settings.

The MSKCC is the most used tool for prognostic stratification of patients with mCCRCCs receiving second-line therapy. Data from the pivotal second-line studies, including the CheckMate 025 trial demonstrating superior efficacy for nivolumab over everolimus, or the phase III study METEOR, which showed that cabozantinib significantly improved OS and PFS compared to everolimus [41,42].

Furthermore, a large study by Dudani et al. demonstrated that the IMDC score continues to risk stratify patients with mCCRCCs treated with second-line immune checkpoint inhibitors [43].

Our results show that both MSKCC and IMDC have important prognostic value. Patients with favorable risks had significantly longer overall survival compared to those with poor or intermediate scores (HR = 8.907, 2.148–36.935, p = 0.004 for MSKCC; and HR = 1.826, 1.068–3.122, p = 0.028 for IMDC).

The neutrophil-to-lymphocyte ratio (NLR) is a controversial topic. NLR is a marker of systemic inflammatory response with prognostic significance. It reflects a dynamic relationship between neutrophils, an essential part of the innate immune system, and the adaptive immune responses carried out by lymphocytes. A high NLR has been unequivocally associated with adverse prognoses in many cancers. The evidence indicates that pretreatment NLR could predict recurrence, disease progression, and survival outcomes in RCCs patients [44–46].

However, NLR varies widely between patients. The normal range of NLR in healthy adults varies between 1 and 3. Values higher than three are considered pathological [47–50]. Therefore, it is essential to find a cut-off value to mark the lower and the upper limit of NLR associated with prognosis.

We also investigated the optimal cut-off value for NLR. Firstly, we determined the median NLR value for the patients initiating second-line therapy, which was 2.85 ± 2.05. The findings align with the literature, where NLR values between 1.7 and 5 have been reported, with an NLR of 3 being frequently used [51–53]. Secondly, using ROC analysis, we identified the optimal NLR cut-off value of 3. Furthermore, we classified patients into high (>3) NLR and low (<3) NLR. The study reported a negative association between high NLR and OS. Clearly, an increased NLR > 3 was associated with worse outcomes in both univariate and multivariate analyses (p = 0.005). Therefore, we hypothesized that adding NLR to the established risk models might help improve their prognostic accuracy.

Accumulating evidence has established the role of NLR as a biomarker of increased immune activation in different disorders, including infections, autoimmune diseases, cardiovascular diseases, and metabolic syndromes [54,55]. Even if most studies have explored the prognostic value of NLR in cancer patients with an optimal cut-off value above 3, a grey zone of NLR values between two and three may serve as an early warning of endothelial dysfunction, chronic vascular inflammation, and atherosclerosis [56–58]. The chronic vascular inflammatory process plays a role in the pathogenesis of atherosclerosis, hypertension, diabetes, and obesity. Prior evidence has shown that NLR is a reliable biomarker to predict cardiovascular risk. In patients with coronary artery disease, an NLR over 2.13 independently predicted myocardial damage [59].

Furthermore, NLR has also been investigated as a marker of disease activity and a predictor of relapse in autoimmune diseases [60,61]. For example, D'Amico et al. demonstrated that an NLR > 1 in patients with multiple sclerosis strongly predicted disease activity and aggressive evolution [62]. In conclusion, even an NLR value lower than three may be an unfavorable prognostic factor for RCCs patients with underlying comorbidities

Although growing evidence showed that immune checkpoint inhibitors, targeted therapy combinations, and late lines of therapy improve clinical outcomes, whether increased treatment duration improves survival remains unclear. Prolonged treatment time for mCCRCCs has an important positive impact on cancer-specific survival. But despite recent therapeutic advances in mCCRCCs treatments, challenges remain in managing progressive or recurrent disease. About 10 to 25% of mCCRCCs patients experience rapidly progressive disease on first-line therapy [39]. An established and accepted definition of a rapidly progressive mCCRCCs still needs to be found.

For example, Chang and his colleagues defined rapid disease progression as occurring within one month of systemic therapy in mCCRCCs [63]. Another study on real-world treatment patterns in mCCRCCs patients by Bersanelli described rapid progression as the time from the start of the first-line to the beginning of second-line treatment $\leq$ 24 weeks [64].

This study also examined whether treatment duration in first or subsequent therapeutic lines is associated with survival.

Chen VJ et al. defined the term "clinical benefit" as receiving a therapeutic line for at least three months. They found no statistically significant difference in clinical benefit between first- and second-line treatment, but their data confirm that receiving therapy for at least three months correlated with a longer overall survival [26]. Our results also suggest that patients with over 3 months of clinical benefit had a statistically higher median overall survival in both first-line (22.25 months versus 7.62 months) and second-line (25.22 months versus 12.33 months) therapy. Unfortunately, we found no statistically significant correlation between the surgical approach and clinical benefit in first and second-line therapy.

Histopathological features can predict prognosis and may help to stratify patients to receive appropriate therapy. The concept of sarcomatoid RCCs was first comprehensively described by Farrow and colleagues in 1968 under the spectrum of renal sarcomas [65]. However, in 2012, the International Society of Urological Pathology (ISUP) grading system for renal cell carcinoma established that sarcomatoid differentiation is an uncommon histological transformation that can occur in most histological subtypes of RCCs, giving them aggressive biological features and poor prognoses [66,67]. Due to the rarity of these tumors and their exclusion from most studies, limited data exist on the current standard of care associated with their treatment [68–71]. Most clinical evidence is therefore based on case reports or subgroup analyses.

A systematic review and meta-analysis showed that immune checkpoint inhibitors were associated with remarkable clinical efficacy in patients with sarcomatoid RCCs [72].

However, agents targeting the vascular endothelial growth factor pathway did not report a significant improvement in patient outcomes [73]. Therefore, regardless of the PD-L1 level, the preferred options are immunotherapy-based regimens.

In this current study, we evaluated the pathological features and the subsequent treatment-related outcomes of patients with sarcomatoid RCCs. In total, seven patients (9.5%) had sarcomatoid features. Only two patients with sarcomatoid differentiation received second-line therapy with a modest response rate and poor clinical benefit (14.86 months OS).

Naturally, the single-center retrospective observational analysis has several limitations, including susceptibility to random errors. The main weaknesses of this study are the retrospective nature and the relatively small sample size of patients. There was also a relatively short follow-up period and a lack of external validation. The second-line therapy administered was not well-balanced, preventing a direct comparison of the efficacy of each regimen. Furthermore, the patients did not receive novel drugs approved in mCCRCCs, which have demonstrated survival benefits. However, we propose that the analysis generates hypotheses and may serve as a basis for investigating these potential prognostic factors in larger studies, particularly in later-line treatment settings.

5. Conclusions

No consistent reports regarding prognostic factors in second-line therapy for mCCRCCs had been published prior to this study. Therefore, this study contributed valuable new insights to the literature by shedding light on the prognostic role of biomarkers and emphasizing the importance of risk models in mCCRCCs.

The results indicated that second-line therapy is more effective in patients with favorable IMDC and MSKCC prognostic scores, NLR < 3, and no sarcomatoid differentiation. Moreover, patients with clinical benefit > 3 months on a therapeutic line experienced significantly longer overall survival.

Large-scale research studies are needed to confirm these observations and to increase our understanding of the best therapeutic sequence in mCCRCCs patients.

Supplementary Materials: The following supporting information can be downloaded at: https://www.mdpi.com/article/10.3390/diagnostics13142430/s1, Table S1: The clinical and biological parameters of patients with mCCRCC.

Author Contributions: Conceptualization, C.B. and A.I.P.; methodology, C.N.; software, I.M.S.; validation C.F.P.; formal analysis, C.B.; investigation, A.I.P.; resources, C.N.; data curation, A.I.P. writing—original draft preparation, C.B.; writing—review and editing, C.N.; visualization, A.I.P.; supervision, C.B. and C.N.; project administration, A.I.P. All authors have read and agreed to the published version of the manuscript.

Funding: This research received no external funding.

Institutional Review Board Statement: The study was conducted in accordance with the Declaration of Helsinki and approved by the Institutional Review Board of Elias Emergency University Clinic Hospital (no. 7170/12 January 2023).

Informed Consent Statement: Informed consent was obtained from all subjects involved in the study. Written informed consent has been obtained from the patients to publish this paper.

Data Availability Statement: Data supporting this study are included within the article and Supplementary Materials.

Conflicts of Interest: The authors declare no conflict of interest.

References

1. Moch, H.; Amin, M.B.; Berney, D.M. The 2022 World Health Organization Classification of Tumours of the Urinary System and Male Genital Organs-Part A: Renal, Penile, and Testicular Tumours. *Eur. Urol.* **2022**, *82*, 458–468. [CrossRef] [PubMed]
2. Alzubaidi, A.N.; Sekoulopoulos, S.; Pham, J. Incidence and Distribution of New Renal Cell Carcinoma Cases: 27-Year Trends from a Statewide Cancer Registry. *J. Kidney Cancer VHL* **2022**, *9*, 7–12. [CrossRef] [PubMed]
3. Saad, A.M.; Gad, M.M.; Al-Husseini, M.J. Trends in Renal-Cell Carcinoma Incidence and Mortality in the United States in the Last 2 Decades: A SEER-Based Study. *Clin. Genitourin. Cancer* **2019**, *17*, 46–57.e5. [CrossRef] [PubMed]
4. Wang, Z.; Yin, Y.; Wang, J. Standardized Incidence Rate, Risk and Survival Outcomes of Second Primary Malignancy among Renal Cell Carcinoma Survivors: A Nested Case-Control Study. *Front. Oncol.* **2021**, *11*, 716741. [CrossRef]
5. Sung, H.; Ferlay, J.; Siegel, R.L. Global Cancer Statistics 2020: GLOBOCAN Estimates of Incidence and Mortality Worldwide for 36 Cancers in 185 Countries. *CA Cancer J. Clin.* **2021**, *71*, 209–249. [CrossRef] [PubMed]
6. Zhan, X.; Chen, T.; Liu, Y. Trends in cause of death among patients with renal cell carcinoma in the United States: A SEER-based study. *BMC Public. Health* **2023**, *23*, 770. [CrossRef]
7. Padala, S.A.; Barsouk, A.; Thandra, K.C. Epidemiology of Renal Cell Carcinoma. *World J. Oncol.* **2020**, *11*, 79–87. [CrossRef]
8. Motzer, R.J.; Jonasch, E.; Agarwal, N. Kidney Cancer, Version 3.2022, NCCN Clinical Practice Guidelines in Oncology. *J. Natl. Compr. Cancer Netw.* **2022**, *20*, 71–90. [CrossRef]

9. Mitsogiannis, I.C.; Mitsogianni, M.; Papathanassiou, M.A. Current Options for Second-Line Systemic Therapy in Metastatic Renal Cell Carcinoma. *J. Kidney Cancer VHL* **2022**, *9*, 29–40. [CrossRef]
10. Buczek, M.; Escudier, B.; Bartnik, E. Resistance to tyrosine kinase inhibitors in clear cell renal cell carcinoma: From the patient's bed to molecular mechanisms. *Biochim. Biophys. Acta* **2014**, *1845*, 31–41. [CrossRef]
11. Iacovelli, R.; Ciccarese, C.; Procopio, G. Current evidence for second-line treatment in metastatic renal cell carcinoma after progression to immune-based combinations. *Cancer Treat. Rev.* **2022**, *105*, 102379. [CrossRef] [PubMed]
12. Makhov, P.; Joshi, S.; Ghatalia, P. Resistance to systemic therapies in clear cell renal cell carcinoma: Mechanisms and management strategies. *Mol. Cancer Ther.* **2018**, *17*, 1355–1364. [CrossRef]
13. Sorbellini, M. A postoperative prognostic nomogram predicting recurrence for patients with conventional clear cell renal cell carcinoma. *J. Urol.* **2005**, *173*, 48. [CrossRef] [PubMed]
14. Zisman, A. Improved prognostication of renal cell carcinoma using an integrated staging system. *J. Clin. Oncol.* **2001**, *19*, 1649. [CrossRef]
15. Frank, I. An outcome prediction model for patients with clear cell renal cell carcinoma treated with radical nephrectomy based on tumor stage, size, grade and necrosis: The SSIGN score. *J. Urol.* **2002**, *168*, 2395. [CrossRef] [PubMed]
16. Karakiewicz, P.I. Multi-institutional validation of a new renal cancer-specific survival nomogram. *J. Clin. Oncol.* **2007**, *25*, 1316. [CrossRef]
17. Rosiello, G. Head-to-head comparison of all the prognostic models recommended by the European Association of Urology Guidelines to predict oncologic outcomes in patients with renal cell carcinoma. *Urol. Oncol.* **2022**, *40*, 271.e19. [CrossRef]
18. Motzer, R.J.; Bacik, J.; Murphy, B.A. Interferon-alfa as a comparative treatment for clinical trials of new therapies against advanced renal cell carcinoma. *J. Clin. Oncol. Off. J. Am. Soc. Clin. Oncol.* **2002**, *20*, 289–296. [CrossRef]
19. Okita, K. Impact of Disagreement Between Two Risk Group Models on Prognosis in Patients with Metastatic Renal-Cell Carcinoma. *Clin. Genitourin. Cancer* **2019**, *17*, e440. [CrossRef]
20. Heng, D.Y.; Xie, W.; Regan, M.M. External validation and comparison with other models of the International Metastatic Renal-Cell Carcinoma Database Consortium prognostic model: A population-based study. *Lancet Oncol.* **2013**, *14*, 141–148. [CrossRef]
21. Heng, D.Y.; Xie, W.; Regan, M.M.; Warren, M.A.; Golshayan, A.R.; Sahi, C.; Eigl, B.J.; Ruether, J.D.; Cheng, T.; North, S.; et al. Prognostic factors for overall survival in patients with metastatic renal cell carcinoma treated with vascular endothelial growth factor-targeted agents: Results from a large, multicenter study. *J. Clin. Oncol.* **2009**, *27*, 5794–5799. [CrossRef] [PubMed]
22. Massari, F. Addition of Primary Metastatic Site on Bone, Brain, and Liver to IMDC Criteria in Patients with Metastatic Renal Cell Carcinoma: A Validation Study. *Clin. Genitourin. Cancer* **2021**, *19*, 32. [CrossRef] [PubMed]
23. Levy, A.; Menard, J.; Albiges, L. Second line treatment of metastatic renal cell carcinoma: The Institut Gustave Roussy experience with targeted therapies in 251 consecutive patients. *Eur. J. Cancer* **2013**, *49*, 1898–1904. [CrossRef] [PubMed]
24. Eggers, H.; Ivanyi, P.; Hornig, M.; Grunwald, V. Predictive factors for second-line therapy in metastatic renal cell carcinoma: A retrospective analysis. *J. Kidney Cancer VHL* **2017**, *4*, 8–15. [CrossRef] [PubMed]
25. Zanwar, S.; Joshi, A.; Noronha, V. Patterns of care and outcomes for second-line targeted therapy in metastatic renal cell carcinomas: A registry-based analysis. *Indian J. Cancer* **2016**, *53*, 579–582.
26. Chen, V.J.; Hernandez-Meza, G.; Agrawal, P.; Zhang, C.A.; Xie, L.; Gong, C.L.; Hoerner, C.R.; Srinivas, S.; Oermann, E.K.; Fan, A.C. Time on Therapy for at Least Three Months Correlates with Overall Survival in Metastatic Renal Cell Carcinoma. *Cancers* **2019**, *11*, 1000. [CrossRef]
27. Méjean, A.; Escudier, B.; Thezenas, S.; Beauval, J.-B.; Geoffroy, L.; Bensalah, K.; Thiery-Vuillemin, A.; Cormier, L.; Lang, H.; Guy, L. CARMENA: Cytoreductive nephrectomy followed by sunitinib versus sunitinib alone in metastatic renal cell carcinoma—Results of a phase III noninferiority trial. *J. Clin. Oncol.* **2018**, *36*, LBA3. [CrossRef]
28. Buonacera, A.; Stancanelli, B.; Colaci, M.; Malatino, L. Neutrophil to Lymphocyte Ratio: An Emerging Marker of the Relationships between the Immune System and Diseases. *Int. J. Mol. Sci.* **2022**, *23*, 3636. [CrossRef]
29. Huang, Z.; Fu, Z.; Huang, W.; Huang, K. Prognostic value of neutrophil-to-lymphocyte ratio in sepsis: A meta-analysis. *Am. J. Emerg. Med.* **2020**, *38*, 641–647. [CrossRef]
30. Kumar, R.; Geuna, E.; Michalarea, V. The neutrophil-lymphocyte ratio and its utilisation for the management of cancer patients in early clinical trials. *Br. J. Cancer* **2015**, *112*, 1157–1165. [CrossRef]
31. Naszai, M.; Kurjan, A.; Maughan, T.S. The prognostic utility of pre-treatment neutrophil-to-lymphocyte-ratio (NLR) in colorectal cancer: A systematic review and meta-analysis. *Cancer Med.* **2021**, *10*, 5983–5997. [CrossRef] [PubMed]
32. Yu, Y.; Qian, L.; Cui, J. Value of neutrophil-to-lymphocyte ratio for predicting lung cancer prognosis: A meta-analysis of 7219 patients. *Mol. Clin. Oncol.* **2017**, *7*, 498–506. [CrossRef] [PubMed]
33. Blum, K.A.; Gupta, S.; Tickoo, S.K.; Chan, T.A.; Russo, P.; Motzer, R.J.; Karam, J.A.; Hakimi, A.A. Sarcomatoid renal cell carcinoma Biology, natural history and management. *Nat. Rev. Urol.* **2020**, *17*, 659–678. [CrossRef] [PubMed]
34. Shaker, O.; Ayeldeen, G.; Abdelhamid, A. Sarcomatoid-associated gene risk index for clear cell renal cell carcinoma. *Front. Genet.* **2022**, *13*, 985641. [CrossRef]
35. Dutcher, J.P.; Flippot, R.; Fallah, J.; Escudier, B. On the Shoulders of Giants: The Evolution of Renal Cell Carcinoma Treatment-Cytokines, Targeted Therapy, and Immunotherapy. *Am. Soc. Clin. Oncol. Educ. Book* **2020**, *40*, 418–435. [CrossRef]
36. Motzer, R.J.; Mazumdar, M.; Bacik, J.; Berg, W.; Amsterdam, A.; Ferrara, J. Survival and prognostic stratification of 670 patients with advanced renal cell carcinoma. *J. Clin. Oncol.* **1999**, *17*, 2530–2540. [CrossRef]

37. Mekhail, T.M.; Abou-Jawde, R.M.; BouMerhi, G.; Malhi, S.; Wood, L.; Elson, P.; Bukowski, R. Validation and extension of the memorial sloan-kettering prognostic factors model for survival in patients with previously untreated metastatic renal cell carcinoma. *J. Clin. Oncol.* **2005**, *23*, 832–841. [CrossRef]
38. Ko, J.J.; Xie, W.; Kroeger, N.; Lee, J.L.; Rini, B.I.; Knox, J.J.; Bjarnason, G.A.; Srinivas, S.; Pal, S.K.; Yuasa, T.; et al. The International Metastatic Renal Cell Carcinoma Database Consortium model as a prognostic tool in patients with metastatic renal cell carcinoma previously treated with first-line targeted therapy: A population-based study. *Lancet Oncol.* **2015**, *16*, 293–300. [CrossRef]
39. Wells, J.C.; Stukalin, I.; Norton, C.; Srinivas, S.; Lee, J.L.; Donskov, F.; Bjarnason, G.A.; Yamamoto, H.; Beuselinck, B.; Rini, B.I.; et al. Third-line targeted therapy in metastatic renal cell carcinoma: Results from the International Metastatic Renal Cell Carcinoma Database Consortium. *Eur. Urol.* **2017**, *71*, 204–209. [CrossRef]
40. Lakomy, R.; Poprach, A.; Bortlicek, Z.; Melichar, B.; Chloupkova, R.; Vyzula, R.; Zemanova, M.; Kopeckova, K.; Svoboda, M.; Slaby, O.; et al. Utilization and efficacy of second-line targeted therapy in metastatic renal cell carcinoma: Data from a national registry. *BMC Cancer* **2017**, *17*, 880. [CrossRef]
41. Motzer, R.J.; Escudier, B.; McDermott, D.F.; George, S.; Hammers, H.J.; Srinivas, S.; Tykodi, S.S.; Sosman, J.A.; Procopio, G.; Plimack, E.R.; et al. Nivolumab versus everolimus in advanced renal-cell carcinoma. *N. Engl. J. Med.* **2015**, *373*, 1803–1813. [CrossRef] [PubMed]
42. Choueiri, T.K.; Escudier, B.; Powles, T.; Mainwaring, P.N.; Rini, B.I.; Donskov, F.; Hammers, H.; Hutson, T.E.; Lee, J.L.; Peltola, K.; et al. Cabozantinib versus everolimus in advanced renal-cell carcinoma. *N. Engl. J. Med.* **2015**, *373*, 1814–1823. [CrossRef] [PubMed]
43. Dudani, S.; Gan, C.L.; Wells, C.; Bakouny, Z.; Dizman, N.; Pal, S.K.; Wood, L.; Kollmannsberger, C.K.; Szabados, B.; Powles, T.; et al. Application of IMDC criteria across first-line (1L) and second-line (2L) therapies in metastatic renal-cell carcinoma (mRCC): New and updated benchmarks of clinical outcomes. *J. Clin. Oncol.* **2020**, *38* (Suppl. 15), 5063. [CrossRef]
44. Zahorec, R. Neutrophil-to-lymphocyte ratio, past, present and future perspectives. *Bratisl. Lek. Listy.* **2021**, *122*, 474–488. [CrossRef]
45. Boissier, R.; Campagna, J.; Branger, N.; Karsenty, G.; Lechevallier, E. The prognostic value of the neutrophil-lymphocyte ratio in renal oncology: A review. *Urol. Oncol.* **2017**, *35*, 135–141. [CrossRef]
46. Gago-Dominguez, M.; Matabuena, M.; Redondo, C.M.; Patel, S.P.; Carracedo, A.; Ponte, S.M.; Martínez, M.E.; Castelao, J.E. Neutrophil to lymphocyte ratio and breast cancer risk: Analysis by subtype and potential interactions. *Sci. Rep.* **2020**, *10*, 13203, Erratum in *Sci. Rep.* **2020**, *10*, 20641. [CrossRef]
47. Arda, E.; Yuksel, I.; Cakiroglu, B.; Akdeniz, E.; Cilesiz, N. Valuation of Neutrophil/Lymphocyte Ratio in Renal Cell Carcinoma Grading and Progression. *Cureus* **2018**, *10*, e2051. [CrossRef]
48. Parosanu, A.I.; Pirlog, C.F.; Slavu, C.O.; Stanciu, I.M.; Cotan, H.-T.; Vrabie, R.C.; Popa, A.-M.; Olaru, M.; Iaciu, C.; Bratu, L.I.; et al. The Prognostic Value of Neutrophil-to-Lymphocyte Ratio in Patients with Metastatic Renal Cell Carcinoma. *Curr. Oncol.* **2023**, *30*, 2457–2464. [CrossRef]
49. Wen, R.M.; Zhang, Y.J.; Ma, S.; Xu, Y.L.; Chen, Y.S.; Li, H.L.; Bai, J.; Zheng, J.N. Preoperative Neutrophil to Lymphocyte Ratio as a Prognostic Factor in Patients with Non-metastatic Renal Cell Carcinoma. *Asian Pac. J. Cancer Prev.* **2015**, *16*, 3703–3708. [CrossRef]
50. Nunno, V.D.; Mollica, V.; Gatto, L.; Santoni, M.; Cosmai, L.; Porta, C.; Massari, F. Prognostic impact of neutrophil-to-lymphocyte ratio in renal cell carcinoma: A systematic review and meta-analysis. *Immunotherapy* **2019**, *11*, 631–643. [CrossRef]
51. Shao, Y.; Wu, B.; Jia, W. Prognostic value of pretreatment neutrophil-to-lymphocyte ratio in renal cell carcinoma: A systematic review and meta-analysis. *BMC Urol.* **2020**, *20*, 90. [CrossRef] [PubMed]
52. Lalani, A.A.; Xie, W.; Martini, D.J.; Steinharter, J.A.; Norton, C.K.; Krajewski, K.M.; Duquette, A.; Bossé, D.; Bellmunt, J.; Van Allen, E.M.; et al. Change in Neutrophil-to-lymphocyte ratio (NLR) in response to immune checkpoint blockade for metastatic renal cell carcinoma. *J. Immunother. Cancer* **2018**, *6*, 5. [CrossRef] [PubMed]
53. Basu, A.; Phone, A.; Bice, T. Change in neutrophil to lymphocyte ratio (NLR) as a predictor of treatment failure in renal cell carcinoma patients: Analysis of the IROC (Investigating RCC Outcomes) cohort. *J. Clin. Oncol.* **2021**, *39* (Suppl. 6), 344. [CrossRef]
54. Cordeiro, M.D.; Ilario, E.N.; Abe, D.K.; Carvalho, P.A.; Muniz, D.Q.B.; Sarkis, A.S.; Coelho, R.F.; Guimarães, R.M.; Haddad, M.V.; Nahas, W.C. Neutrophil-to-Lymphocyte Ratio Predicts Cancer Outcome in Locally Advanced Clear Renal Cell Carcinoma. *Clin. Genitourin. Cancer* **2022**, *20*, 102–106. [CrossRef]
55. Pichler, M.; Hutterer, G.C.; Stoeckigt, C.; Chromecki, T.F.; Stojakovic, T.; Golbeck, S.; Eberhard, K.; Gerger, A.; Mannweiler, S.; Pummer, K.; et al. Validation of the pre-treatment neutrophil-lymphocyte ratio as a prognostic factor in a large European cohort of renal cell carcinoma patients. *Br. J. Cancer* **2013**, *108*, 901–907. [CrossRef] [PubMed]
56. Chung, D.; Lee, K.O.; Choi, J.W.; Kim, N.K.; Kim, O.J.; Kim, S.H.; Oh, S.H.; Kim, W.C. Blood Neutrophil/Lymphocyte Ratio Is Associated with Cerebral Large-Artery Atherosclerosis but Not with Cerebral Small-Vessel Disease. *Front. Neurol.* **2020**, *11*, 1022. [CrossRef]
57. Fu, X.; Liu, H.; Huang, G.; Dai, S.S. The emerging role of neutrophils in autoimmune-associated disorders: Effector, predictor, and therapeutic targets. *MedComm* **2021**, *2*, 402–413. [CrossRef]
58. Balta, S.; Celik, T.; Mikhailidis, D.P.; Ozturk, C.; Demirkol, S.; Aparci, M.; Iyisoy, A. The Relation between Atherosclerosis and the Neutrophil-Lymphocyte Ratio. *Clin. Appl. Thromb. Hemost.* **2016**, *22*, 405–411. [CrossRef]
59. Sharma, K.; Patel, A.K.; Shah, K.H.; Konat, A. Is Neutrophil-to-Lymphocyte Ratio a Predictor of Coronary Artery Disease in Western Indians? *Int. J. Inflam.* **2017**, *2017*, 4136126. [CrossRef]

60. Erre, G.L.; Paliogiannis, P.; Castagna, F.; Mangoni, A.A.; Carru, C.; Passiu, G.; Zinellu, A. Meta-analysis of neutrophil-to-lymphocyte and platelet-to-lymphocyte ratio in rheumatoid arthritis. *Eur. J. Clin. Investig.* **2019**, *49*, e13037. [CrossRef]
61. Bisgaard, A.K.; Pihl-Jensen, G.; Frederiksen, J.L. The neutrophil-to-lymphocyte ratio as disease actvity marker in multiple sclerosis and optic neuritis. *Mult. Scler. Relat Disord.* **2017**, *18*, 213–217. [CrossRef] [PubMed]
62. D'Amico, E.; Zanghì, A.; Romano, A.; Sciandra, M.; Palumbo, G.A.M.; Patti, F. The Neutrophil-to-Lymphocyte Ratio is Related to Disease Activity in Relapsing Remitting Multiple Sclerosis. *Cells* **2019**, *8*, 1114. [CrossRef] [PubMed]
63. Chang, S.; Liaw, C.L.; Huang, J.S. Procedure Triggers Rapid Progression of Renal Cell Carcinoma. *J. Cancer Res. Pract.* **2014**, *1*, 57–62. [CrossRef]
64. Bersanelli, M.; Iacovelli, R.; Buti, S.; Houede, N.; Laguerre, B.; Procopio, G.; Lheureux, S.; Fischer, R.; Negrier, S.; Ravaud, A.; et al. Metastatic Renal Cell Carcinoma Rapidly Progressive to Sunitinib: What to Do Next? *Eur. Urol. Oncol.* **2021**, *4*, 274–281 [CrossRef]
65. Farrow, G.M.; Harrison, E.G., Jr.; Utz, D.C. Sarcomas and sarcomatoid and mixed malignant tumors of the kidney in adults *Cancer* **1968**, *22*, 556–563. [CrossRef]
66. de Peralta-Venturina, M.; Moch, H.; Amin, M.; Tamboli, P.; Hailemariam, S.; Mihatsch, M.; Javidan, J.; Stricker, H.; Ro, J.Y.; Amin, M.B. Sarcomatoid differentiation in renal cell carcinoma: A study of 101 cases. *Am. J. Surg. Pathol.* **2001**, *25*, 275–284 [CrossRef]
67. Cheville, J.C.; Lohse, C.M.; Zincke, H. Sarcomatoid renal cell carcinoma: An examination of underlying histologic subtype and an analysis of associations with patient outcome. *Am. J. Surg. Pathol.* **2004**, *28*, 435–441. [CrossRef]
68. Tully, K.H.; Berg, S.; Paciotti, M. The Natural History of Renal-Cell Carcinoma with Sarcomatoid Differentiation, a Stage-by-Stage Analysis. *Clin. Genitourin. Cancer* **2023**, *21*, 63–68. [CrossRef]
69. Pichler, R.; Compérat, E.; Klatte, T.; Pichler, M.; Loidl, W.; Lusuardi, L.; Schmidinger, M. Renal Cell Carcinoma with Sarcomatoid Features: Finally New Therapeutic Hope? *Cancers* **2019**, *11*, 422. [CrossRef]
70. Candelario, N.; Geiger, C.; Flaig, T. Sarcomatoid Renal Cell Carcinoma: The Present and Future of Treatment Paradigms. *Kidney Cancer* **2021**, *5*, 167–179. [CrossRef]
71. Kyriakopoulos, C.E.; Chittoria, N.; Choueiri, T.K. Outcome of patients with metastatic sarcomatoid renal cell carcinoma: Results from the International Metastatic Renal Cell Carcinoma Database Consortium. *Clin. Genitourin. Cancer* **2015**, *13*, e79–e85 [CrossRef] [PubMed]
72. Iacovelli, R.; Ciccarese, C.; Bria, E. Patients with sarcomatoid renal cell carcinoma—Re-defining the first-line of treatment: A meta-analysis of randomised clinical trials with immune checkpoint inhibitors. *Eur. J. Cancer* **2020**, *136*, 195–203. [CrossRef] [PubMed]
73. Tannir, N.M.; Signoretti, S.; Choueiri, T.K. Efficacy and Safety of Nivolumab Plus Ipilimumab versus Sunitinib in First-line Treatment of Patients with Advanced Sarcomatoid Renal Cell Carcinoma. *Clin. Cancer Res.* **2021**, *27*, 78–86. [CrossRef] [PubMed]

diagnostics

MDPI

Article

A Novel Two-Gene Expression-Based Prognostic Score in Malignant Pleural Mesothelioma

Velizar Shivarov [1,*], Georgi Blazhev [2] and Angel Yordanov [3]

1 Department of Experimental Research, Medical University Pleven, 5800 Pleven, Bulgaria
2 Department of Genetics, Faculty of Biology, St. Kliment Ohridski Sofia University, 1164 Sofia, Bulgaria
3 Department of Gynaecological Oncology, Medical University Pleven, 5800 Pleven, Bulgaria
* Correspondence: vshivarov@abv.bg; Tel.: +359-887881766

Abstract: Background: Malignant pleural mesothelioma (MPM) is a rare cancer type with an increasing incidence worldwide. We aimed to develop a rational gene expression-based prognostic score in MPM using publicly available datasets. Methods: We developed and validated a two-gene prognostic score (2-PS) using three independent publicly available gene expression datasets. The 2-PS was built using the Robust Likelihood-Based Survival Modeling with Microarray Data method. Results: We narrowed down the model building to the analysis of 179 genes, which have been shown previously to be of importance to MPM development. Our statistical approach showed that a model including two genes (GOLT1B and MAD2L1) was the best predictor for overall survival (OS) ($p < 0.0001$). The binary score based on the median of the continuous score stratified the patients into low and high score groups and also showed statistical significance in uni- and multivariate models. The 2-PS was validated using two independent transcriptomic datasets. Furthermore, gene set enrichment analysis using training and validation datasets showed that high score patients had distinct gene expression profiles. Our 2-PS also showed a correlation with the estimated infiltration by some immune cell fractions such as CD8+ T cells and M1/2 macrophages. Finally, 2-PS correlated with sensitivity or resistance to some commonly used chemotherapeutic drugs. Conclusion: This is the first study to demonstrate good performance of only two-gene expression-based prognostic scores in MPM. Our initial approach for features selection allowed for an increased likelihood for the predictive value of the developed score, which we were also able to demonstrate.

Keywords: gene expression; prognosis; prediction; score; mesothelioma

Citation: Shivarov, V.; Blazhev, G.; Yordanov, A. A Novel Two-Gene Expression-Based Prognostic Score in Malignant Pleural Mesothelioma. *Diagnostics* **2023**, *13*, 1556. https://doi.org/10.3390/diagnostics13091556

Academic Editors: Yuli Huang, Yong Yuan and Peisong Chen

Received: 26 March 2023
Revised: 16 April 2023
Accepted: 24 April 2023
Published: 26 April 2023

1. Introduction

Primary malignant pleural mesothelioma (MPM) is a rare malignancy with an aggressive course and a dismal prognosis for 5-year overall survival of less than 10%. It is currently considered a premier example of a chemically-induced cancer due to occupational or environmental factors because the major risk factor remains a history of asbestos exposure. The elucidation of asbestos exposure as a primary risk factor for MPM led to its restricted use worldwide, which resulted in a slight decline in the incidence of MPM in developed Western countries. However, most of the cases due to occupational exposure develop with a latency of approximately 20 years or even longer, which suggests that we may still expect to witness a peak or a steady state in MPM incidence in various regions worldwide beyond 2020 [1]. Histologically, MPM is classified into three major subtypes, namely epithelioid, sarcomatoid, and biphasic subtypes. Histological subtyping has prognostic value, with epithelioid MPM having a much better prognosis in comparison to sarcomatoid and biphasic cases. Recent molecular profiling studies, however, suggested that the clonal structure of MPM is much more complex and actually consists of subpopulations of the two major histological subtypes (epithelioid and sarcomatoid) in a variable proportion. At a genomic level, MPM is characterized by a relatively low mutational burden [2]. The recurrently mutated genes are mainly tumor suppressor genes such as *BAP1*, *CDKNA2*, *NF2*, *SETD2*, and *TP53* [3]. Unfortunately, most patients with MPM present initially with inoperable disease and are assessed for fitness for systemic therapy and symptoms control approaches. Since the mid-2000s, the mainstay of frontline chemotherapy is cisplatin drug in combination with pemetrexed, with some evidence that anti-angiogenic therapy may add benefit to this doublet. Immune checkpoint inhibitors were shown to provide survival benefits in relapsed settings [4] and gained regulatory approval. Phase 3 results also favor immune checkpoint inhibitors in comparison to chemotherapy in first-line settings [5,6].

In the last two decades, a number of studies addressed the development of gene expression-based prognostic scores. Because of the relatively small number of cases in each study and the diverse patient profiles of each study, the wide applicability of each of those is debatable. Additionally, it will be of greater value if any prognostic score has some predictive power as well. Therefore, the MPM field still requires the identification of reliable prognostic or predictive biomarkers. These become easier to address more than ever before because of the recent advances in genomic technologies and the possibilities to integrate multilayer omics data.

Here we aimed for the development of a novel gene expression-based prognostic score in MPM with potential predictive power. We further aimed to integrate of gene expression and epigenetic data to interrogate the underlying biological features of the newly defined prognostic subgroups of the disease. For the purposes of this study, we used, as a training set, the publicly available dataset of MPM cases that had been profiled as part of The Cancer Genome Atlas (TCGA). As validation datasets, we used two other datasets of gene expression data from MPM reported in the last decade. We were able to develop a 2-gene expression-based score, which showed independent prognostic power in both training and validation datasets. Furthermore, the score has some predictive power as demonstrated by the integration of gene expression and drug sensitivity data from MPM cell lines.

2. Materials and Methods

2.1. Datasets

We identified 3 datasets of whole transcriptome analysis of at least 50 MPM patients performed in the last decade as outlined in a recent review [7]. As a training dataset, we used RNA-Seq data for 87 MPM cases from the TCGA project [3]. TCGA data were downloaded as RPKM and log10 transformed from the Genomics Data Commons Portal (https://portal.gdc.cancer.gov/ (accessed on 25 March 2023)). As validation datasets, we used data from one study using RNA sequencing (Bueno) [8] of 211 MPM cases and one study with a planar gene expression array (Blum) with 67 MPM cases [9]. Bueno RNA-seq data were downloaded as raw counts from the European Genome-phenome Archive (EGA) from a study, EGAS000010015631, conducted by Bueno et al. (EGAD00001001915, $n = 211$) sequenced via Illumina HiSeq 2000 technology. Paired-end reads were trimmed using Trim Galore v. 0.6.3 to remove Illumina adapters. Paired-end fastq files were subjected to a quality control procedure using FastQC v. 0.72. fastq files were then to reference the human genome hg38 build using HISAT2 v. 2.1 and subsequently annotated to the gene level using featureCounts v 1.6.4. Raw counts data were subsequently voom transformed and quantile normalized using *limma* package for R. Blum dataset data were downloaded as raw .cel files from the Array Express server with the accession number E-MTAB-6877. They were RMA transformed and quantile normalized using *limma* package and analyzed as described below. The overall analytical approach is summarized in Figure 1.

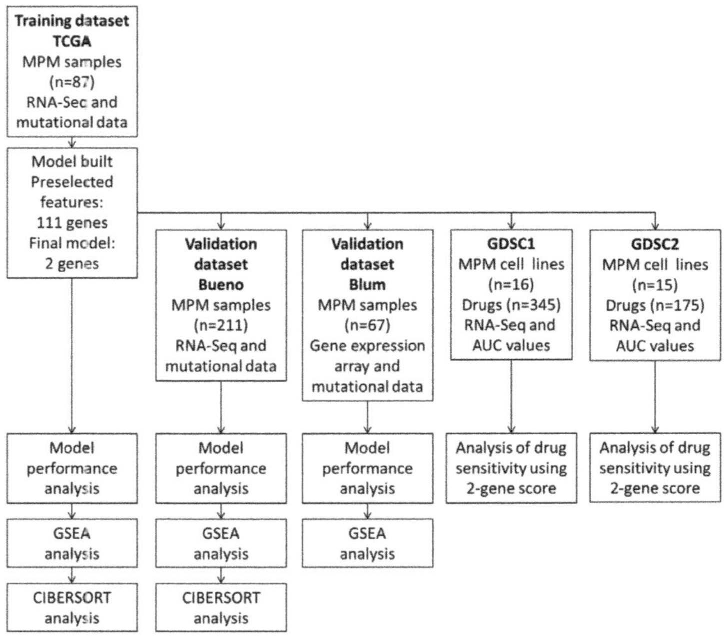

Figure 1. Summary of the analytical approach in the study using different datasets.

2.2. Model Building

In order to reduce the number of features (genes) that we start with in the model build, we decided to select only genes on which MPM cells lines have already been reported to be dependent as part of the DepMap project (https://depmap.org/portal/ (accessed on 25 March 2023)) [10–13]. We identified a total of 179 genes that were reported to confer significant dependence of MPM in at least one RNA interference screen included in the DepMap project (Supplementary Table S1). We subsequently used RNA-Seq data for 87 MPM cases included in the TCGA project as a training dataset to build a model based on the expression of survival-associated selected genes. We filtered the RPKM expression values of the identified 179 genes and applied the Robust Likelihood-Based Survival Modeling with Microarray Data [14], which was implemented through the *rbsurv* package for the R statistical environment. This technique utilizes the partial likelihood of the Cox model and functions through the generation of multiple gene models. It also divides the input dataset into training and validation sets and performs multiple cross-validations of a series of gene models so that it finally provides the optimal model based on the Akaike Information Criterion (AIC). The AIC measure does not have any biological meaning. It is interpreted as a statistical criterion to select the best model. The lower the value of the criterion, the better the statistical model performs for the given dataset. The *rbsurv* package automatically selects and proposes the best multivariate model with the lowest AIC. Cox regression coefficients for the genes included in the model for both datasets were obtained using the *survival* package for R. A total continuous score was calculated for each sample through weighted summation of the gene expression values according to the formula: $\text{Scorei} = \sum \text{wj} * \text{xij}$ where xij is the log-transformed expression value for the gene j in patient i, and wj is the weight assigned to probe set j (here wj was the Cox regression coefficient from the univariate analysis in the training set). The total score was calculated for each patient sample in the training and validation datasets. To build a binary score (high vs. low), we defined a cut-off value specific for each cohort corresponding to the median of the continuous score for the respective cohort. Performance of the median cut-off score was assessed using Receiver Operating Characteristics (ROC) curves implemented through the web-based graphical user interface of the Cutoff Finder package for R (https://molpathoheidelberg.shinyapps.io/CutoffFinder_v1/ (accessed on 25 March 2023)) [15]. Univariate and multivariate analyses for correlation of the continuous and discrete scores with overall survival were performed for each of the datasets using the *survminer* package for R. We have previously used an identical approach for the development of a microRNA-expression based prognostic score in acute myeloid leukemia (AML) [16].

2.3. Gene Set Enrichment Analysis

Gene set enrichment analysis (GSEA) was performed using the stand-alone version GSEA 4.0.3 (Boston, MA, USA) developed by Broad Institute (http://www.broadinstitute.org/gsea/index.jsp (accessed on 25 March 2023)) [17]. The cancer hallmarks collection of oncogenic gene ontology signatures from the Molecular Signatures Database (MSigDB) (https://www.gsea-msigdb.org/gsea/msigdb/human/collections.jsp#H (accessed on 25 March 2023)) was used as the reference gene-sets to test for enrichment in the gene expression profiles of high score versus low score patients from each cohort.

2.4. Cibersort

We downloaded the CIBERSORT-inferred immune signatures [18] for TCGA mesothelioma cases, which were determined as part of the The Immune Landscape of Cancer study (https://gdc.cancer.gov/about-data/publications/panimmune (accessed on 25 March 2023)) [19]. We used the CIBERSORTx portal (https://cibersortx.stanford.edu/index.php (accessed on 25 March 2023)) [20] to estimate the immune cell fractions of the samples from the Bueno dataset. Pearson correlations between continuous score and each immune cell fraction were calculated using the *cor* function from the stats package for R. Estimated correlation coefficients were presented graphically using the *corrplot* package for R.

2.5. Drug Sensitivity Analysis

RNA-Seq data for 16 MPM cells lines included in the Sanger Genomics of Drug Sensitivity in Cancer Project (GDSC) [21,22] were downloaded from the ArrayExpress server with the accession number E-MTAB-3983 (https://www.ebi.ac.uk/biostudies/arrayexpress/studies/E-MTAB-3983 (accessed on 25 March 2023)). FRPKM gene expression values were log2 transformed and the 2-PS was calculated for each MPM cell line using the regression coefficient as described above. The GDSC1 analysis dataset included data for 16 MPM cell lines treated with 345 unique compounds. The GDSC2 analysis dataset included data for 15 MPM cell lines treated with 175 drugs. We analyzed the correlation between the 2-PS of the analyzed cell lines and the Area Under the Curve (AUC) of the dose–response analysis of each of the drugs in the two datasets. ANOVA models were used to identify significant correlations, with a two-sided p-value of 0.05 being considered significant.

2.6. Common Statistical Procedures

All statistical procedures were performed using the R v. 4.2.2 environment for statistical computing. The chi-squared test was used for the assessment of independence in the distribution of categorical variables. A two-sided t-test for independent samples was used to compare the means of normally distributed continuous variables. The Wilcoxon–Mann–Whitney test was used to compare the medians of continuous variables without normal distribution. For all statistical tests, an alpha level of 0.05 was considered statistically significant.

3. Results

3.1. Building and Initial Performance of a Two-Gene Prognostic Score (2-PS)

We applied the Robust Likelihood-Based Survival Modeling with Microarray Data to the training dataset (TCGA dataset) with genes that MPM cell lines were shown to be dependent on (Supplementary Table S1) (Figure 1). We chose such an approach for two main reasons. Firstly, we were able to reduce the number of features (genes) to be tested in the prognostic model. Secondly, this approach, in our view, increases the likelihood that the prognostic model built may also have some predictive power. The algorithm selected a best performing prognostic model with AIC of consisting of two genes, *GOLT1B* and *MAD2L1*. The estimated Cox regression coefficients (ln(HR)) for *GOLT1B* and *MAD2L1* were 1.403 and 0.945, respectively. The continuous score for each sample in each dataset was calculated as the sum of expression values for each of the genes in the model multiplied by the regression coefficient. In univariate analysis, the continuous score was prognostic for the OS (Cox regression $p = 9.69 \times 10^{-10}$). We further defined a binary score using as a cut-off the median of the continuous score for all samples. In univariate analysis, the binary score also showed significant prognostic value (Cox regression $p = 2.85 \times 10^{-6}$ and Figure 2A) with an Area Under the Curve value of the Receiver Operator Characteristics (ROC) analysis of 0.67 (Figure 2B). This AUC value is distinct from 0.5 suggesting there exists a true difference in survival between the two groups of patients [23]. As it is below 0.7, it is expected that the binary score may not be the only contributing prognostic factor in this cohort. Therefore, we further evaluated the performance of the binary score in a multivariate model with age, sex, stage, histology, and mutational status as covariates. Notably, the binary score retained independent prognostic power (Cox regression $p = 7.34 \times 10^{-6}$, Figure 2C).

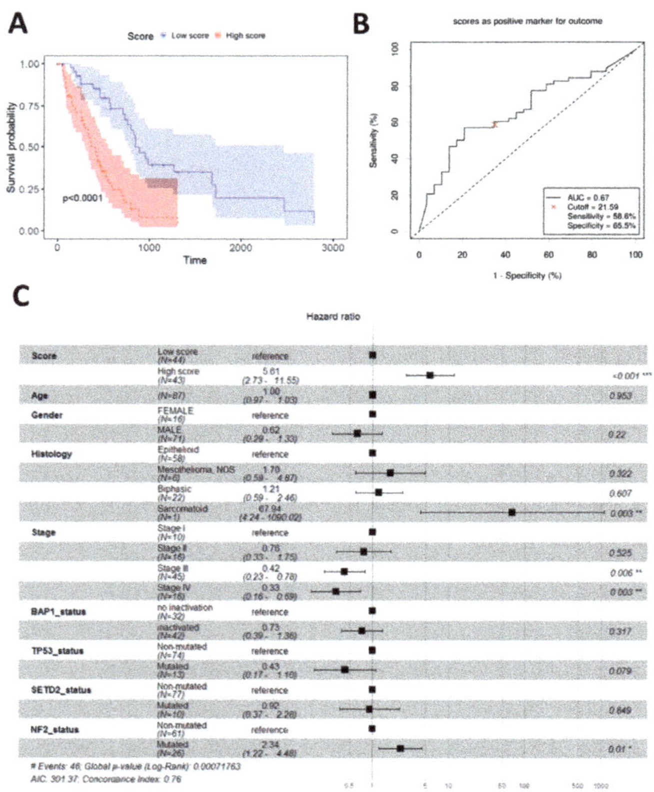

Figure 2. Performance of the 2-gene PS in the TCGA cohort. (**A**) Overall survival in low vs. high score patients; (**B**) ROC analysis using the median value of the continuous score to define low and high score patients; (**C**) Multivariate prognostic model including 2-PS and other demographic and clinical parameters. Significance notations: "*"—<0.05, "**"—<0.01, "***"—<0.001.

3.2. Validation of the 2-PS

In order to validate our 2-gene prognostic score, we used two recent publicly available datasets with either RNA-Seq data ($n = 211$) (Bueno) or planar expression arrays ($n = 67$) (Blum). The estimated continuous score in the Bueno dataset showed a clear prognostic value (Cox regression $p = 1.22 \times 10^{-7}$). The score was further converted to a binary one using the median of the continuous score as a cut-off. Analogous to the data training dataset, the binary score in this validation dataset also had prognostic power (Cox regression $p = 1.4 \times 10^{-5}$, Figure 3A) with an AUC of the ROC analysis of 0.75 (Figure 3B), which is considered acceptable for a diagnostic test [23]. Furthermore, in the extensive multivariate model for the Bueno dataset, the binary score was still of independent prognostic value (Cox regression $p = 0.03$, Figure 3C).

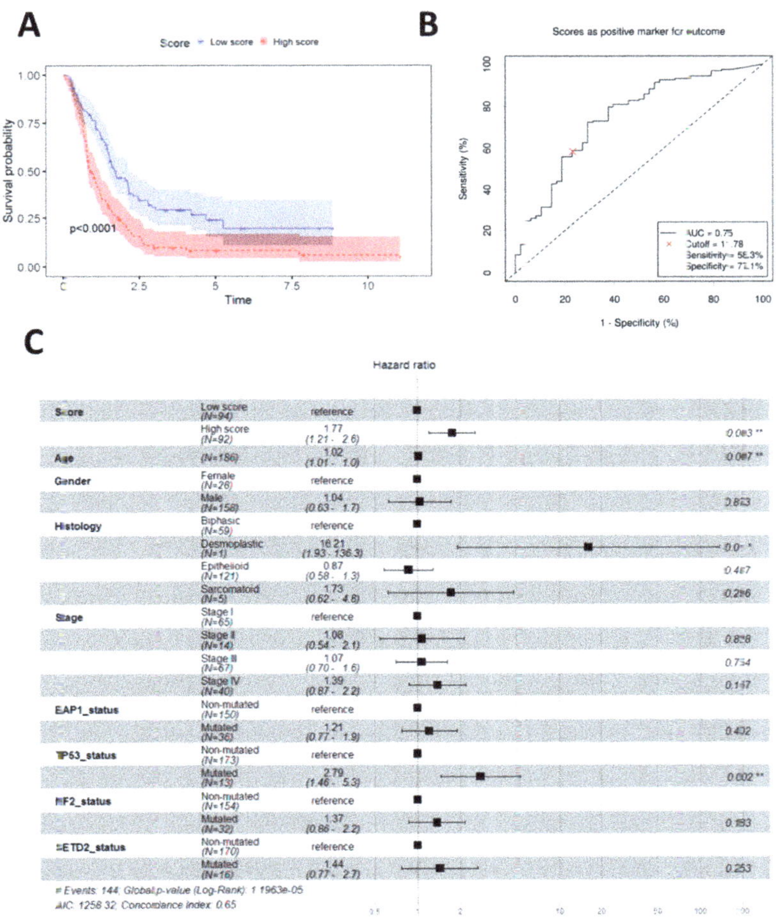

Figure 3. Performance of the 2-gene PS in the Bueno cohort. (**A**) Overall survival in low vs. high score patients; (**B**) ROC analysis using the median value of the continuous score to define low and high score patients; (**C**) Multivariate prognostic model including 2-PS and other demographic and clinical parameters. Significance notations: "*"—<0.05, "**"—<0.01.

Following the same procedure, we analyzed the performance of the estimated continuous score in the Blum dataset. In multivariate analysis, the continuous model was a significant prognostic factor ($p = 1.13 \times 10^{-5}$). The same also held true for the binary model defined by the cut-off of the median for the continuous score (Cox regression $p = 0.0014$, Figure 4A). The AUC in the ROC analysis, in that case, was 0.85 (Figure 4B), which is considered excellent performance for a diagnostic test [23]. Finally, a multivariate model for the Blum dataset was built using binary score, sex, age, stage, and histology, albeit without mutational data, as those were not publicly available. The model demonstrated the independent prognostic value of the binary score (Cox regression $p = 0.00671$, Figure 4C).

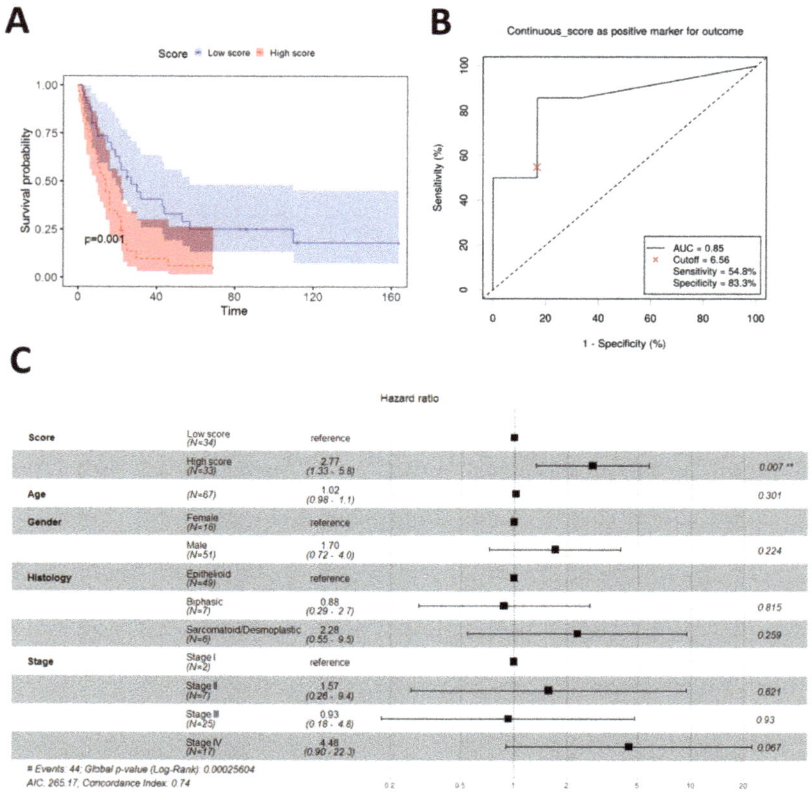

Figure 4. Performance of the 2-gene PS in the Blum cohort. (**A**) Overall survival in low vs. high score patients; (**B**) ROC analysis using the median value of the continuous score to define low and high score patients; (**C**) Multivariate prognostic model including 2-PS and other demographic and clinical parameters. Significance notations: "**"—<0.01.

3.3. Gene Set Enrichment Analysis

Based on the observation that our novel 2-PS performed similarly well in both training and the two validation datasets, we hypothesized that the score may correlate with specific gene expression signatures. Therefore, we performed GSEA using predefined cancer hallmark signatures from the MSig database. For each of the datasets, we obtained a number of signatures enriched in the high score patients' subgroups as follows: TCGA ($n = 37$), Bueno ($n = 34$), Blum ($n = 34$). There was a more significant overlap between the overexpressed signatures in the TCGA and Bueno datasets and slightly less so between any of those two and the Blum dataset. However, there were a total of 25 signatures that were commonly overexpressed in high score patients from the three cohorts (Figure 5A). Most of them were related to DNA repair and DNA damage response (Figure 5B).

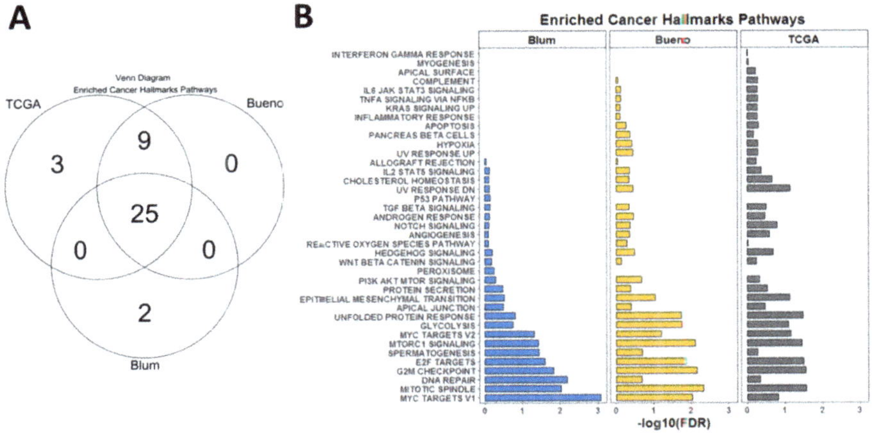

Figure 5. Summary of the GSEA analysis in all three cohorts. (**A**) Venn diagram showing the overlap of enriched cancer hallmark pathways in high score patients from the three cohorts. (**B**) List of all enriched cancer hallmark pathways in high score patients from the three cohorts at a false discovery rate (FDR) of below 0.05.

3.4. Correlation with Immune Signatures

Infiltrating immune cells are a major player in the immune response against cancer and may be used as prognostic and predictive markers. We therefore questioned whether 2-PS-defined groups of MPM patients would also have distinct underlying profiles in terms of the microenvironment. To address this question, we analyzed whether the 2-PS correlated with specific immune cell subtype infiltration in MPM. We used the inferred infiltrating immune cell fractions using the CIBERSORT algorithm using TCGA (Figure 6A) and Bueno (Figure 6B) datasets. Notably for both datasets the continuous prognostic score showed a positive correlation with CD8+ T cell fraction as well as with M1 and M2 macrophage fractions.

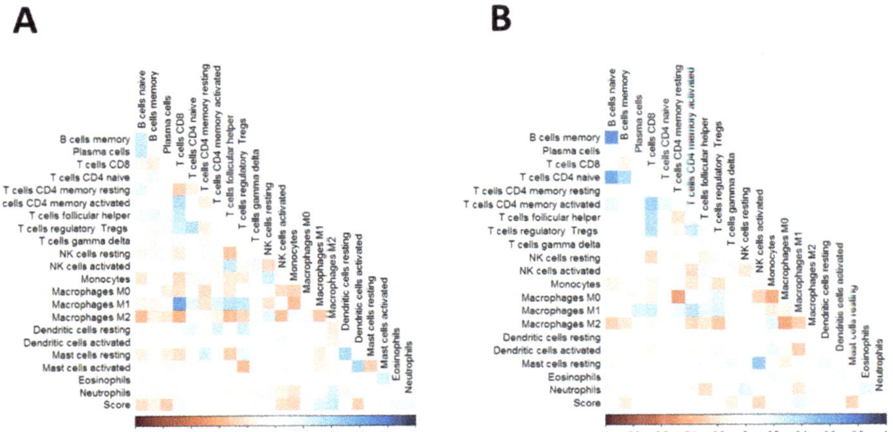

Figure 6. Plots of correlation matrices between 2-gene prognostic score and the estimated immune cell population contents using CIBERSORT. (**A**) TCGA dataset; (**B**) Bueno dataset. Color codes represent Pearson correlation coefficients.

3.5. Potential Predictive Power of the 2-PS

We specifically developed our score focusing on genes for which there had been a demonstration of dependency using some knock-out screen experiments. It was therefore reasonable to accept that the score may have some potential predictive power. The most straightforward approach to provide some preliminary evidence in that regard was to use drug sensitivity screens data from MPM cell lines. We calculated the 2-gene prognostic score for each of the cell MPM cell lines included in the Genomics of Drug Sensitivity in Cancer project GDSC1 part (n = 16) and GDSC2 part (n = 15) [21,22]. We subsequently tested the correlation between the 2-PS and the sensitivity to each of the drugs tested in both projects using AUC values. The AUC-defined response to 11 drugs from the GDSC1 set showed a significant correlation with the 2-PS of the tested mesothelioma cell lines (Figure 7A); whereas for the GDSC2 set, the number of such significant correlations was 18 (Figure 7B). Interestingly, this analysis revealed a correlation of the 2-PS with response to commonly used drugs in mesothelioma management such as cisplatin (R = −0.51, p = 0.046), gemcitabine (R = 0.69, p = 0.019), and vinblastine (R = 0.63, p = 0.037).

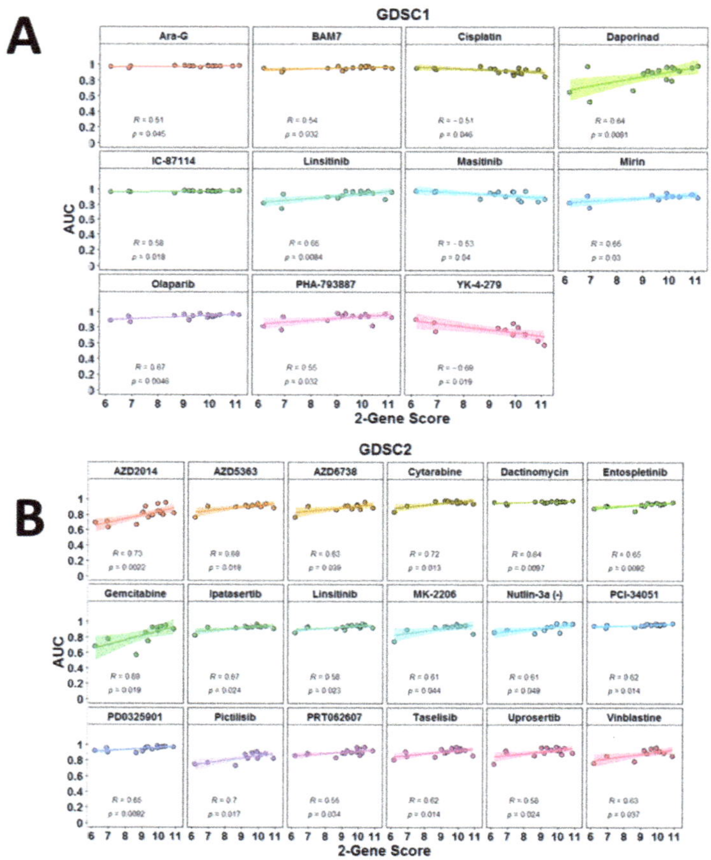

Figure 7. Linear correlations between 2-gene prognostic score and sensitivity of MPM cell lines (as measured by Area Under the Curve (AUC)) to different compounds. (**A**) Data from the analysis using the GDSC1 dataset. (**B**) Data from the analysis using the GDSC2 dataset. Correlation coefficients are from Pearson correlation.

4. Discussion

A number of studies proposed gene expression-based prognostic models in MPM [3,8,9,24–31]. They differ significantly in their approaches for feature selection, training and validation datasets, the number of genes included in the final model, as well as in the performance in different MPM cohorts. Additionally, the predictive value of each proposed score remains largely unexplored. Here we implemented a novel approach in features selection by limiting the number of genes to be tested for inclusion in an MPM prognostic model only to genes for which it has been that MPM cell lines have been sensitive to their knock-down. We then applied the RBSURV approach to the TCGA dataset and built a two-gene prognostic model, which showed moderate prognostic power as a continuous or binary score in both univariate and multivariate models in three different MPM cohorts. This moderate prognostic power is in trade-off with the minimal number of genes included in the prognostic score, avoiding over-fitting of the model by the inclusion of a higher number of features. The limited number of genes in our 2-PS may further allow simple validation using low-throughput techniques such as quantitative RT-PCR or immunohistochemistry. We tested the performance on continuous and binary scores using univariate and multivariate Cox regression analyses in three independent cohorts (two from the USA and one from Europe). The definition of cut-offs for binary scores cannot be directly applied to prospective studies as we used a dataset-specific median to define binary scores provided that the datasets we used were from different gene expression profiling platforms and had different pre-processing steps. Ideally, the model must be validated in a prospective fashion using a simple, readily reproducible technique providing a uniform standardized read-out of gene expression. The two genes included in our model are not widely studied in MPM.

The *MAD2L1* gene encodes the mitotic arrest deficient 2 like 1, coding for the respective protein, which is an integral part of the mitotic spindle assembly checkpoint and ensures that all chromosomes are properly aligned at the metaphase plate before the cell can proceed to anaphase [32]. MAD2L1 was recently found overexpressed in several MPM cell lines at the mRNA and protein level [33]. This study conforms with a previous one which demonstrated higher MAD2L1 protein expression (both nuclear and cytoplasmic) in MPM cell lines as compared to normal mesothelium [34]. Interestingly, according to the latter study, the total *MAD2L1* mRNA expression level did not correlate with the overall survival of 80 MPM patients [34]. However, the same study showed that higher nuclear MAD2L1 expression determined using immunohistochemistry correlated with a shorter overall survival [34]. Notably, a recent study showed that BRCA1 in the mesothelioma leads to the co-depletion of MAD2L1 mRNA and protein [35]. Additionally, loss of BRCA1/MAD2L1 was associated with resistance to vinorelbine ex vivo, and the survival was shorter for patients lacking BRCA1/MAD2L1 expression in comparison to those with double-positive tumors [35]. This observation can explain the fact that our 2-PS correlated with resistance to vinblastine (mitotic spindle assembly inhibitor) and olaparib (PARP inhibitor) in MPM cell lines (Figure 7). In addition, among the top enriched pathways in the GSEA analysis of all three cohorts were pathways directly involving mechanisms of DNA replication such as the following pathways: "Mitotic spindle", "G2M checkpoint", and "DNA repair" (Figure 5).

The *GOLT1B* gene encodes for the human vesicle transport protein (Golgi Transport 1B) GOT1B protein [36]. GOLT1B might be overexpressed in various tumors because of the amplification of the chromosome 12p region [37]. Recent studies show that overexpression of GOLT1B in breast and colorectal cancer might be associated with poorer outcomes due to the promotion of immune evasion [38,39]. Consistent with that, we found that in high 2-PS MPM, patients from all three cohorts in our analysis, "Epithelial mesenchymal transition", "Apical junction", and "Protein secretion pathways", were significantly enriched in the gene expression profiles (Figure 5).

Recent reports regarding the role of GOLT1B in immune evasion let us investigate whether our 2-PS correlated with the estimated fractions of immune cells within the tumor tissue. We used the now standard deconvolution algorithm to obtain those fractions and were able to demonstrate that 2-PS correlated with the CD8+ T cells and M1/2 macrophage content (Figure 6). Using a similar approach, Blum et al. demonstrated that epithelioid-like morphology and transcriptomic profile correlated with an estimated fraction of CD8+ T cells [9]. Nguen et al. also demonstrated that inferred infiltrating immune cell fractions can be combined with genomic parameters to develop prognostic models in MPM [40]. Finally, another recent study showed that markers for higher levels of systemic inflammation correlated with shorter overall survival in MPM patients [41]. Our observations in the context of those studies obviously suggest that immune-based markers are to be included in prognostic schemes for MPM patients. In addition, it is rational to expect that they may have predictive power for the success of immune-checkpoint inhibitors (ICIs)-based therapy in MPMs [4,6].

However, even in the era of ICIs, combinations with conventional chemotherapy or targeted therapy may yield additional clinical benefits in MPM. Therefore, we further evaluated our 2-PS as a possible marker to predict the sensitivity of MPM cell lines to small molecule drugs. Interestingly, 2-PS inversely correlated with AUC values for cisplatin, suggesting that it may predict higher sensitivity to cisplatin. The opposite observation was made for two other common chemotherapeutics, gemcitabine and vinblastine, suggesting that our 2-PS can predict resistance to those two. These findings also suggest that the performance of the 2-PS in various datasets might be highly dependent on the therapeutic approach used in any cohort and its further evaluation need to focus on uniformly treated MPM patients.

5. Conclusions

In sum, here we demonstrated the development of a 2-gene expression-based prognostic score in MPM with initial filtration of features based on predicted gene dependency of MPM. Our score was further validated in two independent cohorts. Furthermore, it obviously defines patient subgroups with specific gene profile expressions, underlying immune surveillance mechanisms, and drug sensitivity. Our 2-PS can be tested in a prospective fashion using readily available pathological techniques such as RT-PCR and immunohistochemistry. Finally, our approach to the development of the 2-PS can be applied to other cancer types.

Supplementary Materials: The following supporting information can be downloaded at: https://www.mdpi.com/article/10.3390/diagnostics13091556/s1.

Author Contributions: V.S. proposed research, collected and analyzed data, supervised and coordinated work, and wrote the manuscript. G.B. collected data. A.Y. analyzed data. All authors have read and agreed to the published version of the manuscript.

Funding: Velizar Shivarov's work on cancer immunology is partially supported by NSF (Bulgaria) project KP-06-H41/2.

Institutional Review Board Statement: This study was an exempt from institutional review board assessment as it uses only publicly available data.

Informed Consent Statement: Not applicable.

Data Availability Statement: All data used in this study are publicly available and the sources are referenced in the main text accordingly.

Conflicts of Interest: The authors declare no conflict of interest.

References

1. Alpert, N.; van Gerwen, M.; Taioli, E. Epidemiology of mesothelioma in the 21st century in Europe and the United States, 40 years after restricted/banned asbestos use. *Transl. Lung Cancer Res.* **2020**, *9* (Suppl. 1), S28. [CrossRef]
2. Shao, C.; Li, G.; Huang, L.; Pruitt, S.; Castellanos, E.; Frampton, G.; Carson, K.R.; Snow, T.; Singal, G.; Fabrizio, D. Prevalence of high tumor mutational burden and association with survival in patients with less common solid tumors. *JAMA Netw. Open* **2020**, *3*, e2025109. [CrossRef]

3. Hmeljak, J.; Sanchez-Vega, F.; Hoadley, K.A.; Shih, J.; Stewart, C.; Heiman, D.; Tarpey, P.; Danilova, L.; Drill, E.; Gibb, E.A. Integrative molecular characterization of malignant pleural mesothelioma. *Cancer Discov.* **2018**, *8*, 1548–1565. [CrossRef]
4. Fennell, D.A.; Ewings, S.; Ottensmeier, C.; Califano, R.; Hanna, G.G.; Hill, K.; Danson, S.; Steele, N.; Nye, M.; Johnson, L. Nivolumab versus placebo in patients with relapsed malignant mesothelioma (CONFIRM): A multicentre, double-blind, randomised, phase 3 trial. *Lancet Oncol.* **2021**, *22*, 1530–1540. [CrossRef]
5. Peters, S.; Scherpereel, A.; Cornelissen, R.; Oulkhouir, Y.; Greillier, L.; Kaplan, M.; Talbot, T.; Monnet, I.; Hiret, S.; Baas, P. First-line nivolumab plus ipilimumab versus chemotherapy in patients with unresectable malignant pleural mesothelioma: 3-year outcomes from CheckMate 743. *Ann. Oncol.* **2022**, *33*, 488–499. [CrossRef] [PubMed]
6. Yap, T.A.; Nakagawa, K.; Fujimoto, N.; Kuribayashi, K.; Guren, T.K.; Calabrò, L.; Shapira-Frommer, R.; Gao, B.; Kao, S.; Matos, I. Efficacy and safety of pembrolizumab in patients with advanced mesothelioma in the open-label, single-arm, phase 2 KEYNOTE-158 study. *Lancet Respir. Med.* **2021**, *9*, 613–621. [CrossRef] [PubMed]
7. Wadowski, B.; De Rienzo, A.; Bueno, R. The molecular basis of malignant pleural mesothelioma. *Thorac. Surg. Clin.* **2020**, *30*, 383–393. [CrossRef] [PubMed]
8. Bueno, R.; Stawiski, E.W.; Goldstein, L.D.; Durinck, S.; De Rienzo, A.; Modrusan, Z.; Gnad, F.; Nguyen, T.T.; Jaiswal, B.S.; Chirieac, L.R. Comprehensive genomic analysis of malignant pleural mesothelioma identifies recurrent mutations, gene fusions and splicing alterations. *Nat. Genet.* **2016**, *48*, 407–416. [CrossRef]
9. Blum, Y.; Meiller, C.; Quetel, L.; Elarouci, N.; Ayadi, M.; Tashtanbaeva, D.; Armenoult, L.; Montagne, F.; Tranchant, R.; Renier, A. Dissecting heterogeneity in malignant pleural mesothelioma through histo-molecular gradients for clinical applications. *Nat. Commun.* **2019**, *10*, 1333. [CrossRef]
10. Yu, C.; Mannan, A.M.; Yvone, G.M.; Ross, K.N.; Zhang, Y.-L.; Marton, M.A.; Taylor, B.R.; Crenshaw, A.; Gould, J.Z.; Tamayo, P. High-throughput identification of genotype-specific cancer vulnerabilities in mixtures of barcoded tumor cell lines. *Nat. Biotechnol.* **2016**, *34*, 419–423. [CrossRef]
11. Meyers, R.M.; Bryan, J.G.; McFarland, J.M.; Weir, B.A.; Sizemore, A.E.; Xu, H.; Dharia, N.V.; Montgomery, P.G.; Cowley, G.S.; Pantel, S. Computational correction of copy number effect improves specificity of CRISPR–Cas9 essentiality screens in cancer cells. *Nat. Genet.* **2017**, *49*, 1779–1784. [CrossRef] [PubMed]
12. Tsherniak, A.; Vazquez, F.; Montgomery, P.G.; Weir, B.A.; Kryukov, G.; Cowley, G.S.; Gill, S.; Harrington, W.F.; Pantel, S.; Krill-Burger, J.M. Defining a cancer dependency map. *Cell* **2017**, *170*, 564–576.e16. [CrossRef] [PubMed]
13. Boehm, J.S.; Golub, T.R. An ecosystem of cancer cell line factories to support a cancer dependency map. *Nat. Rev. Genet.* **2015**, *16*, 373–374. [CrossRef] [PubMed]
14. Cho, H.J.; Yu, A.; Kim, S.; Kang, J.; Hong, S.M. Robust likelihood-based survival modeling with microarray data. *J. Stat. Softw.* **2009**, *29*, 1–16. [CrossRef]
15. Budczies, J.; Klauschen, F.; Sinn, B.V.; Gyorffy, B.; Schmitt, W.D.; Darb-Esfahani, S.; Denkert, C. Cutoff Finder: A comprehensive and straightforward Web application enabling rapid biomarker cutoff optimization. *PLoS ONE* **2012**, *7*, e51862. [CrossRef]
16. Shivarov, V.; Dolnik, A.; Lang, K.M.; Krönke, J.; Kuchenbauer, F.; Paschka, P.; Gaidzik, V.I.; Döhner, H.; Schlenk, R.F.; Döhner, K. MicroRNA expression-based outcome prediction in acute myeloid leukemia: Novel insights through cross-platform integrative analyses. *Haematologica* **2016**, *101*, e454. [CrossRef]
17. Subramanian, A.; Tamayo, P.; Mootha, V.K.; Mukherjee, S.; Ebert, B.L.; Gillette, M.A.; Paulovich, A.; Pomeroy, S.L.; Golub, T.R.; Lander, E.S.; et al. Gene set enrichment analysis: A knowledge-based approach for interpreting genome-wide expression profiles. *Proc. Natl. Acad. Sci. USA* **2005**, *102*, 15545–15550. [CrossRef]
18. Newman, A.M.; Liu, C.L.; Green, M.R.; Gentles, A.J.; Feng, W.; Xu, Y.; Hoang, C.D.; Diehn, M.; Alizadeh, A.A. Robust enumeration of cell subsets from tissue expression profiles. *Nat. Methods* **2015**, *12*, 453–457. [CrossRef]
19. Thorsson, V.; Gibbs, D.L.; Brown, S.D.; Wolf, D.; Bortone, D.S.; Yang, T.-H.O.; Porta-Pardo, E.; Gao, G.F.; Plaisier, C.L.; Eddy, J.A. The immune landscape of cancer. *Immunity* **2018**, *48*, 812–830.e14. [CrossRef]
20. Newman, A.M.; Steen, C.B.; Liu, C.L.; Gentles, A.J.; Chaudhuri, A.A.; Scherer, F.; Khodadoust, M.S.; Esfahani, M.S.; Luca, B.A.; Steiner, D. Determining cell type abundance and expression from bulk tissues with digital cytometry. *Nat. Biotechnol.* **2019**, *37*, 773–782. [CrossRef]
21. Iorio, F.; Knijnenburg, T.A.; Vis, D.J.; Bignell, G.R.; Menden, M.P.; Schubert, M.; Aben, N.; Gonçalves, E.; Barthorpe, S.; Lightfoot, H. A landscape of pharmacogenomic interactions in cancer. *Cell* **2016**, *166*, 740–754. [CrossRef] [PubMed]
22. Yang, W.; Soares, J.; Greninger, P.; Edelman, E.J.; Lightfoot, H.; Forbes, S.; Bindal, N.; Beare, D.; Smith, J.A.; Thompson, I.R. Genomics of Drug Sensitivity in Cancer (GDSC): A resource for therapeutic biomarker discovery in cancer cells. *Nucleic Acids Res.* **2012**, *41*, D955–D961. [CrossRef]
23. Mandrekar, J.N. Receiver operating characteristic curve in diagnostic test assessment. *J. Thorac. Oncol.* **2010**, *5*, 1315–1316. [CrossRef]
24. Pass, H.I.; Liu, Z.; Wali, A.; Bueno, R.; Land, S.; Lott, D.; Siddiq, F.; Lonardo, F.; Carbone, M.; Draghici, S. Gene expression profiles predict survival and progression of pleural mesothelioma. *Clin. Cancer Res.* **2004**, *10*, 849–859. [CrossRef]
25. De Reynies, A.; Jaurand, M.-C.; Renier, A.; Couchy, G.; Hysi, I.; Elarouci, N.; Galateau-Sallé, F.; Copin, M.-C.; Hofman, P.; Cazes, A. Molecular classification of malignant pleural mesothelioma: Identification of a poor prognosis subgroup linked to the epithelial-to-mesenchymal transition. *Clin. Cancer Res.* **2014**, *20*, 1323–1334. [CrossRef]

26. Gordon, G.J.; Jensen, R.V.; Hsiao, L.-L.; Gullans, S.R.; Blumenstock, J.E.; Richards, W.G.; Jaklitsch, M.T.; Sugarbaker, D.J.; Bueno, R. Using gene expression ratios to predict outcome among patients with mesothelioma. *J. Natl. Cancer Inst.* **2003**, *95*, 598–605. [CrossRef] [PubMed]
27. Gordon, G.J.; Rockwell, G.N.; Jensen, R.V.; Rheinwald, J.G.; Glickman, J.N.; Aronson, J.P.; Pottorf, B.J.; Nitz, M.D.; Richards, W.G.; Sugarbaker, D.J. Identification of novel candidate oncogenes and tumor suppressors in malignant pleural mesothelioma using large-scale transcriptional profiling. *Am. J. Pathol.* **2005**, *166*, 1827–1840. [CrossRef] [PubMed]
28. Gordon, G.J.; Dong, L.; Yeap, B.Y.; Richards, W.G.; Glickman, J.N.; Edenfield, H.; Mani, M.; Colquitt, R.; Maulik, G.; Van Oss, B. Four-gene expression ratio test for survival in patients undergoing surgery for mesothelioma. *J. Natl. Cancer Inst.* **2009**, *101*, 678–686. [CrossRef]
29. Zhou, J.-G.; Zhong, H.; Zhang, J.; Jin, S.-H.; Roudi, R.; Ma, H. Development and validation of a prognostic signature for malignant pleural mesothelioma. *Front. Oncol.* **2019**, *9*, 78. [CrossRef]
30. Bai, Y.; Wang, X.; Hou, J.; Geng, L.; Liang, X.; Ruan, Z.; Guo, H.; Nan, K.; Jiang, L. Identification of a five-gene signature for predicting survival in malignant pleural mesothelioma patients. *Front. Genet.* **2020**, *11*, 899. [CrossRef]
31. López-Ríos, F.; Chuai, S.; Flores, R.; Shimizu, S.; Ohno, T.; Wakahara, K.; Illei, P.B.; Hussain, S.; Krug, L.; Zakowski, M.F. Global gene expression profiling of pleural mesotheliomas: Overexpression of aurora kinases and P16/CDKN2A deletion as prognostic factors and critical evaluation of microarray-based prognostic prediction. *Cancer Res.* **2006**, *66*, 2970–2979. [CrossRef]
32. Musacchio, A.; Hardwick, K.G. The spindle checkpoint: Structural insights into dynamic signalling. *Nat. Rev. Mol. Cell Biol.* **2002**, *3*, 731–741. [CrossRef]
33. Morani, F.; Bisceglia, L.; Rosini, G.; Mutti, L.; Melaiu, O.; Landi, S.; Gemignani, F. Identification of overexpressed genes in malignant pleural mesothelioma. *Int. J. Mol. Sci.* **2021**, *22*, 2738. [CrossRef] [PubMed]
34. Suraokar, M.; Nunez, M.; Diao, L.; Chow, C.; Kim, D.; Behrens, C.; Lin, H.; Lee, S.; Raso, G.; Moran, C. Expression profiling stratifies mesothelioma tumors and signifies deregulation of spindle checkpoint pathway and microtubule network with therapeutic implications. *Ann. Oncol.* **2014**, *25*, 1184–1192. [CrossRef] [PubMed]
35. Busacca, S.; O'Regan, L.; Singh, A.; Sharkey, A.J.; Dawson, A.G.; Dzialo, J.; Parsons, A.; Kumar, N.; Schunselaar, L.M.; Guppy, N. BRCA1/MAD2L1 Deficiency Disrupts the Spindle Assembly Checkpoint to Confer Vinorelbine Resistance in MesotheliomaBRCA1/MAD2L1 Expression Predicts Response to Vinorelbine. *Mol. Cancer Ther.* **2021**, *20*, 379–388. [CrossRef]
36. Conchon, S.; Cao, X.; Barlowe, C.; Pelham, H.R. Got1p and Sft2p: Membrane proteins involved in traffic to the Golgi complex. *EMBO J.* **1999**, *18*, 3934–3946. [CrossRef] [PubMed]
37. Bourdon, V.; Naef, F.; Rao, P.H.; Reuter, V.; Mok, S.C.; Bosl, G.J.; Koul, S.; Murty, V.V.; Kucherlapati, R.S.; Chaganti, R. Genomic and expression analysis of the 12p11-p12 amplicon using EST arrays identifies two novel amplified and overexpressed genes. *Cancer Res.* **2002**, *62*, 6218–6223.
38. Liu, T.; Liu, B.; Liu, Y.; Feng, X.; Jiang, X.; Long, J.; Gao, Q.; Yang, Z. Vesicle transporter GOLT1B mediates the cell membrane localization of DVL2 and PD-L2 and promotes colorectal cancer metastasis. *Cancer Cell Int.* **2021**, *21*, 287. [CrossRef]
39. Liu, J.; Zhang, W.; Cai, W.; Chen, Y.; Cai, X.; Tang, D.; Tang, M.; Dai, Y. Multi-omics analyses revealed GOLT1B as a potential prognostic gene in breast cancer probably regulating the immune microenvironment. *Front. Oncol.* **2022**, *11*, 5871. [CrossRef]
40. Nguyen, T.T.; Lee, H.-S.; Burt, B.M.; Amos, C.I.; Cheng, C. A combination of intrinsic and extrinsic features improves prognostic prediction in malignant pleural mesothelioma. *Br. J. Cancer* **2022**, *127*, 1691–1700. [CrossRef]
41. Fournel, L.; Charrier, T.; Huriet, M.; Iaffaldano, A.; Lupo, A.; Damotte, D.; Arrondeau, J.; Alifano, M. Prognostic impact of inflammation in malignant pleural mesothelioma: A large-scale analysis of consecutive patients. *Lung Cancer* **2022**, *166*, 221–227. [CrossRef] [PubMed]

diagnostics

MDPI

Article

Diagnostic and Prognostic Nomograms for Hepatocellular Carcinoma Based on PIVKA-II and Serum Biomarkers

Shu An [1,†], Xiaoxia Zhan [1,†], Min Liu [1], Laisheng Li [1,*] and Jian Wu [2,*]

1 Department of Laboratory Medicine, The First Affiliated Hospital, Sun Yat-sen University, Guangzhou 510080, China
2 Center of Hepato-Pancreato-Biliary Surgery, The First Affiliated Hospital, Sun Yat-sen University, Guangzhou 510080, China
* Correspondence: lilaish@mail.sysu.edu.cn (L.L.); wujian3@mail.sysu.edu.cn (J.W.); Tel./Fax: +86-20-28823350-8464 (L.L.); +86-20-28823350-8214 (J.W.)
† These authors contributed equally to this work.

Abstract: Background: The aim of the present study was to develop an improved diagnostic and prognostic model for HBV-associated HCC by combining AFP with PIVKA-II and other potential serum/plasma protein biomarkers. Methods: A total of 578 patients, including 352 patients with HBV-related HCC, 102 patients with HBV-associated liver cirrhosis (LC), 124 patients with chronic HBV, and 127 healthy subjects (HS), were enrolled in the study. The serum levels of AFP PIVKA-II, and other laboratory parameters were collected. Univariate and multivariate logistic regression and Cox regression analyses were performed to identify independent diagnostic and prognostic factors, respectively. The diagnostic efficacy of the nomogram was evaluated using receiver operator curve (ROC) analysis and the prognostic performance was measured by Harrell's concordance index (C-index). Results: AFP and PIVKA-II levels were significantly increased in HBV-related HCC, compared with those in HBV-associated LC and chronic HBV participants ($p < 0.05$ and $p < 0.001$, respectively). The diagnostic nomogram, which included age, gender, AFP, PIVKA-II, prothrombin time (PT), and total protein (TP), discriminated patients with HBV-HCC from those with HBV-LC or chronic HBV with an AUC of 0.970. In addition, based on the univariate and multivariate Cox regression analysis, PIVKA-II, γ-glutamyl transpeptidase, and albumin were found to be significantly associated with the prognosis of HBV-related HCC and were incorporated into a nomogram. The C-index of the nomogram for predicting 3-year survival in the training and validation groups was 0.75 and 0.78, respectively. The calibration curves for the probability of 3-year OS showed good agreement between the nomogram prediction and the actual observation in the training and the validation groups. Furthermore, the nomogram had a higher C-index (0.74) than that of the Child–Pugh grade (0.62), the albumin–bilirubin (ALBI) score (0.64), and Barcelona Clinic Liver Cancer (0.56) in all follow-up cases. Conclusion: Our study suggests that the nomograms based on AFP, PIVKA-II, and potential serum protein biomarkers showed a better performance in the diagnosis and prognosis of HCC, which may help to guide therapeutic strategies and assess the prognosis of HCC.

Keywords: hepatocellular carcinoma; PIVKA-II; AFP; nomogram; diagnosis; prognosis

Citation: An, S.; Zhan, X.; Liu, M.; Li, L.; Wu, J. Diagnostic and Prognostic Nomograms for Hepatocellular Carcinoma Based on PIVKA-II and Serum Biomarkers. *Diagnostics* **2023**, *13*, 1442. https://doi.org/10.3390/diagnostics13081442

Academic Editor: Gian Paolo Caviglia

Received: 27 February 2023
Revised: 6 April 2023
Accepted: 7 April 2023
Published: 17 April 2023

1. Introduction

Hepatocellular carcinoma (HCC) is one of the most prevalent cancers and one of the leading causes of cancer-related death globally, with an estimated 905,677 new cases and 830,180 new deaths in 2020 [1]. The dominant histologic type of primary liver cancer is hepatocellular carcinoma, accounting for nearly 75–90% of cases [2]. The majority of cases of liver cirrhosis and hepatocellular carcinoma (HCC) are attributed to persistent infections with the hepatitis B virus (HBV) and hepatitis C virus (HCV) [3]. Moreover, the high number of chronic HBV carriers are responsible for the high prevalence of HCC in China [4]. Unfortunately, lack of surveillance and inadequate early diagnosis account for the poor prognosis and high mortality of HCC [5,6]. Therefore, the accurate and effective diagnosis of HCC is essential to rapidly determine potentially curative therapies, such as liver resection or transplantation [6,7].

Currently, abdominal ultrasound with serum alpha-fetoprotein (AFP) assessment is recommended for the early detection of HCC according to the guidelines [6,8]. Significant heterogeneity in tumor size, equipment, and research experience compromises the efficiency of abdominal ultrasound in HCC detection [9,10]. In addition, HBV and cirrhosis patients with serum AFP levels up to 20 ng/mL at 12 months after entecavir treatment are most likely to develop into HCC [11,12]. However, approximately 30% of patients with liver cancer are always negative for serum AFP [13]. AFP detection has a sensitivity of only 46% to 59% for clinical HCC diagnosis and only 40% for preclinical prediction [14,15]. Therefore, the sensitivity and specificity of these methods in the screening and diagnosis of small and early HCC remain inefficient [16]. There is an urgent need to explore new strategies for the identification of high-risk groups for liver cancer and for the screening of HCC patients, especially those with early-stage HCC, AFP-negative HCC, and micro-liver cancer in the subclinical stage [6].

Protein-induced vitamin K absence or antagonist II (PIVKA-II) is widely recognized as a reliable biomarker for the diagnosis, prognosis, treatment response, and recurrence monitoring of HCC [17–19]. Studies have demonstrated that PIVKA-II levels of 40 mAU/mL are indicative of early HCC with sensitivity and specificity rates of 64% and 89%, respectively [20]. Furthermore, PIVKA-II shows a significant diagnostic value for AFP-negative liver cancer [21,22]. Although PIVKA-II is noted as a highly specific biomarker for HCC, the limited sensitivity of PIVKA-II raises concerns about its efficacy as a surveillance biomarker, as alcohol-related liver disease, obstructive jaundice, and cholestasis can also influence the expression of PIVKA-II [23,24]. Therefore, combination studies of AFP and PIVKA-II have been recommended to improve the sensitivity and specificity of HCC screening [15].

Several serum biomarkers have been shown to assist in the diagnosis and prognosis of HCC, including liver enzyme indicators (aspartate aminotransferase (AST), alanine aminotransferase (ALT), alkaline phosphatase (ALKP), and γ-glutamyl transpeptidase (GGT)), or indicators reflecting liver metabolism (total bilirubin (TBIL)) and protein synthesis function (total protein (TP), albumin (ALB), and prothrombin time (PT)) [25–30]. To fully explore and harness the potential benefits of different biomarkers, it is necessary to identify potential serum/plasma protein biomarkers that can be combined with AFP and PIVKA-II to facilitate HCC detection and surveillance. The present study aimed at optimizing AFP and PIVKA-II related diagnostic and prognostic models by integrating potential serum/plasma protein biomarkers in HCC.

2. Materials and Methods

2.1. Data Collection

From January 2014 to December 2016, individuals with HBV infection, HBV-related liver cirrhosis (HBV-LC), and HBV-HCC, as well as healthy subjects (HS) from the First Affiliated Hospital of Sun Yat-sen University (FAH-SYSU), were enrolled in the training cohort. From June 2017 to December 2018, individuals with HBV, HBV-LC, HBV-HCC, and HS from FAH-SYSU were recruited for the validation cohort. The exclusion criteria were as follows: (a) history of previous treatment (hepatic resection, liver transplant, trans-arterial chemoembolization, radiofrequency, anti-angiogenetic drugs, and warfarin therapy); (b) Child–Pugh C; (c) obstructive jaundice; (d) estimated creatinine clearance <30 mL/min; (e) diagnosis of second extrahepatic neoplasia; and (f) metastasis. HBV infection was defined as hepatitis B surface antigen (HBsAg) positivity within the previous 6 months. HBV-HCC was defined as HCC with HBV infection, excluding alcoholic liver disease or hepatitis C virus infection [6]. Patients with HCC met the diagnostic criteria for HCC, including imaging evidence (ultrasound, computed tomography, and MRI) and histopathological confirmation [6]. Liver cirrhosis was diagnosed based on clinical parameters, including histologic examination, laboratory tests, and radiologic or endoscopic evidence of cirrhosis [31]. HS were blood donors with no history of chronic liver disease or gastrointestinal malignancy. The study was approved by the Institutional Review Boards at FAH-SYSU and informed consent was obtained from each participant

2.2. Laboratory Methods

Peripheral blood was collected from each participant and centrifuged at 800× g for 10 min. The serum was aliquoted and immediately frozen at −80 °C until testing. Serum concentrations of AFP, PIVKA-II, carcinoembryonic antigen (CEA), and carbohydrate antigen 199 (CA199) were determined using the ARCHITECT immunoassay according to protocol (Abbott Diagnostics). Clinical laboratory test results, including biochemical indices, blood routine results, and coagulation function results, were collected from routine clinical practice. The clinical laboratory test results included glutamic-oxalacetic transaminase (AST), glutamic-pyruvic transaminase (ALT), albumin (ALB), total protein (TP), lactate dehydrogenase (LDH), total bilirubin (TBIL), direct bilirubin (DBIL), γ-glutamyl transpeptidase (GGT), and HBV surface antigen (HBsAg). Blood routine indexes included white blood cell (WBC), neutrophil (NET), lymphocyte (LY), neutrophil-to-lymphocyte ratio (NLR), red blood cell (RBC), red cell distribution width (RDW), hemoglobin (Hb), platelet count (PLT), and mean platelet volume (MPV). Indicators of the coagulation function included prothrombin time (PT), thrombin time (TT), activated partial thromboplastin time (APTT), fibrinogen (FIB), and international normalized ratio (INR). All of the biochemical indices, blood routine indexes, and coagulation function results were acquired via standard automated laboratory methods and utilizing commercially available kits according to the manufacturer's protocols.

2.3. Statistical Analysis

For demographic data, categorical variables were expressed as a mean ± standard deviation (SD), range, or ratio. Univariate analysis was performed using the Mann−Whitney U-test, chi-squared test, or Fisher's exact test, as appropriate. Variables with right-skewed distributions, including AFP and PVIKA-II, were log-transformed before logistic or Cox regression analysis. OS was defined as the interval from the date of treatment to the date of patient death or lost follow-up. Univariate and multivariate logistic regression and Cox regression analyses were performed to identify independent diagnostic and prognostic factors, respectively. Variables with $p < 0.05$ were included in the multivariate regression analyses. The nomogram was constructed based on the results of the multivariate analysis utilizing the rms package. The discriminative performance of the diagnostic nomogram model was examined using area under the curve (AUC) analysis [32]. In addition, the final variables for the construction of the prognostic nomogram were selected using the backward step-down method based on the Akaike information criterion (AIC) [33]. To evaluate the discriminative ability of the prognostic nomogram, Harrell's concordance index (C-index) and receiver operating characteristic (ROC) curve were used. Calibration curves were plotted to evaluate the predictive accuracy of the prognostic nomogram. Furthermore, the discriminatory ability of the prognostic nomograms was compared with the HCC prognostic model of Barcelona Clinic Liver Cancer, ALBI score, and Child−Pugh grade by analyzing the ROC curves. The results were considered statistically significant if the p value was less than 0.05. Statistical analysis was accomplished using the SPSS software version 25.0 and R (version 3.4.2).

3. Results

3.1. Characteristics of Patients

A total of 705 subjects were included in the study. The clinical laboratory characteristics of the patients with HBV-related HCC, HBV-associate liver cirrhosis (LC), chronic HBV infection, and healthy subjects are shown in Table 1. Consistent with previous studies, the serum concentrations of liver function biomarkers AST and ALT were significantly higher in patients with HCC than in other patients ($p < 0.001$). Serum levels of ALB, TP,

PLT, and Hb were lower in the HCC and cirrhosis groups than in the hepatitis and normal control groups. In addition, TBIL was increased moderately in the HCC group, but was significantly elevated in the LC group. The time of PT was the longest in the cirrhosis group, and PLT and Hb were dramatically decreased in the cirrhosis group.

Table 1. Clinical characteristics of the study population.

Parameters	HCC (N = 352)	Liver Cirrhosis (N = 102)	HBV (N = 124)	Healthy Subjects (N = 127)
Age, years mean (SD)	54 (11.8)	55.7 (12.7)	36.3 (10.8)	50.1 (10.7)
Gender, n (%)				
Male	309 (87.8%)	67 (65.7%)	83 (66.9%)	69 (54.3%)
Female	43 (12.2%)	35 (34.3%)	41 (33.1%)	58 (45.7%)
AST (U/L) mean (SD)	92.2 (135.2)	79.2 (91.9)	40.0 (59.5)	24.1 (6.1)
ALT (U/L) mean (SD)	79.4 (190.2)	52.1 (56.5)	45.5 (120.8)	24.0 (13.6)
TBIL, μmol/L mean (SD)	32.5 (57.0)	60.7 (96.2)	17.0 (14.5)	12.4 (3.9)
ALB (g/L) mean (SD)	36.3 (5.4)	32.7 (6.6)	45.6 (3.0)	44.5 (3.8)
TP (g/L) mean (SD)	66.5 (8.5)	66.2 (9.0)	75.4 (4.7)	73.1 (4.8)
PLT ($\times 10^9$/L) mean (SD)	174.3 (90.3)	124.2 (87.8)	214.5 (47.9)	244.6 (53.8)
Hb (g/L) mean (SD)	129.5 (23.6)	106.7 (27.4)	145.2 (16.7)	146.1 (14.8)
PT, s mean (SD)	13.2 (2.8)	15.7 (4.5)	12.8 (2.0)	13.0 (1.1)

Categorical and continuous variables are presented as frequencies or mean (SD), respectively. AST: aspartate aminotransferase; ALT: alanine aminotransferase; TBIL: total bilirubin; ALB: albumin; TP: total protein; PLT: platelets; Hb: hemoglobin; PT: prothrombin time.

3.2. AFP and PIVKA-II Distribution among Disease Groups

As depicted in Figure 1, there were significant differences between the HCC group and the other groups in the serum concentration of PIVKA-II and AFP ($p < 0.001$ and $p < 0.05$, respectively). Among 352 patients with HCC, 125 (35.51%) were AFP negative when the positive screening value of AFP was defined as >20 ng/mL [34]. Similarly, when the positive screening value of PIVKA-II was defined as >40 mAU/mL [35], 84.37% of patients with HCC had increased serum levels of PIVKA-II. Notably, elevated levels of PIVKA-II were observed in 100 (80.0%) AFP negative patients. Furthermore, the sensitivity for AFP in combination with PIVKA-II to predict HCC was 92.89% (Supplementary Table S1). These results suggest that PIVKA-II is a complementary biomarker to AFP in differentiating HBV-HCC.

3.3. Diagnostic Factors of HCC Based on Logistic Regression

The results based on the univariate and multivariate logistic regression analysis to assess the predictive value of biomarkers for HCC are shown in Table 2. The results of the univariate regression analysis indicate that night variables, including age, gender, AFP, PIVKA-II, AST, ALT, ALB, PT, and TP, were significant predictors of HCC when compared with HBV infection or liver cirrhosis. In the multivariate logistic regression analysis, increasing age (OR: 1.08, 95% CI 1.050–1.116; $p < 0.001$), male gender (OR: 2.662, 95% CI 1.159–6.117; $p = 0.021$), elevated AFP (OR: 8.291, 95% CI 4.366–15.743; $p < 0.001$), increased PIVKA-II (OR: 12.231, 95% CI 5.853–25.559; $p < 0.001$), decreased TP (OR: 0.912, 95% CI 0.871–0.954; $p < 0.001$), and prolonged PT (OR: 0.839, 95% CI 0.752–0.935; $p = 0.002$) were independent predictors of HCC.

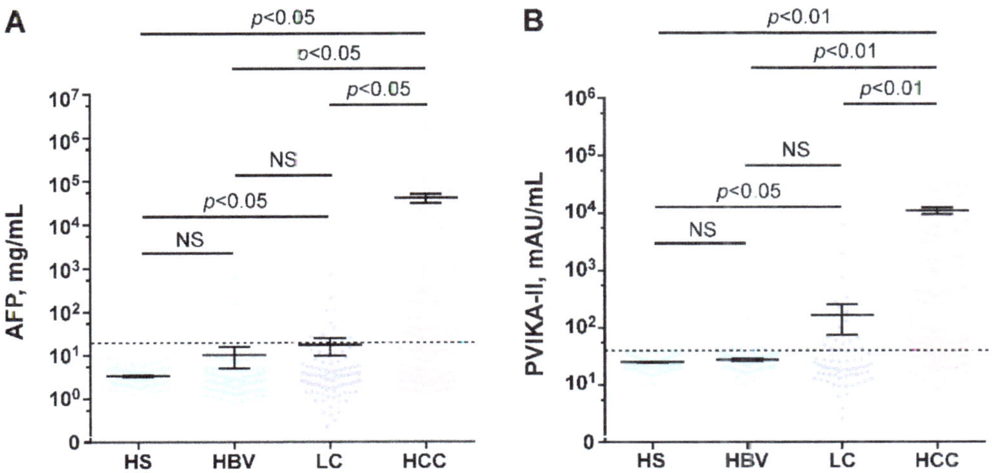

Figure 1. Serum AFP and PIVKA-II concentrations in different populations. (**A**) Serum AFP levels in healthy subjects (HS), chronic HBV infections (HBV), HBV-associate liver cirrhosis (LC), and patients with HBV-related HCC. (**B**) Serum PIVKA-II levels in healthy subjects (HS), chronic HBV infection (HBV), HBV-associated liver cirrhosis (LC), and patients with HBV-related HCC. AFP reference line: 20 ng/mL and PIVKA-II reference line: 40 mAU/mL.

Table 2. Univariate and multivariate logistic regression analysis between HCC and liver cirrhosis or HBV infection groups.

Parameters	Univariate Analysis		Multivariate Analysis	
	OR (95% CI)	***p* Value**	**OR (95% CI)**	***p* Value**
Age	1.052 (1.037–1.066)	<0.001	1.082 (1.05–1.116)	<0.001
Gender (male)	3.641 (2.388–5.551)	<0.001	2.662 (1.159–6.117)	0.021
Log_{10}AFP	6.572 (4.515–9.566)	<0.001	8.291 (4.366–15.743)	<0.001
Log_{10}PVIKA-II	17.056 (9.366–31.059)	<0.001	12.231 (5.853–25.559)	<0.001
AST	1.004 (1.002–1.007)	0.001	0.996 (0.989–1.004)	0.339
ALT	1.004 (1.001–1.006)	0.01	1.009 (0.999–1.019)	0.078
TBIL	0.999 (0.996–1.002)	0.426		
ALB	0.923 (0.899–0.948)	<0.001	1.073 (0.998–1.153)	0.055
TP	0.933 (0.912–0.954)	<0.001	0.912 (0.871–0.954)	<0.001
PLT	1.00 (0.998–1.002)	0.968		
Hb	1.002 (0.996–1.009)	0.487		
PT	0.918 (0.862–0.977)	0.007	0.839 (0.752–0.935)	0.002

Based on the results of the multivariate regression analysis, risk factors including age, gender, AFP, PIVKA-II, PT, and TP were incorporated into a nomogram, referred to as the

APPT grade. According to the nomogram, log-transformed AFP and PIVKA-II values had the most significant impact on early HCC diagnosis, followed by PT, TP, age, and sex. The discriminative power of the nomogram was evaluated using ROC curves and the AUC of the diagnostic nomogram was 0.970 (Figure 2). The ROC analysis indicated that the diagnostic nomogram demonstrated an excellent ability to discriminate HCC from CHB and LC.

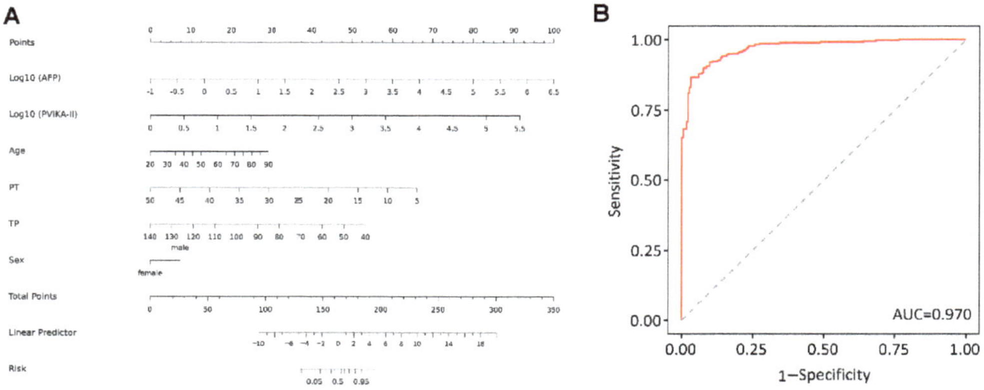

Figure 2. Diagnostic nomogram for patients with HCC. (**A**) A diagnostic nomogram for differentiating HCC cases from chronic HBV and LC. (**B**) AUROC for the diagnostic nomogram in differentiating HCC cases from chronic HBV and LC.

3.4. Clinical Characteristics of HCC Patients Enrolled in Survival Analysis

A total of 241 HCC patients were successfully followed up for 3 years. Correspondingly, in our data, 106 patients (44%) died during the 3-year follow-up. Non-survivors had significantly lower levels of ALB and higher levels of PIVKA-II, CA199, RDW, AST, GGT, TBIL, DBIL, PT, and FIB (Table 3, $p < 0.05$).

Table 3. Characteristics of HCC patients according to 3-year mortality.

Parameters	Survivor (n = 135)	Non-Survivor (n = 106)	*p* Value
Age (years) Median (IQR)	55.0 (48.0–63.0)	54.5 (48–65.0)	0.730 [a]
Gender Male (%)	117 (86.7%)	93 (87.7%)	0.806 [b]
Child-Pugh grade Number (%)			<0.001 [b]
A	120 (88.9%)	69 (65.1%)	
B	15 (11.1%)	37 (34.9%)	
BCLC stage Number (%)			0.092 [b]
A	26 (19.3%)	19 (17.9%)	
B	81 (60.0%)	52 (49.1%)	
C	28 (20.7%)	35 (33.0%)	
HBsAg Positive (%)	108 (80.0%)	86 (81.1%)	0.826 [b]
AFP (mg/mL) Median (IQR)	49.8 (8.3–719.0)	92.9 (8.5–8406.2)	0.099 [a]
PVIKA-II (mAU/mL) Median (IQR)	296.6 (61.0–1725.3)	1626.3 (123.2–10833.8)	<0.001 [a]
CEA (mg/mL) Median (IQR)	2.8 (1.5–4.2)	2.6 (1.7–4.2)	0.948 [a]

Table 3. *Cont.*

Parameters	Survivor (n = 135)	Non-Survivor (n = 106)	*p* Value
CA199 (mg/mL) Median (IQR)	7.4 (4.1–14.0)	12.5 (5.6–31.8)	0.001 [a]
WBC (×10^9/L) Median (IQR)	7.2 (5.1–10.2)	7.7 (5.5–9.5)	0.858 [a]
LY (×10^9/L) Median (IQR)	1.1 (0.8–1.6)	1.0 (0.7–1.5)	0.240 [a]
NET (×10^9/L) Median (IQR)	5.0 (3.4–7.9)	5.0 (3.1–7.8)	0.918 [a]
NLR Median (IQR)	4.5 (2.7–8.4)	4.6 (2.7–8.9)	0.577 [a]
RBC (×10^9/L) Median (IQR)	4.0 (3.4–4.4)	4.1 (3.4–4.5)	0.230 [a]
Hb (g/L) Median (IQR)	131.0 (117.0–145.5)	130.0 (113.3–142.0)	0.460 [a]
RDW (%) Median (IQR)	13.5 (13.0–14.0)	14.0 (13.0–15.2)	0.049 [a]
PLT (×10^9/L) Median (IQR)	166.0 (113.0–223.0)	163.5 (99.5–243.3)	0.913 [a]
MPV (fL) Median (IQR)	10.0 (9.2–11.1)	10.0 (9.2–10.8)	0.686 [a]
ALT (U/L) Median (IQR)	39.0 (24.0–66.5)	45.0 (32.0–65.8)	0.109 [a]
AST (U/L) Median (IQR)	39.0 (29.0–64.0)	56.5 (37.5–98.5)	<0.001 [a]
GGT Median (IQR)	64.5 (39.2–112.7)	138.5 (72.7–242.7)	<0.001 [a]
LDH Median (IQR)	228.5 (195.5–298.0)	250.5 (206.2–345.5)	0.072 [a]
TBIL (µmol/L) Median (IQR)	16.4 (12.5–24.8)	19.7 (14.1–34.9)	0.004 [a]
DBIL (µmol/L) Median (IQR)	3.8 (2.5–6.8)	5.5 (3.2–12.9)	0.002 [a]
ALB (g/L) Median (IQR)	37.0 (34.8–40.7)	35.2 (31.7–39.0)	0.002 [a]
TP (g/L) Median (IQR)	66.2 (61.3–71.2)	66.9 (61.2–71.6)	0.630 [a]
Cr (µmol/L) Median (IQR)	71.0 (60.0–81.0)	67.0 (56.0–79.0)	0.114 [a]
PT (s) Median (IQR)	12.4 (11.8–13.1)	12.8 (12.1–14.2)	0.004 [a]
INR Median (IQR)	1.1 (1.1–1.2)	1.1 (1.0–1.2)	0.949 [a]
APTT (s) Median (IQR)	29.2 (27.2–33.4)	30.8 (27.6–35.6)	0.066 [a]
TT (s) Median (IQR)	18.3 (17.5–19.1)	17.9 (17.2–19.1)	0.316 [a]
FIB (g/L) Median (IQR)	2.6 (2.1–3.4)	3.1 (2.1–4.0)	0.035 [a]

Categorical and continuous variables are presented as frequencies or medians (IQR), respectively. Differences between groups in categorical and continuous variables are analyzed using the chi-squared test and Mann–Whitney U test, respectively. BCLC: Barcelona Clinic Liver Cancer; HBsAg: HBV surface antigen; CEA: carcinoembryonic antigen; CA199: carbohydrate antigen 199; LY: lymphocyte; NET: neutrophils; NLR: neutrophil-to-lymphocyte ratio; RDW: red cell distribution width; MPV: mean platelet volume; Cr: creatinine; TBIL: total bilirubin; DBIL direct bilirubin; INR: international normalized ratio; APTT: activated partial thromboplastin time; TT: thrombin time; FIB: fibrinogen. [a] Wilcoxon rank sum test; [b] Chi-squared test.

From January 2014 to December 2016, 142 HCC patients were successfully followed up and enrolled in the training group, while from June 2017 to December 2018, 99 HCC patients were followed up and enrolled in the validation group. The characteristics and laboratory indexes of the HCC patients in the training and validation cohorts are shown in Supplementary Table S2. Univariable and multivariable Cox regression analyses were applied to identify the independent prognostic factors for HCC patients in the training cohort (Table 4). Univariate Cox regression analyses revealed that PIVKA-II, GGT, DBIL, FIB, and ALB were significant predictors of survival in HCC patients. Variables with a *p* value of <0.05 in univariable analysis were included in the multivariable analysis. The results of multivariate Cox regression analysis demonstrated that $\log_{10}$PIVKA-II (HR: 1.347, 95% CI: 1.121–1.780), GGT (HR: 1.002, 95% CI: 1.001–1.003), and ALB (HR: 0.932, 95% CI: 0.888–0.979) were independent variables for the prognosis of HCC (Table 4).

Table 4. Univariate and multivariate Cox regression analysis in HCC patients according to 3-year mortality.

Parameters	Univariate Analysis		Multivariate Analysis	
	HR (95% CI)	*p* Value	HR (95% CI)	*p* Value
Log_{10}PVIKA-II	1.573 (1.263–1.958)	<0.001	1.347 (1.053–1.724)	0.018
GGT	1.005 (1.002–1.008)	<0.001	1.002 (1.001–1.003)	0.002
CA199	1.004 (1.002–1.006)	<0.001	1.001 (0.998–1.004)	0.395
AST	1.001 (1.000–1.002)	0.07		
TBIL	1.007 (1.003–1.012)	0.002	1.001 (0.994–1.009)	0.763
DBIL	1.012 (1.005–1.018)	0.001	1.007 (0.999–1.016)	0.263
ALB	0.924 (0.884–0.966)	<0.001	0.932 (0.888–0.979)	0.005
PT	1.082 (0.985–1.188)	0.102		
FIB	1.22 (1.032–1.443)	0.02	1.121 (0.918–1.368)	0.263

3.5. Predictive Potentials of Prognostic Nomogram

The prognostic nomogram was established based on the risk factors PIVKA-II, GGT, and ALB, which were identified by the multivariate analysis. Higher levels of PIVKA-II and GGT and lower levels of ALB during hospitalization were associated with a poorer prognosis for HCC patients. The nomogram based on PIVKA-II, GGT, and ALB was referred to as the PGA grade (Figure 3).

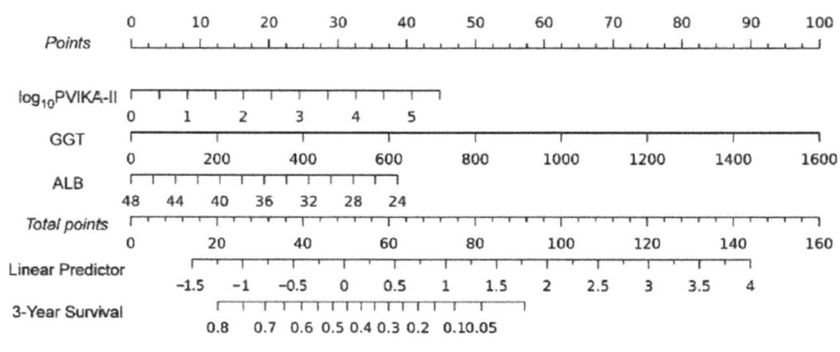

Figure 3. A prognostic nomogram for the survival at the 3-year follow-up in hepatocellular carcinoma.

The discriminative power of the nomogram was assessed using Harrell's concordance index and ROC curves. The C-indexes for the prediction of overall survive in the training and validation groups were 0.75 (95% CI 0.67–0.83) and 0.78 (95% CI 0.68–0.88), respectively (Figure 4). Furthermore, the calibration curves for the probability of 3-year OS demonstrated good agreement between prediction by the nomograms and the actual observation in the training and the validation sets (Figure 4). Therefore, the PGA grade exhibited a great prediction efficiency for the 3-year prognosis of HCC patients.

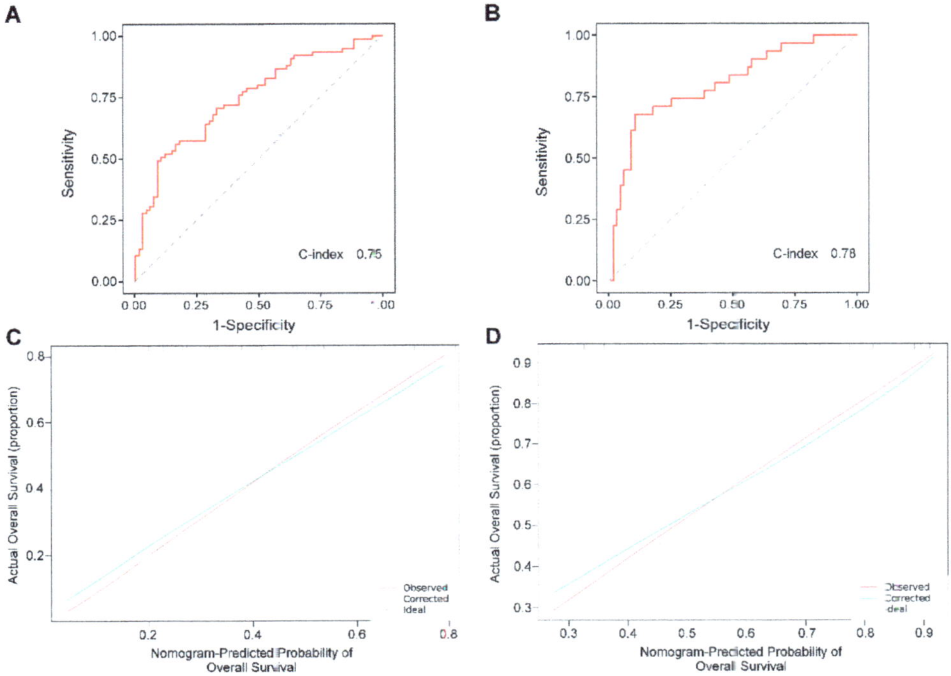

Figure 4. The C-index of the nomogram for the prognostic prediction of OS in the training cohort (**A**) and validation cohort (**B**). The calibration curves of OS based on nomogram prediction and the actual observation in the training cohort (**C**) and validation cohort (**D**).

Child–Pugh grade, ALBI score, and BCLC were utilized to predict the prognosis of HCC. To further evaluate the clinical value of the PGA grade, we compared the performance of the predictive potential of 3-year survival probability in Child–Pugh grade, ALBI score, BCLC, and PGA grade in all follow-up cases. The C-index for OS prediction of the PGA grade was 0.74 (95% CI 0.68–0.80), which was found to be superior to that of the Child–Pugh grade (0.62), ALBI score (0.64), and BCLC (0.56). The result suggests that the PGA grade has better predictive potential than the ALBI score, Child–Pugh grade, and BCLC (Figure 5).

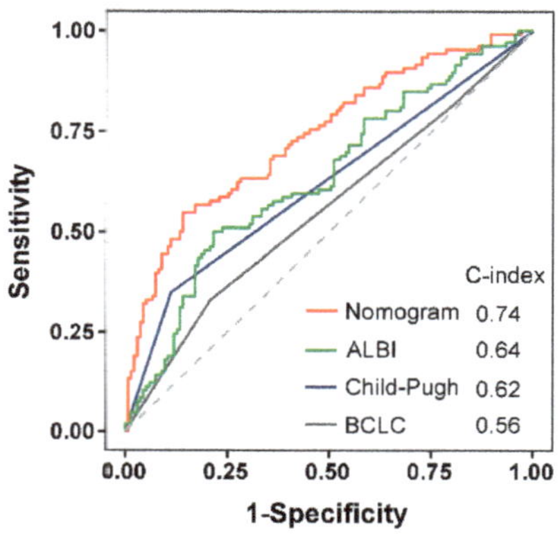

Figure 5. Comparison of predictive accuracy between ALBI score, Child−Pugh grade, BCLC, and the PGA grade. BCLC, Barcelona Clinic Liver Cancer staging; ALBI, albumin−bilirubin score; Child−Pugh, Child−Pugh grade.

4. Discussion

The high incidence rate and mortality of HCC patients remains a major clinical concern. Early intervention based on risk stratification is an effective strategy to improve the survival rate of HCC patients. Clinical and histopathological parameters, including tumor burden, vascular invasion, lymph node, and extrahepatic metastasis, are closely related to the prognosis of HCC [6,36]. As accurate histological diagnosis is not easily available, and ultrasonography is far from excellent, serum biomarker assessments are more objective and accurate at predicting and evaluating the prognosis of HCC [10,14]. Herein, we presented a reliable and easy-to-use model for better diagnosis or prognosis of early HCC by integrating potential serum biomarkers. The diagnostic nomogram based on age, gender, AFP, PIVKA, PT, and TP (refer to APPT grade) can improve the diagnostic efficacy of early HCC. In addition, the prognostic nomogram, constructed by PIVKA, GGT, and ALB (refer to PGA grade) exhibited excellent discrimination between survival and non-survival HCC patients. Moreover, the prognostic nomogram (PGA grade) outperformed the routinely used prognostic models including ALBI grade, BCLC, and CTP classification. A high PGA grade may serve as an effective predictor of survival rate in HCC and support optimal therapeutic selection in HCC.

Collective evidence suggests a positive correlation between AFP and PIVKA-II levels and clinicopathological performance, such as tumor size, tumor differentiation, and vascular invasion [18,21,37]. In the present study, using 20 ng/mL as a cut-off, 35.51% of HCC patients were AFP negative, which is consistent with previous reports indicating that approximately 30% of liver cancer patients were consistently negative for serum AFP [13]. The sensitivity of AFP for HCC in CHB patients was 64.49%. On the other hand, PIVKA-II with a level of up to 40 ng/mL revealed a superior specificity of 84.37% for HCC diagnosis. Furthermore, the combination of AFP and PIVKA-II further increased the specificity to 92.89%. Our results were consistent with the conclusions of other researchers who suggested that the combination of AFP and PIVKA-II can improve the diagnosis of HBV-related HCC [22,37].

The occurrence and development of HCC involves a multi-step evolutionary process from the molecular to the clinical level, characterized by marked abnormalities in liver function, including liver enzymes, metabolism, and protein synthesis function [38]. Emerging evidence suggests that the imbalance of tumor and blood coagulation disorders promotes tumor growth, invasion, and metastasis [39,40]. For example, prolonged prothrombin time (PT) has been associated with aggressive tumor growth and poor survival rates in various cancers, such as lung cancer [39], kidney cancer [41], and early HCC [40]. After performing univariate and multivariate logistic regression analyses, we identified PT as one of the independent risk factors for HCC (OR = 0.839, $p = 0.002$). In addition, hypoproteinemia, especially hypoalbuminemia, suggests poor nutritional status and decreased hepatic synthesis function due to chronic liver disease and HCC [27,30]. According to our study, TP was also an independent risk factor for HCC (OR = 0.912, $p < 0.001$).

Therefore, the objective of this study was to construct a reliable and precise nomogram for the prediction of HCC by integrating biomarkers such as AFP, PIVKA-II, PT, and TP. The AUROC of the diagnostic nomogram was 0.970, indicating the critical role of APPT grade in the early detection of HBV-related HCC. By evaluating the individualized potential to develop HCC, the APPT grade may enable physicians to optimize the implementation and efficiency of a screening surveillance strategy.

Liver function indicators, including GGT, ALB, PT, and TP, are critical serum biomarkers for determining hepatic reserve function, which is essential for the prognosis of HCC. Currently, several prognostic models, such as BCLC [42], albumin−bilirubin (ALBI) grade [43,44], and Child−Turcotte−Pugh (CTP) classification, have been established and validated as effective tools for predicting HCC outcomes [44–46]. Despite the widespread use of CTP scores to evaluate preoperative liver function, the limitations regarding its subjective variables, such as clinical grading of ascites and encephalopathy, have been extensively discussed [6,47]. The albumin−bilirubin (ALBI) score has the specific advantage of being based on statistical and objective evidence [44], but our multivariable Cox analyses showed no correlation between bilirubin levels and worse survival in HCC patients. Although previous studies have reported that HCC patients with higher bilirubin levels had a worse prognosis, our results suggested otherwise.

PIVKA-II is a novel prognostic predictor for HCC as elevated levels are associated with early recurrence, vascular invasion, large HCC size, and poor prognosis [17,22]. However, the elevation of PIVKA-II is not specific to HCC because interfering factors such as taking warfarin, primary gastric adenocarcinoma, vitamin K deficiency, inflammatory bowel disease, intestinal flora imbalance, renal failure, malnutrition, and alcoholic liver disease can lead to elevated serum levels of PIVKA-II in non-HCC patients [48]. In the current work, the prognostic value of PIVKA-II in combination with other liver function parameters, especially GGT and ALB, was evaluated through survival analysis in HCC patients. High levels of GGT protein, which resulted from impaired biliary excretion in gastrointestinal cancer or the secretion of HCC cells, were positively associated with a large tumor size and advanced TNM stage, and were considered as an independent prognostic factor for predicting the survival rate of individuals with AFP-negative HCC [49–51]. Consistent with previous studies, the results of our analysis suggest that the increased GGT level seemed to be a strong risk factor for an unfavorable survival rate in patients with HCC. Although PIVKA-II, GGT, and ALB partially reflect different aspects of HCC including tumor burden vascular invasion, and poor tumor differentiation, respectively, they may complement each other when used in combination. Therefore, the combined use of PIVKA-II, GGT, and ALB could validate hepatic reserve function and increase their predictive probability in the prognosis of HCC. In the present study, when PIVKA-II was used in combination with GGT and ALB (PGA grade), the PGA grade (C-index: 0.74) exhibited an excellent discrimination and good accuracy and outperformed commonly used prognostic models, including ALBI grade (0.64), BCLC (0.56), and CTP classification (0.62).

Furthermore, Xu and colleagues conducted a study to evaluate the predictive efficacy of GGT for prognosis in patients with HCC who underwent liver resection [52]. The study found that elevated GGT levels were significantly associated with a higher risk in this patient population, with an AUC of 0.643. In another study, Park et al. [53] investigated the clinical utility of the response of AFP and PIVKA-II on the prognosis of patients with locally advanced HCC who received local treatment. The combination of AFP and PIVKA-II had a prognostic power of 0.626 for overall survival, which was better than AFP alone (0.592). Although the management of patients may vary among different studies, encompassing variations in therapeutic schedules and etiology, our results suggest that the proposed PGA grade could improve prognostic prediction compared with using

GGT alone or combining AFP and PIVKA. In addition, several nomograms have been developed to predict recurrence and survival in HCC patients who had undergone resection treatment [54–56]. Wang et al. reported that preoperative TACE therapy, microvascular invasion (MVI), AFP, ALBI grade, tumor differentiation, tumor size, intraoperative blood transfusion, and surgical modality were independent risk factors for overall survival in patients with single large and huge HCC who underwent curative resection treatment [54]. Based on these risk factors, a nomogram was developed that achieved high C-indexes (0.86 for overall survival). Compared with these nomograms [54–56], which are based on clinical data such as tumor size and vascular invasion, the PGA grade was found to have less potential for predicting prognosis, possibly due to the lack of clinical characteristics, different treatment methods received by patients and the different histopathological types of liver cancer.

The present study has some limitations that need to be considered. First, the lack of clinical characteristics, such as treatment methods, tumor size, tumor differentiation, and vascular invasion, may have led to confounding deviations. Therefore, further prospective research is warranted, which should include a comprehensive evaluation of both clinical and laboratory data to devise a better method for the diagnosis and prognosis of HCC. Second, this study was conducted at a single center and had a relatively small sample size. Therefore, to draw a precise conclusion, our results need to be validated in a larger, multicenter clinical trial, which should include long-term follow-up evaluations.

In conclusion, the combination of AFP and PIVKA with PT and TP showed the greatest diagnostic ability for HBV-related HCC, while the combination of PIVKA-II with GGT and ALB was efficient at validating the hepatic reserve function and predicting the prognosis of HCC. Therefore, clinicians should select the most appropriate biomarkers for HCC, which can facilitate better assessment and guide appropriate therapeutic strategies for HCC.

Supplementary Materials: The following supporting information can be downloaded at: https://www.mdpi.com/article/10.3390/diagnostics13081442/s1. Table S1: Differences in the rates of HCC diagnosis between AFP and PVIKA-II. Table S2: Characteristics of the patients in the training cohort and validation cohort.

Author Contributions: Conceptualization, L.L. and J.W.; data curation, S.A.; formal analysis, S.A.; methodology, S.A., X.Z. and J.W.; project administration, L.L.; resources, M.L., L.L. and J.W.; supervision, M.L. and J.W.; validation, X.Z.; writing—original draft, S.A. and L.L.; writing—review and editing, X.Z. and M.L. All authors have read and agreed to the published version of the manuscript.

Funding: This work was funded by grants from the National Natural Science Foundation of China (82072384 to L.L. and 82101775 to S.A.). The funder had no role in the study design, data collection and analysis, manuscript preparation, and publish decision.

Institutional Review Board Statement: All experimental protocols in this study involving human subjects were performed in accordance with the ethical standards of the First Affiliated Hospital of Sun Yat-sen University and the national research committee and with the 1964 Declaration of Helsinki and its later amendments or comparable ethical standards.

Informed Consent Statement: All individual subjects in this study provided informed consent.

Data Availability Statement: Data will be made available upon request.

Acknowledgments: We thank all of the participants for their contribution to this work.

Conflicts of Interest: The authors declare that there is no commercial or financial conflict of interest.

References

1. Sung, H.; Ferlay, J.; Siegel, R.L.; Laversanne, M.; Soerjomataram, I.; Jemal, A.; Bray, F. Global Cancer Statistics 2020: GLOBOCAN Estimates of Incidence and Mortality Worldwide for 36 Cancers in 185 Countries. *CA Cancer J. Clin.* **2021**, *71*, 209–249. [CrossRef] [PubMed]
2. Komuta, M. Histological Heterogeneity of Primary Liver Cancers: Clinical Relevance, Diagnostic Pitfalls and the Pathologist's Role. *Cancers* **2021**, *13*, 2871. [CrossRef] [PubMed]

3. Shen, C.; Jiang, X.; Li, M.; Luo, Y. Hepatitis Virus and Hepatocellular Carcinoma: Recent Advances. *Cancers* **2023**, *15*, 533. [CrossRef] [PubMed]
4. Wong, M.C.S.; Huang, J.L.W.; George, J.; Huang, J.; Leung, C.; Eslam, M.; Chan, H.L.Y.; Ng, S.C. The changing epidemiology of liver diseases in the Asia-Pacific region. *Nat. Rev. Gastroenterol. Hepatol* **2019**, *16*, 57–73. [CrossRef]
5. Forner, A.; Reig, M.; Bruix, J. Hepatocellular carcinoma. *Lancet* **2018**, *391*, 1301–1314. [CrossRef] [PubMed]
6. European Association for the Study of the Liver. EASL Clinical Practice Guidelines: Management of hepatocellular carcinoma. *J. Hepatol.* **2018**, *69*, 182–236. [CrossRef]
7. Lee, Y.T.; Fujiwara, N.; Yang, J.D.; Hoshida, Y. Risk stratification and early detection biomarkers for precision HCC screening. *Hepatology* **2022**. [CrossRef]
8. Omata, M.; Cheng, A.L.; Kokudo, N.; Kudo, M.; Lee, J.M.; Jia, J.; Tateishi, R.; Han, K.H.; Chawla, Y.K.; Shiina, S.; et al. Asia-Pacific clinical practice guidelines on the management of hepatocellular carcinoma: A 2017 update. *Hepatol. Int.* **2017**, *11*, 17–370. [CrossRef] [PubMed]
9. Tzartzeva, K.; Obi, J.; Rich, N.E.; Parikh, N.D.; Marrero, J.A.; Yopp, A.; Waljee, A.K.; Singal, A.G. Surveillance Imaging and Alpha Fetoprotein for Early Detection of Hepatocellular Carcinoma in Patients With Cirrhosis: A Meta-analysis. *Gastroenterology* **2018**, *154*, 1706–1718.e1. [CrossRef]
10. Goldberg, D.S.; Taddei, T.H.; Serper, M.; Mehta, R.; Dieperink, E.; Aytaman, A.; Baytarian, M.; Fox, R.; Hunt, K.; Pedrosa, M.; et al. Identifying barriers to hepatocellular carcinoma surveillance in a national sample of patients with cirrhosis. *Hepatology* **2017**, *65*, 864–874. [CrossRef] [PubMed]
11. Kim, G.A.; Seock, C.H.; Park, J.W.; An, J.; Lee, K.S.; Yang, J.E.; Lim, Y.S.; Kim, K.M.; Shim, J.H.; Lee, D.; et al. Reappraisal of serum alpha-foetoprotein as a surveillance test for hepatocellular carcinoma during entecavir treatment. *Liver Int.* **2015**, *35*, 232–239. [CrossRef]
12. Asrih, M.; Lenglet, S.; Mach, F.; Montecucco, F. Alpha-fetoprotein: A controversial prognostic biomarker for small hepatocellular carcinoma. *World J. Gastroenterol.* **2013**, *19*, 328–330. [CrossRef] [PubMed]
13. Agopian, V.G.; Harlander-Locke, M.P.; Markovic, D.; Zarrinpar, A.; Kaldas, F.M.; Cheng, E.Y.; Yersiz, H.; Farmer, D.G.; Hiatt, J.R.; Busuttil, R.W. Evaluation of Patients With Hepatocellular Carcinomas That Do Not Produce alpha-Fetoprotein. *JAMA Surg.* **2017**, *152*, 55–64. [CrossRef] [PubMed]
14. Tsuchiya, N.; Sawada, Y.; Endo, I.; Saito, K.; Uemura, Y.; Nakatsura, T. Biomarkers for the early diagnosis of hepatocellular carcinoma. *World J. Gastroenterol.* **2015**, *21*, 10573–10583. [CrossRef] [PubMed]
15. Kim, D.Y.; Toan, B.N.; Tan, C.K.; Hasan, I.; Setiawan, L.; Yu, M.L.; Izumi, N.; Huyen, N.N.; Chow, P.K.; Mohamed, R.; et al. Utility of Combining PIVKA-II and AFP in the Surveillance and Monitoring of Hepatocellular Carcinoma in the Asia-Pacific Region. *Clin. Mol. Hepatol.* **2023**, *29*, 277–292. [CrossRef]
16. Trevisani, F.; Garuti, F.; Neri, A. Alpha-fetoprotein for Diagnosis, Prognosis, and Transplant Selection. *Semin Liver. Dis.* **2019**, *39*, 163–177. [CrossRef] [PubMed]
17. Wang, B.L.; Tan, Q.W.; Gao, X.H.; Wu, J.; Guo, W. Elevated PIVKA-II is associated with early recurrence and poor prognosis in BCLC 0-A hepatocellular carcinomas. *Asian Pac. J. Cancer Prev.* **2014**, *15*, 6673–6678. [CrossRef]
18. Pote, N.; Cauchy, F.; Albuquerque, M.; Voitot, H.; Belghiti, J.; Castera, L.; Puy, H.; Bedossa, P.; Paradis, V. Performance of PIVKA-II for early hepatocellular carcinoma diagnosis and prediction of microvascular invasion. *J. Hepatol.* **2015**, *62*, 848–854. [CrossRef] [PubMed]
19. Wu, J.; Xiang, Z.; Bai, L.; He, L.; Tan, L.; Hu, M.; Ren, Y. Diagnostic value of serum PIVKA-II levels for BCLC early hepatocellular carcinoma and correlation with HBV DNA. *Cancer Biomark.* **2018**, *23*, 235–242. [CrossRef]
20. Yang, Y.; Li, G.; Lu, Z.; Liu, Y.; Kong, J.; Liu, J. Progression of Prothrombin Induced by Vitamin K Absence-II in Hepatocellular Carcinoma. *Front Oncol.* **2021**, *11*, 726213. [CrossRef]
21. Si, Y.Q.; Wang, X.Q.; Fan, G.; Wang, C.Y.; Zheng, Y.W.; Song, X.; Pan, C.C.; Chu, F.L.; Liu, Z.F.; Lu, B.R.; et al. Value of AFP and PIVKA-II in diagnosis of HBV-related hepatocellular carcinoma and prediction of vascular invasion and tumor differentiation. *Infect. Agents Cancer* **2020**, *15*, 70. [CrossRef]
22. Seo, S.I.; Kim, H.S.; Kim, W.J.; Shin, W.G.; Kim, D.J.; Kim, K.H.; Jang, M.K.; Lee, J.H.; Kim, J.S.; Kim, H.Y.; et al. Diagnostic value of PIVKA-II and alpha-fetoprotein in hepatitis B virus-associated hepatocellular carcinoma. *World J. Gastroenterol.* **2015**, *21*, 3928–3935. [CrossRef] [PubMed]
23. Strople, J.; Lovell, G.; Heubi, J. Prevalence of subclinical vitamin K deficiency in cholestatic liver disease. *J. Pediatr. Gastroenterol. Nutr.* **2009**, *49*, 78–84. [CrossRef]
24. Chen, J.; Wu, G.; Li, Y. Evaluation of Serum Des-Gamma-Carboxy Prothrombin for the Diagnosis of Hepatitis B Virus-Related Hepatocellular Carcinoma: A Meta-Analysis. *Dis. Markers* **2018**, *2018*, 8906023. [CrossRef]
25. Yang, J.G.; He, X.F.; Huang, B.; Zhang, H.A.; He, Y.K. Rule of changes in serum GGT levels and GGT/ALT and AST/ALT ratios in primary hepatic carcinoma patients with different AFP levels. *Cancer Biomark* **2018**, *21*, 743–746. [CrossRef]
26. Zhang, C.H.; Ni, X.C.; Chen, B.Y.; Qiu, S.J.; Zhu, Y.M.; Luo, M. Combined preoperative albumin-bilirubin (ALBI) and serum gamma-glutamyl transpeptidase (GGT) predicts the outcome of hepatocellular carcinoma patients following hepatic resection. *J. Cancer* **2019**, *10*, 4836–4845. [CrossRef] [PubMed]
27. Bagirsakci, E.; Sahin, E.; Atabey, N.; Erdal, E.; Guerra, V.; Carr, B.I. Role of Albumin in Growth Inhibition in Hepatocellular Carcinoma. *Oncology* **2017**, *93*, 136–142. [CrossRef]

28. Xia, F.; Ndhlovu, E.; Liu, Z.; Chen, X.; Zhang, B.; Zhu, P. Alpha-Fetoprotein+Alkaline Phosphatase (A-A) Score Can Predict the Prognosis of Patients with Ruptured Hepatocellular Carcinoma Underwent Hepatectomy. *Dis. Markers* **2022**, *2022*, 9934189. [CrossRef] [PubMed]
29. Mao, M.; Wang, X.; Song, Y.; Sheng, H.; Han, R.; Lin, W.; Dai, S. Novel Prognostic Scores Based on Plasma Prothrombin Time and Fibrinogen Levels in Patients With AFP-Negative Hepatocellular Carcinoma. *Cancer Control* **2020**, *27*, 1073274820915520. [CrossRef]
30. Haruki, K.; Shiba, H.; Saito, N.; Horiuchi, T.; Shirai, Y.; Fujiwara, Y.; Furukawa, K.; Sakamoto, T.; Yanaga, K. Risk stratification using a novel liver functional reserve score of combination prothrombin time-international normalized ratio to albumin ratio and albumin in patients with hepatocellular carcinoma. *Surgery* **2018**, *164*, 404–410. [CrossRef]
31. Schuppan, D.; Afdhal, N.H. Liver cirrhosis. *Lancet* **2008**, *371*, 838–851. [CrossRef] [PubMed]
32. Yang, T.; Xing, H.; Wang, G.; Wang, N.; Liu, M.; Yan, C.; Li, H.; Wei, L.; Li, S.; Fan, Z.; et al. A Novel Online Calculator Based on Serum Biomarkers to Detect Hepatocellular Carcinoma among Patients with Hepatitis B. *Clin Chem.* **2019**, *65*, 1543–1553. [CrossRef] [PubMed]
33. Moons, K.G.; Altman, D.G.; Reitsma, J.B.; Ioannidis, J.P.; Macaskill, P.; Steyerberg, E.W.; Vickers, A.J.; Ransohoff, D.F.; Collins, G.S. Transparent Reporting of a multivariable prediction model for Individual Prognosis or Diagnosis (TRIPOD): Explanation and elaboration. *Ann. Intern. Med.* **2015**, *162*, W1–W73. [CrossRef]
34. Park, H.; Park, J.Y. Clinical significance of AFP and PIVKA-II responses for monitoring treatment outcomes and predicting prognosis in patients with hepatocellular carcinoma. *Biomed Res. Int.* **2013**, *2013*, 310427. [CrossRef] [PubMed]
35. Viggiani, V.; Palombi, S.; Gennarini, G.; D'Ettorre, G.; De Vito, C.; Angeloni, A.; Frati, L.; Anastasi, E. Protein induced by vitamin K absence or antagonist-II (PIVKA-II) specifically increased in Italian hepatocellular carcinoma patients. *Scand. J. Gastroenterol.* **2016**, *51*, 1257–1262. [CrossRef]
36. Llovet, J.M.; Kelley, R.K.; Villanueva, A.; Singal, A.G.; Pikarsky, E.; Roayaie, S.; Lencioni, R.; Koike, K.; Zucman-Rossi, J.; Finn, R.S. Hepatocellular carcinoma. *Nat. Rev. Dis. Primers.* **2021**, *7*, 6. [CrossRef] [PubMed]
37. Feng, H.; Li, B.; Li, Z.; Wei, Q.; Ren, L. PIVKA-II serves as a potential biomarker that complements AFP for the diagnosis of hepatocellular carcinoma. *BMC Cancer* **2021**, *21*, 401. [CrossRef]
38. Shahini, E.; Pasculli, G.; Solimando, A.G.; Tiribelli, C.; Cozzolongo, R.; Giannelli, G. Updating the Clinical Application of Blood Biomarkers and Their Algorithms in the Diagnosis and Surveillance of Hepatocellular Carcinoma: A Critical Review. *Int. J. Mol. Sci.* **2023**, *24*, 4286. [CrossRef]
39. Tas, F.; Kilic, L.; Serilmez, M.; Keskin, S.; Sen, F.; Duranyildiz, D. Clinical and prognostic significance of coagulation assays in lung cancer. *Respir. Med.* **2013**, *107*, 451–457. [CrossRef]
40. Wang, X.P.; Mao, M.J.; He, Z.L.; Zhang, L.; Chi, P.D.; Su, J.R.; Dai, S.Q.; Liu, W.L. A retrospective discussion of the prognostic value of combining prothrombin time(PT) and fibrinogen(Fbg) in patients with Hepatocellular carcinoma. *J. Cancer* **2017**, *8*, 2079–2087. [CrossRef]
41. Bian, Z.; Meng, J.; Niu, Q.; Jin, X.; Wang, J.; Feng, X.; Che, H.; Zhou, J.; Zhang, L.; Zhang, M.; et al. Prognostic Role of Prothrombin Time Activity, Prothrombin Time, Albumin/Globulin Ratio, Platelets, Sex, and Fibrinogen in Predicting Recurrence-Free Survival Time of Renal Cancer. *Cancer Manag. Res.* **2020**, *12*, 8481–8490. [CrossRef] [PubMed]
42. Reig, M.; Forner, A.; Rimola, J.; Ferrer-Fabrega, J.; Burrel, M.; Garcia-Criado, A.; Kelley, R.K.; Galle, P.R.; Mazzaferro, V.; Salem, R.; et al. BCLC strategy for prognosis prediction and treatment recommendation: The 2022 update. *J. Hepatol.* **2022**, *76*, 681–693. [CrossRef] [PubMed]
43. Hiraoka, A.; Kumada, T.; Michitaka, K.; Toyoda, H.; Tada, T.; Ueki, H.; Kaneto, M.; Aibiki, T.; Okudaira, T.; Kawakami, T.; et al. Usefulness of albumin-bilirubin grade for evaluation of prognosis of 2584 Japanese patients with hepatocellular carcinoma. *J. Gastroenterol. Hepatol.* **2016**, *31*, 1031–1036. [CrossRef]
44. Demirtas, C.O.; D'Alessio, A.; Rimassa, L.; Sharma, R.; Pinato, D.J. ALBI grade: Evidence for an improved model for liver functional estimation in patients with hepatocellular carcinoma. *JHEP Rep.* **2021**, *3*, 100347. [CrossRef]
45. Zhang, J.; Zhao, L.; Zhou, Y.; Ding, J.; Zhang, Q.; Jing, X. The comparison between albumin-bilirubin grade and Child-Pugh grade for assessing the prognosis of hepatocellular carcinoma after thermal ablation: A propensity score-matched analysis. *Transl. Cancer Res.* **2022**, *11*, 2523–2535. [CrossRef]
46. Huang, F.; Gao, J. Modified Child-Pugh grade vs albumin-bilirubin grade for predicting prognosis of hepatocellular carcinoma patients after hepatectomy. *World J. Gastroenterol.* **2020**, *26*, 749–758. [CrossRef]
47. European Association for the Study of the Liver; European Organisation for Research and Treatment of Cancer. EASL-EORTC clinical practice guidelines: Management of hepatocellular carcinoma. *J. Hepatol.* **2012**, *56*, 908–943. [CrossRef] [PubMed]
48. Kang, K.H.; Kim, J.H.; Kang, S.H.; Lee, B.J.; Seo, Y.S.; Yim, H.J.; Yeon, J.E.; Park, J.J.; Kim, J.S.; Bak, Y.T.; et al. The influence of alcoholic liver disease on serum PIVKA-II levels in patients without hepatocellular carcinoma. *Gut. Liver* **2015**, *9*, 224–230. [CrossRef]
49. Wang, X.; Mao, M.; He, Z.; Zhang, L.; Li, H.; Lin, J.; He, Y.; Dai, S.; Hu, W.; Liu, W. Development and Validation of a Prognostic Nomogram in AFP-negative hepatocellular carcinoma. *Int. J. Biol. Sci.* **2019**, *15*, 221–228. [CrossRef]
50. Li, S.; Xu, W.; Liao, M.; Zhou, Y.; Weng, J.; Ren, L.; Yu, J.; Liao, W.; Huang, Z. The Significance of Gamma-Glutamyl Transpeptidase to Lymphocyte Count Ratio in the Early Postoperative Recurrence Monitoring and Prognosis Prediction of AFP-Negative Hepatocellular Carcinoma. *J. Hepatocell Carcinoma.* **2021**, *8*, 23–33. [CrossRef]

51. Wu, S.J.; Lin, Y.X.; Ye, H.; Xiong, X.Z.; Li, F.Y.; Cheng, N.S. Prognostic value of alkaline phosphatase, gamma-glutamyl transpeptidase and lactate dehydrogenase in hepatocellular carcinoma patients treated with liver resection. *Int. J. Surg.* **2016**, *36*, 143–151. [CrossRef] [PubMed]
52. Xu, X.S.; Wan, Y.; Song, S.D.; Chen, W.; Miao, R.C.; Zhou, Y.Y.; Zhang, L.Q.; Qu, K.; Liu, S.N.; Zhang, Y.L.; et al. Model based on gamma-glutamyltransferase and alkaline phosphatase for hepatocellular carcinoma prognosis. *World J. Gastroenterol.* **2014**, *20*, 10944–10952. [CrossRef] [PubMed]
53. Park, H.; Kim, S.U.; Park, J.Y.; Kim, D.Y.; Ahn, S.H.; Chon, C.Y.; Han, K.H.; Seong, J. Clinical usefulness of double biomarkers AFP and PIVKA-II for subdividing prognostic groups in locally advanced hepatocellular carcinoma. *Liver Int.* **2014**, *34*, 313–321. [CrossRef]
54. Wang, J.C.; Hou, J.Y.; Chen, J.C.; Xiang, C.L.; Mao, X.H.; Yang, B.; Li, Q.; Liu, Q.B.; Chen, J.; Ye, Z.W.; et al. Development and validation of prognostic nomograms for single large and huge hepatocellular carcinoma after curative resection. *Eur. J. Cancer* **2021**, *155*, 85–96. [CrossRef] [PubMed]
55. Wang, X.H.; Liu, Q.B.; Xiang, C.L.; Mao, X.H.; Yang, B.; Li, Q.; Zhou, Q.F.; Li, S.Q.; Zhou, Z.G.; Chen, M.S. Multi-institutional validation of novel models for predicting the prognosis of patients with huge hepatocellular carcinoma. *Int. J. Cancer* **2021**, *149*, 127–138. [CrossRef] [PubMed]
56. Zhou, Y.W.; Li, Q.F.; Chen, Y.Y.; Wang, K.; Pu, D.; Chen, X.R.; Li, C.H.; Jiang, L.; Wang, Y.; Li, Q.; et al. Clinicopathologic features, treatment, survival, and prognostic factors of combined hepatocellular and cholangiocarcinoma: A nomogram development based on SEER database and validation in multicenter study. *Eur. J. Surg. Oncol.* **2022**, *48*, 1559–1566. [CrossRef] [PubMed]

diagnostics

MDPI

Article

Identification and Application of a Novel Immune-Related lncRNA Signature on the Prognosis and Immunotherapy for Lung Adenocarcinoma

Zhimin Zeng [1,2,†], Yuxia Liang [1,2,†], Jia Shi [1,2], Lisha Xiao [1,2], Lu Tang [1,2], Yubiao Guo [1,2,*], Fengjia Chen [1,2,*] and Gengpeng Lin [1,2,*]

1 Division of Pulmonary and Critical Care Medicine, The First Affiliated Hospital of Sun Yat-sen University, Zhongshan Second Road No. 58, Guangzhou 510080, China
2 Institute of Pulmonary Diseases, Sun Yat-sen University, Guangzhou 510275, China
* Correspondence: guoyb@mail.sysu.edu.cn(Y.G.); chenfjia@mail.sysu.edu.cn (F.C.); lingp6@mail.sysu.edu.cn (G.L.)
† These authors contributed equally to this work.

Abstract: Background: Long non-coding RNA (lncRNA) participates in the immune regulation of lung cancer. However, limited studies showed the potential roles of immune-related lncRNAs (IRLs) in predicting survival and immunotherapy response of lung adenocarcinoma (LUAD). Methods: Based on The Cancer Genome Atlas (TCGA) and ImmLnc databases, IRLs were identified through weighted gene coexpression network analysis (WGCNA), Cox regression, and Lasso regression analyses. The predictive ability was validated by Kaplan–Meier (KM) and receiver operating characteristic (ROC) curves in the internal dataset, external dataset, and clinical study. The immunophenoscore (IPS)-PD1/PD-L1 blocker and IPS-CTLA4 blocker data of LUAD were obtained in TCIA to predict the response to immune checkpoint inhibitors (ICIs). The expression levels of immune checkpoint molecules and markers for hyperprogressive disease were analyzed. Results: A six-IRL signature was identified, and patients were stratified into high- and low-risk groups. The low-risk had improved survival outcome ($p = 0.006$ in the training dataset, $p = 0.010$ in the testing dataset, $p < 0.001$ in the entire dataset), a stronger response to ICI ($p < 0.001$ in response to anti-PD-1/PD-L1, $p < 0.001$ in response to anti-CTLA4), and higher expression levels of immune checkpoint molecules ($p < 0.001$ in PD-1, $p < 0.001$ in PD-L1, $p < 0.001$ in CTLA4) but expressed more biomarkers of hyperprogression in immunotherapy ($p = 0.002$ in MDM2, $p < 0.001$ in MDM4). Conclusion: The six-IRL signature exhibits a promising prediction value of clinical prognosis and ICI efficacy in LUAD. Patients with low risk might gain benefits from ICI, although some have a risk of hyperprogressive disease.

Keywords: immune-related lncRNA; ddPCR; prognosis; immunotherapy; lung adenocarcinoma

Citation: Zeng, Z.; Liang, Y.; Shi, J.; Xiao, L.; Tang, L.; Guo, Y.; Chen, F.; Lin, G. Identification and Application of a Novel Immune-Related lncRNA Signature on the Prognosis and Immunotherapy for Lung Adenocarcinoma. *Diagnostics* **2022**, *12*, 2891. https://doi.org/10.3390/diagnostics12112891

Academic Editors: Yuli Huang, Yong Yuan and Peisong Chen

Received: 20 September 2022
Accepted: 16 November 2022
Published: 21 November 2022

1. Introduction

Non-small cell lung cancer (NSCLC) remains the second-most commonly diagnosed cancer (with an estimated prevalence of 11.4%) and the leading cause of cancer death (with an estimated prevalence of 18%) in the global population, but incidence declines and survival increases in the United States of America. The tobacco epidemic, ambient air pollution, the diagnosis period, and late-stage treatments likely contribute to the remarkable difference [1]. Lung adenocarcinoma (LUAD), as one of the common and severe lung cancers, accounts for approximately 50% of NSCLC [2] and has an average 5-year survival rate of 21% given to the missed diagnosis in an early stage and advanced cancer refractory to traditional treatments [1]. A large proportion of patients still cannot benefit from current conventional chemotherapy and targeted treatments because of the resistance, which leads to a relatively high recurrence rate in LUAD [3].

Cancer immunologic and immune therapeutic advances seem promising for gaining a survival benefit for LUAD. Increasing studies focused on tumor microenvironments reported that infiltrating immune cells and modulating immune pathways have prominent effects on the progression and aggressiveness of LUAD [4]. Besides, immune checkpoint inhibitors (ICIs) directed against programmed cell death-1 (PD-1) and its ligand (PD-L1) have revolutionized the treatment of advanced LUAD without targetable mutations [5]. Unfortunately, the overall response to ICI is modestly low, and a paradoxical acceleration of tumor growth, defined as "hyperprogressive disease (HPD)," happens in a subset of patients with NSCLC treated with ICI [6]. Thus, the molecular signature relevant to tumor immunology is needed to be recognized as prognostic biomarkers to optimize personalized medicine and improve long-term survival.

Along with advances in immunology, researchers studied the critical regulatory ability of long non-coding RNA (lncRNA) in different phases of cancer immunity, such as antigen release and presentation, immune activation, immune cell migration, infiltration, and killing of tumor cells [7]. Besides, the immune signature can be a conspicuous marker to evaluate the overall survival (OS) in patients with LUAD [8]. However, few studies have comprehensively considered prognostic immune-related lncRNAs (IRLs) and their roles in predicting the efficacy of ICI treatment.

In this study, according to the IRL-based risk model, the subtypes of LUAD were identified to evaluate prognosis, immune cell infiltration, therapeutic benefit, and HPD during immune checkpoint blockade via integrative bioinformatics.

2. Materials and Methods

2.1. Data Download

Expression, phenotype, and survival data were downloaded from The Cancer Genome Atlas (TCGA (RRID:SCR_003193)) cohort of the UCSC Xena database (https://xenabrowser.net/ accessed on 16 September 2020), in which 513 LUAD samples were obtained as an entire dataset after removing 13 samples with the missing phenotype (Table S1). Then, gene symbol names were retrieved from the human gtf file in the Ensembl database (http://www.ensembl.org/info/data/ftp/index.html accessed on 23 April 2020). A total of 3547 LUAD-associated IRLs were acquired from Lnc_Immunecell_Sig and Lnc_Pathways_Sig files in the immLnc database (http://bio-bigdata.hrbmu.edu.cn/ImmLnc/ accessed on 28 March 2021) [9]. The RNA sequencing data and relevant clinical characteristics of GSE120622 of patients with LUAD were downloaded from the Gene Expression Omnibus (GEO) (https://www.ncbi.nlm.nih.gov/geo/ accessed on 4 July 2022) [10]. Besides, the version 22 reference genome data, immunophenoscore (IPS) data [11], and target mRNAs of IRLs were obtained from The Gencode database(https://www.gencodegenes.org/ accessed on 27 April 2020), The Cancer Immunome Atlas database(TCIA, https://tcia.at/home accessed on 10 May 2021) [11], and starbase3.0 (http://starbase.sysu.edu.cn/ accessed on 6 April 2020) database.

2.2. Patients' Samples

To verify the risk score calculated from the TCGA database, we collected 40 samples of lung tissue punctured by thoracoscope and relevant clinical information as a retrospective case-control study in the First Affiliated Hospital of Sun Yat-sen University between October 2020 and June 2021. In accordance with the International Association for the Study of Lung Cancer tumor–node–metastasis classification [12], 30 and 10 cases were diagnosed with LUAD and nonlung cancer, respectively. The clinical characteristics of patients with LUAD are shown in Table S1. This study was approved and supervised by the Research Ethics Committee of the First Affiliated Hospital of Sun Yat-sen University (No. (2022)049).

2.3. Weighted Gene Coexpression Network Analysis (WGCNA)

The WGCNA (RRID:SCR_003302) R package (version 1.69) was used to analyze the co-expression network of IRLs [13]. Specifically, the screening criterion was R-square

> 0.85, and the soft thresholding power of four was selected. By using the power of four and a merged module threshold < 0.25, highly correlated clusters were merged into similar modules. Modules were generated, and the hierarchical clustering dendrogram was plotted.

2.4. Efficacy Analysis of Risk Score

All samples were randomly divided into TCGA training (360 samples) and testing (153 samples) datasets in accordance with the ratio of 7:3. Combining survival data, univariate Cox regression was performed on the training dataset by using R package survival (version 3.2-7) and R package survminer (version 0.4.8), and a p value < 0.05 was set for screening significantly differentially expressed genes [14]. Afterward, the Lasso regression was used to further narrow differentially expressed genes via the R package glmnet (version 4.0-2), in which the minimal lambda was obtained by a cross-validation procedure and then used to fit the Lasso model [15]. According to $\text{RScore}_i = \sum_{j=1}^{n} \exp_{ji} \times \beta_j$ (exp represents gene expression; β represents coefficients of genes identified by Lasso regression; and i and j represent each sample and each gene), we calculated the risk score of each sample and divided all patients with LUAD into high- and low-risk groups by using the median risk score [16]. Combined with survival data, the Kaplan–Meier (KM) curve was plotted between high- and low-risk groups with a p value < 0.05. The receiver operating characteristic (ROC) curves and area under the ROC curves (AUC) were drawn and calculated to estimate the 5-year survival probability [17].

2.5. Analysis of Stability and Independence

To validate the stability of the risk score, all patients were stratified into different subgroups on the basis of age (<60 vs. $\geq$60 years), gender (male vs. female), and TNM stage (stage I + II vs. stage III + IV). The survival status between high- and low-risk groups in KM curves was determined (p value < 0.05 defined as significant). We incorporated several parameters, including age, gender, M stage, N stage, T stage, and risk score, into univariate and multivariate Cox regression analyses.

2.6. Construction of a Nomogram

By using the R package rms (version 6.1-0) and survival (version 3.2-7), the prognostic nomogram was plotted with clinical parameters after establishing the Cox proportional hazards model and calculating survival probability [18]. The ROC curve was used to validate and predict the nomogram.

2.7. Assessment of Immune Microenvironment

The gene expression *signature* matrix of LM22 was obtained (CIBERSORT, RRID:SCR_016955) to estimate the proportions of 22 types of infiltrating immune cells, and expression levels were calculated in all samples by using the R package CIBERSORT (version 1.03) [19]. The R package estimate (version 1.0.13) was used to calculate the stromal score, immune score, ESTIMATE score, and tumor purity of all samples to compare the difference between high- and low-risk groups. Besides, univariate and multivariate Cox regression analyses were used to analyze the clinical traits and risk scores of LUAD.

2.8. Evaluation of Response to ICIs

IPS-PD1/PD-L1 blocker and IPS-CTLA4 blocker data on LUAD from TCGA were obtained in TCIA for predicting patients' responses to ICI in high- and low-risk groups. In addition, the expression levels of immune checkpoint molecules and markers for HPD were compared in these two groups.

2.9. Gene Set Enrichment Analysis

The Hallmark pathway enrichment analysis was performed using gene expression profiling data from different groups with LUAD by a GESA software (version 3.0) from

Sangerbox (http://vip.sangerbox.com/home.html accessed on 27 July 2021) [20] In the GSEA runs, gene-set sizes were set to between 5 and 5000 and parameters were set to 1000. A p value < 0.05 and an FDR < 0.25 were considered statistically significant.

2.10. Droplet Digital PCR

The Digital PCR MicroDrop-100 and Reagents (Kitforevergen) were used for droplet digital PCR (ddPCR) experiments. Primers are shown in Table S3. The cycle parameters used were as follows: 95 °C for 10 min and 95 °C for 30 s; 40 cycles of 60 °C for 30 s, and 72 °C for 1 min, and 72 °C for 30 s and final heating at 16 °C. The QuantaSoft software was used to analyze the number of copies of each sample.

2.11. Statistical Analysis

The R software (version 4.0.5) or GraphPad Prism (version 7.0) was used for statistical computing and graphics. For continuous variables, a t-test was used if data were normally distributed, *whereas the Mann–Whitney test* or Wilcoxon was performed for data that did not follow a normal distribution. For categorical variables, the chi-square test was performed. The univariate Cox regression analysis was used to examine potential risk factors, and the multivariate Cox regression analysis was further carried out for covariates whose p value < 0.05 in the univariate analysis. The KM curve was plotted to analyze the differences in survival by the log-rank test. Relationships between modules and traits were analyzed by the Pearson correlation. A p value < 0.05 represented statistical significance in all analyses.

3. Results

3.1. Identification of Prognostic IRLs

Integrative bioinformatics was conducted to explore the prognosis, immune cell infiltration, therapeutic benefit, and HPD of LUAD. To obtain relevant IRLs from core modules in LUAD, we acquired 513 TCGA–LUAD samples with complete clinical information (Table S1) and 3547 LUAD-associated IRLs. The WGCNA was then conducted to categorize the expression pattern of IRLs in TCGA–LUAD samples into 11 similar modules via the hierarchical clustering dendrogram at the appropriate soft threshold power of four (Figure 1A). Module–trait relationships were analyzed by correlating the 11 modules with clinical characteristics. We found that the brown co-expression network, including 174 IRLs, had negative associations with TNM stage and clinical events (Figure 1B).

To further narrow significant prognosis-related IRLs, we identified 18 candidate IRLs (p < 0.05) by investigating the association between the expression levels of 174 IRLs and survival information in the TCGA training dataset by using univariate Cox regression (Figure 1C). Afterward, six prognostic IRLs were identified as the key immune signature by Lasso regression with a minimal λ value (Figure 1D) and clustered in heatmaps (Figure S1). Based on the coefficients of the six IRLs, an optimized prognostic model for LUAD was constructed by calculating the risk score as follows: risk score = AC104971.3 × (−0.1592) + FAM215A × (−0.0697) + AC021678.2 × (−0.0330) + LINC02413 × (−0.0228) + AL161781.2 × (−0.0207) + LY86-AS1 × (−0.0051).

3.2. Predictive Capability and Sensitivity of the Risk Score Model

Aiming to assess the predictive value of the constructed prognostic model with six IRLs, patients were divided into low-risk and high-risk groups according to the median risk score of −0.46. The KM curve was plotted to compare the survival time in the TCGA training dataset, revealing that the median survival time of the high-risk group was shorter than that of the low-risk group (p = 0.006, Figure 2A). Afterward, KM curves were also drawn in the TCGA testing dataset (p = 0.010), TCGA entire dataset (p < 0.001), and GSE120622 (p = 0.040) to validate the predictive ability of the prognostic model (Figure 2B–D). These KM curves showed that patients with LUAD in the high-risk group had a worse prognosis than those in the low-risk group. The ROC curves of the 5-year OS indicated that the risk score was essential in predicting prognosis in patients with LUAD (TCGA training dataset: AUC = 0.64; TCGA testing dataset: AUC = 0.64; TCGA entire dataset: AUC = 0.64; GSE120622: AUC = 0.62; Figure S2), which indicated a predictive ability based on the risk score.

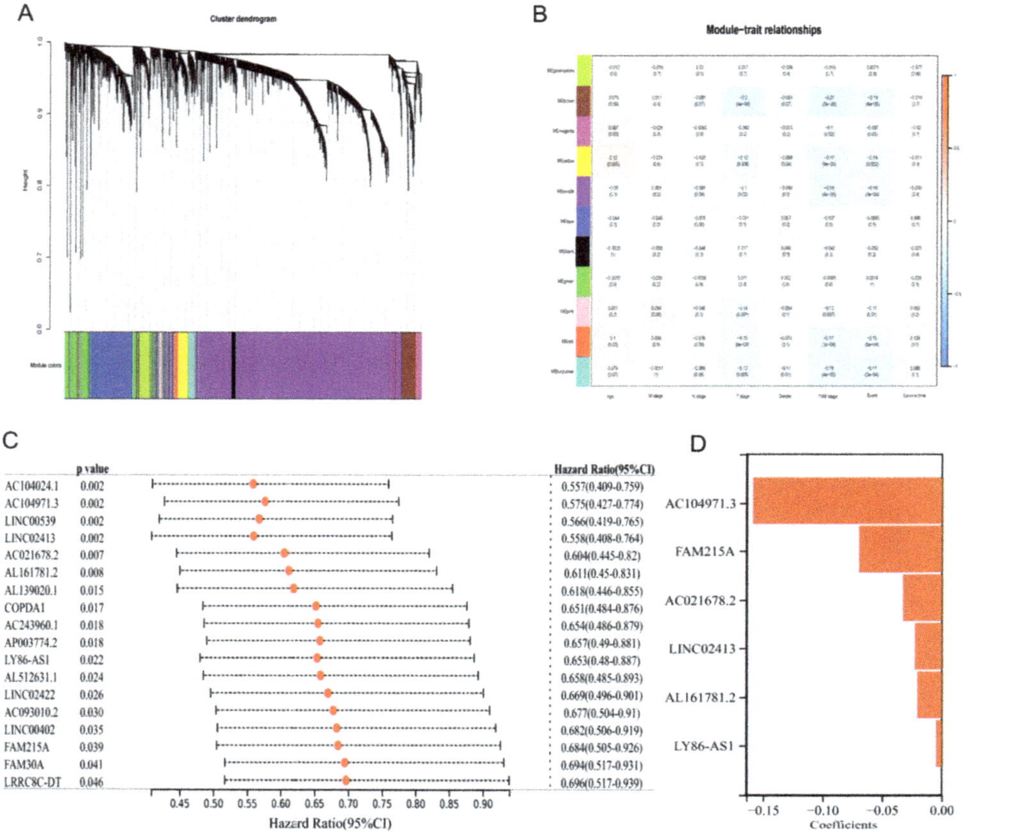

Figure 1. Six identified immune-related lncRNAs (IRLs) after weighted gene coexpression network analysis (WGCNA) and regression analyses. (**A**) Hierarchical clustering tree on IRL's coexpression network. (**B**) A heatmap of modules and clinical traits, including age, M stage, N stage, T stage, gender, TNM stage, clinical event, and survival time. The numbers in each box and parenthesis represented correlation coefficient and p value. (**C**) Eighteen prognosis-related IRLs screened by univariate Cox regression. (**D**) Coefficients of six prognostic IRLs identified by the Lasso model.

Furthermore, we verified the risk score by directly detecting the absolute quantification of these six IRLs via ddPCR in LUAD samples and controls from a case-control study in south China. Consistent with the results from the TCGA and GEO datasets, patients with LUAD and high-risk scores exhibited poor outcomes in our clinical study (Table 1). Except for the predictive power of the IRL signature, we wondered whether these six IRLs expressed differently between LUAD and normal controls. Results showed that AC104971.3 ($p < 0.01$), AC021678.2 ($p < 0.01$), LINC02413 ($p < 0.05$), AL161781.2 ($p < 0.05$), and LY86-AS1 ($p < 0.01$) were significantly downregulated in clinical LUAD tumor tissues, and that FAM215A expression was not significantly different but slightly increased in tumor samples (Figure S3). Besides, patients with LUAD from the TCGA dataset, GEO dataset, and clinical study were divided into high- and low-risk groups to avoid baseline bias. No difference was observed between the two groups in gender and age (Table S2).

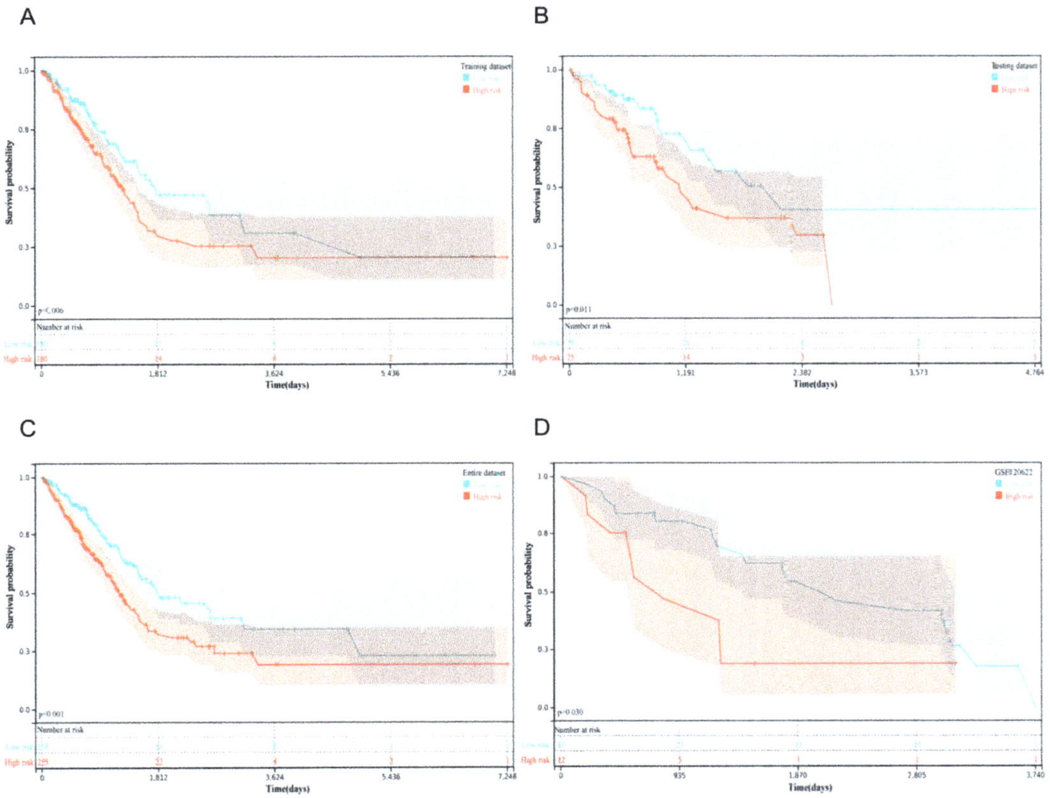

Figure 2. Survival prognosis of the risk score model. Kaplan–Meier (KM) survival curves of patients with LUAD in high- and low-risk groups from (**A**) TCGA training dataset, (**B**) TCGA testing dataset, (**C**) TCGA entire dataset, and (**D**) GSE120622.

Table 1. Severity of patients with lung adenocarcinoma (LUAD) and a low-risk score vs. patients with LUAD and a high-risk score in the case-control study.

	Low Risk	High Risk	*p* Value
TNM			
Stage I + II	9	1	0.005 **
Stage III + IV	6	14	
T stage			
1–2	9	4	0.139
3–4	6	11	
M stage			
0	11	2	0.003 **
1	4	13	
N stage			
0	9	2	0.021 *
1–3	6	13	
Performance Status			
0	13	6	0.021 *
1–4	2	9	

Note: ** $p < 0.01$ is considered statistically significant. * $p < 0.05$ is considered statistically significant.

Additionally, to prove the stability and independence of the risk score based on IRLs, we plotted KM curves on patients with LUAD under different subgroups divided by baseline characteristics and performed univariate and multivariate Cox regression analysis on the risk score and various clinicopathological characteristics. Results showed that patients in the low-risk group had a high chance of survival at a different age ($p = 0.019$ in age ≤ 60 years, $p = 0.001$ in age > 60 years), TNM stages ($p = 0.046$ in stage I + II, $p = 0.011$ in stage III + IV), and male patients ($p = 0.001$ in male patients; Figure 3A–F). Univariate and multivariate Cox regression analyses demonstrated that the risk score ($p = 0.004$) and TNM stage ($p = 0.007$) had a remarkable predictive capability when considering potential risk factors (Figure 3G,H). Hence, the established risk score had an independent and reliable prognostic performance in predicting patients with LUAD.

3.3. Construction of an IRL Signature-Based Nomogram

A nomogram was established to visualize the above independent factors, including risk score and TNM stage (Figure 4A), in which the ROC curve showed that the AUC of the 5-year survival probability in the nomogram was 0.75 (Figure 4B). This finding suggested that the complex nomogram integrating IRLs and clinical characteristics could be effective in predicting the survival status of LUAD. Besides, we compared several IRL prognostic models and relevant nomograms of LUAD in a published paper [21–24]. Similarly, the nomogram AUC values of these four TCGA entire datasets were more than 0.70, but the nomogram AUC of GSE120622 was between 0.58 and 0.75 (Figure S4).

3.4. Immune Landscape and Efficacy of ICI

In order to characterize the immune environment of patients with LUAD, the proportions of infiltrating immune cells were compared between low- and high-risk groups in all samples using the CIBERSORT and LM22 signature matrices. The low-risk group had higher percentages of naive B cells, plasma cells, CD8+ T cells, and activated memory CD4+ T cells but lower percentages of M0 macrophages, M2 macrophages, and activated dendritic cells than the high-risk group (Figure 5A). Additionally, the low-risk group was found to have higher stromal scores ($p < 0.001$), immune scores ($p < 0.001$), and ESTIMATE scores ($p < 0.001$), but lower tumor purity ($p < 0.001$) than the high-risk group (Figure 5B–E).

To explore the ability of the IRL signature in the prediction of immunotherapeutic sensitivity for LUAD patients, IPS values, which were calculated based on immunogenicity from the TCIA database, were analyzed in the risk model. The outcome showed that the potentials of the low-risk group to respond to anti-PD-1/PD-L1 ($p < 0.001$) and anti-cytotoxic T lymphocyte-associated antigen-4 (CTLA4, $p < 0.001$) treatment were higher than those of the high-risk group (Figure 6A,B). In accordance with these results, we found that the expression levels of PD-1 ($p < 0.001$), PD-L1 ($p < 0.001$), and CTLA4 ($p < 0.001$) were relatively increased in the low-risk group (Figure 6C–E). Thus, ICI treatment might be effective for patients with LUAD with low-risk scores. However, the expression profiles of the amplification of murine double minute (MDM) 2 and 4 ($p = 0.002$, $p < 0.001$), which were markers for HPD, were modestly elevated in low-risk patients (Figure S5), although no difference in DNA methyltransferase 3 alpha (DNMT3A), Cyclin D1, or Fibroblast Growth Factor (FGF) 3/4/19 was observed.

After conducting gene set enrichment analysis (GSEA) on the mRNAs of all LUAD samples, we found that more pathways in the low-risk group were upregulated compared with those in the high-risk group (Figure S6). Compared with the high-risk group, the low-risk group had significantly enriched V-Ki-Ras2 Kirsten rat sarcoma viral oncogene homolog (KRAS) signaling, interferon-gamma response, interleukin (IL) 2–signal transduction, and activator of transcription (STAT) 5 signaling. Moreover, class II major histocompatibility complex transactivator (CIITA) and interferon alpha and beta receptor subunit 2, as target mRNAs of AC104971.3 and LY86-AS1, took part in the interferon-gamma response, implying that IRLs might regulate the relevant pathways via target mRNAs and play an essential role in the pathogenesis of LUAD.

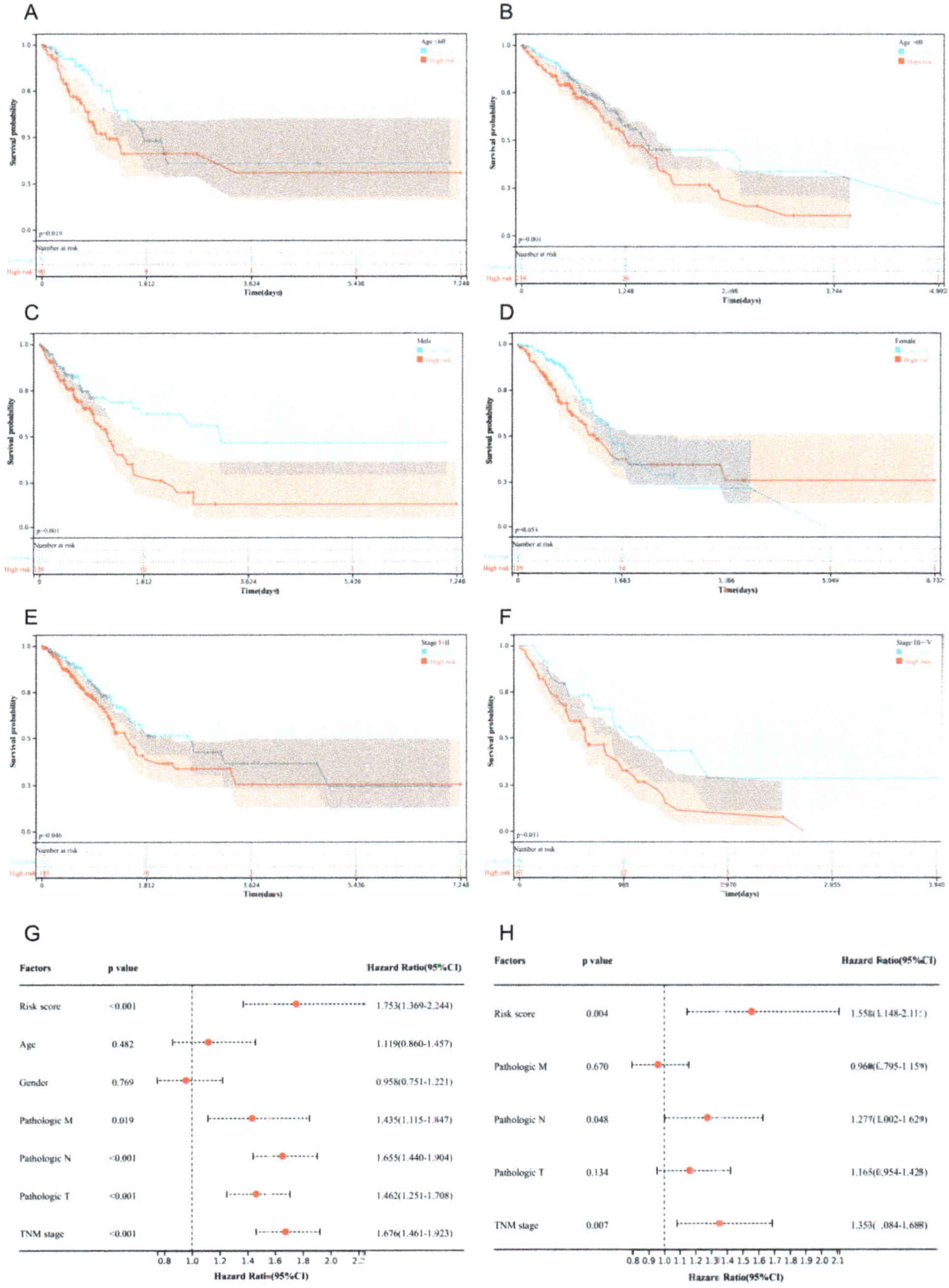

Figure 3. Stability and independence of the prognostic risk model. Survival times of patients among different (**A**,**B**) age, (**C**,**D**) gender, and (**E**,**F**) TNM stage subgroups. (**G**) Univariate and (**H**) multivariate Cox regression analyses on risk score, age, gender, M, N, T, and TNM stages.

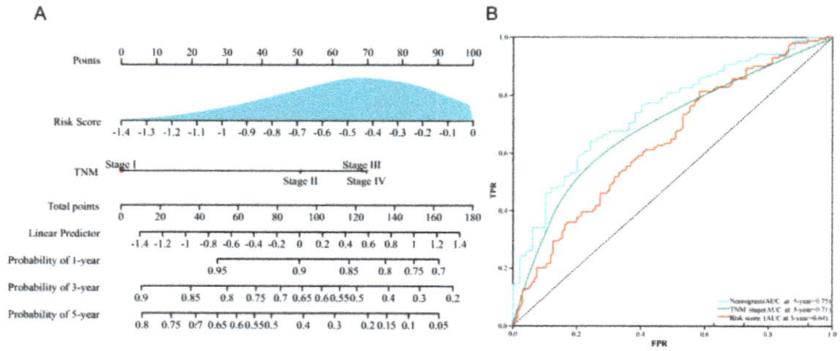

Figure 4. Nomogram integrating risk score and clinical features. (**A**) Nomogram for predicting 1-, 3-, and 5-year overall survival rates of patients with LUAD. (**B**) ROC curves of the risk score, TNM stage, and nomogram. AUC, areas under the ROC curve. FPR, false positive rate. TPR, true positive rate.

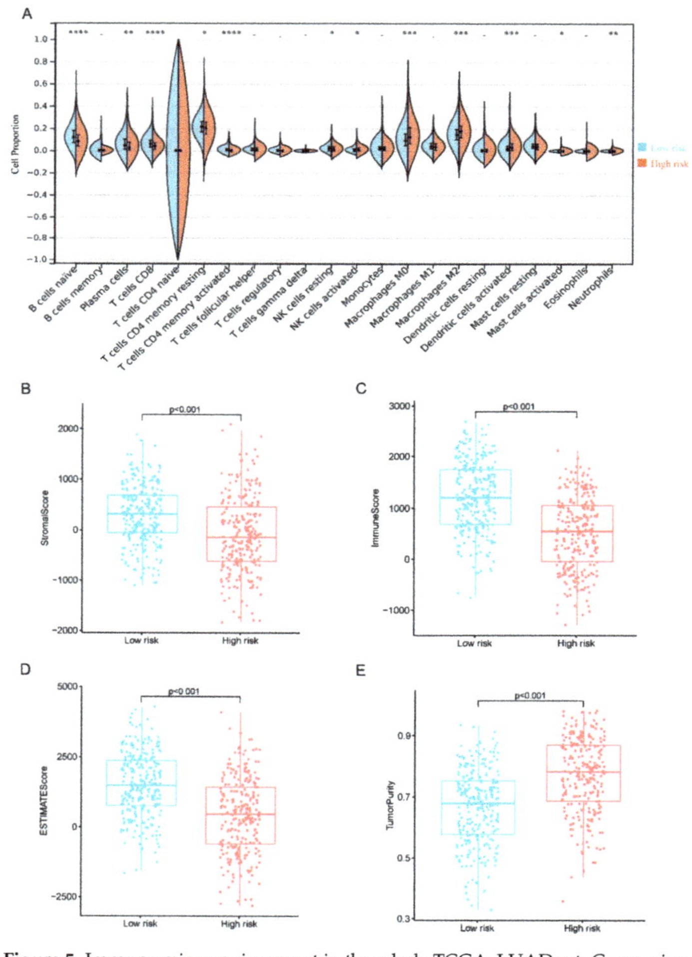

Figure 5. Immune microenvironment in the whole TCGA–LUAD set. Comparison of (**A**) 22 immune cell proportion, (**B**) stromal score, (**C**) immune score, (**D**) ESTIMATE score, and (**E**) tumor purity. Note: **** means $p < 0.0001$. *** means $p < 0.001$. ** means $p < 0.01$. * means $p < 0.05$. - means no significant.

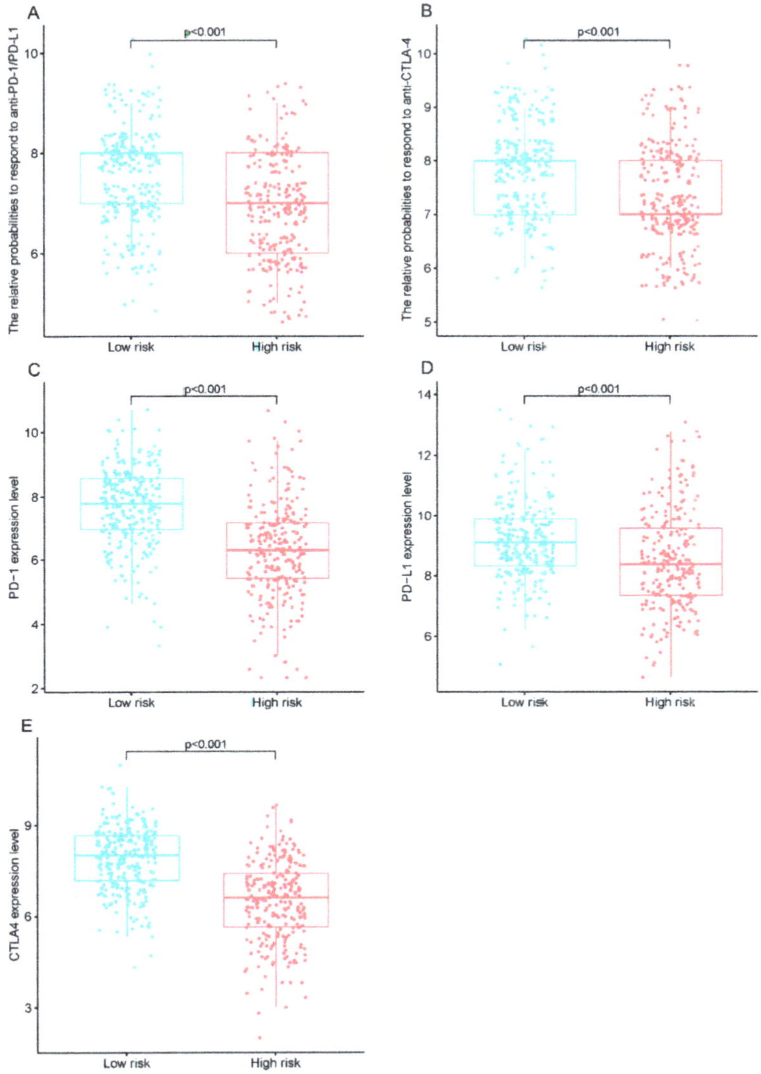

Figure 6. Predictive response to immune checkpoint inhibitor (ICI) treatment in patients with LUAD. Relative probabilities of responding to (**A**) anti-PD-1/PD-L1 and (**B**) anti-CTLA4 treatments. The mRNA levels of (**C**) PD-1, (**D**) PD-L1, and (**E**) CTLA4.

4. Discussion

Infiltrating immune cells in the tumor microenvironment are critical in cancer progression [25], and the quantitative evaluation of tumor immune infiltrates is still a major challenge using the traditional immunohistochemistry immunoscoring approach. Additionally, ICI has become the first-line treatment for advanced LUAD that is refractory to targeted therapy [5]. Nowadays. PD-L1 expression, tumor somatic mutation burden, mismatch repair deficiency, and microsatellite instability have been widely applied to predict ICI efficacy. However, several concerns, including the poor uniformity of detection technologies and different cutoff values for positivity across clinical trials. have limited its utility [26]. In recent years, IRLs have been proven to be indispensable in tumor progression and oncogenic pathways by regulating gene expression and can be used as a potential prognostic biomarker for cancer [27].

Immunologic features, which actively participate in cancer development, can more effectively predict patients' survival than traditional intrinsic features of tumors [28]. With the dramatic development of gene sequencing technology, molecular profiling-based signatures to infer immune infiltration have become a reality. Several previous studies constructed risk models with novel IRLs that can predict the prognosis in patients with LUAD. In most of these studies, IRLs are obtained through different expressions combined with Pearson correlation or interaction prediction with immune-related mRNA expression profiles from databases. Another study selected lncRNAs that are upregulated in immune cell lines but downregulated in NSCLC cell lines as tumor-infiltrating IRLs [29]. LncRNAs are suggested to participate in the immune response by regulating the expression of target mRNAs and interacting with chromatin, proteins, and miRNA in various ways [30,31]. Hence, only a few IRLs have been found to play a role in LUAD so far, and high-throughput methods for the identification of lncRNAs affecting immune activity are still largely unknown.

Therefore, our study of several novel IRLs is an essential complement to identifying their roles in immune regulation and immunotherapy targets in LUAD. Here, we assessed lncRNAs directly related to immune response, which are systematically identified via a computational algorithm and represent immune pathways and distinct immune cell types in the immLnc database. Afterward, we extracted IRLs and relevant clinical information from TCGA–LUAD to construct a risk signature via integrative analysis and then divided patients with LUAD into low- and high-risk groups. Comparing the survival outcomes of patients classified by the six IRLs, we supposed that the risk score model and integrated nomogram have a reliable and stable prediction performance. Moreover, cytotoxic T cells, Th1 helper cells, B cells, and plasma cells eliminate tumor cells in the antitumor immune milieu, whereas specific macrophages and regulatory T cells can accelerate immune escape and tumor growth in the protumorigenic immune milieu [32]. Protective CD8+ T cell responses can be induced by activated dendritic cells in the inflammation of normal tissue [33], whereas an immune reaction to lung cancer in the presence of mature dendritic cells (activated or not) is necessary to organize cytotoxic T cells, which are associated with a good clinical outcome and response to therapeutics [34]. Similarly, we found that patients with low-risk immune signatures have long OS; increased B cells, plasma cells, CD8+ T cells, and CD4+ T cells; and minimal macrophages and activated dendritic cells infiltrated in tumors.

Some similar published papers studied IRLs by performing coexpression analysis between lncRNAs from TCGA and immune-related genes from MsigDB, ImmPort, or the GSEA database, of which only three papers were validated by GEO microarray data [22,23] or clinical studies [21]. We retrieved IRLs for LUAD from the ImmLnc database, which was established by integrating tumor purity estimation, GSEA, and powerful algorithms [9], and uninvestigated IRLs were found by directly retrieving IRLs from ImmLnc and TCGA and integrating bioinformatics. The risk model and nomogram presented here were validated by GEO sequencing data and our clinical study, suggesting that the IRL signature is available in Chinese patients with LUAD.

An effective IRL-based model for patient selection before ICI treatment in LUAD has not been studied yet. A considerable proportion of patients with NSCLC have a poor response to immunotherapy despite the high expression of immune checkpoint molecules [35]. Thus, developing comprehensive predictive biomarkers is indispensable. The complex interaction between tumor immune infiltrates and the immunotherapy response affects NSCLC [36]. For instance, more CD8+ T cell infiltration in lung cancer tissue is associated with a superior treatment response from pembrolizumab treatment [37]. In our model, patients in the low-risk group have increased IPS values for PD1/PD-L1 and CTLA4 blockers, and the expression of immune checkpoint molecules may be associated with improved sensitivity to ICI treatment, suggesting that their tumors are in a preactivated immune status. Thus, six IRLs can be useful for choosing suitable immunotherapy.

Additionally, HPD is a novel pattern of tumor progression, with unexpected and rapid tumor growth, poor prognosis of patients, and high rates of fatality, which have limited the clinical application of ICI [6]. Considering the limitations of ICI due to HPD incidence, valid biomarkers are urgently needed to predict the occurrence of HPD to improve ICI efficacy. To date, several tumor cell biomarkers, including MDM2/4, epidermal growth factor receptor mutation, DNMT3A, and FGF3/4/19, have been shown to be associated with HPD [38]. We found that the expression profiles of MDM2 and MDM4, which regulate p53 and apoptotic responses to cellular injuries when overexpressed [39], are modestly elevated in low-risk patients. This finding implies that patients with a minimal IRL signature may gain benefits from ICI treatment, but some patients may have a potential risk of HPD. Hence, multiple factors influencing the efficacy of ICI should be comprehensively considered to optimize treatment regimens.

To explore the possible underlying mechanisms, we conducted GSEA on targeted mRNAs of IRLs. KRAS signaling, interferon-gamma response, and IL2–STAT5 signaling, as potential positive predictors of antitumor immunity [40], are enriched in the low-risk group. CIITA, the target mRNA of AC104971, plays an important role in the interferon-gamma response. Consistent with our result, the loss of CIITA converts lung cancer from anti-PD-1-sensitive to anti-PD-1-resistant [41]. Our study suggests that these IRLs may regulate target mRNAs and play a functional role in the sensitivity to ICI treatment for LUAD.

Even with the above promising findings, some limitations remain in this study. First, the prognostic model is established by public databases, which may increase the bias. Even though the model has been validated with the internal database, external database, and single-center case-control study, the optimal cut-off value and predictive capability of the six IRLs demand further confirmation in prospective clinical trials with complete survival time based on a large sample size. Second, the functions of the six IRLs have not been validated. Thus, functional and mechanistic experiments are needed to support our findings.

In conclusion, the six-IRL signature is a promising biomarker for prognosis prediction and facilitates the management of immunotherapy in LUAD. Patients with low risk might gain benefits from ICI, although some have a risk of hyperprogressive disease.

Supplementary Materials: The following supporting information can be downloaded at: https://www.mdpi.com/article/10.3390/diagnostics12112891/s1, Table S1: Clinicopathological characteristics of LUAD patients in TCGA, GEO datasets and clinical cohort; Table S2: Baseline clinical characteristics of LUAD patients with low and high risk in TCGA, GEO datasets, and clinical cohorts; Table S3: Primer sequence of lncRNAs; Figure S1: Expression profiles of the prognostic immune signature; Figure S2: ROC curves of risk score with IRLs; Figure S3: Absolute quantification of six IRLs in a case-control study; Figure S4: Comparison with other reported IRL risk score models and relevant nomograms; Figure S5: Biomarkers for hyperprogression in response to ICI treatment; Figure S6: Top 10 hallmark pathways of GSEA for patients with LUAD and low risk score vs. patients with LUAD and high risk score.

Author Contributions: G.L., F.C. and Y.G. conceived and supervised the project. Z.Z. designed and performed essential data analyses and wrote the initial manuscript draft. Y.L., J.S., L.X. and L.T. participated in data downloading and collection. All authors edited the manuscript. All authors have read and agreed to the published version of the manuscript.

Funding: This work was supported by the Natural Science Foundation of Guangdong Province (2021A1515012027), National Natural Science Foundation of China (82200029) and the Guangdong Finance Foundation for Industrial Technology Research and Development (20160907)

Institutional Review Board Statement: The study was conducted in accordance with the Declaration of Helsinki and approved by the Ethics Committee of the First Affiliated Hospital of Sun Yat-sen University for studies involving humans (No. (2022)049 on 28 January 2022).

Informed Consent Statement: Not applicable.

Data Availability Statement: The data that support the findings of this study are openly available in the TCGA cohort of the UCSC Xena database (https://xenabrowser.net/ accessed on 16 September 2020), the GEO database (https://www.ncbi.nlm.nih.gov/geo/ accessed on 4 July 2022), and the immLnc database (http://bio-bigdata.hrbmu.edu.cn/ImmLnc/ accessed on 28 March 2021). The version 22 references that genome data, IPS data, and target mRNAs of IRLs were obtained from the Gencode database (https://www.gencodegenes.org/ accessed on 27 April 2020), The Cancer Immunome Atlas database (https://tcia.at/home accessed on 10 May 202), and the starbase3.0 database (http://starbase.sysu.edu.cn/ accessed on 6 April 2020).

Conflicts of Interest: The authors declare that they have no known competing financial interest or personal relationships that could have appeared to influence the work reported in this paper.

References

1. Sung, H.; Ferlay, J.; Siegel, R.L.; Laversanne, M.; Soerjomataram, I.; Jemal, A.; Bray, F. Global Cancer Statistics 2020: GLOBOCAN Estimates of Incidence and Mortality Worldwide for 36 Cancers in 185 Countries. *CA Cancer J. Clin.* **2021**, *71*, 209–249. [CrossRef] [PubMed]
2. Chen, Z.; Fillmore, C.M.; Hammerman, P.S.; Kim, C.F.; Wong, K.K. Non-small-cell lung cancers: A heterogeneous set of diseases. *Nat. Rev. Cancer* **2014**, *14*, 535–546. [CrossRef] [PubMed]
3. Lewis, J.; Gillaspie, E.A.; Osmundson, E.C.; Horn, L. Before or After: Evolving Neoadjuvant Approaches to Locally Advanced Non-Small Cell Lung Cancer. *Front. Oncol.* **2018**, *8*, 5. [CrossRef] [PubMed]
4. Chi, A.; He, X.; Hou, L.; Nguyen, N.P.; Zhu, G.; Cameron, R.B.; Lee, J.M. Classification of Non-Small Cell Lung Cancer's Tumor Immune Micro-Environment and Strategies to Augment Its Response to Immune Checkpoint Blockade. *Cancers* **2021**, *13*, 2924. [CrossRef] [PubMed]
5. Santarpia, M.; Aguilar, A.; Chaib, I.; Cardona, A.F.; Fancelli, S.; Laguia, F.; Bracht, J.W.P.; Cao, P.; Molina-Vila, M.A.; Karachaliou, N.; et al. Non-Small-Cell Lung Cancer Signaling Pathways, Metabolism, and PD-1/PD-L1 Antibodies. *Cancers* **2020**, *12*, 1475. [CrossRef]
6. Ferrara, R.; Mezquita, L.; Texier, M.; Lahmar, J.; Audigier-Valette, C.; Tessonnier, L.; Mazieres, J.; Zalcman, G.; Brosseau, S.; Le Moulec, S.; et al. Hyperprogressive Disease in Patients with Advanced Non-Small Cell Lung Cancer Treated with PD-1/PD-L1 Inhibitors or with Single-Agent Chemotherapy. *JAMA Oncol.* **2018**, *4*, 1543–1552. [CrossRef]
7. Yu, W.D.; Wang, H.; He, Q.F.; Xu, Y.; Wang, X.C. Long noncoding RNAs in cancer-immunity cycle. *J. Cellul. Physiol.* **2018**, *233*, 6518–6523. [CrossRef]
8. Li, B.; Cui, Y.; Diehn, M.; Li, R. Development and Validation of an Individualized Immune Prognostic Signature in Early-Stage Nonsquamous Non-Small Cell Lung Cancer. *JAMA Oncol.* **2017**, *3*, 1529–1537. [CrossRef]
9. Li, Y.; Jiang, T.; Zhou, W.; Li, J.; Li, X.; Wang, Q.; Jin, X.; Yin, J.; Chen, L.; Zhang, Y.; et al. Pan-cancer characterization of immune-related lncRNAs identifies potential oncogenic biomarkers. *Nat. Commun.* **2020**, *11*, 1000. [CrossRef]
10. Weng, R.R.; Lu, H.H.; Lin, C.T.; Fan, C.C.; Lin, R.S.; Huang, T.C.; Lin, S.Y.; Huang, Y.J.; Juan, Y.H.; Wu, Y.C.; et al. Epigenetic modulation of immune synaptic-cytoskeletal networks potentiates γδ T cell-mediated cytotoxicity in lung cancer. *Nat. Commun.* **2021**, *12*, 2163. [CrossRef]
11. Charoentong, P.; Finotello, F.; Angelova, M.; Mayer, C.; Efremova, M.; Rieder, D.; Hackl, H.; Trajanoski, Z. Pan-cancer Immunogenomic Analyses Reveal Genotype-Immunophenotype Relationships and Predictors of Response to Checkpoint Blockade. *Cell Rep.* **2017**, *18*, 248–262. [CrossRef]
12. Rami-Porta, R.; Bolejack, V.; Crowley, J.; Ball, D.; Kim, J.; Lyons, G.; Rice, T.; Suzuki, K.; Thomas, C.F., Jr.; Travis, W.D.; et al. The IASLC Lung Cancer Staging Project: Proposals for the Revisions of the T Descriptors in the Forthcoming Eighth Edition of the TNM Classification for Lung Cancer. *J. Thoracic Oncol.* **2015**, *10*, 990–1003. [CrossRef]
13. Langfelder, P.; Horvath, S. WGCNA: An R package for weighted correlation network analysis. *BMC Bioinf.* **2008**, *9*, 559. [CrossRef]
14. Stel, V.S.; Dekker, F.W.; Tripepi, G.; Zoccali, C.; Jager, K.J. Survival analysis II: Cox regression. *Nephron. Clin. Pract.* **2011**, *119*, c255–c260. [CrossRef]
15. Tibshirani, R. The lasso method for variable selection in the Cox model. *Stat. Med.* **1997**, *16*, 385–395. [CrossRef]
16. Yang, S.; Wu, Y.; Deng, Y.; Zhou, L.; Yang, P.; Zheng, Y.; Zhang, D.; Zhai, Z.; Li, N.; Hao, Q.; et al. Identification of a prognostic immune signature for cervical cancer to predict survival and response to immune checkpoint inhibitors. *Oncoimmunology* **2019**, *8*, e1659094. [CrossRef]
17. Heagerty, P.J.; Lumley, T.; Pepe, M.S. Time-dependent ROC curves for censored survival data and a diagnostic marker. *Biometrics* **2000**, *56*, 337–344. [CrossRef]
18. Iasonos, A.; Schrag, D.; Raj, G.V.; Panageas, K.S. How to build and interpret a nomogram for cancer prognosis. *J. Clin. Oncol.* **2008**, *26*, 1364–1370. [CrossRef]
19. Newman, A.M.; Liu, C.L.; Green, M.R.; Gentles, A.J.; Feng, W.; Xu, Y.; Hoang, C.D.; Diehn, M.; Alizadeh, A.A. Robust enumeration of cell subsets from tissue expression profiles. *Nat. Methods* **2015**, *12*, 453–457. [CrossRef]

20. Subramanian, A.; Tamayo, P.; Mootha, V.K.; Mukherjee, S.; Ebert, B.L.; Gillette, M.A.; Paulovich, A.; Pomeroy, S.L.; Golub, T.R.; Lander, E.S.; et al. Gene set enrichment analysis: A knowledge-based approach for interpreting genome-wide expression profiles. *Proc. Natl. Acad. Sci. USA* **2005**, *102*, 15545–15550. [CrossRef]
21. Zhang, B.; Wang, R.; Li, K.; Peng, Z.; Liu, D.; Zhang, Y.; Zhou, L. An Immune-Related lncRNA Expression Profile to Improve Prognosis Prediction for Lung Adenocarcinoma: From Bioinformatics to Clinical Word. *Front. Oncol.* **2021**, *11*, 671341. [CrossRef] [PubMed]
22. Cao, K.; Liu, M.; Ma, K.; Jiang, X.; Ma, J.; Zhu, J. Prediction of prognosis and immunotherapy response with a robust immune-related lncRNA pair signature in lung adenocarcinoma. *Cancer Immunol. Immunother.* **2022**, *71*, 1295–1311. [CrossRef] [PubMed]
23. Zhuang, J.; Chen, Z.; Chen, Z.; Chen, J. Liu, M.; Xu, X.; Liu, Y.; Yang, S.; Hu, Z.; He, F. Construction of an immune-related lncRNA signature pair for predicting oncologic outcomes and the sensitivity of immunosuppressor in treatment of lung adenocarcinoma. *Respir. Res.* **2022**, *23*, 123. [CrossRef] [PubMed]
24. Miao, H.; Chen, D.; Li, R.; Hu, J.; Chen, Y.; Xu, C.; Wen, Z. Identification of an immune-related six-long noncoding RNA signature as a novel prognosis biomarker for adenocarcinoma of lung. *Biosci. Rep.* **2021**, *41*, 2444. [CrossRef] [PubMed]
25. Jochems, C.; Schlom, J. Tumor-infiltrating immune cells and prognosis: The potential link between conventional cancer therapy and immunity. *Exp. Biol. Med.* **2011**, *236*, 567–579. [CrossRef]
26. Hellmann, M.D.; Ciuleanu, T.E.; Pluzanski, A.; Lee, J.S.; Otterson, G.A.; Audigier-Valette, C.; Minenza, E.; Linardou, H.; Burgers, S.; Salman, P.; et al. Nivolumab plus Ipilimumab in Lung Cancer with a High Tumor Mutational Burden. *New Engl. J. Med.* **2018**, *378*, 2093–2104. [CrossRef]
27. Li, Y.; Zhao, J.; Yu, S.; Wang, Z.; He, X.; Su, Y.; Guo, T.; Sheng, H.; Chen, J.; Zheng, Q.; et al. Extracellular Vesicles Long RNA Sequencing Reveals Abundant mRNA, circRNA, and lncRNA in Human Blood as Potential Biomarkers for Cancer Diagnosis. *Clin. Chem.* **2019**, *65*, 798–808. [CrossRef]
28. Guo, D.; Wang, M.; Shen, Z.; Zhu, J. A new immune signature for survival prediction and immune checkpoint molecules in lung adenocarcinoma. *J. Trans. Med.* **2020**, *18*, 123. [CrossRef]
29. Sun, J.; Zhang, Z.; Bao, S.; Yan, C.; Hou, P.; Wu, N.; Su, J.; Xu, L.; Zhou, M. Identification of tumor immune infiltration-associated lncRNAs for improving prognosis and immunotherapy response of patients with non-small cell lung cancer. *J. Immunother. Cancer* **2020**, *8*, 110. [CrossRef]
30. Guo, Z.; Wang, Y.H.; Xu, H.; Yuan, C.S.; Zhou, H.H.; Huang, W.H.; Wang, H.; Zhang, W. LncRNA linc00312 suppresses radiotherapy resistance by targeting DNA-PKcs and impairing DNA damage repair in nasopharyngeal carcinoma. *Cell Death Disease* **2021**, *12*, 69. [CrossRef]
31. Brahic, M.; Bureau, J.F.; Michiels, T. The genetics of the persistent infection and demyelinating disease caused by Theiler's virus. *Annual Review Microbiol.* **2005**, *59*, 279–298. [CrossRef]
32. Taube, J.M.; Galon, J.; Sholl, L.M.; Rodig, S.J.; Cottrell, T.R.; Giraldo, N.A.; Baras, A.S.; Patel, S.S.; Anders, R.A.; Rimm, D.L.; et al. Implications of the tumor immune microenvironment for staging and therapeutics. *Modern Pathol.* **2018**, *31*, 214–234. [CrossRef]
33. McKay, P.F.; Cizmeci, D.; Aldon, Y.; Maertzdorf, J.; Weiner, J.; Kaufmann, S.H.; Lewis, D.J.; van den Berg, R.A.; Del Giudice, G.; Shattock, R.J. Identification of potential biomarkers of vaccine inflammation in mice. *eLife* **2019**, *8*, 46149. [CrossRef]
34. Goc, J.; Germain, C.; Vo-Bourgais, T.K.; Lupo, A.; Klein, C.; Knockaert, S.; de Chaisemartin, L.; Ouakrim, H.; Becht, E.; Alifano, M.; et al. Dendritic cells in tumor-associated tertiary lymphoid structures signal a Th1 cytotoxic immune contexture and license the positive prognostic value of infiltrating CD8+ T cells. *Cancer Res.* **2014**, *74*, 705–715. [CrossRef]
35. Pu, X.; Wu, L.; Su, D.; Mao, W.; Fang, B. Immunotherapy for non-small cell lung cancers: Biomarkers for predicting responses and strategies to overcome resistance. *BMC Cancer* **2018**, *18*, 1082. [CrossRef]
36. Varn, F.S.; Tafe, L.J.; Amos, C.I.; Cheng, C. Computational immune profiling in lung adenocarcinoma reveals reproducible prognostic associations with implications for immunotherapy. *Oncoimmunology* **2018**, *7*, e1431084. [CrossRef]
37. Garon, E.B.; Hellmann, M.D.; Rizvi, N.A.; Carcereny, E.; Leighl, N.B.; Ahn, M.J.; Eder, J.P.; Balmanoukian, A.S.; Aggarwal, C.; Horn, L.; et al. Five-Year Overall Survival for Patients with Advanced Non–Small-Cell Lung Cancer Treated with Pembrolizumab: Results From the Phase I KEYNOTE-001 Study. *J. Clin. Oncol.* **2019**, *37*, 2518–2527. [CrossRef]
38. Singavi, A.K.; Menon, S.; Kilari, D.; Alqwasmi, A.; Ritch, P.S.; Thomas, J.P.; Martin, A.L.; Oxencis, C.; Ali, S.; George, B. 1140PD-Predictive biomarkers for hyper-progression (HP) in response to immune checkpoint inhibitors (ICI)–analysis of somatic alterations (SAs). *Ann. Oncol.* **2017**, *28*, v405. [CrossRef]
39. Hou, H.; Sun, D.; Zhang, X. The role of MDM2 amplification and overexpression in therapeutic resistance of malignant tumors. *Cancer Cell Int.* **2019**, *19*, 216. [CrossRef]
40. Dong, Z.Y.; Zhong, W.Z.; Zhang, X.C.; Su, J.; Xie, Z.; Liu, S.Y.; Tu, H.Y.; Chen, H.J.; Sun, Y.L.; Zhou, Q.; et al. Potential Predictive Value of TP53 and KRAS Mutation Status for Response to PD-1 Blockade Immunotherapy in Lung Adenocarcinoma. *Clin. Cancer Res.* **2017**, *23*, 3012–3024. [CrossRef]
41. Johnson, A.M.; Bullock, B.L.; Neuwelt, A.J.; Poczobutt, J.M.; Kaspar, R.E.; Li, H.Y.; Kwak, J.W.; Hopp, K.; Weiser-Evans, M.C.M.; Heasley, L.E.; et al. Cancer Cell-Intrinsic Expression of MHC Class II Regulates the Immune Microenvironment and Response to Anti-PD-1 Therapy in Lung Adenocarcinoma. *J. Immunol.* **2020**, *204*, 2295–2307. [CrossRef] [PubMed]

diagnostics

MDPI

Article

Radiomics Based on Nomogram Predict Pelvic Lymphnode Metastasis in Early-Stage Cervical Cancer

Xueming Xia [1,2,†], Dongdong Li [3,†], Wei Du [2,†], Yu Wang [4,†], Shihong Nie [2], Qiaoyue Tan [2] and Qiheng Gou [1,2,*,†]

1 Department of Head and Neck Oncology, Cancer Center, West China Hospital, Sichuan University, Chengdu 610041, China
2 Department of Radiotherapy, Cancer Center, West China Hospital, Sichuan University, Chengdu 610041, China
3 Department of Network Engineering, School of Computer Science and Engineering, South China University of Technology, Guangzhou 510641, China
4 Department of Radiology, The Affiliated Hospital of Southwest Medical University, Luzhou 402103, China
* Correspondence: gouqiheng513@wchscu.cn
† These authors contributed equally to this work.

Abstract: The accurate prediction of the status of PLNM preoperatively plays a key role in treatment strategy decisions in early-stage cervical cancer. The aim of this study was to develop and validate a radiomics-based nomogram for the preoperative prediction of pelvic lymph node metastatic status in early-stage cervical cancer. One hundred fifty patients were enrolled in this study. Radiomics features were extracted from T2-weighted MRI imaging (T2WI). Based on the selected features, a support vector machine (SVM) algorithm was used to build the radiomics signature. The radiomics-based nomogram was developed incorporating radiomics signature and clinical risk factors. In the training cohort (AUC = 0.925, accuracy = 81.6%, sensitivity = 70.3%, and specificity = 92.0%) and the testing cohort (AUC = 0.839, accuracy = 74.2%, sensitivity = 65.7%, and specificity = 82.8%), clinical models that combine stromal invasion depth, FIGO stage, and MTD perform poorly. The combined model had the highest AUC in the training cohort (AUC = 0.988, accuracy = 95.9%, sensitivity = 92.0%, and specificity = 100.0%) and the testing cohort (AUC = 0.922, accuracy = 87.1%, sensitivity = 85.7%, and specificity = 88.6%) when compared to the radiomics and clinical models. The study may provide valuable guidance for clinical physicians regarding the treatment strategies for early-stage cervical cancer patients.

Keywords: cervical cancer; pelvic lymph node metastasis; radiomics; nomogram

Citation: Xia, X.; Li, D.; Du, W.; Wang, Y.; Nie, S.; Tan, Q.; Gou, Q. Radiomics Based on Nomogram Predict Pelvic Lymphnode Metastasis in Early-Stage Cervical Cancer. *Diagnostics* **2022**, *12*, 2446. https://doi.org/10.3390/diagnostics12102446

Academic Editor: Romaric Loffroy

Received: 1 September 2022
Accepted: 6 October 2022
Published: 10 October 2022

1. Introduction

Cervical cancer is the fourth most common malignant tumor and the fourth leading cause of cancer death in women worldwide [1]. Due to the promotion of tumor screening, an increasing number of early cervical cancers have been discovered in recent years [2,3]. In the European Society for Medical Oncology (ESMO) Clinical Practice Guidelines on cervical cancer, radical hysterectomy with bilateral lymph node dissection [with or without sentinel lymph node (SLN)], carried out by laparotomy, was regarded as standard treatment in patients with the International Federation of Gynecology and Obstetrics (FIGO) stage IA2, IB and IIA [4,5]. However, nearly 30% of early-stage cervical cancer patients have pelvic lymph node metastasis (PLNM), which is an extremely important factor affecting treatment decisions [6,7]. For cervical cancer patients with PLNM, concurrent chemoradiation is preferred instead of surgical treatment, according to ESMO Clinical Practice Guidelines [8]. A considerable number of patients have received unnecessary radical hysterectomies, which may lead to serious complications and reduce their quality of life [4,5]. In addition, PLNM is a poor prognostic indicator for recurrence and metastasis in early-stage cervical cancer [9,10]. Therefore, the accurate prediction of the status of PLNM preoperatively plays a key role in treatment strategy decisions.

Magnetic resonance imaging (MRI) has been widely recommended as the optimal imaging equipment for preoperative staging and lymph node exploration of cervical cancer [11]. However, traditional MRI based on morphological assessment, such as lymph node size and morphology, has only a low sensitivity of 75% for lymph node status [12]. Due to the presence of inflammatory hyperplasia and micrometastatic lymph nodes, the efficacy of traditional MRI in differentiating lymph node status is difficult to satisfy [13]. Low diagnostic efficiency may result in many cervical cancer patients being understaged, which affects clinical decision-making [14].

Radiomics, which can extract quantitative features from digital medical images and convert them into mineable high-dimensional data, plays an important role in personalized clinical decision-making [15]. Radiomics analysis can be used in the diagnosis, prognosis, and prediction of curative effects in a variety of tumor types by developing appropriate model refinement features. This method has shown a huge potential for predicting lymph node metastatic status in a wide range of tumor types in a cost-effective and non-invasive way [16–18]. T2-weighted MRI imaging has also been widely utilized in the staging of cervical cancer [19].

Therefore, the purpose of this research was to develop and validate a radiomics-based nomogram for the preoperative prediction of pelvic lymph node metastatic status in early cervical cancer.

2. Materials and Methods

2.1. Patient

This retrospective study was approved by the hospital ethics committee, and the requirement for informed consent was waived. We retrospectively collected patients diagnosed with cervical cancer with biopsy-proven cervical carcinoma receiving initial treatment with surgery in our hospital from February 2017 to October 2019. The inclusion criteria were as follows: (i) patients receiving radical hysterectomy and pelvic lymphadenectomy; (ii) pathologically diagnosed cervical cancer; (iii) pretreatment MRI scan available in our hospital; (iv) primary cancer lesions visible on sagittal T2WI; and (v) available clinical characteristics. The exclusion criteria were as follows: (i) patients receiving other treatment before surgery, including neoadjuvant chemotherapy, radiotherapy, or conization; (ii) absence of preoperative MRI scan in our hospital; (iii) poor MRI image quality, for example, indistinct MRI image of the cervical structure; and (iv) rare pathological types of cervical tumor, for example, mucoepidermoid carcinoma [20]. The MRI images were reviewed by two radiologists with 7 and 9 years of experience.

In total, 150 patients were enrolled. Among these patients, 35 patients with PLNM and 115 patients without PLNM were treated surgically and confirmed pathologically. Baseline clinicopathologic characteristics, including age, FIGO stage, maximal tumor diameter (MTD), histological subtype, and stromal invasion, were derived from medical records. The patient selection process is shown in Figure 1.

2.2. MRI Acquisition Protocol and Tumor Segmentation

The MRI scans were performed on each patient before the initial surgical treatment. A T MRI scanner was used (Achieva 3.0 T, Philips, Amsterdam, The Netherlands), which was equipped with a 16-channel abdominal coil. Abdomen and pelvic MRI examinations were performed on patients. To avoid the loss of image information, the Digital Imaging and Communications in Medicine (DICOM) images from Picture Archiving were acquired without any compression or downsampling. The conventional protocol included T2-weighted imaging on the axial, sagittal, and coronal planes, T1-weighted imaging, fat-suppressed T1-weighted imaging, fat-suppressed T2-weighted imaging, and diffusion-weighted imaging (b = 0.800 s/mm^2) on the axial planes. The scanning parameters for the fat-suppressed turbo spin echo (TSE) T2-weighted images were as follows: TR/TE:

4854/85 ms, FOV = 300 × 300 mm; matrix = 232 × 171; slice thickness/gap: 5/1 mm; NEX = 2.

All of the MRI images were loaded into 3D slicer software and then manually three-dimensionally segmented (open-source software; https://download.slicer.org/, accessed on 1 January 2022). All of the manual segmentations of the primary tumor region were contoured as the region of interest (ROI) on axial T2WI, and the coronal and sagittal MRI images were used for guiding the segmentation of the ROIs in the cross-sectional plane. Segmentation was performed by a radiologist who had 7 years of experience in gynecological MR imaging and resolved any uncertainty through consultation with another radiologist who had 9 years of experience. The representative images of the lesions are shown in Figure 2. The workflow of the radiomics analysis is presented in Figure 3.

2.3. Radiomic Feature Extraction

Following the hand segmentation of the tumor, the radiographic characteristics of the tumor were extracted using open-source software called Pyradiomics. To generate a consistent normal-distribution image distribution, standardized techniques on T2WI pictures were applied. We collected 1688 radiological characteristics from seven imaging types on the T2WI of the tumor. Tumor size (e.g., volume), shape (e.g., circumference, diameter), grayscale cooccurrence matrices (e.g., energy, contrast, entropy), grayscale run-length matrices, and grayscale dependency matrices are all quantified using these characteristics. Python 3.7 is used to implement all functionalities. The workflow of the radiomic feature extraction is presented in Figure 3.

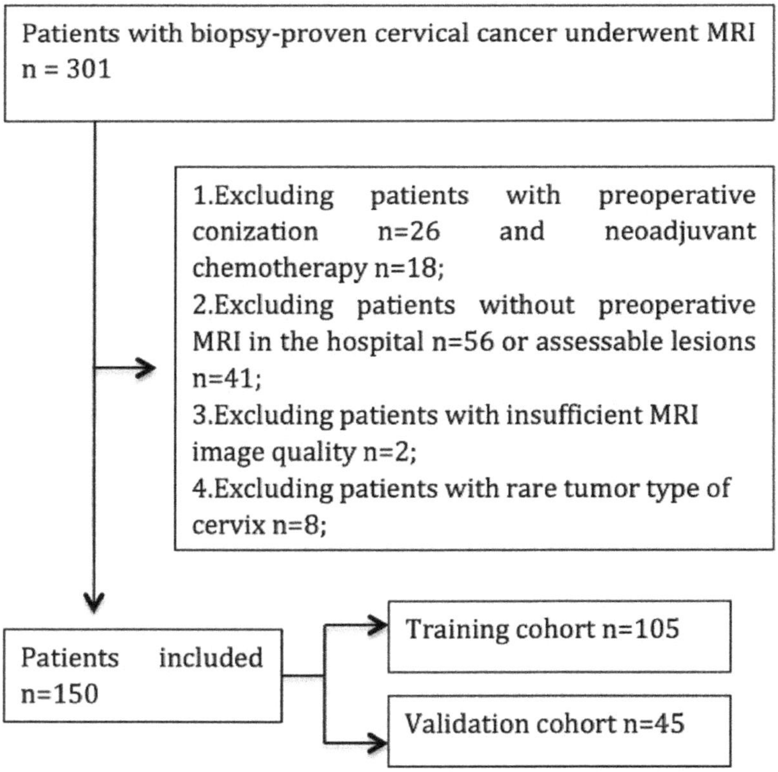

Figure 1. Flow chart of patient enrollment in this study.

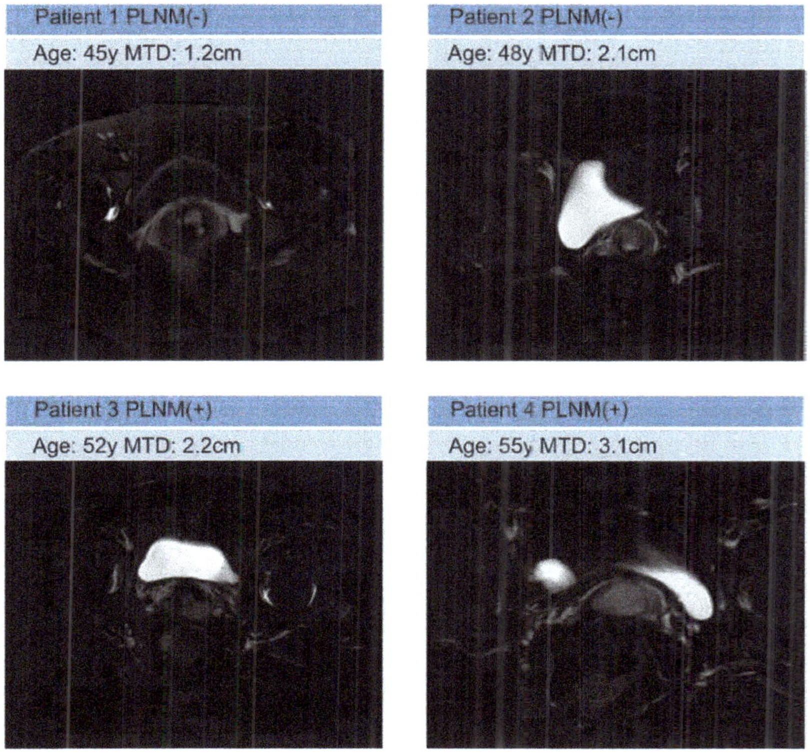

Figure 2. Representative MRI images in PLMN(−) and PLMN(+) patients.

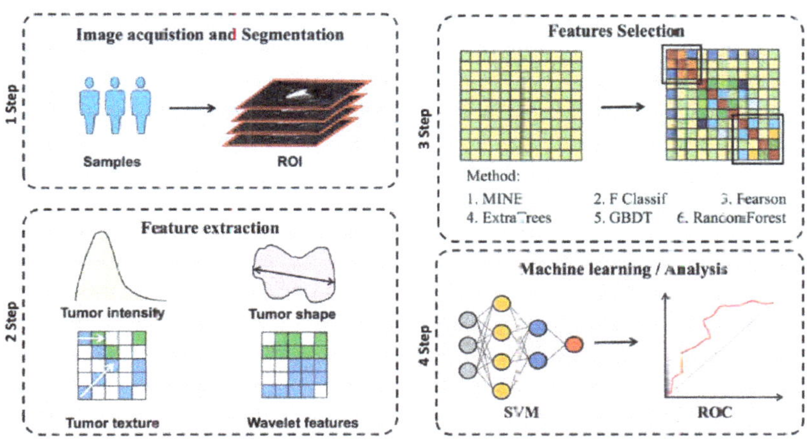

Figure 3. Radiomics analysis workflow. The study's subject underwent T2WI image collection. Radiomics performed feature extraction after manually marking the ROI region of the tumor lesion. The efficient feature combination was filtered using one of two feature extraction methods: filtered or embedded. For further statistical analysis, ROC curves were used.

2.4. Radiomics Feature Selection and Development of the Radiomics Model

The goal of feature selection is to choose some of the most useful features from the original features to minimize dimensionality and increase the model's generalization ability, as well as to speed up the fitting process. We have chosen two types of feature selection methods: filtered and embedded. MINE, Pearson's initial relation number, and ANOVA F value (F Classif) are chosen in the filtered scheme, whereas ExtraTree, GBDT, and Random Forest are chosen in the embedded scheme.

Because we used Pyradiomics to extract a vast number of features from the images, these features are not fully associated with the prevalence of the disease. Therefore, we chose the most effective feature extraction approach among six alternative feature filtering strategies to implement feature extraction. We also evaluated the validation of Chalkidou et al. [21], in which research showed that a stable model might be built using 10–15 characteristics. The outcomes of the experiments reveal that not all datasets will benefit from such a selection.

We use a support vector machine model to implement the sample classification when we have finished selecting features. The RBF kernel is also chosen in the SVM model, and the hyperparameter C is chosen to optimize the AUC value under the training set.

To develop the clinical prediction model, the stromal invasion depth, MTD, and FIGO phases were examined. To determine the sample classes, the SVM model is employed. The logistic regression approach with forward stepwise selection was used to create the combined model. In the combined model, the radiomics signature and clinical risk variables were incorporated. The combined model was presented as a radiomics-based nomogram to make it a more user-friendly tool for the preoperative prediction of PLMN status. The formula of the radiomics signature of the final radiomics model is shown in Supplementary S1.

2.5. Assessment of Predictive Models

The performance of these prediction models was evaluated in the training cohort before being verified in the validation cohort using the receiver operating characteristic (ROC) curve. The agreement between the nomogram prediction probabilities of the PLMN status and actual results was assessed using a calibration curve.

2.6. Statistical Analysis

All of the statistical analyses were conducted with R 3.4.1 and Python 3.7. The independent-sample t-test of independent samples was used to assess the significance of age and MTD between the training cohort and the validation cohort. The chi-squared test or Fisher's exact test was used to evaluate the significance of categorical variables such as FIGO stage, histology type, stromal invasion, lymphovascular invasion (LVSI), and nerve invasion (NI) between the training and validation cohorts. Two-tailed p-values less than 0.05 were considered statistically significant.

3. Results

3.1. Patients' Clinicopathologic Characteristics

Between February 2017 and October 2019, a total of 301 patients who underwent surgery for cervical cancer were enrolled. According to the inclusion and exclusion criteria, 151 patients were excluded. Finally, 150 patients fulfilled the eligibility criteria and were enrolled in the following analysis (Figure 1).

The patient characteristics are summarized in Table 1. The distribution of clinical characteristics (age, FIGO stage, MTD) and pathological characteristics (histology type, stromal invasion depth, LVSI, nerve invasion) were balanced between the training and validation cohorts. The MTD, stromal invasion depth, and LVSI status showed a significant difference between the patients with and without PLNM metastasis in the training cohorts. The LVSI status and NI status showed a significant difference in the validation cohort, as shown in Table 1.

Table 1. Characteristics of the included patients.

Characteristics	Training Cohort (n = 104)			Validation Cohort (n = 45)			p *
	PLNM(+) (n = 25)	PLNM(−) (n = 80)	p	PLNM(+) (n = 10)	PLNM(−) (n = 35)	p	
Age, years			0.095			0.312	0.520
Mean	47.12	46.66		43.30	46.60		
FIGO Stage (N, %)			0.787			0.208	0.871
IA	0	2		1	0		
IB	15	52		5	22		
IIA	10	26		4	13		
MTD, cm			0.007			0.220	0.815
Mean	3.95	2.77		2.97	3.07		
Histology (N, %)			0.331			0.657	0.664
SCC	19	67		10	30		
AC	4	11		0	4		
ASC	2	2		0	1		
Stromal Invasion			<0.001			0.261	0.443
Deep 1/3	21	28		6	10		
Middle 1/3	2	24		2	12		
Superficial 1/3	2	28		2	13		
LVSI status			<0.001			0.010	0.519
Positive	2	57		1	21		
Negative	23	23		9	14		
NI status			0.627			0.016	0.185
Positive	23	76		6	33		
Negative	2	4		4	2		

Note: p is derived from the chi-squared test or Fisher's exact test between patients with and without PLNM in the training and validation cohorts, respectively. p * represents the difference in each clinicopathological variable between the training and validation cohorts. Abbreviations: PLNM: pelvic lymph node metastasis; MTD: maximal tumor diameter; LVSI: lymphovascular invasion; SCC: squamous cell carcinoma; AC: adenocarcinoma; ASC: adenosquamous carcinoma; NI: nerve invasion.

3.2. Feature Selection and Performance of the Clinical Model and Radiomics Model

We collected a total of 1688 characteristics, with 744 associated with wavelets, 372 associated with LBP, 186 associated with squares, and 107 associated with the original. The exponential, the gradient, and the logarithmic functions each have 93 characteristics. Furthermore, feature filtering was accomplished utilizing two types of methods: filtered and embedded. Overall, the filtered technique beats the embedded extraction strategy, ignoring the fact that the embedded scheme varies significantly with different feature extraction ratios.

Although the AUC increases by over 90% as the number of features embedded grows, the outcomes are more heavily influenced by the number of chosen features. In contrast, when the number of features grows, the filtered solution maintains a consistent rising trend. MINE outperforms F Classification and Pearson in regard to feature selection, with a rank of 138. The MINE approach is focused on a feature number of 138 for later trials (Figure 4).

In the training cohort (AUC = 0.925, accuracy = 81.6%, sensitivity = 70.3%, and specificity = 92.0%) and the testing cohort (AUC = 0.839, accuracy = 74.2%, sensitivity = 65.7%, and specificity = 82.8%), the clinical models that combined stromal invasion depth, FIGO stage, and MTD performed poorly (Table 2). In both the training (AUC = 0.975, accuracy = 91.8 percent, sensitivity = 92.0 percent, and specificity = 91.7 percent) and testing cohorts (AUC = 0.852, accuracy = 77.1 percent, sensitivity = 82.9 percent, and specificity = 71.4 percent), the radiomics model outperformed the clinical model significantly, as depicted in the ROC curves (Figure 5).

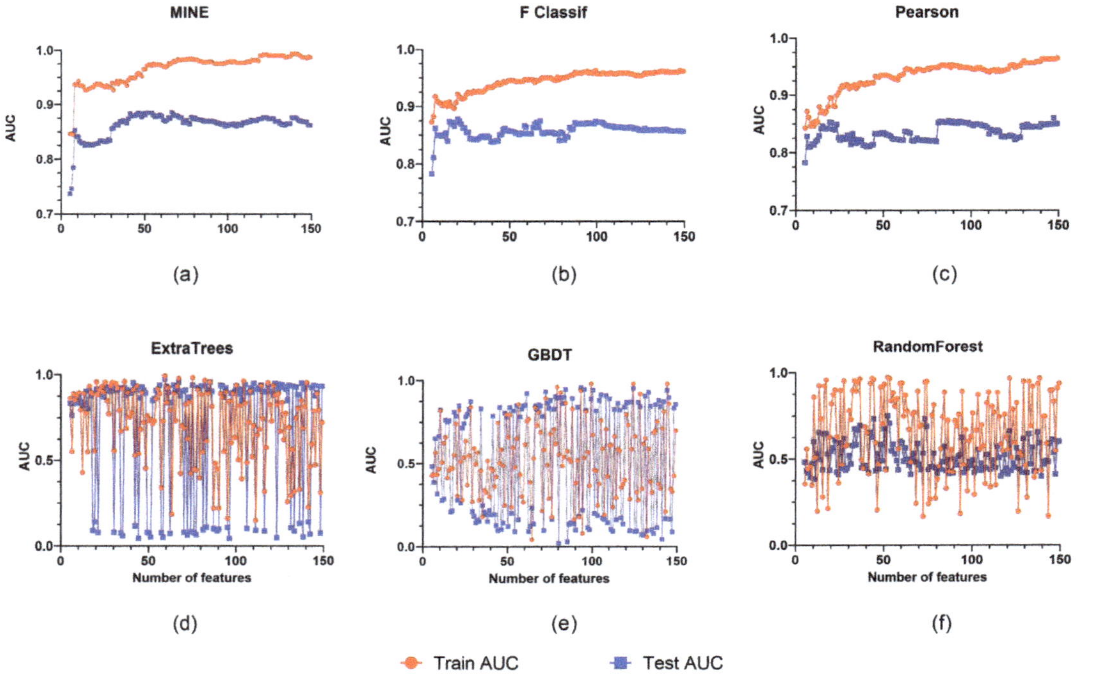

Figure 4. Six different feature screening approaches are illustrated, with the horizontal coordinate showing the number of features selected and the vertical coordinate showing the classifier's AUC. The filtered scheme includes (**a**–**c**). The embedded scheme includes (**d**–**f**).

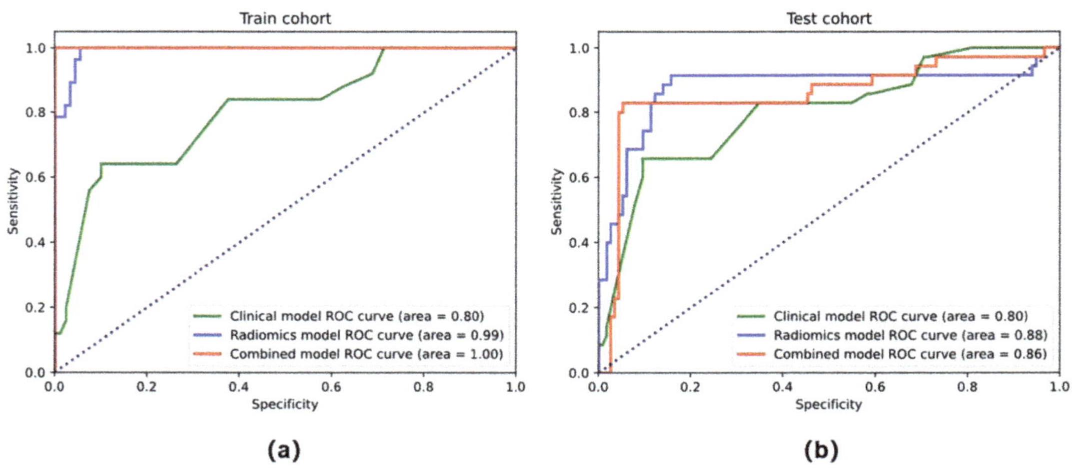

Figure 5. The ROC curves of the clinical model, radiomics models, and combined model in the training cohort (**a**) and testing cohort (**b**).

Table 2. Performance of models.

		Training Cohort				Testing Cohort			
		AUC	ACC (%)	SEN (%)	SPE (%)	AUC	ACC (%)	SEN (%)	SPE (%)
Clinical	stage	0.832	0.816	0.708	0.920	0.767	0.771	0.657	0.886
	pathology	0.476	0.531	0.08	0.96	0.527	0.485	0.08	0.886
	diameter	0.712	0.714	0.75	0.68	0.728	0.729	0.771	0.685
	Stage+pathology+diameter	0.925	0.816	0.703	0.920	0.839	0.742	0.657	0.828
Radiomics model		0.975	0.918	0.920	0.917	0.852	0.771	0.829	0.714
Combined model		0.988	0.959	0.920	1.00	0.922	0.871	0.857	0.886

3.3. *Performance of the Combined Model and the Radiomics Nomogram*

Tumor stage, tumor infiltration depth, and radiomics signature were chosen throughout the creation of the integrated model. The combined model had the highest AUC in the training cohort (AUC = 0.988, accuracy = 95.9%, sensitivity = 92.0%, and specificity = 100.0%) and the testing cohort (AUC = 0.922, accuracy = 87.1%, sensitivity = 85.7%, and specificity = 88.6%) when compared to the radiomics and clinical models (Table 2 and Figure 5). The combined model has a better performance than the clinical model in both the training and test cohorts. The radiomics-based nomogram and the combined model are shown in Figure 6. The calibration curves of the radiomics-based nomogram demonstrated satisfactory agreement between the predictive and observational possibility of the PLMN status in both the training and validation cohorts.

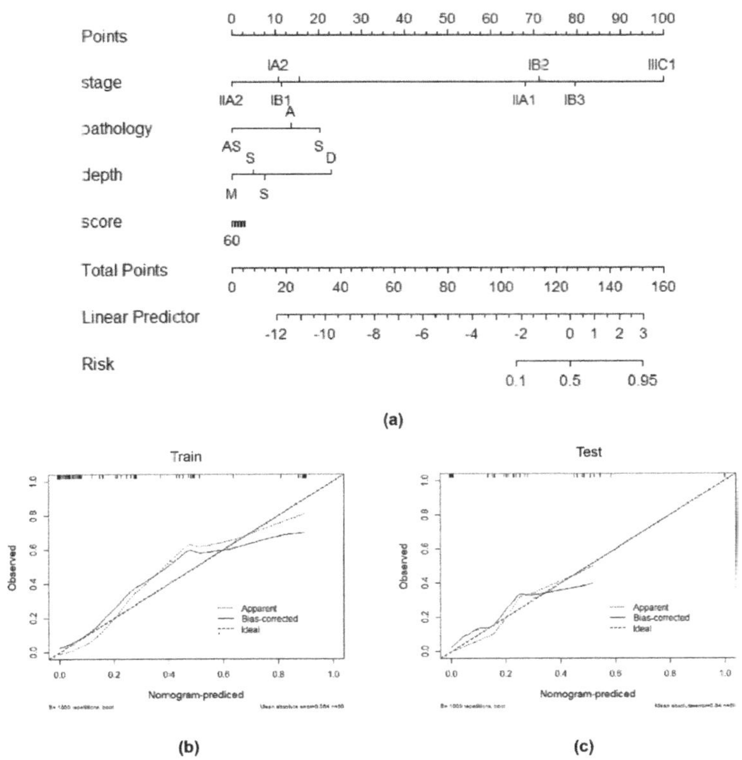

Figure 6. The radiomics-based nomogram (**a**). Calibration curves in the training cohort (**b**) and validation cohort (**c**). A closer fit to the diagonal line indicates a better evaluation.

4. Discussion

For patients with early-stage cervical cancer, pelvic lymph node (LN) status is one of the most important factors taken into account when making clinical decisions regarding surgery or radical chemoradiotherapy. Many inspection methods have been made to enhance the accurate evaluation of LN status before surgery, such as sentinel lymph node (SLN) biopsy, magnetic resonance imaging (MRI), and positron emission tomography (PET)-computed tomography (CT). SLN biopsy can acquire optimal diagnostic efficiency, but the absence of a standard surgical technique and the surgeon's training and experience may influence the outcome of SLN biopsy [22]. In addition, SLN biopsy is an invasive approach and is not routinely applied. MRI has widely been used in preoperative staging evaluation and lymph node detection of cervical cancer. Although MRI has the potential to identify LNM status according to the MRI morphological appearances of PLNM, such as size and shape, its efficiency and sensitivity in diagnosing PLNM are unsatisfactory [23]. Several previous studies have shown that a considerable proportion of cervical cancer patients were misclassified according to morphologic criteria on MRI images [24,25]. Although PET-CT shows favorable performance and is superior to MRI, considering the availability of PET-CT devices and high inspection costs, its wide application has a long way to go [26].

In this study, we successfully developed and validated a noninvasive individualized radiomics-based nomogram integrating the radiomics signature and clinicopathologic factors for predicting PLNM in patients with early-stage cervical cancer before surgery. The proposed radiomics nomogram exhibited a favorable performance in discriminating lymph node status in the training and validation sets. This radiomics nomogram also showed more predictive efficacy than the traditional diagnostic criteria of PLNM.

Chalkidou and colleagues proposed a solution to the problem of picking a limited number of finite features [21], which is the focus of this study. The results of the studies reveal that only a subset of datasets fit the criteria, not all of them. As part of our related work on lymphovascular space invasion, we substituted the study population in the original dataset in this experiment. It was not possible to achieve a decent result in Lasso's technique based on the theory presented by Chalkidou et al. To achieve this, two broad categories of feature selection strategies were chosen for validation, and all experiments consistently indicated that even the same dataset might have a significant influence on the selected features when the experimental aim is changed.

The findings of the experiments indicate that the tree-based feature extraction approach may not be suited for feature extraction in imagingomics. The tree scheme is not ideal for processing continuous features, and the continuity of the final derived features is quite visible. Second, the tree-based algorithm's performance tends to be poor when processing data with substantial feature correlation, and the generated features are highly correlated. The other set of approaches, on the other hand, makes better use of data correlation and hence beats the tree-based strategy in the experimental comparison. MINE is the most effective of these approaches since it finds not only the linear but also the nonlinear correlations between the variables. It is possible that this is why the MINE scheme outperforms the others.

By combining the radiomics signature with easily available, preoperative clinical risk factors, we developed a quantified radiomics nomogram, which is convenient to utilize for clinicians. The radiomics nomogram could generate a personalized probability of PLNM for patients before surgery, which could provide more information for clinicians to make treatment decisions. This individualized treatment is in line with the trend of personalized precision medicine [27].

In our study, some clinical characteristics were independently correlated with LNM status, including MRI-reported LN status and FIGO stage. These clinical characteristics were quite similar to those from other previous studies based on conventional MRI analytical methods, indicating that the LNM status was closely related to the total status of the primary tumor [24,25,28]. Recently, some studies [29–31] developed combined models for predicting LNM status in cervical cancer patients based on clinical and histological information and MRI images. These studies utilizing the radiomics method showed comparable prediction performance with ours. Compared with these previous studies, our study provides an easier visual tool for doctors to evaluate the PLNM status preoperatively and noninvasively.

There were several limitations in our study. First, we performed radiomics analysis on the T2WI sequence. During the segmentations, we excluded the T1WI for the unsatisfactory performance in displaying the lesions. Therefore, we did not continue the segmentations on T1WI. We also excluded DWI due to the low signal-to-noise ratio and motion artifacts. In future research, we will combine or compare more MRI sequences to improve diagnostic efficiency. Second, the cervical lesions that were difficult to recognize or invisible on MRI were excluded, which may lead to potential selection biases. Third, due to the relatively small number of cases in this single-center retrospective study, a larger sample size, multiple centers, and prospective datasets are needed to optimize the performance of the radiomics model in the future. Finally, all of the manual segmentations of the tumor tissues on axial T2WI and contrast-enhanced T1-weighted spin–echo images were performed by a radiologist who had 5 years of experience in gynecological MR imaging, and each segmentation was validated by a senior radiologist who had 12 years of experience. Therefore, the final segmentations were revised and verified by two experienced radiologists. However, the concordance between the two observers was not measured. In the future, we will perform the analysis of concordance to provide more precise data. Automatic tumor segmentation and extraction with machine learning could be further explored [32].

5. Conclusions

This study provides an effective and noninvasive tool for the individualized preoperative prediction of PLNM status. This radiomics-based nomogram would aid the selection of the optimal therapeutic strategy and clinical decision-making for individuals. The study may provide valuable guidance for clinical physicians regarding treatment strategies for early-stage cervical cancer patients.

Supplementary Materials: The following supporting information can be downloaded at https://www.mdpi.com/article/10.3390/diagnostics12102446/s1, Supplementary S1: The formula of the radiomics signature.

Author Contributions: Conceptualization, W.D.; Data curation, X.X. and Y.W.; Formal analysis, X.X., D.L., S.N. and Q.T.; Methodology, D.L. and Y.W.; Software, D.L.; Supervision, W.D.; Validation, W.D.; Writing—original draft, X.X., D.L., Y.W. and Q.T.; Writing—review & editing, Q.C. All authors have read and agreed to the published version of the manuscript.

Funding: This work was supported by National Natural Science Foundation of China(81902723).

Institutional Review Board Statement: The study was conducted in accordance with the Declaration of Helsinki and approved by the Ethics Committee of the Affiliated Hospital of Southwest Medical University (protocol code 2010-092 and date of approval 10 December 2019).

Informed Consent Statement: Patient consent was waived because all data and images in this study were anonymous.

Data Availability Statement: Raw data can be shared from the first author if there is a reasonable request.

Conflicts of Interest: The authors declare no conflict of interest.

References

1. Bray, F.; Ferlay, J.; Soerjomataram, I.; Siegel, R.L.; Torre, L.A.; Jemal, A. Global cancer statistics 2018: GLOBOCAN estimates of incidence and mortality worldwide for 36 cancers in 185 countries. *CA Cancer J. Clin.* **2018**, *68*, 394–424. [CrossRef] [PubMed]
2. Quinn, M.A.; Benedet, J.L.; Odicino, F.; Maisonneuve, P.; Beller, U.; Creasman, W.T.; Heintz, A.P.; Ngan, H.Y.; Pecorelli, S Carcinoma of the cervix uteri. FIGO 26th Annual Report on the Results of Treatment in Gynecological Cancer. *Int. J. Gynaecol. Obstet.* **2006**, *95* (Suppl. S1), S43–S103. [CrossRef]
3. Greggi, S.; Scaffa, C. Surgical management of early cervical cancer: The shape of future studies. *Curr. Oncol. Rep.* **2012**, *14*, 527–534. [CrossRef] [PubMed]

4. Ramirez, P.T.; Frumovitz, M.; Pareja, R.; Lopez, A.; Vieira, M.; Ribeiro, R.; Buda, A.; Yan, X.; Shuzhong, Y.; Chetty, N.; et al. Minimally Invasive versus Abdominal Radical Hysterectomy for Cervical Cancer. *N. Engl. J. Med.* **2018**, *379*, 1895–1904. [CrossRef] [PubMed]
5. Melamed, A.; Margul, D.J.; Chen, L.; Keating, N.L.; Del Carmen, M.G.; Yang, J.; Seagle, B.L.L.; Alexander, A.; Barber, E.L.; Rice, L.W.J.O.; et al. Survival After Minimally Invasive Radical Hysterectomy for Early-stage Cervical Cancer. *JAMA Oncol.* **2019**, *74*, 84–85. [CrossRef]
6. Biewenga, P.; van der Velden, J.; Mol, B.W.; Stalpers, L.J.; Schilthuis, M.S.; van der Steeg, J.W.; Burger, M.P.; Buist, M.R. Prognostic model for survival in patients with early stage cervical cancer. *Cancer* **2011**, *117*, 768–776. [CrossRef]
7. Bats, A.S.; Frati, A.; Mathevet, P.; Orliaguet, I.; Querleu, D.; Zerdoud, S.; Leblanc, E.; Gauthier, H.; Uzan, C.; Deandreis, D.; et al. Contribution of lymphoscintigraphy to intraoperative sentinel lymph node detection in early cervical cancer: Analysis of the prospective multicenter SENTICOL cohort. *Gynecol. Oncol.* **2015**, *137*, 264–269. [CrossRef]
8. Marth, C.; Landoni, F.; Mahner, S.; McCormack, M.; Gonzalez-Martin, A.; Colombo, N.; Committee, E.G. Cervical cancer: ESMO Clinical Practice Guidelines for diagnosis, treatment and follow-up. *Ann. Oncol.* **2017**, *28*, iv72–iv83. [CrossRef]
9. Diab, Y. Sentinel Lymph Nodes Mapping in Cervical Cancer a Comprehensive Review. *Int. J. Gynecol. Cancer* **2017**, *27*, 154–158. [CrossRef] [PubMed]
10. Ferrandina, G.; Anchora, L.P.; Gallotta, V.; Fagotti, A.; Vizza, E.; Chiantera, V.; De Iaco, P.; Ercoli, A.; Corrado, G.; Bottoni, C.; et al. Can We Define the Risk of Lymph Node Metastasis in Early-Stage Cervical Cancer Patients? A Large-Scale, Retrospective Study. *Ann. Surg. Oncol.* **2017**, *24*, 2311–2318. [CrossRef] [PubMed]
11. Balcacer, P.; Shergill, A.; Litkouhi, B. MRI of cervical cancer with a surgical perspective: Staging, prognostic implications and pitfalls. *Abdom. Radiol.* **2019**, *44*, 2557–2571. [CrossRef] [PubMed]
12. Khiewvan, B.; Torigian, D.A.; Emamzadehfard, S.; Paydary, K.; Salavati, A.; Houshmand, S.; Shamchi, S.P.; Werner, T.J.; Aydin, A.; Roy, S.G.; et al. Update of the role of PET/CT and PET/MRI in the management of patients with cervical cancer. *Hell. J. Nucl. Med.* **2016**, *19*, 254–268. [CrossRef] [PubMed]
13. Cibula, D.; Zikan, M.; Slama, J.; Fischerova, D.; Kocian, R.; Germanova, A.; Burgetova, A.; Dusek, L.; Dundr, P.; Gregova, M.; et al. Risk of micrometastases in non-sentinel pelvic lymph nodes in cervical cancer. *Gynecol. Oncol.* **2016**, *143*, 83–86. [CrossRef] [PubMed]
14. Wu, Q.; Zheng, D.; Shi, L.; Liu, M.; Wang, M.; Shi, D. Differentiating metastatic from nonmetastatic lymph nodes in cervical cancer patients using monoexponential, biexponential, and stretched exponential diffusion-weighted MR imaging. *Eur. Radiol.* **2017**, *27*, 5272–5279. [CrossRef]
15. Gillies, R.J.; Kinahan, P.E.; Hricak, H. Radiomics: Images Are More than Pictures, They Are Data. *Radiology* **2016**, *278*, 563–577. [CrossRef] [PubMed]
16. Song, L.; Yin, J. Application of Texture Analysis Based on Sagittal Fat-Suppression and Oblique Axial T2-Weighted Magnetic Resonance Imaging to Identify Lymph Node Invasion Status of Rectal Cancer. *Front. Oncol.* **2020**, *10*, 1364. [CrossRef]
17. Zhou, X.; Yi, Y.; Liu, Z.; Zhou, Z.; Lai, B.; Sun, K.; Li, L.; Huang, L.; Feng, Y.; Cao, W.; et al. Radiomics-Based Preoperative Prediction of Lymph Node Status Following Neoadjuvant Therapy in Locally Advanced Rectal Cancer. *Front. Oncol.* **2020**, *10*, 604. [CrossRef] [PubMed]
18. Yang, Y.S.; Feng, F.; Qiu, Y.J.; Zheng, G.H.; Ge, Y.Q.; Wang, Y.T. High-resolution MRI-based radiomics analysis to predict lymph node metastasis and tumor deposits respectively in rectal cancer. *Abdom. Radiol.* **2021**, *46*, 873–884. [CrossRef] [PubMed]
19. Bhatla, N.; Berek, J.S.; Fredes, M.C.; Denny, L.A.; Grenman, S.; Karunaratne, K.; Kehoe, S.T.; Konishi, I.; Olawaiye, A.B.; Prat, J.; et al. Revised FIGO staging for carcinoma of the cervix uteri. *Int. J. Gynaecol. Obstet.* **2019**, *145*, 129–135. [CrossRef]
20. Cree, I.A.; White, V.A.; Indave, B.I.; Lokuhetty, D. Revising the WHO classification: Female genital tract tumours. *Histopathology* **2020**, *76*, 151–156. [CrossRef]
21. Chalkidou, A.; O'Doherty, M.; Marsden, P. False Discovery Rates in PET and CT Studies with Texture Features: A Systematic Review. *PLoS ONE* **2015**, *10*, e0124165. [CrossRef]
22. Lecuru, F.; Mathevet, P.; Querleu, D.; Leblanc, E.; Morice, P.; Darai, E.; Marret, H.; Magaud, L.; Gillaizeau, F.; Chatellier, G.; et al. Bilateral negative sentinel nodes accurately predict absence of lymph node metastasis in early cervical cancer: Results of the SENTICOL study. *J. Clin. Oncol.* **2011**, *29*, 1686–1691. [CrossRef]
23. Bhosale, P.R.; Iyer, R.B.; Ramalingam, P.; Schmeler, K.M.; Wei, W.; Bassett, R.L.; Ramirez, P.T.; Frumovitz, M. Is MRI helpful in assessing the distance of the tumour from the internal os in patients with cervical cancer below FIGO Stage IB2? *Clin. Radiol.* **2016**, *71*, 515–522. [CrossRef]
24. Kim, S.H.; Lee, H.J.; Kim, Y.W. Correlation between tumor size and surveillance of lymph node metastasis for IB and IIA cervical cancer by magnetic resonance images. *Eur. J. Radiol.* **2012**, *81*, 1945–1950. [CrossRef]
25. Zhang, W.; Zhang, J.; Yang, J.; Xue, H.; Cao, D.; Huang, H.; Wu, M.; Cui, Q.; Chen, J.; Lang, J.; et al. The role of magnetic resonance imaging in pretreatment evaluation of early-stage cervical cancer. *Int. J. Gynecol. Cancer* **2014**, *24*, 1292–1298. [CrossRef]
26. Lv, K.; Guo, H.M.; Lu, Y.J.; Wu, Z.X.; Zhang, K.; Han, J.K. Role of 18F-FDG PET/CT in detecting pelvic lymph-node metastases in patients with early-stage uterine cervical cancer: Comparison with MRI findings. *Nucl. Med. Commun.* **2014**, *35*, 1204–1211. [CrossRef]
27. Balachandran, V.P.; Gonen, M.; Smith, J.J.; DeMatteo, R.P. Nomograms in oncology: More than meets the eye. *Lancet Oncol.* **2015**, *16*, e173–e180. [CrossRef]

28. Choi, H.J.; Ju, W.; Myung, S.K.; Kim, Y. Diagnostic performance of computer tomography, magnetic resonance imaging, and positron emission tomography or positron emission tomography/computer tomography for detection of metastatic lymph nodes in patients with cervical cancer: Meta-analysis. *Cancer Sci.* **2010**, *101*, 1471–1479. [CrossRef] [PubMed]
29. Wang, T.; Gao, T.; Yang, J.; Yan, X.; Wang, Y.; Zhou, X.; Tian, J.; Huang, L.; Zhang, M. Preoperative prediction of pelvic lymph nodes metastasis in early-stage cervical cancer using radiomics nomogram developed based on T2-weighted MRI and diffusion-weighted imaging. *Eur. J. Radiol.* **2019**, *114*, 128–135. [CrossRef]
30. Kan, Y.; Dong, D.; Zhang, Y.; Jiang, W.; Zhao, N.; Han, L.; Fang, M.; Zang, Y.; Hu, C.; Tian, J.; et al. Radiomic signature as a predictive factor for lymph node metastasis in early-stage cervical cancer. *J. Magn. Reason. Imaging* **2019**, *49*, 304–310. [CrossRef]
31. Yu, Y.Y.; Zhang, R.; Dong, R.T.; Hu, Q.Y.; Yu, T.; Liu, F.; Luo, Y.H.; Dong, Y. Feasibility of an ADC-based radiomics model for predicting pelvic lymph node metastases in patients with stage IB-IIA cervical squamous cell carcinoma. *Br. J. Radiol.* **2019**, *92*, 20180986. [CrossRef]
32. Lin, Y.C.; Lin, C.H.; Lu, H.Y.; Chiang, H.J.; Wang, H.K.; Huang, Y.T.; Ng, S.H.; Hong, J.H.; Yen, T.C.; Lai, C.H.; et al. Deep learning for fully automated tumor segmentation and extraction of magnetic resonance radiomics features in cervical cancer. *Eur. Radiol.* **2020**, *30*, 1297–1305. [CrossRef]

diagnostics

MDPI

Review

Nectins and Nectin-like Molecules in Colorectal Cancer: Role in Diagnostics, Prognostic Values, and Emerging Treatment Options: A Literature Review

Jakub Kobecki [1,2], Paweł Gajdzis [3,4], Grzegorz Mazur [5] and Mariusz Chabowski [1,2,*]

1 Department of Surgery, 4th Military Teaching Hospital, 5 Weigla Street, 50-981 Wroclaw, Poland
2 Division of Anaesthesiological and Surgical Nursing, Department of Nursing and Obstetrics, Faculty of Health Science, Wroclaw Medical University, 5 Bartla Street, 51-618 Wroclaw, Poland
3 Department of Pathomorphology, 4th Military Teaching Hospital, 5 Weigla Street, 50-981 Wroclaw, Poland
4 Department of Clinical Pathology, Wroclaw Medical University, 213 Borowska Street, 50-556 Wroclaw, Poland
5 Department of Internal Medicine, Occupational Diseases, Hypertension and Clinical Oncology, Wroclaw Medical University, 213 Borowska Street, 50-556 Wroclaw, Poland
* Correspondence: mariusz.chabowski@gmail.com; Tel.: +48-261-660-247; Fax: +48-261-660-245

Abstract: In 2020, colorectal cancer was the third most common type of cancer worldwide with a clearly visible increase in the number of cases each year. With relatively high mortality rates and an uncertain prognosis, colorectal cancer is a serious health problem. There is an urgent need to investigate its specific mechanism of carcinogenesis and progression in order to develop new strategies of action against this cancer. Nectins and Nectin-like molecules are cell adhesion molecules that take part in a plethora of essential processes in healthy tissues as well as mediating substantial actions for tumor initiation and evolution. Our understanding of their role and a viable application of this in anti-cancer therapy has rapidly improved in recent years. This review summarizes the current data on the role nectins and Nectin-like molecules play in colorectal cancer.

Keywords: Nectin; Necl; cancer; colorectal; diagnostics

Citation: Kobecki, J.; Gajdzis, P.; Mazur, G.; Chabowski, M. Nectins and Nectin-like Molecules in Colorectal Cancer: Role in Diagnostics, Prognostic Values, and Emerging Treatment Options: A Literature Review. *Diagnostics* **2022**, *12*, 3076. https://doi.org/10.3390/diagnostics12123076

Academic Editors: Yuli Huang, Yong Yuan and Peisong Chen

Received: 24 October 2022
Accepted: 4 December 2022
Published: 7 December 2022

1. Introduction

Colorectal cancer (CRC) is a serious worldwide health problem with over 1.9 million new cases estimated in 2020 [1]. Globally, CRC is the third most commonly diagnosed cancer in both sexes combined [2]. It is estimated that CRC is the second most common cause of cancer-related mortality and accounts for approximately 930 thousand deaths around the world in 2020 [1,3]. Increase in incidence, altogether with predictive models for future years, prove that CRC is a growing burden for patients and national healthcare systems [4]. A significant shift toward the growing incidence among younger adults (<50 y.o.) makes those predictions even more worrisome [5]. The COVID-19 pandemic led to weighty health care delivery disruptions also in the matter of the prevention and treatment of patients with CRC [6,7]. The association between the risk of a more advanced stage of disease and the COVID-19 pandemic was reported in patients undergoing CRC when compared with pre-pandemic cases [8]. Limited access to health care and a CRC screening slowdown emphasized the importance of research in new diagnostic tools. Colorectal cancer is a nonhomogeneous group in which over 90% of cases are adenocarcinomas [9]. Long-term survival and cure rates have not improved significantly in recent years, especially in more advanced stages of the disease [9–11]. Carcinogenesis involves an accumulation of irreversible genetic mutations and epigenetic alterations. Within the same tumor, there are genetically distinguished cell populations coexisting. Epigenetic alterations vary from "hard-wired" and stable during cell division to transient and changing with every few divisions [12]. Epigenetic alterations, leading to developmental genes' reactivation, normally silent in somatic tissues, have been observed in CRC specimens [13]. Not every genetical mutation and epigenetic alternation lead to specific phenotype change. Phenotypic plasticity has been described, also in CRC cells, showing that there is an ability to alter transcriptome without underlying genetic or epigenetic heritable mutation [14]. All the above mentioned sum up into a complex intra-tumor heterogeneity and are closely connected in cancer evolution. Clonal selection and evolution are at the base of treatment resistance [13]. Approximately 60% of all CRCs are sporadic–they develop without any known family history or any obvious genetic cancer syndrome (e.g., FAP or HNPCC) [15,16]. One-in-four CRCs has a hereditary component, as family and twin studies show. Yet, environmental factors play a role in carcinogenesis in those cases. Only 5% of CRC is attributed to well-characterized, high-penetrance syndromes [17].

The prognosis for patients with CRC is strongly determined by tumor invasiveness and the development of local or distant metastases. The presence of advanced locally or disseminated disease is a strong negative prognostic for a cure and long-term survival. Surgical treatment remains at the base of a curative approach [10]. Big differences in overall survival, even within same-stage groups, place the search for valuable prognostic factors in the spotlight of clinical research [18,19]. Among many others, cell surface molecules are being investigated for their clinical relevance (e.g., carcino-embryonic antigen). Cell adhesion molecules (CAMs) play a role in diagnostics and remain a field of interest in pathophysiology and the development of new treatment strategies in CRC [20–23]. With modern improvements we are on the verge of introducing a wide array of prognostic and predictive tools into clinical practice, which will help treatment decision making [24]. Advanced statistical tools can help with navigation through numerous biomarkers, combining individual and distinct features into more relevant groups–e.g., consensus molecular subtypes classification of CRC as a prognostic and predictive instrument for clinical decisions [25–27] Factors, once associated only with poor prognosis, lead researchers to new therapy development (e.g., BRAF, HER2) [28]. The new methods, such as circulating tumor DNA (ctDNA) allow more precise and sooner detection of the residual disease or relapse [29–31]. Nectins and nectin-like molecule seem to be worthy research candidates for future clinical implementation.

2. General Characteristics of Nectins and Nectin-like Molecules

Nectin and Nectin-like molecules (Necls) belong to immunoglobulin-like (Ig-like) transmembrane CAMs that play a significant role in cell-to-cell, calcium-independent adhesion formation, cell motility, proliferation, differentiation, polarization, and apoptosis. First discovered in 1986 as poliovirus receptor-related proteins (PRR proteins); then, after amino acid sequencing and further studies, they were identified as members of the immunoglobulin superfamily (IgSF). Finally, after discovering CAMs properties, a change in nomenclature was made: Nectins derived from latin "necto" i.e., "to bind" [32] (Table 1). Divided into four members of Nectins and five members of Necls, the superfamily gained attention due to their functions in immune response (e.g., TIGIT-related), being virion-entry cell receptors and tumor suppressors and oncoproteins [33]. Several experiments have shown lesser tumorogenic potential cells with positive expression of selected Nectins and Necls [34–36]. By contrast, the loss of expression of Nectins and Necls has been associated with a higher aggressiveness in several cancer cells e.g., Nectin-3 in pancreatic neuroendocrine tumor [37]. A specific mechanism of the mentioned tumor suppression is not well elucidated, whether contact inhibition is a main process is not clear. Studies about the evolution of Nectins and Necls have showed Nectin-like protein 5 to possess characteristics that are intermediate between these two groups. Moreover, sequence conservations of Nectins across species suggests evolutionary constraints due to the critical roles played by these CAMs [38]. Despite being diverse in terms of function, as well as phylogenetic distribution, members of the IgSF family share structural similarities, such as an extracellular domain consisting of three Ig-like loops, a transmembrane domain, and a short cytoplasmic tail [39]. On account of their ability to establish both, homophilic and heterophilic interactions with other CAMs (such as cadherins), Nectins perform a crucial role in the initiation of the formation of adherence junctions (AJs) [38].

Table 1. The Nectin nomenclature and viral cell-entry mediation examples.

Molecule	Other Names	Viral Cell-Entry Mediation
Nectin-1	CD111, PVRL-1, PRR-1, HVEC	HSV-1 [1], HSV-2 [2], PRV [3], BoHV-1 [4]
Nectin-2	CD112, PRVL-2, PRR-2, HVEB	HSV-1 [1], HSV-2 [1], PRV [3], HHV-6B [5]
Nectin-3	CD113, PRVL-3, PRR-3	-
Nectin-4	PVRL-4, PRR-4,	MeV [6]
Nectin-like molecule 5	CD155, PVR, Tage4	PV [7]

[1] Herpes simplex virus 1, [2] herpes simplex virus 2, [3] pseudorabies virus, [4] bovine herpes virus-1, [5] human herpesvirus 6B, [6] measles virus, [7] poliovirus.

2.1. Homophilic and Heterophicic Interactions

Nectins form transient and weak clusters of homo-cis dimers on the cell surface, which then trans-interacts with extracellular regions of opposing cell membrane Nectin molecules. It was demonstrated that Nectins are able to form both homophilic and heterophilic trans-interaction; however, heterophilic interactions are selective—for example, Nectin-1 interacts with both Nectin-3 and Nectin-4, but there is no interaction with Nectin-2 [40]. The selectiveness of heterophilic interactions is probably due to conserved charged residue localized in the center of the adhesive face of each Nectin. Therefore, only Nectins with opposed charged residues form heterophilic trans-interactions. Additionally, the transient and weak nature of homo-cis dimers can be, at least partially, explained by the repulsion of identical electrostatic charges [41].

Recruitment of E-cadherin to Nectin-based cell–cell adhesion sites and the subsequent activation of Cdc42 and Rac small G proteins enhances the formation of cadherin-based AJs through the reorganization of the actin cytoskeleton [42].

2.2. Nectin Spots

The C-terminus of the Nectins cytoplasmic domain determines their ability to interact with afadin and PAR3 [39]. Additionally, Nectins initiate afadin- and cadherin-independent cell adhesion apparatus that is not linked with the cytoskeleton. This type of apparatus is observed as dots or short lines under an immunofluorescence microscope and is named Nectin Spots [43].

The cytoplasmic tails of Necls differ from those of Nectins and interact with various scaffolding proteins, such as MAGUK or TCTEX1—a subunit of the dynein motor complex. Necls are ubiquitously expressed and have a greater variety of functions than Nectins [44].

2.3. Crosstalk between Cells and Extracellular Matrix

Many studies on MDCK cells have proven that Nectins and Necls promote crosstalk between cell–cell junctions and cell–extracellular matrix (ECM) junctions [42,44]. Extracellular regions of Nectins interact with integrins during the formation of AJs. Integrins are responsible for interactions with ECM proteins. The integrin-mediated cell–ECM junctions positively or negatively regulate the formation and stability of cell–cell junctions through protein tyrosine kinases associated with integrins (e.g., FAK, c-Src). Interaction between Nectins is responsible for integrin inactivation and, therefore, the stabilization of AJs [45]. The breakdown of AJs leads to the disruption of tight junctions (TJs) and desmosomes and the loss of epithelial cell integrity by affecting subsequently cell–ECM junctions. This process may lead to epithelial-to-mesenchymal cell transition (EMT). In cancerous cells, such increased motility is responsible for pathological implications, i.e., invasion—EMT and metastasing—cell migration [46–48].

2.4. Interactions with Growth Factors

Growth factors are responsible for cell proliferation regulation. In mature tissues, that process is transduced by a complex network of molecules. Interactions between Nectins, Necls, and growth factor receptors play a significant role in this network. For example, Nectins interact with the platelet derived growth factor (PDGF) receptor through the extracellular region and regulates the PDGF-induced activation of Akt and the inhibition of apoptosis in cooperation with afadin at PI3K activation. Experiments on Nectin-3 or afadin knockdown resulted in attenuated phosphorylation of Akt as well as decreased activity of PI3K enhanced by the PDGF receptor in NIH3T3 cell lines. Furthermore, embryoid bodies derived from afadin displayed an enormous number of apoptotic cells [45,49].

2.5. Tissue Distribution

Nectins differ in tissue distribution, where Nectin-4 normally occurs in placenta and embryonic tissues and Nectins-1 to -3 are commonly expressed in healthy adult tissues [50,51]. Many studies have shown an overexpression of Nectin-4 in numerous cancerous tumors, such as breast, pancreatic, urotheliary, ovarian, squamous cell carcinoma, and colorectal cancer [52,53] (Table 2). Necl-5 has a rather low expression in healthy adult organs but is overexpressed in many cancers. Necl-1 is specifically expressed in neural tissue whereas other members are ubiquitously expressed in tissues [39].

Table 2. Expression of Nectins and Necls in selected cancers.

Molecule	Colorectal Cancer	Breast Cancer	Urothelial Cancer
Nectin-1	↑ *	↑ *	- ***/↑ *
Nectin-2	↑ *	↑ *	NED ****
Nectin-3	NED ***	↑ */↓ **	NED ****
Nectin-4	↑ *	↑ */↓ **	↑ *
Nectin-like molecule 5	↑ *	↑ *	↑ *

↑ * = high expression, ↓ ** = low expression, - *** = standard expression, **** = not enough data.

2.6. Nectins in Disease

Studies have shown that mutations in Nectin and Necls coding genes can lead to many diseases, such as cancer, ectodermal dysplasia, Alzheimer's disease, stress-related mental disorders, viral infections, and cataracts [43]. Several Nectin-1 mutations cause disorders in ectodermally derived structures, such as lip/palate clefts, dental anomalies, and hypertrichosis—known as, e.g., Zlotogor–Ogur syndrome (CLPED1) or Margarita Islands Syndrome (CLPED2) [45]. Studies on the biology and pathology of Nectins and Necls has led to important innovations and clinical applications, such as FDA-approved therapy for treating locally advanced or metastatic urothelial carcinoma antibody–drug conjugate: enfortumab vendotin uses Nectin-4 as a target molecule for entering tumor cells [54,55].

3. Nectins and Necls in Cancer

3.1. Nectin-1

Nectin-1 is known as CD111, PRR-1, PVRL-1, and expressed under normal conditions on cells of the gastrointestinal tract, liver, gallbladder, female reproductive organs, and skin [50,51]. It is probably best studied as an entry receptor for the human herpes virus 1 (HSV-1) infection (Table 1). Nectin-1 has gained additional attention due to its interactions with CD96 (known also as Tactile) [56,57]. Abnormal expression of Nectin-1 can be observed in tumors of epithelial origin, such as cervical squamous cell carcinomas, cancer-associated fibroblasts of pancreatic ductal adenocarcinomas, CRCs and gastric cancers, or malignant transformations of keratinocytes. The observed expression varies from low to high and the published data are not consistent in this matter [40] (Table 2). CD96/Nectin-1 interactions have not yet been fully explored, although studies have shown that cells with Nectin-1 expression have been more susceptible to NK-mediated cell toxicity compared to those with no expression [56]. It can be speculated that, if soluble ligands circulating in serum can competitively inhibit NK-cell activation, they can also induce the internalization and degradation of activating receptors on cell surfaces. Colon cancer might be a promising target for NK cell-based adoptive immunotherapy, and there is a need to investigate the role of Nectin-1 in further studies [58]. HSV-1 is a well-studied virus and was used as the first rationally designed replication-competent onco-lytic virus (RCOV). Moreover, a soluble variable domain of Nectin-1 was successfully used in an experiment as an "adaptor" for increasing the efficacy of an HSV-1 infection in CHO cells with no Nectin-1 expression. Based on these findings, cancerous cells, which display enhanced Nectin-1 availability, may serve as a receptor for HSV-1 viral oncolysis [59].

Nectin-1 tends to be a possible prognostic factor in the disease-free survival of patients with CRC. Research conducted by Tampakis et al. showed that Nectin-1 is strongly expressed in the cytoplasm of CRC cells in comparison to adjacent cells. Nectin-1 expression in colorectal cancer is associated with a significantly worse three-year progression-free survival, therefore, identifying a group of patients at high risk of an early recurrence of the disease [60]. These findings correlate with studies on pancreatic ductal adenocarcinoma patients. Yamada et al. reported that diffuse Nectin-1 expression in the cancer-associated fibroblasts of pancreatic ductal adenocarcinoma patients is associated with invasion, metastasis, and shorter survival [61]. It is important to note that colorectal endometriosis, which represents the most aggressive form of endometriosis, is characterized by both an increased expression of Nectin-1 and a decreased expression of Necl-2 [62].

3.2. Nectin-2

Nectin-2 (also known as CD112 and PVRL-2) is expressed in a vast number of adult tissues [51]. There is an overexpression of Nectin-2 in tumors of epithelial origin, such as squamous cell carcinomas and in adenocarcinomas (e.g., colorectal, esophageal, lung, pancreatic, and gallbladder cancer) (Table 2). Besides involvement in cell–cell adhesion, Nectin-2 interacts with immune cells by binding to different immune receptors, including CD226 (DNAM-1, DNAX accessory molecule-1), T-cell immunoreceptor Ig tyrosine (TIGIT)-based inhibition motif (ITIM) domain, and CD112R [63,64]. Coupling with receptors of CD8+ T-cells and NKs to modulate immune functioning by mediating immune-activating or inhibitory signaling in leukocytes. Necl-5 binds to the same immune receptors as Nectin-2, and it has been proven in MCA-induced tumors of Necl-5-deficient mice that Nectin-2 upregulation compensates Necl-5 absence in immune surveillance. In this study, Nagumo et al. pointed out that there is some sort of modulation of expression between Nectin-2, Necl-5, TIGIT, DNAM-1, and CD96 since they did not observe any difference in tumor growth in the Necl-5-deficient mice model compared to wild phenotypes [65]. Nectin-2 has been found to play an important role in the process of the NK cell-mediated killing of colon adenocarcinoma cells. A functional blockage of Nectin-2 with a specific antibody led to the significant inhibition of NK cell cytotoxicity to a colon cancer cell [58]. Platelet cloaking of circulating tumor cells and releasing transforming growth factor beta (TGFß) from platelets is an evasion method to hide from immune surveillance [66]. Cluxton et al. reported that cancer cells cloaked by platelets had a significantly reduced expression of Nectin-2 and Necl-5 on the tumor cell surface. Simultaneously, TGFß mediated the CD266 downregulation on NK cell surface. This suggests that the "immune decoy" mechanism is mediated by platelets. Platelet cloaking actively disrupts the CD226/CD96–Nectin-2/Necl-5 axis of circulating cancer cells recognition and plays a significant role in metastatic cascade [67].

An estimated 3–5% of patients with diagnosed CRC display HER2 amplification. In recent years, this has emerged as an actionable therapeutic target. There are treatments targeted at HER2 in clinical trials, e.g., with trastuzumab-deruxtecan ADC in metastatic CRC [68,69]. Nectin-2 has an affinity with the inhibitory immune receptor CD112R present in T-cells and NK cells [70]. Trastuzumab has limited action on cells with low HER2 expression. Antibody blockage of CD112R-enhanced NK cell cytokine production when NK cells were incubated with trastuzumab-coated breast cancer cells [71]. Thus, Nectin-2 and its receptors can be utilized to improve the therapeutic level of trastuzumab and can possibly be used in cases with lower HER2 expression. The complex and not yet fully explored TIGIT-CD96-CD112R-CD-226 axis is a promising target for a next generation of immunotherapy in cancer [57,64].

PVRIG binds with high affinity to Nectin-2 and suppresses T-cell function. This may lead to the presence of exhausted T-cell phenotypes in the tumor microenvironment (TME) that is unable to perform effective cytotoxicity. Whelan et al. found that PVRIG-Nectin-2 axis blockage enhanced cytokine production and the cytotoxic function of T-cells [72]. Immune-checkpoint inhibitors targeting the PVRIG-Nectin-2 axis in Nectin-2 positive cancers may be evaluated as potential drugs.

What is more, Nectin-2 is the target of post-translational modifications. Ubiquitination not only decreases Nectin-2 surface expression by targeting the protein for degradation but also by promotes Nectin-2 intracellular retention. Upon the inhibition of the ubiquitin pathway, increased Nectin2 surface expression renders tumor cells more efficiently recognized and lysed by NK cells by mediating the CD226-dependent co-stimulation of both NK and CD8+ T cells [73].

An interesting finding was described by Bekes et al. in a study on endothelial permeability and Nectin-2 downregulation. This study observed that patients with more advanced stages of ovarian cancer displayed an overexpression of Nectin-2 in cancer cells, a lower expression of Nectin-2 in peritoneal endothelium cells, and a significantly higher concentration of circulating vascular endothelial growth factor (VEGF). These findings advocate a thesis that the downregulation of Nectin-2 is responsible for increased endothelial permeability, which is VEGF driven [74]. These findings are in line with a study carried out by Russo et al. However, increased vascular coverage in the retina and spleen of Nectin-2-deficient mice was assumed to be caused rather by a defect in contact inhibition than other angiogenic factors (e.g., VEGF). Moreover, it has been demonstrated that Nectin-2 is involved in the homecoming process of T-cell entry into the spleen. Both T-cell entry routes into the spleen (through the red pulp and the marginal zone) were impaired in the Nectin-2-deficient model [75].

Karabulut et al. measured serum levels of Nectin-2 in 140 patients with diagnosed CRC. It was found that serum Nectin-2 levels in patients with both metastatic and non-metastatic CRC have a diagnostic value since there was a significant difference in baseline serum Nectin-2 levels between the whole group patients and the healthy control group ($p < 0.001$; for all, non-metastatic (stage II or III), and metastatic patients). Patients with elevated serum Nectin-2 concentrations had significantly less favorable progression-free survival rates compared with those with lower levels (median 5.8 v 9.1 months, respectively, $p = 0.04$). However, the results of this study did not show a statistically significant relationship between serum Nectin-2 concentration and overall survival or chemotherapy responsiveness [76].

This finding is in line with results obtained by researchers in different types of cancers. Liang et al. tested the significance of Nectin- 2 in pancreatic ductal adenocarcinomas. It has been shown that Nectin-2 expression is significantly correlated with clinical progression, as indicated by large tumor size and lymph node metastasis. Moreover, positive Nectin-2 expression is correlated with shorter survival in the cited study [77]. Nectin-2 also acts as an entry receptor for the Herpes simplex virus by interacting with viral glycoprotein D [40,78]. The role of Nectin-2 in colorectal cancer needs to be further elucidated.

3.3. Nectin-3

Nectin-3, also known as PVRL-3, PRR3, or CD113, plays a significant role in organ development (e.g., ocular, inner ear, and cerebral cortex development) and is widely expressed in healthy adult tissues (e.g., endocrine, gastrointestinal tissues, testis) (Table 1). Nectin-3 has the most versatile skill in terms of trans-interactions with other family members [39,40]. It is the only known Nectin that is expressed on the T-cell surface. Nectin-3 plays a significant role in T-lymphocytes extravasation. It trans-interacts with Nectin-2 on endothelial cells (localized near high endothelial venules) and facilitates the transendothelial migration of immune cells to secondary immune organs or surrounding tissues [79]. A similar mechanism may apply to malignant cells during disease dissemination [80]. Nectin-3 was identified as a mandatory C. difficile receptor for TcdB-mediated cytotoxicity as it is highly expressed in the colon. It may serve as a drug-target to prevent pseudomembranous colitis symptoms in C. difficile infections [81]. There are no specific studies on the clinical application of Nectin-3 expression in CRC. From the published data, we know that Nectin-3 expression is upregulated in lung adenocarcinomas, ovarian, and nasopharyngeal carcinomas (Table 2). Minawa et al. presented a study in which the membranous expression of Nectin-3 (normally absent in healthy lung tissues) was found to be associated with a poor prognosis for lung adenocarcinoma patients. Surprisingly patients who showed the membranous expression of Nectin-3 together with e-cadherin co-location had a better overall survival rate than in the case of the separate localization of both molecules within a tumor cell. In cases where Nectin-3 was expressed on a cell membrane with no expression of e-cadherin, the overall outcome was the worst of all patient groups. Thus, the membranous Nectin-3 that does not have a physiological function (the recruitment of E-cadherin) may contribute to increased tumor malignancy [82]. Zhao et al. established Nectin-3 and NRXN3 (both are members of CAMs) as downstream target genes of the zinc finger protein 582 (ZNF582). The hypermetylation of ZNF582 in nasopharyngeal carcinoma (NPC) is associated with higher migration, invasion, and metastasis. The restoration of ZNF582 led to the downregulation of Nectin-3 expression and the upregulation of NRXN3. Subsequent knock-out experiments and an in vivo model confirmed that Nectin-3 acts as an oncogene in NPC. This study also elucidated a new way of regulating Nectin-3 expression by ZNF582 [83]. Similarly, Xu et al. reported that Nectin-3 overexpression in ovarian cancer is associated with a worse overall survival rate. As an oncogene, Nectin-3 contributes to tumor progression in ovarian cancer. These results indicate that the expression of Nectin-3 upregulates the expression matrix metalloproteinases (MMPs 1 and 2) and leads to enhanced migration and invasion in OC cells by inducing ECM degradation in the area surrounding the tumor [84].

Meanwhile, Nectin-3 is downregulated in pancreatic adenocarcinoma and neuroendocrine tumors along with breast cancer. In both cases, the loss of Nectin-3 expression is associated with higher malignancy and a poorer prognosis. Martin et al. conducted a study in which induced overexpression of Nectin-3 in breast cancer cell lines reduced their motility and invasion properties. Established TJs, measured by trans epithelial resistance, also appeared to be "tighter". Nectin-3 may be a key component in the formation of cell junctions and a putative suppressor molecule to the invasion and metastases of breast cancer cells [85].

Covering results were obtained by Hirabayashi et al. in their study on Nectin-3 expression in pancreatic endocrine tumors (PanNETs). Loss of Nectin-3 expression in PanNETs was associated with larger tumor size, a higher grade, lymphatic involvement, a higher Ki67 labeling index, an advanced pT-factor, lymph node metastasis, an advanced tumor stage, nonfunctioning tumors, and shorter disease-free survival. The authors presumed that the higher rate of cell proliferation may be promoted by the disruption of trans-interaction between Nectin-3 and Necl-5 [37,86]. In the case of pancreatic adenocarcinomas, reports on Nectin-3 expression are in line with previously cited studies, where diffuse expression in cancer cells was associated with a favorable prognosis [87].

More than 90% of microsatellite instability (MSI) colon cancers cells carry TGF-ß-receptor type II (TGFBR2) mutations. Lee et al. reported that the reconstitution of TGFBR2 in HCT116 colorectal cancer cells led to an increased sialylation of Nectin-3. However, overall synthesis and expression appeared to not be affected [88]. Available studies on Nectin-3 are not consistent and suggest that the role of Nectin-3 may be dependent on the histopathological type and location of the tumor. Further research to evaluate the role of Nectin-3 in CRC is needed.

3.4. Nectin-4

This is probably the most extensively researched Nectin family member in cancer and disease in general, and in breast cancer in particular. Nectin-4, known also as PVRL-4 or PRR4, is strongly expressed in fetal tissues during development (Table 1). There is little or no expression in adult tissues, besides the placenta, throat, bladder, breast, stomach, esophagus, salivary gland (ducts), and skin (epidermis and sweat glands) [40,89,90]. The importance of Nectin-4 during embryogenesis can be illustrated with the example of ectodermal dysplasia syndactyly associated with missense mutations of the PVLR4 gene. Disturbed trans-interactions between Nectin-1 and Nectin-4 causing Rac1 pathway alteration and delayed AJs formation in mutant cells are responsible for phenotypic presentations of EDSS1 and CLEPED1 [45,91,92]. By contrast, it was observed that Nectin-4 is overexpressed in various types of tumors (e.g., colorectal, gastric, esophageal, urotherial, breast, ovarian, hepatocellular, non-small cell lung carcinoma, and renal papillary cell) [90,93–99] (Table 2) Its role as an oncogene is being investigated.

Nectin-4 has been identified as a biomarker of cancer stem cells (CSCs). CSCs have been recognized as the root of cancers' initiation and the resistance of cancer cells to conventional chemo- and radiotherapies; hence, they are critical in the metastasis, recurrence, and thus, the disease-free survival of, e.g., colorectal cancer [100–102]. Siddharth et al. alleged that Nectin-4 is a CSC biomarker in the breast cancer model. Nectin-4 deletion inhibited the invasion of EMT/TME, a WNT-signaling cascade and an anchorage-independent growth [103].

Colon cancer cells exposed to 5-fluorouracil (5-FU, a core drug in CRC chemotherapy worldwide) increased endogenous Nectin-4 expression. The 5-FU sensitivity is inversely related to Nectin-4 expression in CRC cell line studies. Thus, it has been proposed that Nectin-4 is one of the factors related to 5-FU resistance. Nectin-4 coupling with afadin and subsequent cell growth induction through the Pi3k/Akt axis is a putative mechanism of resistance to 5-FU therapy in CRC cells. A combination of BCNU and resveratrol-induced apoptosis in 5-FU resistant colon cancer cells by decreasing Nectin-4 expression [104].

It has been experimentally demonstrated by Siddharth et al. that Nectin-4 is responsible for the induction of WNT/β-Catenin signaling via the Pi3k/Akt axis and promotes cancer stem cell proliferation as well as EMT and metastasing [103]. The self-renewal properties of circulating cancer cells induced by Nectin-4 expression are considered responsible for tumor aggressiveness and may play a role in EMT–TME transitions. A greater expression of Nectin-4 was observed in secondary tumors [89]. Nectin-4 has been associated with virtually every stage of tumor progression and the dissemination of disease. This statement is strongly supported by studies reporting Nectin-4 overexpression correlating with disease advancement and a worse prognosis in other cancer types [52,89,93,94,96,103,105–109]. Nectin-4 appears to increase with tumor grade (e.g., based on the tumor-nodes-metastasis classification). Measured overexpression was highest in relapsed tumors [89,107].

Nectin-4 has an impact on the cancer microenvironment. ADAM17 (a disintegrin and metalloproteinase 17)/TACE (a TNFα-converting enzyme) overexpressed in many cancers, including CRC, is capable of Nectin-4 cleavage under hypoxic conditions. As a result, the endodomain and ectodomain of Nectin-4 are released in a process called shedding [110].

Siddharth et al. has investigated the role of the soluble Nectin-4 ectodomain in the tumor microenvironment. It induces angiogenesis by direct interaction with endothelial integrin-ß4 via Src/PI3K/Akt/iNOS cascade [111]. Chatterjee et al. have provided experimental evidence that the Nectin-4 endodomain physically interacts with IMPORTIN-α (KARYOPHERIN-α2) and is translocated and accumulated in a nucleus, which activates DNA repair and enhances cell proliferation. In addition, nano formulated quinacrine (NQC) inhibits the action of both shedded Nectin-4 domains in in vitro assays [112]. There has been demonstrated an anti-angiogenic effect of curcumin and veliparib (a

PARP inhibitor) through the deregulation of Nectin-4 in alleged DNA repair inhibition as well [113]. Taken together, ADAM-17 and Nectin-4 are putative targets for anticancer therapies [110,111,113–115].

Kedashiro et al. reported that Nectin-4 and p95-ErbB2 (one of the trastuzumab-resistant HER2 receptor splice variants) cooperatively activated the Hippo signaling-dependent SOX2 gene expression in the T47D serum-free suspension breast cancer cell line. It subsequently enhances cell proliferation in an anchorage-independent manner [116,117]. It has been outlined that the SOX2-β-catenin/Beclin1/autophagy signaling axis enhances chemoresistance, induces EMT, and is responsible for CSC properties in CRC cells [118]. Surviving away from ECM and the ability to proliferate in blood is one of the key features of malignant cells. Clusters of circulating tumor cells have been identified as well from the blood samples of CRC patients [99]. The mechanisms mentioned above and potential Nectin-4 involvement in CRC cell lines are yet to be elucidated.

Additionally, Nectin-4 tends to be involved in vasculogenic mimicry (VM)—the phenomenon of fluid conducting, microcirculatory channels lined by nonendothelial cells. VM channels are generated by pluripotent embryonic stem cells, highly invasive tumor cells, and the ECM in aggressive primary and metastatic tumors in order to provide a sufficient blood supply [119,120]. Zhang et al. reported a strong association between Nectin-4 mRNA expression, metastases, and an advanced disease stage in CRC patients. They also observed an association between VM, Nectin-4, and integrin ß-1 (ITGB1) presence. Additionally, Nectin-4, ITGB1 and VM were significantly associated with metastases and TNM stage, which features highly invasive CRCs. All the three parameters can be utilized as prognostic factors for CRC [97].

Sethy et al. indicated that Nectin-4 is also responsible for the promotion of lymphangiogenesis. Their study was conducted on specimens collected from breast cancer patients. Nectin-4 was found to have a predominant role in promotion of tumor-induced lymphangiogenesis by increasing lymphatic vessel density (LVD) and activating the chemokine axis (CXCR4/CXCL12) [107]. These mechanisms were also investigated in other cancer types [121]. Moreover, LVD in peritumoral tissues are associated with local recurrence and DSF in CRC [122]. Whether Nectin-4 induces lymphangiogenesis in CRC is a matter for future research.

Furthermore, Nectin-4 serves as a measles virus (MV) entry receptor [123]. It gives an opportunity to utilize the oncolytic properties of MV in the treatment of Nectin-4-induced cancer. In the available literature, MV was reported to act as natural cancer killer cells for Burkitt's lymphoma, Hodgkin's disease, and squamous cell carcinoma [124]. Sugiyama et al. utilized the natural killing effect of MV in the treatment of induced breast cancer in mice. Their findings showed that recombinant MV specifically binds to Nectin-4 expressed on the surface of cells and additionally that it can reveal anti-cancer effects in several other malignancies, including CRC [125].

Antibody-based cancer therapies target specific antigens on cancer cells to deliver a highly cytotoxic payload to tumor sites by harnessing the exquisite specificity of monoclonal antibodies (mAb) as a delivery vehicle [90,126]. There are many (>200 at the time of submission) antibody–drug conjugates (ADCs) in clinical trials. Enfortumab vedotin (sold under brand name PADCEV) is an ADC consisting of a human anti-Nectin-4 antibody linked to the cytotoxic microtubule-disrupting agent monomethyl auristatin E (MMAE). It is an FDA-approved therapeutic for locally advanced or metastatic urothelial carcinoma. The activity of this ADC is under investigation in other cancers known to express Nectin-4 [127].

Likewise, bicycle toxin conjugates (BTCs) are a new class of anticancer agents that allow efficient and targeted delivery of toxin payloads into tumors. Bicyclic peptide is conjugated to a cytotoxic agent via a cleavable linker. The molecules of BTCs are much smaller than ADCs and exhibit better tissue distribution and tumor cell penetration [128]. The recently discovered BTC8009 is targeted against Nectin-4. Linked to MMAE, it has demonstrated at least equal efficacy as enfortumab vedotin in rodent models. In addition, the new BTC is also a putative anti-cancer drug against other Nectin-4-expressing cancers [129].

Thus, based on current knowledge, Nectin-4 expression could be used as a prognostic and potentially predictive factor in CRC. Molecule itself can be a putative target for the development of new therapeutics. Nectin-4 can be utilized for new target-specific probes to improve tumor visualization and metastases detection in single-photon emission computerized tomography (SPECT) [130].

3.5. Nectin-like Molecule 5

Nectin-like Molecule 5 (Necl-5), also known as polio-virus-receptor (PVR), Tage4, or CD155, is phylogenetically more closely related to Nectins than to other Nectin-like molecules (Table 1). It has probably diverged from Nectin-2 [38]. It trans-interacts with Nectin-3 on neighboring cells and also mediates cell–ECM junctions by binding to vitronectin. Necl-5 takes part in the contact inhibition mechanism. When there is no cell–cell contact, Necl-5 prevents Sprouty2 (Spry2) phosphorylation. After cells come into contact with each other, Necl-5 is downregulated by endocytosis following transient trans-interaction with Nectin-3. Unprotected Spry2 is tyrosine-phosphorylated by c-Src, which is activated by the PDGF receptor in response to PDGF, and subsequently inhibits PDGF-induced Ras signaling for cell proliferation [39,49]. Upregulation of Necl-5 in cancerous transformed cells that exceed the rate of internalization during cell–cell contact has been proven in studies to increase cell proliferation and hence tumor progression [44]. Necl-5 colocalizes with integrin on leading edges and takes part in growth factor-induced lamellipodia formation [49]. Necl-5 is upregulated through the Sonic hedgehog pathway as well as in Ras-mutated cells and allegedly induces cancer cells proliferation by inter alia shortening the G0/G1 phase [131].

Through interactions with CD226 and TIGIT, molecules present on leukocytes, Necl-5 together with Nectin-2 is a key regulator in cell-mediated immune response [57,132].

By binding to TIGIT, Necl-5 induces immunosuppression by the inhibition of NK cell and CD8+ T-cell cytotoxicity [133]. At the same time, Necl-5 has an affinity to the DNAM-1 molecule, which enhances immunological response by recognizing and killing tumors [134].

Necl-5 is known to be overexpressed on various types of malignant cells, including CRC [135,136] (Table 2). Zheng et al. evaluated Necl-5 in CRC cell lines under different conditions. They observed increased apoptosis, inhibited colony formation ability, and cell cycle arrest in the G1 phase in CRC cells after Necl-5 knockdown. In addition, the authors observed reduced expression of some cell-invasion-related molecules, such as FAK, Src, and MMP-2. Necl-5 knockdown inhibited Akt phosphorylation. Taken together, their findings support previous studies on that subject—Necl-5 attributes to tumor progression, invasion, and metastases in CRC cell lines and may be considered an anti-apoptotic factor in CRC [135]. Morimoto et al. indicated that Necl-5 augments the metastasis of cancer cells, including CRC, to the lungs. Necl-5 mAb blockage reduced secondary tumor formation in lungs by 60% in a mice model [46]. The authors suggested that cancer cells with high Necl-5 expression attach to CD226-expressing platelets. The mentioned process leads to platelet cloaking and enhances immune evasion of cancerous cells. Cell aggregates were arrested in pulmonary capillaries where extravasation and metastases formation take place [46,67].

There are many examples of the importance of Necl-5 in carcinogenesis, CRC progression, and dissemination. Necl-5 is also involved in immune surveillance and can act both as a tumor inducer and suppressor. Therefore, research on the application of Necl-5 in diagnostics and treatment strategies is recommended.

4. Conclusions and Future Perspectives

Nectins and Nectin-like molecules takes part in a plethora of essential processes in healthy tissues as well as mediating substantial actions in tumor initiation and evolution. Most of the crucial mechanisms of CRC carcinogenesis cannot be explained by a single factor, but nonetheless, Nectins and Nectin-like molecules are surely an important part of the entire picture. Our understanding of their role and their viable application in anti-cancer therapy is far from complete. Additional studies are needed to evaluate reliable Nectin and

Necls expression patterns in CRC cells. Research into potential spatial distribution and sidedness are necessary to elucidate the role of Nectins in different CRC clinical presentations. There is a possible place for Nectins and Necls in a future standard biomarkers array. Nectin-4 tends to be the most achievable therapeutic target for novel CRC treatment options with completion of present and upcoming clinical trials. Based on the broad spectrum of available papers, Nectins and Necls may serve as diagnostic (e.g., PET probes), prognostic (e.g., OS and DFS), and predictive (selection of targeted-therapy regimens) tools in CRC.

Funding: This paper received no external funding.

Conflicts of Interest: The authors declare no conflict of interest.

References

1. Sung, H.; Ferlay, J.; Siegel, R.L.; Laversanne, M.; Soerjomataram, I.; Jemal, A.; Bray, F. Global Cancer Statistics 2020: GLOBOCAN Estimates of Incidence and Mortality Worldwide for 36 Cancers in 185 Countries. *CA Cancer J. Clin.* **2021**, *71*, 209–249. [CrossRef] [PubMed]
2. Arnold, M.; Abnet, C.C.; Neale, R.E.; Vignat, J.; Giovannucci, E.L.; McGlynn, K.A.; Bray, F. Global Burden of 5 Major Types of Gastrointestinal Cancer. *Gastroenterology* **2020**, *159*, 335–349.e15. [CrossRef] [PubMed]
3. Siegel, R.L.; Miller, K.D.; Fuchs, H.E.; Jemal, A. Cancer Statistics, 2022. *CA Cancer J. Clin.* **2022**, *72*, 7–33. [CrossRef] [PubMed]
4. Morgan, E.; Arnold, M.; Gini, A.; Lorenzoni, V.; Cabasag, C.J.; Laversanne, M.; Vignat, J.; Ferlay, J.; Murphy, N.; Bray, F. Global Burden of Colorectal Cancer in 2020 and 2040: Incidence and Mortality Estimates from GLOBOCAN. *Gut* **2022**. gutjnl-2022-327736. [CrossRef]
5. Sifaki-Pistolla, D.; Poimenaki, V.; Fotopoulou, I.; Saloustros, E.; Mavroudis, D.; Vamvakas, L.; Lionis, C. Significant Rise of Colorectal Cancer Incidence in Younger Adults and Strong Determinants: 30 Years Longitudinal Differences between under and over 50s. *Cancers* **2022**, *14*, 4799. [CrossRef]
6. del Vecchio Blanco, G.; Calabrese, E.; Biancone, L.; Monteleone, G.; Paoluzi, O.A. The Impact of COVID-19 Pandemic in the Colorectal Cancer Prevention. *Int. J. Colorectal Dis.* **2020**, *35*, 1951–1954. [CrossRef]
7. Czeisler, M.É.; Marynak, K.; Clarke, K.E.N.; Salah, Z.; Shakya, I.; Thierry, J.M.; Ali, N.; McMillan, H.; Wiley, J.F.; Weaver, M.D.; et al. Delay or Avoidance of Medical Care Because of COVID-19–Related Concerns—United States, June 2020. *MMWR Morb. Mortal. Wkly. Rep.* **2020**, *69*, 1250–1257. [CrossRef]
8. Rottoli, M.; Gori, A.; Pellino, G.; Flacco, M.E.; Martellucci, C.; Spinelli, A.; Poggioli, G. Colorectal Cancer Stage at Diagnosis Before vs During the COVID-19 Pandemic in Italy. *JAMA Netw. Open* **2022**, *5*, e2243119. [CrossRef]
9. Keum, N.N.; Giovannucci, E. Global Burden of Colorectal Cancer: Emerging Trends, Risk Factors and Prevention Strategies. *Nat. Rev. Gastroenterol. Hepatol.* **2019**, *16*, 713–732. [CrossRef]
10. Kuipers, E.J.; Grady, W.M.; Lieberman, D.; Seufferlein, T.; Sung, J.J.; Boelens, P.G.; van de Velde, C.J.H.; Watanabe, T. Colorectal Cancer. *Nat. Rev. Dis. Primers* **2015**, *1*, 1–25. [CrossRef]
11. Arnold, M.; Sierra, M.S.; Laversanne, M.; Soerjomataram, I.; Jemal, A.; Bray, F. Global Patterns and Trends in Colorectal Cancer Incidence and Mortality. *Gut* **2017**, *66*, 683–691. [CrossRef] [PubMed]
12. Marusyk, A.; Janiszewska, M.; Polyak, K. Intratumor Heterogeneity: The Rosetta Stone of Therapy Resistance. *Cancer Cell* **2020**, *37*, 471–484. [CrossRef] [PubMed]
13. Heide, T.; Househam, J.; Cresswell, G.D.; Spiteri, I.; Lynn, C.; Mossner, M.; Kimberley, C.; Fernandez-Mateos, J.; Chen, B.; Zapata, L.; et al. The Co-Evolution of the Genome and Epigenome in Colorectal Cancer. *Nature* **2022**, *611*, 733–743. [CrossRef] [PubMed]
14. Househam, J.; Heide, T.; Cresswell, G.D.; Spiteri, I.; Kimberley, C.; Zapata, L.; Lynn, C.; James, C.; Mossner, M.; Fernandez-Mateos, J.; et al. Phenotypic Plasticity and Genetic Control in Colorectal Cancer Evolution. *Nature* **2022**, *611*, 744–753. [CrossRef]
15. Carethers, J.M.; Jung, B.H. Genetics and Genetic Biomarkers in Sporadic Colorectal Cancer. *Gastroenterology* **2015**, *149*, 1177–1190.e3. [CrossRef]
16. Jasperson, K.W.; Tuohy, T.M.; Neklason, D.W.; Burt, R.W. Hereditary and Familial Colon Cancer. *Gastroenterology* **2010**, *138*, 2044–2058. [CrossRef]
17. Jiao, S.; Peters, U.; Berndt, S.; Brenner, H.; Butterbach, K.; Caan, B.J.; Carlson, C.S.; Chan, A.T.; Chang-Claude, J.; Chanock, S.; et al. Estimating the Heritability of Colorectal Cancer. *Hum. Mol. Genet.* **2014**, *23*, 3898–3905. [CrossRef]
18. Çetin, D.A.; Yildirim, M.; Yakan, S.; Çiyiltepe, H.; Aydoğan, S. Effects of Prognostic Factors on Overall and Disease-Free Survival in Patients with Stage I–III Colorectal Cancer. *Arch. Med. Sci.-Civiliz. Dis.* **2016**, *1*, 131–138. [CrossRef]
19. Zygulska, A.L.; Pierzchalski, P. Novel Diagnostic Biomarkers in Colorectal Cancer. *Int. J. Mol. Sci.* **2022**, *23*, 852. [CrossRef]
20. Haier, J.; Nasralla, M.; Nicolson, G.L. Cell Surface Molecules and Their Prognostic Values in Assessing Colorectal Carcinomas. *Ann. Surg.* **2000**, *231*, 11–24. [CrossRef]
21. Holubec, L.; Topolcan, O.; Finek, J.; Holdenrieder, S.; Stieber, P.; Pesta, M.; Pikner, R.; Holubec Sen, L.; Sutnar, A.; Liska, V.; et al. Markers of Cellular Adhesion in Diagnosis and Therapy Control of Colorectal Carcinoma. *Anticancer Res.* **2005**, *25*, 1597–1601. [PubMed]

22. Seo, K.J.; Kim, M.; Kim, J. Prognostic Implications of Adhesion Molecule Expression in Colorectal Cancer. *Int. J. Clin. Exp. Pathol.* **2015**, *8*, 4148–4157. [PubMed]
23. Sluiter, N.; de Cuba, E.; Kwakman, R.; Kazemier, G.; Meijer, G.; te Velde, E.A. Adhesion Molecules in Peritoneal Dissemination: Function, Prognostic Relevance and Therapeutic Options. *Clin. Exp. Metastasis* **2016**, *33*, 401–416. [CrossRef] [PubMed]
24. Sveen, A.; Kopetz, S.; Lothe, R.A. Biomarker-Guided Therapy for Colorectal Cancer: Strength in Complexity. *Nat. Rev. Clin. Oncol.* **2020**, *17*, 11–32. [CrossRef] [PubMed]
25. Eide, P.W.; Bruun, J.; Lothe, R.A.; Sveen, A. CMScaller: An R Package for Consensus Molecular Subtyping of Colorectal Cancer Pre-Clinical Models. *Sci. Rep.* **2017**, *7*, 16618. [CrossRef]
26. Hoorn, S.T.; de Back, T.R.; Sommeijer, D.W.; Vermeulen, L. Clinical Value of Consensus Molecular Subtypes in Colorectal Cancer: A Systematic Review and Meta-Analysis. *J. Natl. Cancer Inst.* **2022**, *114*, 503–516. [CrossRef]
27. Okita, A.; Takahashi, S.; Ouchi, K.; Inoue, M.; Watanabe, M.; Endo, M.; Honda, H.; Yamada, Y.; Ishioka, C. Consensus Molecular Subtypes Classification of Colorectal Cancer as a Predictive Factor for Chemotherapeutic Efficacy against Metastatic Colorectal Cancer. *Oncotarget* **2018**, *9*, 18698–18711. [CrossRef]
28. Afrăsânie, V.A.; Marinca, M.V.; Alexa-Stratulat, T.; Gafton, B.; Păduraru, M.; Adavidoaiei, A.M.; Miron, L.; Rusu, C. KRAS, NRAS, BRAF, HER2 and Microsatellite Instability in Metastatic Colorectal Cancer-Practical Implications for the Clinician. *Radiol. Oncol.* **2019**, *53*, 265–274. [CrossRef]
29. Luo, H.; Zhao, Q.; Wei, W.; Zheng, L.; Yi, S.; Li, G.; Wang, W.; Sheng, H.; Pu, H.; Mo, H.; et al. Circulating Tumor DNA Methylation Profiles Enable Early Diagnosis, Prognosis Prediction, and Screening for Colorectal Cancer. *Sci. Transl. Med.* **2020**, *12*, 1–12. [CrossRef]
30. To, Y.H.; Degeling, K.; Kosmider, S.; Wong, R.; Lee, M.; Dunn, C.; Gard, G.; Jalali, A.; Wong, V.; IJzerman, M.; et al. Circulating Tumour DNA as a Potential Cost-Effective Biomarker to Reduce Adjuvant Chemotherapy Overtreatment in Stage II Colorectal Cancer. *Pharmacoeconomics* **2021**, *39*, 953–964. [CrossRef]
31. Henriksen, T.V.; Tarazona, N.; Frydendahl, A.; Reinert, T.; Gimeno-Valiente, F.; Carbonell-Asins, J.A.; Sharma, S.; Renner, D.; Hafez, D.; Roda, D.; et al. Circulating Tumor DNA in Stage III Colorectal Cancer, beyond Minimal Residual Disease Detection, toward Assessment of Adjuvant Therapy Efficacy and Clinical Behavior of Recurrences. *Clin. Cancer Res.* **2022**, *28*, 507–517. [CrossRef] [PubMed]
32. Takahashi, K.; Nakanishi, H.; Miyahara, M.; Mandai, K.; Satoh, K.; Satoh, A.; Nishioka, H.; Aoki, J.; Nomoto, A.; Mizoguchi, A.; et al. Nectin/PRR: An Immunoglobulin-like Cell Adhesion Molecule Recruited to Cadherin-Based Adherens Junctions through Interaction with Afadin, a PDZ Domain–Containing Protein. *J. Cell Biol.* **1999**, *145*, 539–549. [CrossRef] [PubMed]
33. Sakisaka, T.; Takai, Y. Biology and Pathology of Nectins and Nectin-like Molecules. *Curr. Opin. Cell Biol.* **2004**, *16*, 513–521. [CrossRef]
34. della Salda, L.; Massimini, M.; Romanucci, M.; Palmieri, C.; Perillo, A.; Grieco, V.; Malatesta, D.; Spinillo, M.A.; Passantino, G.; Dondi, F.; et al. Nectin-4 and P63 Immunohistochemical Expression in Canine Prostate Tumourigenesis. *Vet. Comp. Oncol.* **2019**, *17*, 298–307. [CrossRef] [PubMed]
35. Kawanishi, A.; Hirabayashi, K.; Yamada, M.; Takanashi, Y.; Hadano, A.; Kawaguchi, Y.; Nakagohri, T.; Nakamura, N.; Mine, T. Clinicopathological Significance of Necl-4 Expression in Pancreatic Ductal Adenocarcinoma. *J. Clin. Pathol.* **2017**, *70*, 619–624. [CrossRef] [PubMed]
36. Raveh, S.; Gavert, N.; Spiegel, I.; Ben-Ze'ev, A. The Cell Adhesion Nectin-like Molecules (Necl) 1 and 4 Suppress the Growth and Tumorigenic Ability of Colon Cancer Cells. *J. Cell Biochem.* **2009**, *108*, 326–336. [CrossRef]
37. Hirabayashi, K.; Tajiri, T.; Bosch, D.E.; Morimachi, M.; Miyaoka, M.; Inomoto, C.; Nakamura, N.; Yeh, M.M. Loss of Nectin-3 Expression as a Marker of Tumor Aggressiveness in Pancreatic Neuroendocrine Tumor. *Pathol. Int.* **2020**, *70*, 84–91. [CrossRef]
38. Duraivelan, K.; Samanta, D. Tracing the Evolution of Nectin and Nectin-like Cell Adhesion Molecules. *Sci. Rep.* **2020**, *10*, 9434. [CrossRef]
39. Mandai, K.; Rikitake, Y.; Mori, M.; Takai, Y. *Nectins and Nectin-like Molecules in Development and Disease*, 1st ed.; Elsevier Inc.: Amsterdam, The Netherlands, 2015; Volume 112.
40. Duraivelan, K.; Samanta, D. Emerging Roles of the Nectin Family of Cell Adhesion Molecules in Tumour-Associated Pathways. *Biochim. Biophys. Acta Rev. Cancer* **2021**, *1876*, 188589. [CrossRef]
41. Duraivelan, K.; Dash, S.; Samanta, D. An Evolutionarily Conserved Charged Residue Dictates the Specificity of Heterophilic Interactions among Nectins. *Biochem. Biophys. Res. Commun.* **2021**, *534*, 504–510. [CrossRef]
42. Miyoshi, J.; Takai, Y. Nectin and Nectin-like Molecules: Biology and Pathology. *Am. J. Nephrol.* **2007**, *27*, 590–604. [CrossRef] [PubMed]
43. Mizutani, K.; Takai, Y. Nectin Spot: A Novel Type of Nectin-Mediated Cell Adhesion Apparatus. *Biochem. J.* **2016**, *473*, 2691–2715. [CrossRef] [PubMed]
44. Takai, Y.; Miyoshi, J.; Ikeda, W.; Ogita, H. Nectins and Nectin-like Molecules: Roles in Contact Inhibition of Cell Movement and Proliferation. *Nat. Rev. Mol. Cell Biol.* **2008**, *9*, 603–615. [CrossRef] [PubMed]
45. Ogita, H.; Takai, Y. Cross-Talk Among Integrin, Cadherin, and Growth Factor Receptor: Roles of Nectin and Nectin-Like Molecule. *Int. Rev. Cytol* **2008**, *265*, 1–54. [CrossRef]

46. Morimoto, K.; Satoh-Yamaguchi, K.; Hamaguchi, A.; Inoue, Y.; Takeuchi, M.; Okada, M.; Ikeda, W.; Takai, Y.; Imai, T. Interaction of Cancer Cells with Platelets Mediated by Necl-5/Poliovirus Receptor Enhances Cancer Cell Metastasis to the Lungs. *Oncogene* **2008**, *27*, 264–273. [CrossRef]
47. Paschos, K.A.; Majeed, A.W.; Bird, N.C. Natural History of Hepatic Metastases from Colorectal Cancer-Pathobiological Pathways with Clinical Significance. *World J. Gastroenterol.* **2014**, *20*, 3719–3737. [CrossRef]
48. Pretzsch, E.; Bösch, F.; Neumann, J.; Ganschow, P.; Bazhin, A.; Guba, M.; Werner, J.; Angele, M. Mechanisms of Metastasis in Colorectal Cancer and Metastatic Organotropism: Hematogenous versus Peritoneal Spread. *J. Oncol.* **2019**, *2019*, 7407190. [CrossRef]
49. Kajita, M.; Ikeda, W.; Tamaru, Y.; Takai, Y. Regulation of Platelet-Derived Growth Factor-Induced Ras Signaling by Poliovirus Receptor Necl-5 and Negative Growth Regulator Sprouty2. *Genes Cells* **2007**, *12*, 345–357. [CrossRef]
50. Thul, P.J.; Lindskog, C. The Human Protein Atlas: A Spatial Map of the Human Proteome. *Protein Sci.* **2018**, *27*, 233–244. [CrossRef]
51. Uhlén, M.; Fagerberg, L.; Hallström, B.M.; Lindskog, C.; Oksvold, P.; Mardinoglu, A.; Sivertsson, Å.; Kampf, C.; Sjöstedt, E.; Asplund, A.; et al. Tissue-Based Map of the Human Proteome. *Science* **2015**, *347*, 1260419. [CrossRef]
52. Miyake, M.; Miyamoto, T.; Shimizu, T.; Ohnishi, S.; Fujii, T.; Nishimura, N.; Oda, Y.; Morizawa, Y.; Hori, S.; Gotoh, D.; et al. Tumor Expression of Nectin-1–4 and Its Clinical Implication in Muscle Invasive Bladder Cancer: An Intra-Patient Variability of Nectin-4 Expression. *Pathol. Res. Pract.* **2022**, *237*, 154072. [CrossRef] [PubMed]
53. Tanaka, Y.; Murata, M.; Oda, Y.; Furue, M.; Ito, T. Nectin Cell Adhesion Molecule 4 (Nectin4) Expression in Cutaneous Squamous Cell Carcinoma: A New Therapeutic Target? *Biomedicines* **2021**, *9*, 355. [CrossRef] [PubMed]
54. Liu, Y.; Han, X.; Li, L.; Zhang, Y.; Huang, X.; Li, G.; Xu, C.; Yin, M.; Zhou, P.; Shi, F.; et al. Role of Nectin-4 Protein in Cancer (Review). *Int. J. Oncol.* **2021**, *59*, 93. [CrossRef] [PubMed]
55. Chatterjee, S.; Sinha, S.; Kundu, C.N. Nectin Cell Adhesion Molecule-4 (NECTIN-4): A Potential Target for Cancer Therapy. *Eur. J. Pharmacol.* **2021**, *911*, 174516. [CrossRef] [PubMed]
56. Holmes, V.M.; de Motes, C.M.; Richards, P.T.; Roldan, J.; Bhargava, A.K.; Orange, J.S.; Krummenacher, C. Interaction between Nectin-1 and the Human Natural Killer Cell Receptor CD96. *PLoS ONE* **2019**, *14*, e0212443. [CrossRef] [PubMed]
57. Jin, H.-S.; Park, Y. Hitting the Complexity of the TIGIT-CD96-CD112R-CD226 Axis for next-Generation Cancer Immunotherapy. *BMB Rep.* **2021**, *54*, 2–11. [CrossRef] [PubMed]
58. Zhang, Z.; Su, T.; He, L.; Wang, H.; Ji, G.; Liu, X.; Zhang, Y.; Dong, G. Identification and Functional Analysis of Ligands for Natural Killer Cell Activating Receptors in Colon Carcinoma. *Tohoku J. Exp. Med.* **2012**, *226*, 59–68. [CrossRef]
59. Yu, Z.; Chan, M.K.; O-Charoenrat, P.; Eisenberg, D.P.; Shah, J.P.; Singh, B.; Fong, Y.; Wong, R.J. Enhanced Nectin-1 Expression and Herpes Oncolytic Sensitivity in Highly Migratory and Invasive Carcinoma. *Clin. Cancer Res.* **2005**, *11*, 4889–4897. [CrossRef]
60. Tampakis, A.; Tampaki, E.C.; Nonni, A.; Droeser, R.; Posabella, A.; Tsourouflis, G.; Kontzoglou, K.; Patsouris, E.; von Flüe, M.; Kouraklis, G. Nectin-1 Expression in Colorectal Cancer: Is There a Group of Patients with High Risk for Early Disease Recurrence? *Oncology* **2019**, *96*, 318–325. [CrossRef]
61. Yamada, M.; Hirabayashi, K.; Kawanishi, A.; Hadano, A.; Takanashi, Y.; Izumi, H.; Kawaguchi, Y.; Mine, T.; Nakamura, N.; Nakagohri, T. Nectin-1 Expression in Cancer-Associated Fibroblasts Is a Predictor of Poor Prognosis for Pancreatic Ductal Adenocarcinoma. *Surg. Today* **2017**, *48*, 510–516. [CrossRef]
62. Ballester, M.; Gonin, J.; Rodenas, A.; Bernaudin, J.F.; Rouzier, R.; Coutant, C.; Daraï, E. Eutopic Endometrium and Peritoneal, Ovarian and Colorectal Endometriotic Tissues Express a Different Profile of Nectin-1,-3,-4 and Nectin-like Molecule 2. *Hum. Reprod.* **2012**, *27*, 3179–3186. [CrossRef] [PubMed]
63. Alteber, Z.; Kotturi, M.F.; Whelan, S.; Ganguly, S.; Weyl, E.; Pardoll, D.M.; Hunter, J.; Ophir, E. Therapeutic Targeting of Checkpoint Receptors within the Dnam1 Axis. *Cancer Discov.* **2021**, *11*, 1040–1051. [CrossRef] [PubMed]
64. Fathi, M.; Pustokhina, I.; Kuznetsov, S.V.; Khayrullin, M.; Hojjat-Farsangi, M.; Karpisheh, V.; Jalili, A.; Jadidi-Niaragh, F. T-Cell Immunoglobulin and ITIM Domain, as a Potential Immune Checkpoint Target for Immunotherapy of Colorectal Cancer. *IUBMB Life* **2021**, *73*, 726–738. [CrossRef] [PubMed]
65. Nagumo, Y.; Iguchi-Manaka, A.; Yamashita-Kanemaru, Y.; Abe, F.; Bernhardt, G.N.; Shibuya, A.; Shibuya, K. Increased CD112 Expression in Methylcholanthrene- Induced Tumors in CD155-Deficient Mice. *PLoS ONE* **2014**, *9*, e112415. [CrossRef] [PubMed]
66. Tsuchiya, H.; Shiota, G. Immune Evasion by Cancer Stem Cells. *Regen Ther.* **2021**, *17*, 20–33. [CrossRef] [PubMed]
67. Cluxton, C.D.; Spillane, C.; O'Toole, S.A.; Sheils, O.; Gardiner, C.M.; O'Leary, J.J. Suppression of Natural Killer Cell NKG2D and CD226 Anti-Tumour Cascades by Platelet Cloaked Cancer Cells: Implications for the Metastatic Cascade. *PLoS ONE* **2019**, *14*, e0211538. [CrossRef] [PubMed]
68. Siena, S.; di Bartolomeo, M.; Raghav, K.; Masuishi, T.; Loupakis, F.; Kawakami, H.; Yamaguchi, K.; Nishina, T.; Fakih, M.; Elez, E.; et al. Trastuzumab Deruxtecan (DS-8201) in Patients with HER2-Expressing Metastatic Colorectal Cancer (DESTINY-CRC01): A Multicentre, Open-Label, Phase 2 Trial. *Lancet Oncol.* **2021**, *22*, 779–789. [CrossRef]
69. Ahcene Djaballah, S.; Daniel, F.; Milani, A.; Ricagno, G.; Lonardi, S. HER2 in Colorectal Cancer: The Long and Winding Road From Negative Predictive Factor to Positive Actionable Target. *Am. Soc. Clin. Oncol. Book* **2022**, *42*, 219–232. [CrossRef]
70. Zeng, T.; Cao, Y.; Jin, T.; Tian, Y.; Dai, C.; Xu, F. The CD112R/CD112 Axis: A Breakthrough in Cancer Immunotherapy. *J. Exp. Clin. Cancer Res.* **2021**, *40*, 285. [CrossRef]

71. Xu, F.; Sunderland, A.; Zhou, Y.; Schulick, R.D.; Edil, B.H.; Zhu, Y. Blockade of CD112R and TIGIT Signaling Sensitizes Human Natural Killer Cell Functions. *Cancer Immunol. Immunother.* **2017**, *66*, 1367–1375. [CrossRef]
72. Whelan, S.; Eran, O.; Maya, F.K.; Ofer, L.; Ganguly, S.; Leung, L.; Vaknin, I.; Kumar, S.; Dassa, L.; Hansen, K.; et al. PVRIG and PVRL2 Are Induced in Cancer and Inhibit CD8+ T-Cell Function. *Physiol. Behav.* **2017**, *176*, 498–503. [CrossRef] [PubMed]
73. Molfetta, R.; Milito, N.D.; Zitti, B.; Lecce, M.; Fionda, C.; Cippitelli, M.; Santoni, A.; Paolini, R. The Ubiquitin-Proteasome Pathway Regulates Nectin2/CD112 Expression and Impairs NK Cell Recognition and Killing. *Eur. J. Immunol.* **2019**, *49*, 873–883. [CrossRef] [PubMed]
74. Bekes, I.; Löb, S.; Holzheu, I.; Janni, W.; Baumann, L.; Wöckel, A.; Wulff, C. Nectin-2 in Ovarian Cancer: How Is It Expressed and What Might Be Its Functional Role? *Cancer Sci.* **2019**, *110*, 1872–1882. [CrossRef] [PubMed]
75. Russo, E.; Runge, P.; Jahromi, N.H.; Naboth, H.; Landtwing, A.; Montecchi, R.; Leicht, N.; Hunter, M.C.; Takai, Y.; Halin, C. CD112 Regulates Angiogenesis and T Cell Entry into the Spleen. *Cells* **2021**, *10*, 169. [CrossRef]
76. Karabulut, M.; Gunaldi, M.; Alis, H.; Afsar, C.U.; Karabulut, S.; Serilmez, M.; Akarsu, C.; Seyit, H.; Aykan, N.F. Serum Nectin-2 Levels Are Diagnostic and Prognostic in Patients with Colorectal Carcinoma. *Clin. Transl. Oncol.* **2016**, *18*, 160–171. [CrossRef]
77. Liang, S.; Yang, Z.; Li, D.; Miao, X.; Yang, L.; Zou, Q.; Yuan, Y. The Clinical and Pathological Significance of Nectin-2 and DDX3 Expression in Pancreatic Ductal Adenocarcinomas. *Dis Markers* **2015**, *2015*, 379568. [CrossRef]
78. Samanta, D.; Almo, S.C. Nectin Family of Cell-Adhesion Molecules: Structural and Molecular Aspects of Function and Specificity. *Cell. Mol. Life Sci.* **2015**, *72*, 645–658. [CrossRef]
79. Devilard, E.; Xerri, L.; Dubreuil, P.; Lopez, M.; Reymond, N. Nectin-3 (CD113) Interacts with Nectin-2 (CD112) to Promote Lymphocyte Transendothelial Migration. *PLoS ONE* **2013**, *8*, 1–12. [CrossRef]
80. Madsen, C.D.; Sahai, E. Cancer Dissemination-Lessons from Leukocytes. *Dev. Cell* **2010**, *19*, 13–26. [CrossRef]
81. LaFrance, M.E.; Farrow, M.A.; Chandrasekaran, R.; Sheng, J.; Rubin, D.H.; Lacy, D.B. Identification of an Epithelial Cell Receptor Responsible for Clostridium Difficile TcdB-Induced Cytotoxicity. *Proc. Natl. Acad. Sci. USA* **2015**, *112*, 7073–7078. [CrossRef]
82. Maniwa, Y.; Nishio, W.; Okita, Y.; Yoshimura, M. Expression of Nectin 3: Novel Prognostic Marker of Lung Adenocarcinoma. *Thorac. Cancer* **2012**, *3*, 175–181. [CrossRef] [PubMed]
83. Zhao, Y.; Hong, X.H.; Li, K.; Li, Y.Q.; Li, Y.Q.; He, S.W.; Zhang, P.P.; Li, J.Y.; Li, Q.; Liang, Y.L.; et al. ZNF582 Hypermethylation Promotes Metastasis of Nasopharyngeal Carcinoma by Regulating the Transcription of Adhesion Molecules Nectin-3 and NRXN3. *Cancer Commun.* **2020**, *40*, 721–737. [CrossRef]
84. Xu, F.; Si, X.; Wang, J.; Yang, A.; Qin, T.; Yang, Y. Nectin-3 Is a New Biomarker That Mediates the Upregulation of MMP2 and MMP9 in Ovarian Cancer Cells. *Biomed. Pharmacother.* **2019**, *110*, 139–144. [CrossRef] [PubMed]
85. Martin, T.A.; Lane, J.; Harrison, G.M.; Jiang, W.G. The Expression of the Nectin Complex in Human Breast Cancer and the Role of Nectin-3 in the Control of Tight Junctions during Metastasis. *PLoS ONE* **2013**, *8*, e82696. [CrossRef] [PubMed]
86. Fujito, T.; Ikeda, W.; Kakunaga, S.; Minami, Y.; Kajita, M.; Sakamoto, Y.; Monden, M.; Takai, Y. Inhibition of Cell Movement and Proliferation by Cell-Cell Contact-Induced Interaction of Necl-5 with Nectin-3. *J. Cell Biol.* **2005**, *171*, 165–173. [CrossRef] [PubMed]
87. Izumi, H.; Hirabayashi, K.; Nakamura, N.; Nakagohri, T. Nectin Expression in Pancreatic Adenocarcinoma: Nectin-3 Is Associated with a Poor Prognosis. *Surg. Today* **2015**, *45*, 487–494. [CrossRef]
88. Lee, J.; Warnken, U.; Schnölzer, M.; Gebert, J.; Kopitz, J. A New Method for Detection of Tumor Driver-Dependent Changes of Protein Sialylation in a Colon Cancer Cell Line Reveals Nectin-3 as TGFBR2 Target. *Protein Sci.* **2015**, *24*, 1686–1694. [CrossRef]
89. Sethy, C.; Goutam, K.; Nayak, D.; Pradhan, R.; Molla, S.; Chatterjee, S.; Rout, N.; Wyatt, M.D.; Narayan, S.; Kundu, C.N. Clinical Significance of a Pvrl 4 Encoded Gene Nectin-4 in Metastasis and Angiogenesis for Tumor Relapse. *J. Cancer Res. Clin. Oncol.* **2020**, *146*, 245–259. [CrossRef]
90. Challita-Eid, P.M.; Satpayev, D.; Yang, P.; An, Z.; Morrison, K.; Shostak, Y.; Raitano, A.; Nadell, R.; Liu, W.; Lortie, D.R.; et al. Enfortumab Vedotin Antibody-Drug Conjugate Targeting Nectin-4 Is a Highly Potent Therapeutic Agent in Multiple Preclinical Cancer Models. *Cancer Res.* **2016**, *76*, 3003–3013. [CrossRef]
91. Brancati, F.; Fortugno, P.; Bottillo, I.; Lopez, M.; Josselin, E.; Boudghene-Stambouli, O.; Agolini, E.; Bernardini, L.; Bellacchio, E.; Iannicelli, M.; et al. Mutations in PVRL4, Encoding Cell Adhesion Molecule Nectin-4, Cause Ectodermal Dysplasia-Syndactyly Syndrome. *Am. J. Hum. Genet.* **2010**, *87*, 265–273. [CrossRef]
92. Fortugno, P.; Josselin, E.; Tsiakas, K.; Agolini, E.; Cestra, G.; Teson, M.; Santer, R.; Castiglia, D.; Novelli, G.; Dallapiccola, B.; et al. Nectin-4 Mutations Causing Ectodermal Dysplasia with Syndactyly Perturb the Rac1 Pathway and the Kinetics of Adherens Junction Formation. *J. Investig. Dermatol.* **2014**, *134*, 2146–2153. [CrossRef] [PubMed]
93. Lin, X.; Hu, H.; Pan, Y.; Pan, Y.; Gao, S. The Prognostic Role of Expression of Nectin-4 in Esophageal Cancer. *Med. Sci. Monit.* **2019**, *25*, 10089–10094. [CrossRef] [PubMed]
94. Erturk, K.; Karaman, S.; Dagoglu, N.; Serilmez, M.; Duranyildiz, D.; Tas, F. Serum Nectin-2 and Nectin-4 Are Diagnostic in Lung Cancer: Which Is Superior? *Wien. Klin. Wochenschr.* **2019**, *131*, 419–426. [CrossRef] [PubMed]
95. Zeindler, J.; Soysal, S.D.; Piscuoglio, S.; Ng, C.K.Y.; Mechera, R.; Isaak, A.; Weber, W.P.; Muenst, S.; Kurzeder, C. Nectin-4 Expression Is an Independent Prognostic Biomarker and Associated with Better Survival in Triple-Negative Breast Cancer. *Front. Med.* **2019**, *6*, 200. [CrossRef] [PubMed]

96. Patients, H.; Rodler, S.; Eismann, L.; Schlenker, B.; Casuscelli, J.; Brinkmann, I.; Sendelhofert, A.; Waidelich, R.; Buchner, A.; Stief, C.; et al. Expression of Nectin-4 in Variant Histologies of Bladder Cancer and Its Prognostic Value—Need for Biomarker Testing in High-Risk Patients? *Cancers* **2022**, *14*, 4411.
97. Zhang, J.; Liu, K.; Peng, P.; Li, S.; Ye, Z.; Su, Y.; Liu, S.; Qin, M.; Huang, J. Upregulation of Nectin-4 Is Associated with ITGB1 and Vasculogenic Mimicry and May Serve as a Predictor of Poor Prognosis in Colorectal Cancer. *Oncol. Lett.* **2019**, *18*, 1163–1170. [CrossRef]
98. Zschäbitz, S.; Mikuteit, M.; Stöhr, C.; Herrmann, E.; Polifka, I.; Agaimy, A. Expression of Nectin-4 in Papillary Renal Cell Carcinoma. *Discov. Oncol.* **2022**, *13*, 90. [CrossRef]
99. Pavlova, N.N.; Pallasch, C.; Elia, A.E.; Braun, C.J.; Westbrook, T.F.; Hemann, M.; Elledge, S.J. A Role for PVRL4-Driven Cell-Cell Interactions in Tumorigenesis. *Elife* **2013**, *2*, 358. [CrossRef]
100. Das, P.K.; Islam, F.; Lam, A.K. The Roles of Cancer Stem Cells and Therapy Resistance in Colorectal Carcinoma. *Cells* **2020**, *9*, 1392. [CrossRef]
101. Atashzar, M.R.; Baharlou, R.; Karami, J.; Abdollahi, H.; Rezaei, R.; Pourramezan, F.; Zoljalali Moghaddam, S.H. Cancer Stem Cells: A Review from Origin to Therapeutic Implications. *J. Cell Physiol.* **2020**, *235*, 790–803. [CrossRef]
102. Zhou, Y.; Xia, L.; Wang, H.; Oyang, L.; Su, M.; Liu, Q.; Lin, J.; Tan, S.; Tian, Y.; Liao, Q.; et al. Cancer Stem Cells in Progression of Colorectal Cancer. *Oncotarget* **2018**, *9*, 33403–33415. [CrossRef] [PubMed]
103. Siddharth, S.; Goutam, K.; Das, S.; Nayak, A.; Nayak, D.; Sethy, C.; Wyatt, M.D.; Kundu, C.N. Nectin-4 Is a Breast Cancer Stem Cell Marker That Induces WNT/β-Catenin Signaling via Pi3k/Akt Axis. *Int. J. Biochem. Cell Biol.* **2017**, *89*, 85–94. [CrossRef] [PubMed]
104. Das, D.; Satapathy, S.R.; Siddharth, S.; Nayak, A.; Kundu, C.N. NECTIN-4 Increased the 5-FU Resistance in Colon Cancer Cells by Inducing the PI3K-AKT Cascade. *Cancer Chemother. Pharmacol.* **2015**, *76*, 471–479. [CrossRef] [PubMed]
105. Hao, R.T.; Zheng, C.; Wu, C.Y.; Xia, E.J.; Zhou, X.F.; Quan, R.D.; Zhang, X.H. NECTIN4 Promotes Papillary Thyroid Cancer Cell Proliferation, Migration, and Invasion and Triggers EMT by Activating AKT. *Cancer Manag. Res.* **2019**, *11*, 2565–2578. [CrossRef]
106. Takano, A.; Ishikawa, N.; Nishino, R.; Masuda, K.; Yasui, W.; Inai, K.; Nishimura, H.; Ito, H.; Nakayama, H.; Miyagi, Y.; et al. Identification of Nectin-4 Oncoprotein as a Diagnostic and Therapeutic Target for Lung Cancer. *Cancer Res.* **2009**, *69*, 6694–6703. [CrossRef]
107. Sethy, C.; Goutam, K.; Das, B.; Dash, S.R.; Kundu, C.N. Nectin-4 Promotes Lymphangiogenesis and Lymphatic Metastasis in Breast Cancer by Regulating CXCR4-LYVE-1 Axis. *Vascul. Pharmacol.* **2021**, *140*, 106865. [CrossRef]
108. Deng, H.; Shi, H.; Chen, L.; Zhou, Y.; Jiang, J. Over-Expression of Nectin-4 Promotes Progression of Esophageal Cancer and Correlates with Poor Prognosis of the Patients. *Cancer Cell Int.* **2019**, *19*, 106. [CrossRef]
109. Liu, Y.; Li, G.; Zhang, Y.; Li, L.; Zhang, Y.; Huang, X.; Wei, X.; Zhou, P.; Liu, M.; Zhao, G.; et al. Nectin-4 Promotes Osteosarcoma Progression and Metastasis through Activating PI3K/AKT/NF-KB Signaling by down-Regulation of MiR-520c-3p. *Cancer Cell Int.* **2022**, *22*, 1–18. [CrossRef]
110. Fabre-Lafay, S.; Garrido-Urbani, S.; Reymond, N.; Gonçalves, A.; Dubreuil, P.; Lopez, M. Nectin-4, a New Serological Breast Cancer Marker, Is a Substrate for Tumor Necrosis Factor-α-Converting Enzyme (TACE)/ADAM-17. *J. Biol. Chem.* **2005**, *280*, 19543–19550. [CrossRef]
111. Siddharth, S.; Nayak, A.; Das, S.; Nayak, D.; Panda, J.; Wyatt, M.D.; Kundu, C.N. The Soluble Nectin-4 Ecto-Domain Promotes Breast Cancer Induced Angiogenesis via Endothelial Integrin-B4. *Int. J. Biochem. Cell Biol.* **2018**, *102*, 151–160. [CrossRef]
112. Chatterjee, S.; Kundu, C.N. Nanoformulated Quinacrine Regulates NECTIN-4 Domain Specific Functions in Cervical Cancer Stem Cells. *Eur. J. Pharmacol.* **2020**, *883*, 173308. [CrossRef] [PubMed]
113. Chatterjee, S.; Sinha, S.; Molla, S.; Hembram, K.C.; Kundu, C.N. PARP Inhibitor Veliparib (ABT-888) Enhances the Anti-Angiogenic Potentiality of Curcumin through Deregulation of NECTIN-4 in Oral Cancer: Role of Nitric Oxide (NO). *Cell Signal.* **2021**, *80*, 109902. [CrossRef] [PubMed]
114. Dosch, J.; Ziemke, E.; Wan, S.; Luker, K.; Welling, T.; Hardiman, K.; Fearon, E.; Thomas, S.; Flynn, M.; Rios-Doria, J.; et al. Targeting ADAM17 Inhibits Human Colorectal Adenocarcinoma Progression and Tumor-Initiating Cell Frequency. *Oncotarget* **2017**, *8*, 65090–65099. [CrossRef] [PubMed]
115. Dobert, J.P.; Cabron, A.S.; Arnold, P.; Pavlenko, E.; Rose-John, S.; Zunke, F. Functional Characterization of Colon-Cancer-Associated Variants in Adam17 Affecting the Catalytic Domain. *Biomedicines* **2020**, *8*, 463. [CrossRef] [PubMed]
116. Kedashiro, S.; Sugiura, A.; Mizutani, K.; Takai, Y. Nectin-4 Cis-Interacts with ErbB2 and Its Trastuzumab-Resistant Splice Variants, Enhancing Their Activation and DNA Synthesis. *Sci. Rep.* **2019**, *9*, 18997. [CrossRef]
117. Kedashiro, S.; Kameyama, T.; Mizutani, K.; Takai, Y. Nectin-4 and P95-ErbB2 Cooperatively Regulate Hippo Signaling-Dependent SOX2 Gene Expression, Enhancing Anchorage-Independent T47D Cell Proliferation. *Sci. Rep.* **2021**, *11*, 7344. [CrossRef]
118. Zhu, Y.; Huang, S.; Chen, S.; Chen, J.; Wang, Z.; Wang, Y.; Zheng, H. SOX2 Promotes Chemoresistance, Cancer Stem Cells Properties, and Epithelial–Mesenchymal Transition by β-Catenin and Beclin1/Autophagy Signaling in Colorectal Cancer. *Cell Death Dis.* **2021**, *12*, 449. [CrossRef]
119. Sun, B.; Zhang, D.; Zhao, N.; Zhao, X. Epithelial-to-Endothelial Transition and Cancer Stem Cells: Two Cornerstones of Vasculogenic Mimicry in Malignant Tumors. *Oncotarget* **2017**, *8*, 30502–30510. [CrossRef]
120. Zhang, J.; Qiao, L.; Liang, N.; Xie, J.; Luo, H.; Deng, G.; Zhang, J. Vasculogenic Mimicry and Tumor Metastasis. *J. BUON* **2016**, *21*, 533–541.

121. Cui, K.; Zhao, W.; Wang, C.; Wang, A.; Zhang, B.; Zhou, W.; Yu, J.; Sun, Z.; Li, S. The CXCR4-CXCL12 Pathway Facilitates the Progression of Pancreatic Cancer via Induction of Angiogenesis and Lymphangiogenesis. *J. Surg. Res.* **2011**, *171*, 143–150. [CrossRef]
122. Huang, C.; Chen, Y. Lymphangiogenesis and Colorectal Cancer. *Saudi Med. J.* **2017**, *38*, 237–244. [CrossRef] [PubMed]
123. Lin, L.T.; Richardson, C.D. The Host Cell Receptors for Measles Virus and Their Interaction with the Viral Hemagglutinin (H) Protein. *Viruses* **2016**, *8*, 250. [CrossRef] [PubMed]
124. Li, H.; Peng, K.-W.; Russell, S.J. Oncolytic Measles Virus Encoding Thyroidal Sodium Iodide Symporter for Squamous Cell Cancer of the Head and Neck Radiovirotherapy. *Hum. Gene Ther.* **2012**, *23*, 295–301. [CrossRef] [PubMed]
125. Sugiyama, T.; Yoneda, M.; Kuraishi, T.; Hattori, S.; Inoue, Y.; Sato, H.; Kai, C. Measles Virus Selectively Blind to Signaling Lymphocyte Activation Molecule as a Novel Oncolytic Virus for Breast Cancer Treatment. *Gene Ther.* **2013**, *20*, 338–347. [CrossRef]
126. Heath, E.I.; Rosenberg, J.E. The Biology and Rationale of Targeting Nectin-4 in Urothelial Carcinoma. *Nat. Rev. Urol* **2021**, *18*, 93–103. [CrossRef]
127. Tarantino, P.; Carmagnani Pestana, R.; Corti, C.; Modi, S.; Bardia, A.; Tolaney, S.M.; Cortes, J.; Soria, J.; Curigliano, G. Antibody–Drug Conjugates: Smart Chemotherapy Delivery across Tumor Histologies. *CA Cancer J. Clin.* **2022**, *72*, 165–182. [CrossRef]
128. Eder, M.; Pavan, S.; Bauder-Wüst, U.; van Rietschoten, K.; Baranski, A.C.; Harrison, H.; Campbell, S.; Stace, C.L.; Walker, E.H.; Chen, L.; et al. Bicyclic Peptides as a New Modality for Imaging and Targeting of Proteins Overexpressed by Tumors. *Cancer Res.* **2019**, *79*, 841–852. [CrossRef]
129. Gemma, E.M.; Scott, H.; Chen, L.; van Rietschoten, G.I.-B.K.; Dzionek, K.; Brown, A.; Watcham, S.; White, L.; Park, F.U.; Jeffreyc, P.; et al. Discovery of BT8009: A Nectin-4 Targeting Bicycle Toxin Conjugate for the Treatment of Cancer. *J. Med. Chem* **2022**, *65*, 14337–14347. [CrossRef]
130. Shao, F.; Pan, Z.; Long, Y.; Zhu, Z.; Wang, K.; Ji, H.; Zhu, K.; Song, W.; Song, Y.; Song, X.; et al. Nectin-4-Targeted ImmunoSPECT/CT Imaging and Photothermal Therapy of Triple-Negative Breast Cancer. *J. Nanobiotechnol.* **2022**, *20*, 243. [CrossRef]
131. O'Donnell, J.S.; Madore, J.; Li, X.Y.; Smyth, M.J. Tumor Intrinsic and Extrinsic Immune Functions of CD155. *Semin. Cancer Biol.* **2020**, *65*, 189–196. [CrossRef]
132. Bowers, J.R.; Readler, J.M.; Sharma, P.; Excoffon, K.J.D.A. Poliovirus Receptor: More than a Simple Viral Receptor. *Virus Res.* **2017**, *242*, 1–6. [CrossRef] [PubMed]
133. Gorvel, L.; Olive, D. Targeting the "PVR-TIGIT Axis" with Immune Checkpoint Therapies. *F1000Research* **2020**, *9*, 354. [CrossRef] [PubMed]
134. Molfetta, R.; Zitti, B.; Lecce, M.; Milito, N.D.; Stabile, H.; Fionda, C.; Cippitelli, M.; Gismondi, A.; Santoni, A.; Paolini, R. Cd155: A Multi-Functional Molecule in Tumor Progression. *Int. J. Mol. Sci.* **2020**, *21*, 922. [CrossRef] [PubMed]
135. Zheng, Q.; Wang, B.; Gao, J.; Xin, N.; Wang, W.; Song, X.; Shao, Y.; Zhao, C. CD155 Knockdown Promotes Apoptosis via AKT/Bcl-2/Bax in Colon Cancer Cells. *J. Cell Mol. Med.* **2018**, *22*, 131–140. [CrossRef]
136. Masson, D.; Jarry, A.; Baury, B.; Blanchardie, P.; Laboisse, C.; Lustenberger, P.; Denis, M.G. Overexpression of the CD155 Gene in Human Colorectal Carcinoma. *Gut* **2001**, *49*, 236–240. [CrossRef]

diagnostics

MDPI

Review

Bronchoalveolar Lavage Fluid-Isolated Biomarkers for the Diagnostic and Prognostic Assessment of Lung Cancer

Alexandros Kalkanis [1,*], Dimitrios Papadopoulos [1], Dries Testelmans [1], Alexandra Kopitopoulou [2], Eva Boeykens [2] and Els Wauters [1]

1 Department of Respiratory Diseases, University Hospitals Leuven, Campus Gasthuisberg, 3000 Leuven, Belgium
2 Department of Pulmonology, General Hospital Rivierenland, Campus Bornem, 2880 Bornem, Belgium
* Correspondence: alexandros.kalkanis@uzleuven.be

Abstract: Lung cancer is considered one of the most fatal malignant neoplasms because of its late detection. Detecting molecular markers in samples from routine bronchoscopy, including many liquid-based cytology procedures, such as bronchoalveolar lavage fluid (BALF), could serve as a favorable technique to enhance the efficiency of a lung cancer diagnosis. BALF analysis is a promising approach to evaluating the tumor progression microenvironment. BALF's cellular and non-cellular components dictate the inflammatory response in a cancer-proliferating microenvironment. Furthermore, it is an essential material for detecting clinically significant predictive and prognostic biomarkers that may aid in guiding treatment choices and evaluating therapy-induced toxicities in lung cancer. In the present article, we have reviewed recent literature about the utility of BALF analysis for detecting markers in different stages of tumor cell metabolism, employing either specific biomarker assays or broader omics approaches.

Keywords: biomarkers; bronchoalveolar lavage fluid; lung neoplasm; tumor microenvironment

Citation: Kalkanis, A.; Papadopoulos, D.; Testelmans, D.; Kopitopoulou, A.; Boeykens, E.; Wauters, E. Bronchoalveolar Lavage Fluid-Isolated Biomarkers for the Diagnostic and Prognostic Assessment of Lung Cancer. *Diagnostics* **2022**, *12*, 2949. https://doi.org/10.3390/diagnostics12122949

Academic Editors: Yuli Huang, Yong Yuan and Peisong Chen

Received: 31 October 2022
Accepted: 22 November 2022
Published: 25 November 2022

1. Introduction

Lung cancer is among the most common malignant neoplasms and is associated with a poor prognosis since it is usually diagnosed at an advanced stage [1]. However, recent advances in the field of systemic treatment have changed the landscape of the disease. Much of this progress has been achieved by the discovery and utility of predictive biomarkers, whose detection may guide targeted therapies in an attempt to a more personalized approach to treatment. In cancer research, a biomarker is a biological molecule whose levels can be measured in tissue or fluids and indicate the presence or progress of the disease or its response to specific therapies [2]. Thus, biomarkers may aid in determining the diagnosis, prognosis, and prediction of treatment response in cancer patients. Any component of the cancer cell metabolic pathway, starting from gene alterations and ending with the various metabolites, can be tested as a biomarker according to its biological role. In recent years, the contemporary measurement of multiple substances belonging to the same level in the metabolic cascade has been possible with the use of omics technologies, making feasible the identification of biomarker profiles or signatures [2].

A major problem in employing lung cancer biomarker testing in everyday clinical practice is obtaining adequate tissue samples. Most patients are diagnosed based on small bioptic or even more limited cytological materials obtained during bronchoscopy or transthoracic aspiration. These materials are often depleted during standard histological and histochemical procedures used for diagnosis and tumor subtyping, leaving no space for biomarker testing. Fluids, on the other hand, are promising materials for biomarker evaluation since most substances, from DNA to proteins and metabolites, are being excreted from cancer cells in soluble form. Bronchial or bronchoalveolar lavage fluid (BALF) is obtained by a lung irrigation technique routinely performed during bronchoscopy. It allows the harvesting of cellular and non-cellular contents of the bronchial and alveolar space, serving as an excellent marker of the tumor microenvironment [3]. Its collection requires a minimally invasive procedure that can be repeated with minimum risk, providing a valid and adequate medium for biomarker testing in lung cancer. In addition, due to its proximity to the neoplastic tissue, it could have higher sensitivity for biomarker detection, especially in locally advanced non-metastasized lung cancer, compared to other fluids commonly employed for the same reason, such as plasma or pleural fluid [4].

BALF analysis in lung cancer has multiple uses. Numerous research studies have demonstrated the benefit of employing BALF for the cytological identification of malignant lung neoplasms. Malignant cells may be shed from an adjacent lung tumor into the respective bronchoalveolar space and identified with conventional cytology. Although BALF cytology alone has a modest sensitivity for diagnosing lung cancer (between 29% and 69%), the specificity is exceptionally high (between 90% and 100%) [5]. Furthermore, its combination with other histological techniques seems to increase the diagnostic yield. The diagnostic accuracy of transbronchial lung biopsy in endoscopically non-visible lesions was enhanced when combined with BALF cytology [6]. Secondly, the parenchymal cells of the lungs secrete different types of soluble substances, which can be identified and measured in BALF after removing the cellular component with centrifugation (supernatant). With the advent of novel molecular techniques, it has become possible to amplify nucleic acids that are either extracted from cells or extracellular vesicles (EVs) or detected in cell-free form on BALF and perform a wide array of genomic, epigenomic, and transcriptomic testing. Finally, BALF analysis may aid in identifying various treatment-related adverse events in the lungs, such as pulmonary infections or pneumonitis.

This review aims to perform a brief and comprehensive overview of studies published in the previous decade that focus on the utility of BALF for detecting different forms of diagnostic, predictive, and prognostic biomarkers in lung cancer. Studies were identified by searching the MEDLINE/PubMed electronic database for relative reports dating from 2011 to September 2022. The search strategy included a combination of MeSH terms (biomarkers; bronchoalveolar lavage; lung neoplasm) and keywords (biomarker; [prognostic or predictive] marker; [bronchoalveolar or bronchial or lung] lavage; [bronchial or lung] washing; [lung or pulmonary] [cancer or neoplasm or tumor]). According to our judgement, some older seminal articles are also mentioned based on their influence on the afterward-growing literature on the subject.

2. BALF Biomarkers for the Diagnosis and Prognosis of Lung Cancer

2.1. Genetic Biomarkers

Recent studies have conducted various assessments on genetic markers at a large, genome-wide scale to identify more specific and sensitive molecules for targeted drug delivery methods to cure the disease. A recent study compared tumor-derived mutations in similar plasma and BALF samples from patients with non-small-cell lung cancer (NSCLC). Researchers discovered that BALF cell-free DNA (cfDNA) testing is more accurate than serum for detecting mutations related to lung cancer [7]. Clinical applications of BALF cfDNA testing include the detection of mutations in affected individuals and, perhaps, the detection of lung adenocarcinoma. It is convenient to identify mutations related to tumors by targeted sequencing of BALF cfDNA, and this method seems more robust compared to plasma screening. Bronchial washings are also a viable and practical choice for genome-wide molecular studies. Actionable mutations in cancer genes were detected with a higher than 90% concordance with respective gene mutations in tumor biopsies [8].

2.1.1. EGFR Mutations

With cfDNA collected through the BALF supernatant, a polymerase chain reaction (PCR) technique may identify functional EGFR mutations. This might be a quick and accurate way to diagnose NSCLC simultaneously using DNA analysis and morphology [9]. The tissue biopsy results are compared to the detectability of activating EGFR mutations in

BALF and blood plasma samples in another study [10]. Under this investigation, BALF was found to be significantly more sensitive than liquid biopsy at detecting an EGFR mutation in the same patients (92.3% vs. 38.5%) [10]. A real-time peptide nucleic acid (PNA)-mediated polymerase chain clamping assay was used to detect the T790M EGFR mutation on both BALF and bronchial biopsy specimens of cancer-afflicted persons [11]. In a different, more thorough investigation, 20 patients with lung adenocarcinoma had their EGFR mutations and T790M mutations molecularly tested using cfDNA from BALF. The combination of PNA-mediated clamping PCR and the PANAMutyperTM R EGFR kit with PNA clamping-assisted fluorescence melting curve analysis produced 91.7% concordance with the results from a tumor biopsy [12]. In another study of patients with advanced NSCLC, cytology samples from BALF were used to identify the T790M EGFR mutation. The findings demonstrated that the clinical benefit of osimertinib treatment was predicted by cytology sample EGFR T790M positivity [13]. Isolated EVs from BALF include DNA that can be used for EGFR genotyping by liquid biopsy. Particularly when compared to liquid biopsy cfDNA, BALF EV DNA is tissue specific and incredibly sensitive. Additionally, tissue re-biopsy is less effective than BALF EV DNA at identifying the T790M mutation in individuals who developed resistance to EGFR tyrosine kinase inhibitors (TKIs) [14].

2.1.2. KRAS Mutations

Clinical uses for KRAS mutation analysis in lung disease include its use as a diagnostic indicator for cancer in sputum and BALF samples [15]. In adenocarcinomas of the lung and related sputum and BALF samples, matched KRAS mutations have been found [16]. Furthermore, it has been demonstrated that patients with negative cytological results may benefit from screening for KRAS mutations in BALF cells to help with a lung cancer diagnosis [17]. KRAS and p53 gene mutations in BALF may serve as helpful biomarkers for the diagnosis of peripheral NSCLC, according to the findings of a recent study [18].

2.1.3. ALK Translocations

Reverse transcription-PCR has been used to detect ALK translocations in cytology samples having a 97% concordance with tissue samples [19].

2.2. Epigenetic Biomarkers

DNA methylation markers are frequently employed as biomarkers for early diagnosis or recurrence of cancer. DNA methylation occurs when a methyl group is attached to a cytosine base located in a CpG dinucleotide, which may impact gene transcription. Aberrant DNA methylation can reduce the expression of tumor-suppressor genes and allow cancer cells proliferation. Kim et al. discovered that p16, RASSF1A, H-cadherin, and RAR beta gene methylation in tumors may be valuable biomarkers for the early diagnosis of NSCLC in BALF [20]. Moreover, several other studies also reported the potential application value of p16 promoter methylation in sputum for lung cancer diagnosis [21,22]. A large retrospective cohort showed improved diagnostic efficacy of DNA methylation biomarkers (DNA methylation of p16, TERT, WT1, and RASSF1 gene) in cytological bronchial washings evaluation [23]. When lung cancer patients have bronchial aspirates, the Epi pro-Lung BL Reflex Assay is a powerful and practical diagnostic technique for detecting elevated DNA methylation of the SHOX2 gene locus [24]. In comparison to conventional cytology analysis and serum CEA, the methylation analysis of the SHOX2 and RASSF1A panel in BALF using RT-PCR produced better diagnostic sensitivity (81.0%) and specificity (97.4%) [25]. NSCLC could be found by examining the DNA methylation of 7CpGs (TBX15, PHF11, TOX2, PRR15, TFAP2A, HOXA1, and PDGFRA genes) in bronchial washings [26]. A 5-marker (LHX9, GHSR, HOXA11, PTGER4-2, HOXB4-3) methylation model was recently found to have 70% sensitivity and 82% specificity for discriminating malignant from benign pulmonary nodules [27]. According to a meta-analysis, the detection of DNA methylation of the $p16^{INK4a}$ gene in BALF may be a potential biomarker for NSCLC diagnosis [28].

2.3. Post-Transcriptional Biomarkers

Transcriptome comparison allows for the discovery of genes that are differently expressed in various cell groups or in response to different treatments. Researchers have showed that clusters of deregulated miRNAs were expressed in the BALF of smoking-related diseases such as lung adenocarcinoma and COPD [29]. Similarly, the expression of two anti-apoptotic genes, survivin and livin, through their mRNA levels was observed in lung cancer patients versus patients with benign lung disease, using bronchial aspirates as a sample source [30]. Other studies have broadened the target transcriptomes to other specific panels of miRNAs that were significantly upregulated in patients with lung cancer [31,32]. Cluster analysis of the expression levels of a miRNA panel in BALF samples from NSCLC patients provided a diagnostic sensitivity of 85.7% and specificity of 100% [33]. Moreover, a prediction model of specific miRNA biomarkers from liquid cytological specimens has shown potential in discriminating squamous cell carcinoma from adenocarcinoma [34]. Transcriptome signatures can also be potentially helpful in predicting survival for patients with lung cancer. Patients with a low expression level of a 3-miRNA panel were associated with better overall survival [35]. Evaluation of EVs from plasma and BALF in patients with NSCLC and benign lung diseases revealed significant differences in exosome amount and miRNA content between fluids and patient groups, revealing their specific role as biomarkers [36]. A significant increase in the expression profile of immunoglobulin genes in BALF was found between lung cancer and healthy controls in a recent transcriptomics study, which generated a 53-gene signature that showed a significant correlation with inhibitory checkpoint PDCD1 [37].

2.4. Post-Translational Biomarkers

2.4.1. Proteins

Proteins in BALF can represent the physiological and pathological state of the lung. Proteins that lung cancer cells secrete into BALF may be utilized as biomarkers to detect and evaluate malignancy, aiding lung cancer diagnosis, prognosis, subtyping, and therapy response monitoring. Researchers have since long identified a link between bronchial secretions and serum tumor marker concentrations [38]. Among many proteins, the most specific is napsin A, a renowned immunohistochemical marker for lung cancer diagnosis in affected individuals [39]. In addition, levels of several BALF proteins located in antioxidant, inflammatory, or anti-angiogenic pathways, such as cytokines [40–42] and growth [43–45], pro-angiogenic [46–48], or complement [49–51] factors, have been evaluated as diagnostic or prognostic biomarkers employing antibody-specific immunoassays.

Combining proteomics, mass spectrometry (MS), and affinity chemistry-based methods have significantly improved our understanding of the protein oxidative changes that take place in many biological specimens under varied physiological and pathological circumstances in recent years (Figure 1). Glycoproteins are crucial in the standard processing of biological processes such as cell separation, growth, their close contact with their vicinity and invading of tumor cells into surrounding cells. A cohort study used solid-phase N-proteoglycans extraction, iTRAQ labeling, and liquid chromatography-tandem MS to analyze N-proteoglycans levels in BALF from patients with lung cancer and benign lung diseases. Levels of 8 glycoproteins (neutrophil elastase, integrin alpha-M, cullin-4B, napsin A, lysosome-associated membrane protein 2, cathepsin D, BPI fold-containing family B member 2, and neutrophil gelatinase-associated lipocalin) displayed more than two times rise in cancer BALF compared to benign BALF [52]. According to research by Uribarri et al. the expression levels of APOA1, CO4A, CRP, GSTP1, and SAMP led to a diagnostic panel for lung cancer that was 95% sensitive and 81% specific. The measurement of STMN1 and GSTP1 proteins enabled the two main subtypes of lung cancer (non-small-cell and small-cell) to be distinguished with 57% specificity and 90% sensitivity [53]. Ortea et al. identified different proteins in the BALF of patients with lung cancer, including glutathione S-transferase pi, haptoglobin, and complement C4-A. These different types of proteins could be proved as prominent biomarkers for disease diagnosis [54]. Almatroodi et al. performed a quantitative proteomic evaluative study on BALF of patients with lung adenocarcinoma, and their results indicated the occurrence of 1100 proteins in BALF. Among them, the ratio of over-expressed proteins in lung adenocarcinoma individuals were 33 compared to healthy individuals. S100-A8, thymidine phosphorylase, annexin A2, transglutaminase 2, and annexin A1 were lung cancer individuals' most renowned over-expressed proteins [55]. In another prospective cohort of individuals with suspected lung cancer, proteomic analysis of BALF samples identified 133 proteins that could differentiate cancer and non-cancer patients [56]. Using label-free MS analysis of BALF, Hmmier et al. discovered that 4 proteins (cystatin-C, TIMP1, lipocalin 2, and HSP70/HSPA1A) were significantly overexpressed in lung cancer patients compared to healthy controls [57]. A recent investigation evaluated aberrant protein glycosylation using lectin microarrays in BALF. It revealed 15 lectins that could distinguish between the different lung cancer types and 14 lectins whose levels differed between early and advanced disease stages [58].

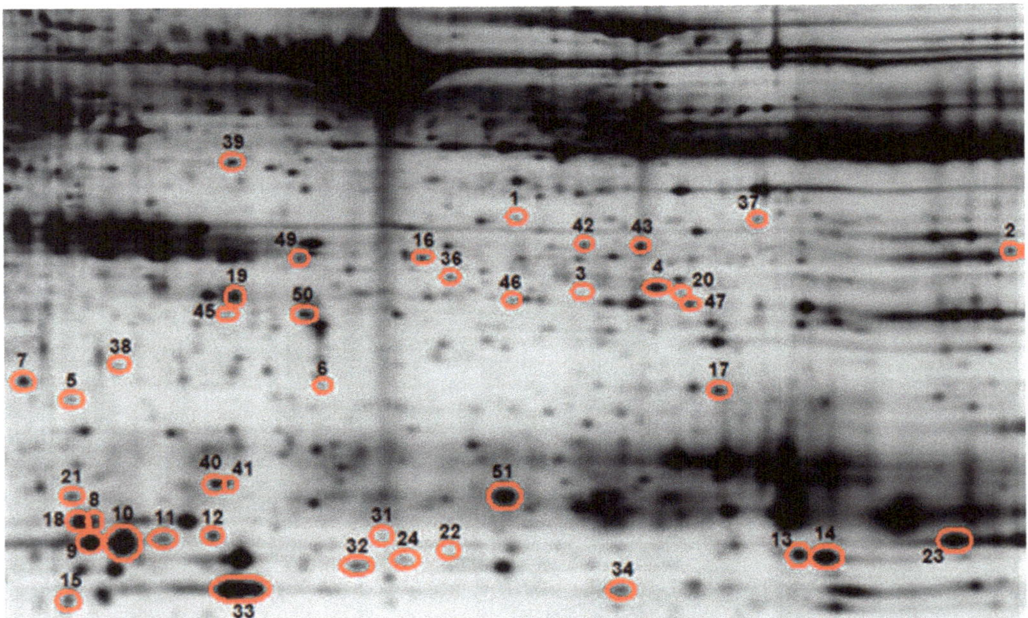

Figure 1. Example of a proteome from a bronchoalveolar lavage fluid sample in a lung cancer patient performed by two-dimensional gel electrophoresis and silver staining. Circled spots represent differentially expressed proteins between lung cancer and controls. Reprinted from [53] with permission from Elsevier.

2.4.2. Cell Epitopes

The combination of immunoassays and flow cytometry in BALF supernatant enables the study of cell epitopes, mainly used in identifying immune cell subpopulations (Figure 2). T cells are essential for antitumor defense, but as cancer progresses, their population changes. Numerous suppressory and regulatory systems can prevent lymphocyte activation and prevent the identification of lung cancer antigens. Various research studies indicated that now lung cancer immunotherapeutic therapy aims to enhance the cytotoxological action of lymphocytes by restricting the activities of suppressor molecules such as cytotoxic T cell antigen 4 (CTLA-4) and programmed death-1 (PD-1) [59]. The examination of BALF revealed that one could easily analyze the different types of immune responses in the tumor microenvironment as well as an alteration in the status of various cells. A higher ratio of cytotoxic CD8+ lymphocytes, neutrophils, T cells, and CTLA-4+ T regulatory cells was found in BALF of the cancer-affected lung compared to the healthy contralateral lung [60]. Moreover, tissue samples of lung adenocarcinoma patients also indicated the presence of M2 polarization of macrophages, which are further associated with the onset of lung cancer [61]. Osińska et al. examined the BALF of 35 patients with lung cancer to figure out the existence of regulatory T cells (Tregs) and their specific role in boosting immunity. Flow cytometry analysis revealed that Treg cells were present in significant proportion in the local environment of lung cancer cells as compared to normal cells. These Tregs are renowned as anticancer defense cells because they have the potential to restrict the normal functioning of NK cells, T-lymphocytes, and dendritic cells, which ultimately enhances the immune tolerance level [62]. In 2018, Hu et al. examined the significant ratio of PD-1+ cells in BALF and peripheral blood in small-cell lung cancer (SCLC) individuals, and their levels dropped after chemotherapy. Therefore, these cells would be potential biomarkers for the diagnosis of diseases as these programmed death cells are recognized as checkpoints in the immune system, and their inhibition could potentially mediate the activation of T cells, ultimately exhibiting antitumor activity [63]. Moreover, another rational study was carried out on BALF as a prognostic agent for patients of lung cancer, and through a cytometric bead array, Hu et al. identified a higher level of interleukin (IL)-10 and IL-10+CD206+CD14+M2-like macrophages in the SCLC affected individuals as compared to NSCLC individuals. These markers were also correlated with advanced stage and reduced survival of SCLC patients [64]. Recently, a cohort study was conducted by Masuhiro et al. to examine the immune profile of tumor microenvironment through BALF of NSCLC individuals according to their response to immunotherapy. They observed a higher level of CXCL9 and a higher ratio of CD56+ cytotoxic T cells in BALF of patients responding to nivolumab compared to those not responding [65].

Some cells disassociate themselves from the primary tumor mass and enter the circulatory system. These cells are known as circulating tumor cells (CTCs) and emerge as a new target that offers clinical insights for the prediction and diagnosis of different types of cancers [66]. Recent studies have demonstrated that the precision and sensitivity of CTCs detected by the chromosomal enumeration probe 8 (CEP8) in lung cancer patients was, correspondingly, 83.3% and 98.6% [67]. Several studies have investigated these distinct CEP8+ CTCs, and the results have demonstrated a substantial correlation between these cells with the detection of lung cancer and also its prognosis with excellent precision and sensitivity [68,69]. Detection of CTCs in serum and BALF showed that CEP8+ CTCs could be used as a supplementary approach for individuals with solitary pulmonary nodules to diagnose lung cancer at an early stage [70].

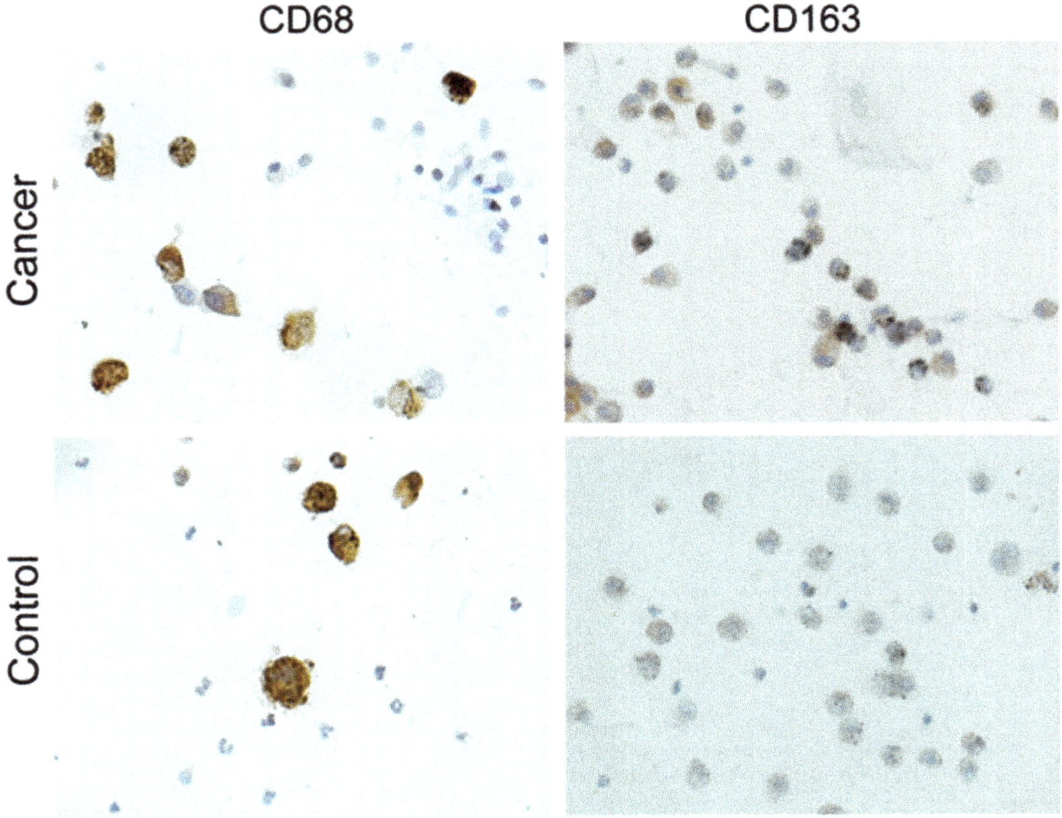

Figure 2. Immunostaining for cell epitopes in macrophages from bronchoalveolar lavage fluid samples in a lung cancer patient and healthy control. Reprinted from [48] with permission from Elsevier.

2.5. Metabolite Biomarkers

Metabolites of tumor cells, such as amino acids, nucleic acids, lipids, and sugars, may reflect cellular metabolic processes that drive tumor formation and progression. Either by using metabolite-specific immunoassays or MS-based metabolomic approaches (Figure 3), recent studies have compared metabolites between lung cancer patients and controls. In an early study, levels of leukotriene B4 and cysteinyl leukotrienes, which are end products of arachidonic acid metabolism, were significantly elevated in BALF of NSCLC patients [71]. In another report, patients with NSCLC had significantly lower ATP and ADP concentrations in BALF than patients with COPD, an effect that was exerted by upregulated expression of P2Y1, P2X4, and P2X7 purinergic receptors [72]. Callejón-Leblic et al. identified 42 altered metabolites in BALF of lung cancer patients compared to non-cancer patients. Based on ROC curve analysis, levels of carnitine, adenine, choline, glycerol, and phosphoric acid show high sensitivity and specificity to distinguish between lung cancer and controls and should be considered as potential biomarkers [73].

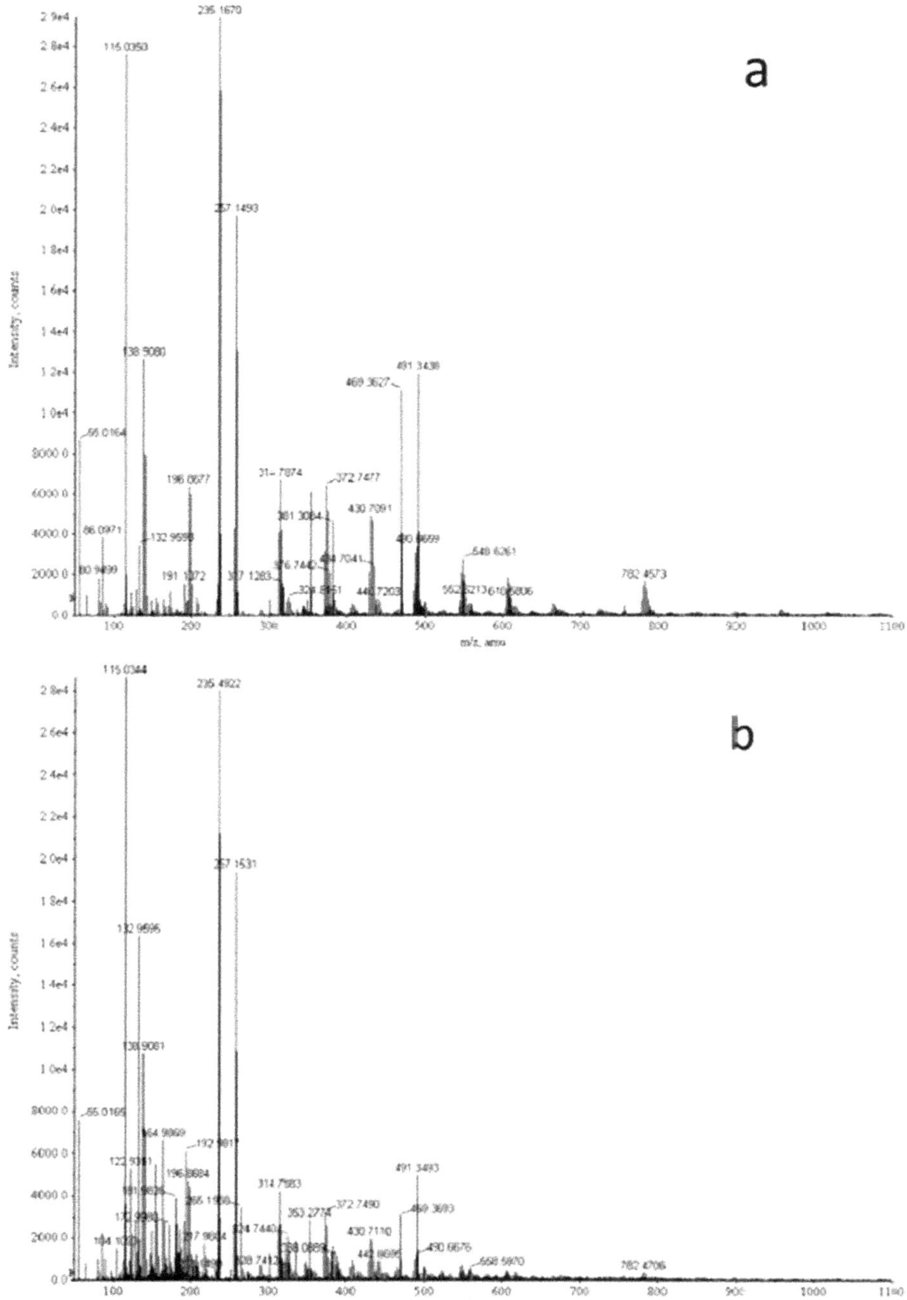

Figure 3. Typical mass spectra from lung cancer patients (**a**) and controls (**b**) by direct infusion triple quadrupole time-of-flight mass spectrometry in bronchoalveolar lavage fluid for the detection of metabolite expressions. Reprinted from [73] with permission from Elsevier.

Certain elements such as essential and non-essential metals (Fe, Cd, Zn, Co, Cu, Pb, V, Cr, etc.) are important for maintaining homeostasis and are involved in several cellular pathways. Excess or deficiency of such elements causes dyshomeostasis and disturbs cellular pathways, ultimately leading to the progression of serious illness [74]. So, identification of levels of such elements in body fluids such as serum and BALF can help to understand the mechanisms involved in the progression of serious illnesses such as lung cancer. As a result, variations in the particular elements' concentrations and the profile of their chemical components can reveal the individual's metabolic and nutritional status and therefore offers a prospective early prognosis of cancer progression. Belén et al. used triple quadrupole inductively coupled plasma MS analysis and identified the presence of metals in body fluids, including BALF, in lung cancer patients and concluded that such metals could be employed as a potent biological marker for more accurate and expeditious identification of lung adenocarcinoma in affected individuals [75].

2.6. Metagenomic Biomarkers

Microbiome generally contributes towards barrier formation, external communication, and immune homeostasis in healthy individuals. The microbiome can also trigger host immune responses against cancer cells. Several studies have shown the interconnectedness between lung cancer and the microbiome. For example, Wang et al. conducted a preliminary analysis of microbiome diversity in saliva and BALF of lung cancer patients [76]. They observed that the microbiome diversity is significantly less than in the healthy samples, indicating a link between cancer and microbiome. It can also be said that certain microbiome genera could act as potential lung cancer biomarkers. In this regard, Cheng et al. characterized microbiomes in BALF for potential associations with lung cancer [77]. The microbiome diversity was compared between diseased and healthy samples using principal coordinate analysis. They observed six different genera that were significantly enriched in BALF of cancer patients: Blautia, Capnocytophaga, Gemmiger, Oscillospira, Sediminibacterium, and TM7-3. Similarly, Lee et al. investigated the microbiome in BALF of the affected group and compared them with benign lung mass [78]. They reported that members of the genera Megasphaera and Veillonella are potential cancer markers as their concentrations are relatively higher in cancer patients than in the control group. In addition, Patnaik et al. observed that the lower airway microbiome could contribute to the recurrence of early-stage NSCLC [79]. They observed that the microbiome diversity differed significantly between patients with recurrence and non-recurrence after surgery. They also observed that the diversity among 16 out of 18 cancer recurrence patients was similar. Zheng et al. performed metagenomic sequencing on BALF samples from cancer patients and healthy controls [80]. Even though the diversity between both samples was comparable, some rare microbiota stood out in lung cancer samples. For example, species such as Bacteroides pyogenes, Chaetomium globosum, Lactobacillus rossiae, Magnetospirillum gryphiswaldense, Paenibacillus odorifer, and Pseudomonas entomophila were common in cancer samples. The results indicate that change in the microbiome's diversity is directly associated with cancer. Moreover, Veillonella was found to increase tumor burden and decrease survival. Furthermore, the microbiome plays a crucial role in cancer cells' response to immunotherapy. For example, Jang et al. analyzed the microbiome's relationship with programmed death-ligand-1 (PD-L1) expression [81]. PD-L1 protein is typically involved in protecting non-cancerous/non-harmful cells from immune cells. The authors also found that the cells with higher PD-L1 expression levels respond better to immunotherapy than the cells with lower PD-L1 expression levels. The population of Veillonella was significantly higher in the group with the higher expression level of PD-L1, while Neisseria was abundant in the group with the lower expression of PD-L1. Moreover, the exact reason for a higher amount of Veillonella in cancer patients is still unknown, but several studies have developed links between different lung-related diseases and the microbiome genera under debate. For example, Veillonella has been found to be involved in pulmonary fibrosis and asthma development [82,83]. As both pulmonary disorders are related to the immune system and lungs, it is also expected that Veillonella will be higher in cancer patients as well.

3. BALF Biomarkers for the Identification of Adverse Events of Lung Cancer Treatment

3.1. Pulmonary Infections

It has been investigated that the risk of developing neutropenia is usually low in the case of lung adenocarcinoma immunotherapeutic procedures compared to chemotherapy [84]. Moreover, long-term immunosuppression therapies with high drug dosage (steroids, TNF-α) are required for immune-related adverse events (irAEs), e.g., colitis and pneumonitis, that ultimately enhance the risk of other infections [85]. Aspergillus fumigatus pneumonia diagnosed after BALF examination has been described in a metastatic melanoma patient treated with systemic corticosteroids and infliximab for an irAE due to ipilimumab [86]. Another report described the detection of Mycobacterium tuberculosis in BALF in a NSCLC patient during third-line nivolumab therapy [87]. Thus, the performance of bronchoscopy and subsequent serological and molecular testing of BALF is an established method for the diagnosis of such infectious complications in this unique patient population [88].

3.2. Therapy-Induced Pulmonary Toxicity

It has been observed that excess application of immunotherapeutic molecules has important side effects; among them, checkpoint inhibitor pneumonitis (CIP) is the most famous. Common clinical presentations of CIP are the onset of dyspnea, hypoxemia, and pulmonary infiltration [89,90]. For advanced NSCLC, the phase I trial of pembrolizumab and nivolumab reported 1 and 3 pneumonitis-related deaths, respectively [91,92]. The extensive examination of BALF showed infiltration and inflammation of lymphocytes in patients with CIP. In 2017, another case study reported T-lymphocytic alveolitis as extrinsic allergic alveolitis in 80% of affected individuals with CD8+ cell predominance [93]. Identifying the dysregulation of immune cell subpopulations in BALF could serve as a promising target for developing therapeutics to minimize adverse effects [94].

Drug-induced sarcoidosis reaction (DISR) can be another after-effect of cancer-related therapies. DISR is a systematic granulomatous reaction that arises after using chemotherapeutic agents and is indistinguishable from sarcoidosis [95]. DISR has been described after treatment with immune checkpoint inhibitors, including PD-L1 inhibitors, PD-1 inhibitors, and anti-CTLA-4 antibodies [96]. Thus, in the absence of biopsy specimens, the surrogate method for the diagnosis of DISR in affected patients is BALF sampling [97]. Montaudie et al. conducted research work, and they presented a DISR-based study that showed 32% lymphocytes along with the enhanced frequency of CD4/CD8 cells [98].

Radiation-induced pneumonitis is an inflammatory process occurring in reaction to chest radiotherapy used to treat thoracic malignancies. Early identification and treatment are crucial, considering the fact that its chronic phase constitutes lung fibrosis. Previous research has detected different types of serum biological markers, including TGF-β, IL-6, TNF-α, and IL-1, that predicted tissue injuries after radiotherapy [99,100]. Later, a series of research studies were performed by Kwang-Joo Park et al. on animal models for the evaluation of radiotherapy-associated injuries of lung tissues, in which the focus was on BALF. Their results revealed that the level of IL-1, NO, TGF-β, and neutrophils were high in BALF [101,102]. Additionally, lymphocytic alveolitis was also found in BALF of affected individuals who took radiotherapy for breast cancer. In most symptomatic cases, a significant rise in CD4+ T lymphocytes was observed, but no increase in CD4/CD8 ratio was found [103]. Crohns et al. reported that radiation therapy resulted in a significant increase in IL-6 in BALF, while elevated IL-8 levels in BALF are associated with worse survival in lung cancer patients receiving radiotherapy [104]. BALF was found to be more effective as compared to serum for the detection of surfactant protein D, which is an important biomarker for radiation-induced pneumonitis in tumor tissues of affected persons [105]. In a cohort study, the profile of cytokines or chemokines was analyzed in

BALF samples from cancer-afflicted individuals before and after radiotherapy. This study showed the overexpression of cytokines, including PAI-1, CD154, IL-1ra, CXCL-1, IL-23, MIF, and IFN-γ, in the lungs of patients with high-grade radiation-induced pneumonitis even before radiotherapy [106].

4. The Value of Repeated BALF Examination in the Course of Lung Cancer

With the advent of modern molecular technologies and extensive research work in biological sciences, highly curable therapies for patients with lung cancer have been recently introduced. These treatment methods wrapped new therapeutic molecules as well as more specific biomarkers for the diagnosis of cancer tissues found in different sections of the lungs more accurately. Apart from their highly beneficial results, the continuous re-evaluation of cancer patients is necessary to gauge the tumor status before and after the application of each line of therapy. It has been proposed that the follow-up should not only be restricted to imaging but repeated histological assessment may be needed when relapses occur after therapeutic interventions. Zarogoulidis et al. performed a case study, and their results suggested that re-biopsy of different lung cancer tissues after the administration of TKIs would detect T790M mutations as well as an altered type of lung cancer [107]. Another cohort study reported that 63% of NSCLC patients showed onset of EGFR mutations and proliferation of cancer after treatment with EGFR-TKIs during re-biopsy, while a 33% frequency of T790M mutation was found [108]. Taking into account the low invasiveness and high sensitivity of BALF testing, a surrogate method of the classic re-biopsy technique could be repeated BALF assessment, not only for the detection of T790M mutations but also for a re-evaluation of biomarkers that may predict cancer recurrence or prognosis and guide further treatment choices.

5. Advantages and Disadvantages of BALF Biomarker Testing in Lung Cancer

Biomarker testing is a central component in nowadays personalized management of lung cancer patients. Many approved drugs for the treatment of stage IV NSCLC are being prescribed according to the detection and quantification of specific molecular markers, such as EGFR, ALK, ROS1, and BRAF mutations and PD-L1 expression. In the case of minimal or no tumor tissue material available for biomarker testing, BALF has the advantage of containing a wide variety of molecules, either cell derived or in soluble form, that resemble the tumor molecular profile. Biomarkers with the highest diagnostic, predictive, and prognostic accuracy from the studies reported above are presented in Table 1. High concordance to tissue markers and better accuracy than plasma in oncogene driver detection makes BALF the best alternative for predictive assessment in lung neoplasms. Panels of BALF-isolated biomarkers of pre- and post-transcriptional gene expression regulation, such as DNA methylation and miRNA expression, have higher diagnostic accuracy than BALF cytology for the detection of lung cancer. Moreover, prediction models derived from proteomic, metabolomic, and microbiomic studies have identified novel markers for the discrimination between cancer types and stages and the prediction of immunotherapy response or tumor progression/recurrence.

Main disadvantages of BALF biomarker testing are the lower yield of tumor cells or tumor nucleic acids compared to tissue samples, which could impede biomarker test performance, and the higher invasiveness compared to plasma sampling. Furthermore, continuous validation of the methylomic, transcriptomic, and proteomic signatures in BALF is needed so that they may assist in the diagnosis of small pulmonary nodules and guide treatment approaches in clinical practice.

Table 1. Overview of lung cancer biomarkers, their measurement methods in bronchoalveolar lavage fluid, their respective diagnostic/predictive roles, and accuracy results from studies included in the review.

Study	Patients	Controls	Processing	Detection	Biomarker	Role	Accuracy
			Genetic/epigenetic biomarkers				
Lee 2020 [4]	*n* = 73 (M 38) Mean age: 65.3 ± 9.8 y Type: AC 65, SQCC 7, other 1 Stage: I 20, II 10, III 10, IV 27	-	Cell-free DNA from BALF supernatant	Droplet digital allele-specific PCR	EGFR L858R mutation	Prediction of tumor molecular profile	AUC 0.96 Acc 95%
					EGFR E19del mutation		AUC 0.86 Acc 85%
Nair 2022 [7]	*n* = 35 (M 18) Age: 40–83 y Type: AC 30, SQCC 2, NSCLC NOS 1, SCC 1, other 1 Stage: I 18, II 6, III 6, IV 5	*n* = 21 (M 12) Age: 46–76 y	Cell-free DNA from BALF supernatant	Deep sequencing	11 gene feature classifier	LC diagnosis	AUC 0.84 Sens 69% Spec 100%
	n = 31	-			Any tumor variant detected	Prediction of tumor molecular profile	Acc 81%
Roncarati 2020 [8]	*n* = 91 (M 60) Age: 47–85 y Type: AC 41, SQCC 31, SCC 11, undefined 7 Stage: I 13, II 7, III 25, IV 43	*n* = 31 (M 21) Age: 42–86 y	Cellular DNA/RNA from BALF cell pellet	Droplet digital methylation-specific PCR	CDH1 methylation	LC diagnosis	Sens 64% Spec 74%
					DLC1 methylation		Sens 37% Spec 94%
					PRPH methylation		Sens 40% Spec 100%
					RASSF1A methylation		Sens 46% Spec 100%
					4-gene methylation panel		AUC 0.93 Sens 97% Spec 74%
		-		Next-generation sequencing	ALK fusions	Prediction of tumor molecular profile	Acc 96%
					BRAF V600E mutation		Acc 100%
					EGFR mutations		Acc 97%
					ERBB2/HER2 mutations		Acc 100%
					KRAS mutations		Acc 90%
					MET mutations		Acc 100%
					ROS1 fusions		Acc 100%
Kawahara 2015 [9]	*n* = 42 (M 23) Age: 42–84 y Type: AC	-	Cell-free DNA from BALF supernatant	Allele-specific PCR/FRET-PHFA	EGFR mutations	Prediction of tumor molecular profile	Acc 47%

Table 1. *Cont.*

Study	Patients	Controls	Processing	Detection	Biomarker	Role	Accuracy
Yanev 2021 [10]	n = 26 (M 13) Mean age: 63.3 y Type: AC	-	Cell-free DNA from BALF supernatant	Allele-specific PCR	EGFR mutations	Prediction of tumor molecular profile	Acc 92%
Park 2017 [12]	n = 20 (M 5) Age: 43–77 y Type: AC Stage: I 1, II 1, III 1, IV 17	-	Cell-free DNA from BALF supernatant	PNA-mediated clamping PCR/ PNA clamping-assisted fluorescence melting curve analysis	EGFR mutations	Prediction of tumor molecular profile	Acc 92%
						Prediction of treatment response (EGFR TKI)	-
Hur 2018 [14]	n = 23	-	EV DNA from BALF EV pellet	PNA-mediated clamping PCR	EGFR mutations	Prediction of tumor molecular profile	Acc 100%
			Cell-free DNA from BALF supernatant				Acc 71%
Ahrendt 1999 [16]	n = 50 Type: AC 25, SQCC 23, other 2 Stage: I 28, II 15, III 7	-	Cellular DNA from BALF cell pellet	Oligonucleotide plaque hybridization	p53 mutations	Prediction of tumor molecular profile	Acc 39%
				Allele-specific PCR	KRAS mutations		Acc 50%
				Methylation-specific PCR	p16 methylation		Acc 63%
				Microsatellite fragment analysis	15 microsatellite markers		Acc 14%
					Combined panel		Acc 53%
Oshita 1999 [17]	n = 20 Type: AC	n = 13	Cellular DNA from BALF cell pellet	Allele-specific PCR	KRAS codon 12 mutation	LC diagnosis	Sens 79% Spec 64%
Li 2014 [18]	n = 48 (M 27) Age: 42–75 y Type: AC 32, SQCC 14, LCC 2 Stage: I 19, II 25, III 4	n = 26 (M 17) Age: 28–70 y	Cellular DNA from BALF cell pellet	PCR single-strand conformation polymorphism	KRAS mutations	LC diagnosis	Sens 38% Spec 92%
						Prediction of tumor molecular profile	Acc 72%
					p53 mutations	LC diagnosis	Sens 44% Spec 96%
						Prediction of tumor molecular profile	Acc 75%

Table 1. *Cont.*

Study	Patients	Controls	Processing	Detection	Biomarker	Role	Accuracy
Li 2014 [18]	*n* = 48 (M 27) Age: 42–75 y Type: AC 32, SQCC 14, LCC 2 Stage: I 19, II 25, III 4	*n* = 26 (M 17) Age: 28–70 y	Cellular DNA from BALF cell pellet	PCR single-strand conformation polymorphism	Combined panel	LC diagnosis	Sens 67% Spec 89%
						Prediction of tumor molecular profile	Acc 70%
Nakamichi 2017 [19]	*n* = 36	-	Cellular RNA from BALF cell pellet	mRNA-specific reverse transcription-PCR	ALK translocations	Prediction of tumor molecular profile	Acc 97%
Kim 2004 [20]	*n* = 85 (M 57) Mean age: 65 ± 17 y Type: AC 31, SQCC 43, other 11 Stage: I 52, II 33	*n* = 127 (M 84) Mean age: 62 ± 14 y	Cellular DNA from BALF cell pellet	Methylation-specific PCR	p16 methylation	LC diagnosis	Sens 16% Spec 94%
					RARβ methylation		Sens 15% Spec 87%
					H-cadherin methylation		Sens 13% Spec 97%
					RASSF1A methylation		Sens 18% Spec 96%
Nikolaidis 2012 [23]	*n* = 139 (M 80) Mean age: 68.4 ± 8.1 y Type: AC 22, SQCC 31, SCC 39, LCC 16, other 20, unknown 11	*n* = 109 (M 63) Mean age: 67.6 ± 8.8 y	Cellular DNA from BALF cell pellet	Methylation-specific PCR	Methylation panel (p16, RASSF1, WT1, TERT)	LC diagnosis	Sens 82% Spec 91%
Dietrich 2012 [24]	*n* = 125 (M 72) Age: 46–85 y Type: AC 26, SQCC 28, NSCLC NOS 9, SCC 40, other 22	*n* = 125 (M 61) Age: 45–86 y	Cellular DNA from BALF cell pellet	Methylation-specific PCR	SHOX2 methylation	LC diagnosis	AUC 0.94 Sens 78% Spec 96%
Zhang 2017 [25]	*n* = 284 (M 212) Age: 31–85 Type: AC 92, SQCC 107, SCC 42, LCC 5, unknown 38 Stage: I 28, II 30, III 133, IV 93	*n* = 38 (M 28) Age: 29–75 y	Cellular DNA from BALF cell pellet	Methylation-specific PCR	Methylation panel (SHOX2, RASSF1A)	LC diagnosis	AUC 0.89 Sens 81% Spec 97%
Um 2018 [26]	*n* = 31	*n* = 10	Cellular DNA from BALF cell pellet	DNA methylation microarray	Methylation panel (TFAP2A, TBX15, PHF11, TOX2, PRR15, PDGFRA, HOXA11)	LC diagnosis	AUC 0.87 Sens 87% Spec 83%
Li 2021 [27]	*n* = 52 (M 33) Type: AC 37, SQCC 10, other 5 Stage: I 34, II 1, III 1, IV 4, unknown 12	*n* = 59 (M 33)	Cellular DNA from BALF cell pellet	Methylation-specific PCR	Methylation panel (LHX9, GHSR, HOXA11, PTGER4-2, HOXB4-3)	LC diagnosis	AUC 0.82 Sens 70% Spec 82%

Table 1. *Cont.*

Study	Patients	Controls	Processing	Detection	Biomarker	Role	Accuracy
			Post-transcriptional biomarkers				
Molina-Pinelo 2014 [29]	*n* = 48 (M 40) Type: AC Stage: I-II 2, III 15, IV 31	*n* = 16 (M 15)	Cellular RNA from BALF cell pellet	MicroRNA microfluidic card array	Four upregulated miRNA clusters (chromosome loci 13q31.3, 7q22.1, Xq26.2, 11q13.1)	LC diagnosis	-
Li 2013 [30]	*n* = 70 (M 52) Mean age: 64 ± 24 y Type: AC 37, SQCC 25, LCC 4, SCC 4 Stage: I 10, II 24, III 25, IV 11	*n* = 26 (M 19) Mean age: 55 ± 19 y	Cellular RNA from BALF cell pellet	mRNA-specific reverse transcription-PCR	Survivin expression ratio > 0.35	LC diagnosis	AUC 0.83 Sens 83% Spec 96%
					Livin expression ratio > 0.3		AUC 0.68 Sens 63% Spec 92%
Rehbein 2015 [31]	*n* = 30 (M 21) Median age 64.5 y	*n* = 30 (M 17) Median age 63.5 y	Cell-free RNA from BALF supernatant	MicroRNA microfluidic card array	Five upregulated miRNAs (U6 snRNA, hsa-miR 1285, hsa-miR 1303, hsa-miR 29a-5p, hsa-miR 650)	LC diagnosis	-
Kim 2018 [32]	*n* = 13 (M 7) Age: 47–72 y Type: AC Stage: I 10, II 3	*n* = 15	EV RNA from BALF EV pellet	miRNA-specific reverse transcription-PCR	miR-126 and Let-7a were significantly upregulated	LC diagnosis	-
Kim 2015 [33]	*n* = 21 (M 17) Age: 46–84 y Type: AC 13, SQCC 5, LCC 3 Stage: I 12, II 9	*n* = 10 (M 8) Age: 30–77 y	Cellular RNA from BALF cell pellet	miRNA-specific reverse transcription-PCR	High expression cluster of a 5-miRNA panel (miR-21, miR-143, miR-155, miR-210, miR-372)	LC diagnosis	Sens 86% Spec 100%
Li 2017 [34]	*n* = 127 (M 82) Age: 66 ± 8 y Type: AC 45, SQCC 82 Stage: I 52, II 38, III-IV 37	-	Cellular RNA from BALF cell pellet	Droplet digital miRNA-specific PCR	2-miRNA (miR-205-5p, miR-944) prediction model	Discrimination of SQCC from AC	AUC 0.997 Sens 95% Spec 97%
Mancuso 2016 [35]	*n* = 50 (M 32) Age: 34–82 y Type: SCC Stage: III 18, IV 32	-	Cellular RNA from BALF cell pellet	miRNA-specific reverse transcription-PCR	Above median expression levels of a 3-miRNA panel (miR-192, miR-200c, miR-205)	Overall survival (worse)	-

Table 1. *Cont.*

Study	Patients	Controls	Processing	Detection	Biomarker	Role	Accuracy
Rodríguez 2014 [36]	n = 30 (M 23) Age: 45–83 y Type: AC 14, SQCC 16	n = 75 (M 46) Age: 18–87 y	EV RNA from BALF EV pellet	MicroRNA real-time PCR array	10 miRNAs were upregulated and 10 down-regulated	LC diagnosis	-
Kuo 2018 [37]	n = 34 (M 19) Mean age: 58.5 ± 12.8 y Type: AC 26, SQCC 8 Stage: III 11, IV 23	n = 14 (M 7) Mean age: 53.3 ± 11.4 y	Cellular RNA from BALF cell pellet	mRNA-specific reverse transcription-PCR	9-gene (SPP1, CEACAM6, MMP7, SLC40A1, IGJ, IGKC, CPA3, YES1, CXCL13) prediction model	LC diagnosis	AUC 0.92
Post-translational biomarkers (proteins, cell epitopes, metabolites)							
Macchia 1987 [38]	n = 37 Type: AC 4, SQCC 23, LCC 3, SCC 7	n = 20	Cell-free BALF supernatant	RIA	CEA	LC diagnosis	Sens 57% Spec 65%
					TPA	LC diagnosis	Sens 65% Spec 20%
					NSE	SCC diagnosis	Sens 71% Spec 90%
					Ferritin	SCC diagnosis	Sens 71% Spec 100%
					CanAg CA-50	SCC diagnosis	Sens 100% Spec 55%
Naumnik 2012 [40]	n = 45 (M 38) Mean age: 61.9 ± 4 y Type: AC 9, SQCC 22, NSCLC NOS 14 Stage: III 18, IV 27	n = 15 (M 13) Mean age: 60.1 ± 5 y	Cell-free BALF supernatant	ELISA	IL-27 (↑)	LC diagnosis	-
					IL-27, IL-29 (↓)	Discrimination of advanced stage	-
Naumnik 2016 [41]	n = 46 (M 46) Mean age: 63 ± 3 y Type: AC 10, SQCC 25, LCC 11 Stage: III 20, IV 26	n = 15 (M 12) Mean age 60 ± 4 y	Cell-free BALF supernatant	ELISA	HGF, IL-22 (↓)	LC diagnosis	-
					IL-22 (↑)	Overall survival (worse)	-
Kontakiotis 2011 [42]	n = 42 (M 42) Age: 43–80 y Type: AC 7, SQCC 22, SCC 10, other 3	n = 16 (M 16) Age: 45–77 y	Cell-free BALF supernatant	ELISA	TNF-α (↑)	LC diagnosis	-
				Colorimetric assay	Total antioxidants, glutathione (↑)	LC diagnosis	-
Jakubowska 2015 [43]	n = 45 (M 38) Mean age: 61.7 ± 8.3 y Type: AC 20, SQCC 22, LCC 3 Stage: III 18, IV 27	n = 15 (M 13) Mean age: 60.1 ± 5.0 y	Cell-free BALF supernatant	ELISA	TGF-β (↑)	LC diagnosis	-

Table 1. *Cont.*

Study	Patients	Controls	Processing	Detection	Biomarker	Role	Accuracy
Chen 2014 [44]	n = 45 (M 28) Mean age: 60.8 ± 1.2 y Type: AC 11, SQCC 18, SCC 10, other 6	n = 33 (M 19) Mean age: 58.2 ± 1.7 y	Cell-free BALF supernatant	ELISA	TGF-β1 >10.85 pg/ml	LC diagnosis	AUC 0.7 Sens 62% Spec 61%
Xiong 2020 [45]	n = 219 (M 150) Mean age: 68.4 ± 18.8 y Type: AC 136, SQCC 43, SCC 35, other 5 Stage: 0 38, I 93, II 50, III 28, IV 10	n = 186 (M 125) Mean age: 40.6 ± 15.5 y	Cell-free BALF supernatant	ELISA	VEGF >234.1 pg/mL, TGF-β >81.8 pg/mL, HGF 44.6 pg/mL (at least 2 positive)	LC diagnosis	AUC 0.81 Sens 82% Spec 61%
Charpidou 2011 [46]	n = 40 (M 37) Age: 45–82 y Type: AC 12, SQCC 19, other 9 Stage: I 3, III 14, IV 23	-	Cell-free BALF supernatant	ELISA	VEGF (↓)	Prediction of treatment response (chemotherapy)	-
					VEGFR1 >53.2 pg/ml	Progression free survival (worse)	-
					VEGFR2 >705.3 pg/ml	Overall survival (worse)	-
Cao 2013 [47]	n = 37 (M 28) Mean age: 55.4 ± 8.4 y Type: AC 15, SQCC 19, SCC 3 Stage: I 23, II 9, III 5	n = 19 (M 12) Mean age: 48.1 ± 9.2 y	Cell-free BALF supernatant	ELISA	VEGF >214 pg/ml	LC diagnosis	AUC 0.86 Sens 82% Spec 84%
Chen 2014 [48]	n = 54 (M 60) Median age: 60 y Type: AC 9, SQCC 36, SCC 9	n = 12 (M 6) Median age: 37 y	Cell-free BALF supernatant	ELISA	IL-8, VEGF (↑)	LC diagnosis	-
Pio 2010 [49]	n = 56 (M 45) Age: 38–83 y Type: AC 12, SQCC 24, LCC 4, SCC 9, other 7	n = 22 (M 14) Age: 30–82 y	Cell-free BALF supernatant	ELISA	Complement factor H >1 μg/mL	LC diagnosis	Sens 62% Spec 77%
					Albumin >17 μg/mL		Sens 68% Spec 71%
Ajona 2013 [50]	n = 50 (M 41) Type: AC 12, SQCC 22, SCC 9, other 7	n = 22 (M 14)	Cell-free BALF supernatant	ELISA	Complement C4-derived fragments	LC diagnosis	AUC 0.73
Ajona 2018 [51]	n = 49 (M 40) Type: AC 12, SQCC 22, SCC 8, other 7	n = 22 (M 14)	Cell-free BALF supernatant	ELISA	Complement C4d	LC diagnosis	AUC 0.80
Li 2013 [52]	n = 18 (M 7) Age: 51–83 y Type: AC	n = 6 (M 3) Age: 18–85 y	Cell-free proteins from BALF supernatant	ELISA	Napsin A >55 ng/mg total protein	LC diagnosis	AUC 0.85 Sens 84% Spec 67%

Table 1. *Cont.*

Study	Patients	Controls	Processing	Detection	Biomarker	Role	Accuracy
Uribarri 2014 [53]	*n* = 204 (M 177) Mean age: 63.0 ± 10.7 y Type: AC 59, SQCC 80, SCC 63, other 2 Stage: I 14, II 4, III 60, IV 109, undefined 17	*n* = 48 (M 38) Mean age: 54.9 ± 14.0 y	Cell-free proteins from BALF supernatant	Fluorescent bead-based immunoassay	5-protein (APOA1, CO4A, CRP, GSTP1, SAMP) prediction model	LC diagnosis	AUC 0.94 Sens 95% Spec 81%
					2-protein (STMN1, GSTP1) prediction model	Discrimination of SCC from NSCLC	AUC 0.80 Sens 90% Spec 57%
Ortea 2016 [54]	*n* = 12 (M 8) Median age: 64 y Type: AC Stage: I-II 2, III-IV 10	*n* = 10 (M 10) Median age: 61	Cell-free proteins from BALF supernatant	Liquid chromatography–mass spectrometry	Discriminant analysis of a 44-protein panel	LC diagnosis	Sens 92% Spec 70%
Almatroodi 2015 [55]	*n* = 8 (M 5) Mean age: 68.1 ± 7.6 y Type: AC Stage: I 2, II 2, III 1, IV 3	*n* = 8 (M 3) Mean age: 60 ± 8.7 y	Cellular proteins from BALF cell pellets	Liquid chromatography–mass spectrometry	33 upregulated proteins	LC diagnosis	-
Carvalho 2017 [56]	*n* = 49 Type: AC 28, SQCC 10, SCC 4, LCC 1, other 6	*n* = 41	Cell-free proteins from BALF supernatant	Liquid chromatography–mass spectrometry	Different spectral count values from all abundant proteins	LC diagnosis	-
					133 differentially expressed proteins		-
Hmmier 2017 [57]	*n* = 26 (M 13) Mean age: 65 y Type: AC 13, SQCC 13 Stage: I-II 15, III-IV 11	*n* = 16 (M 8) Mean age: 56 y	Cell-free proteins from BALF supernatant	Liquid chromatography–mass spectrometry	267 differentially expressed proteins	AC diagnosis	-
					261 differentially expressed proteins	SQCC diagnosis	-
					292 differentially expressed proteins	Discrimination of SQCC from AC	-
Liu 2021 [58]	*n* = 85 (M 60) Type: AC 32, SQCC 32, SCC 21 Stage: I-II 30, III-IV 42, unknown 13	*n* = 33 (M 20)	Cell-free proteins from BALF supernatant	Lectin microarray	3-lectin (ECA, GSL-I, RCA120) prediction model	LC diagnosis	AUC 0.96 Sens 92% Spec 94%
					4-lectin (DBA, STL, UEA-I, BPL) prediction model	Discrimination of AC from other subtypes	AUC 0.62 Sens 71% Spec 59%

Table 1. *Cont.*

Study	Patients	Controls	Processing	Detection	Biomarker	Role	Accuracy
Liu 2021 [58]	n = 85 (M 60) Type: AC 32, SQCC 32, SCC 21 Stage: I-II 30, III-IV 42, unknown 13	n = 33 (M 20)	Cell-free proteins from BALF supernatant	Lectin microarray	1-lectin (PNA) prediction model	Discrimination of AC from other subtypes	AUC 0.69 Sens 80% Spec 67%
					6-lectin (STL, BS-I, PTL-II, SBA, PSA, GNA) prediction model	Discrimination of AC from other subtypes	AUC 0.72 Sens 72% Spec 68%
					6-lectin (MAL-II, LTL, GSL-I, RCA120, PTL-II, PWM) prediction model	Discrimination of early from advanced stage	AUC 0.86 Sens 83% Spec 81%
Kwiecien 2017 [60]	n = 18 (M 12) Age: 50–81 y Type: AC 4, SQCC 9, NSCLC NOS 4 Stage: I 4, II 11, III 3	-	Immune cells from BALF cell pellets	Antibody-specific flow cytometry	% Tregs, CTLA-4$^+$ Tregs (↑)	LC diagnosis (affected vs. healthy lung)	-
Hu 2019 [63]	n = 52 (M 29) Age: 39–73 y Type: NSCLC 26, SCC 26	n = 20 (M 12) Age: 35–75 y	Immune cells from BALF cell pellets	Antibody-specific flow cytometry	% PD-1$^+$ Tph (↓), PD-1$^+$ Tfh/Tph (↑)	SCC diagnosis	-
Hu 2020 [64]	n = 67 (M 46) Age: 39–75 y Type: AC 18, SQCC 17, SCC 32 Stage: 0–IIIA 39, IIIB-IV 28	n = 14 (M 10) Age: 33–71 y	Immune cells from BALF cell pellets	Antibody-specific flow cytometry	Tregs (↑)	LC diagnosis Discrimination of SCC from NSCLC Discrimination of advanced SCC	-
					IL-10$^+$ CD206$^+$ CD14$^+$ M2-like macrophages (↑)	LC diagnosis Discrimination of SCC from NSCLC Discrimination of advanced SCC Overall survival (worse)	-
			Cell-free BALF supernatant	Cytometric bead array	IL-10 (↑)	LC diagnosis Discrimination of SCC from NSCLC Discrimination of advanced SCC Overall survival (worse)	-

Table 1. *Cont.*

Study	Patients	Controls	Processing	Detection	Biomarker	Role	Accuracy
Masuhiro 2022 [65]	n = 12 (M 9) Age: 55–70 y Type: AC 7	-	Cell-free BALF supernatant	Cytometric bead array	CXCL9 (↑)	Prediction of treatment response (immunotherapy)	-
			Bacterial DNA from BALF supernatant	16S rRNA sequencing	Bacterial alpha diversity (↑), *Proteobacteria* (↓), *Bacteroidetes* (↑)		-
	n = 7		Cellular RNA from BALF cell pellets	RNA sequencing	87 genes were upregulated and 28 were downregulated		-
Zhong 2021 [70]	n = 12	n = 6	Tumor cells from BALF cell pellets	Antibody-specific immunostaining + fluorescence in situ hybridization	Circulating tumor cell count ≥2	LC diagnosis	Sens 75% Spec 100%
Schmid 2015 [72]	n = 26 (M 16) Mean age: 60.2 ± 8.3 y Type: AC 20, SQCC 6	n = 21 (M 13) Mean age: 64.7 ± 8.4 y	Cell-free BALF supernatant	BLEIA	ATP, ADP (↑)	LC diagnosis	-
			Cellular RNA from BALF cell pellets	mRNA-specific reverse transcription-PCR	CD39 (↑)	LC diagnosis Discrimination of metastatic disease	-
					P2X4, P2X7, P2Y1 (↑)	Discrimination of metastatic disease	-
Callejón-Leblic 2016 [73]	n = 24 (M 16) Mean age: 66 ± 11 y	n = 31 (M 23) Mean age: 56 ± 13 y	Cell-free metabolites from BALF supernatant	Direct infusion mass spectrometry	Carnitine	LC diagnosis	AUC 0.88
					Adenine		AUC 0.83
					Choline		AUC 0.78
				Gas chromatography–mass spectrometry	Glycerol		AUC 0.89
					Phosphoric acid		AUC 0.79
Callejón-Leblic 2018 [75]	n = 24 (M 20) Mean age: 65 ± 13 y Type: NSCLC 22, SCC 2	n = 31 (M 27) Mean age: 54 ± 14 y	Cell-free elements from BALF supernatant	Inductive coupled plasma mass spectrometry	Mn	LC diagnosis	AUC 0.75
					V/Cu ratio		AUC 0.76

Table 1. *Cont.*

Study	Patients	Controls	Processing	Detection	Biomarker	Role	Accuracy
Suresh 2019 [94]	n = 18 (M 13) Type: NSCLC 15, other 3	-	Immune cells from BALF cell pellets	Antibody-specific flow cytometry	% PD-1^{hi}/CTLA-4^{hi} Tregs (↓), % IL-1RA-expressing B cells (↓), % central memory T cells (↑), % $CD8^+$ TNF-α^{hi} T cells (↑), % IL-$1\beta^{hi}$ monocytes (↑)	Prediction of CIP development	-
			Cell-free BALF supernatant	V-plex immunoassays	IL-1β (↓), IL-8 (↓), MIP-3α (↓), IL-12p40 (↑), IP-10 (↑)		-
Crohns 2010 [104]	n = 36 (M 29) Age: 47–82 y Type: AC 1, SQCC 33, SCC 2 Stage: I 1, III 18, IV 17	n = 36 (M 16) Age: 18–75 y	Cell-free BALF supernatant	ELISA	Il-6 (↑)	LC diagnosis	-
					IL-8 (↑)	Overall survival (worse)	-
Yamagishi 2017 [105]	n = 22 (M 16) Type: AC 8, SQCC 8, SCC 4, unknown 2	-	Cell-free BALF supernatant	ELISA	MMP-9 (↑)	Prediction of radiation pneumonitis	-
					VEGF (↓)		-
			Metagenomic biomarkers				
Wang 2019 [76]	n = 51 (M 31) Type: AC 18, SQCC 19, SCC 14	n = 15 (M 8)	Bacterial DNA from BALF cell pellet	16S rRNA sequencing	Microbial diversity (↓)	LC diagnosis	-
					Treponema		AUC 0.86
Cheng 2020 [77]	n = 32 (M 23) Mean age: 64.3 ± 8.4 y Type: AC 16, SQCC 9, SCC 7 Stage: I 7, III 6, IV 19	n = 22 (M 12) Mean age: 56.5 ± 14.3 y	Bacterial DNA from BALF supernatant	16S rRNA sequencing	10-genera (*f:Pseudomonadaceae, Capnocytophaga, Stenotrophomonas, Microbacterium, Gemmiger, c:TM7-3, Oscillospira, Blautia, Lautropia, Sediminibacterium*) prediction model	LC diagnosis	AUC 0.79

Table 1. *Cont.*

Study	Patients	Controls	Processing	Detection	Biomarker	Role	Accuracy
Lee 2016 [78]	n = 20 (M 13) Median age: 64 y Type: AC 13, SQCC 5, SCC 2 Stage: II 6, III 8, IV 6	n = 8 (M 7) Median age: 58.5 y	Bacterial DNA from BALF supernatant	16S rRNA pyrosequencing	*Veillonella*	LC diagnosis	AUC 0.86
					Megasphaera		AUC 0.78
					Combined panel		AUC 0.89
Patnaik 2021 [79]	n = 36 (M 16) Type: AC 24, SQCC 11, other 1 Stage: I	-	Bacterial DNA from BALF supernatant	16S rRNA sequencing	19-genera microbiome signature	Prediction of recurrence after surgery	AUC 0.77
Zheng 2021 [80]	n = 32 Type: NSCLC Stage: I-II 23, III-IV 9	n = 15	Bacterial DNA from BALF supernatant	16S rRNA sequencing	Differentiated abundance of 19 species	LC diagnosis	-
Jang 2021 [81]	n = 11 (M 9) Median age: 63 y Type: AC 8, SQCC 3 Stage: III 5, IV 6	-	Bacterial DNA from BALF supernatant	16S rRNA sequencing	*Haemophilus influenzae* (↓), *Neisseria perflava* (↓), *Veillonella dispar* (↑)	Prediction of treatment response (immunotherapy)	-

BALF, bronchoalveolar lavage fluid; M, male; AC, adenocarcinoma; SQCC, squamous cell carcinoma; NSCLC; non-small-cell lung cancer; NOS, not otherwise specified; SCC, small-cell carcinoma; LCC, large cell carcinoma; EV, extracellular vesicle; PCR, polymerase chain reaction; FRET-PHFA, fluorescence resonance energy transfer-based preferential homoduplex formation assay; PNA, peptide nucleic acid; RIA, radioimmunoassay; ELISA, enzyme-linked immunosorbent assay; BLEIA, bioluminescent enzyme immunoassay; LC, lung cancer; TKI, tyrosine kinase inhibitor; CIP, checkpoint inhibitor pneumonitis; AUC, area under the ROC curve; Acc, overall accuracy; Sens, sensitivity; Spec, specificity. Note: Upward arrows correspond to increased and downward arrows to decreased detection of the biomarker respectively to predict each outcome. Sensitivity is the probability that a test result will be positive when the disease is present. Specificity is the probability that a test result will be negative when the disease is not present. Overall accuracy is the overall probability that a patient is correctly classified. Area under the ROC curve is the probability that a classifier will distinguish between two classes (disease vs. normal).

6. Conclusions

It can be concluded that BALF is an appropriate medium that may aid in the diagnosis of lung cancer, in assessing prognosis and response to therapy, but also in the early identification of treatment-related adverse events. Furthermore, continuous re-evaluation of the genome and the protein and immune status of lung cancer cells is essential in the course of the disease. Repeated BALF examination is a cost-effective and easily accessible method for evaluating the microenvironment and immune status of cancer-afflicted lung tissues and provides a valid comparison of the disease status before and after treatment.

Author Contributions: Conceptualization, A.K. (Alexandros Kalkanis); methodology, A.K. (Alexandros Kalkanis) and D.P.; data curation, A.K. (Alexandros Kalkanis), D.P., D.T. and A.K. (Alexandra Kopitopoulou); writing—original draft preparation, A.K. (Alexandros Kalkanis) and D.P.; writing—review and editing, D.T., A.K. (Alexandra Kopitopoulou), E.B. and E.W.; supervision, E.W. All authors have read and agreed to the published version of the manuscript.

Funding: This research received no external funding.

Institutional Review Board Statement: Not applicable.

Informed Consent Statement: Not applicable.

Conflicts of Interest: The authors declare no conflict of interest.

References

1. Howlader, N.; Noone, A.; Krapcho, M.; Miller, D.; Brest, A.; Yu, M.; Ruhl, J.; Tatalovich, Z.; Mariotto, A.; Lewis, D.; et al. *SEER Cancer Statistics Review, 1975–2018*; National Cancer Institute: Bethesda, MD, USA, 2021.
2. Henry, N.L.; Hayes, D.F. Cancer biomarkers. *Mol. Oncol.* **2012**, *6*, 140–146. [CrossRef]
3. Domagala-Kulawik, J. The relevance of bronchoalveolar lavage fluid analysis for lung cancer patients. *Expert Rev. Respir. Med.* **2020**, *14*, 329–337. [CrossRef]
4. Lee, S.H.; Kim, E.Y.; Kim, T.; Chang, Y.S. Compared to plasma, bronchial washing fluid shows higher diagnostic yields for detecting EGFR-TKI sensitizing mutations by ddPCR in lung cancer. *Respir. Res.* **2020**, *21*, 142. [CrossRef] [PubMed]
5. Wongsurakiat, P.; Wongbunnate, S.; Dejsomritrutai, W.; Charoenratanakul, S.; Tscheikuna, J.; Youngchaiyud, P.; Pushpakom, R.; Maranetra, N.; Nana, A.; Chierakul, N.; et al. Diagnostic value of bronchoalveolar lavage and postbronchoscopic sputum cytology in peripheral lung cancer. *Respirology* **1998**, *3*, 131–137. [CrossRef] [PubMed]
6. Binesh, F.; Pirdehghan, A.; Mirjalili, M.R.; Samet, M.; Majomerd, Z.A.; Akhavan, A. Comparative assessment of the diagnostic value of transbronchial lung biopsy and bronchoalveolar lavage fluid cytology in lung cancer. *Asian Pac. J. Cancer Prev.* **2015**, *16*, 201–204. [CrossRef] [PubMed]
7. Nair, V.S.; Hui, A.B.-Y.; Chabon, J.J.; Esfahani, M.S.; Stehr, H.; Nabet, B.Y.; Zhou, L.; Chaudhuri, A.A.; Benson, J.; Ayers, K.; et al. Genomic profiling of bronchoalveolar lavage fluid in lung cancer. *Cancer Res.* **2022**, *82*, 2838. [CrossRef]
8. Roncarati, R.; Lupini, L.; Miotto, E.; Saccenti, E.; Mascetti, S.; Morandi, L.; Bassi, C.; Rasio, D.; Callegari, E.; Conti, V.; et al. Molecular testing on bronchial washings for the diagnosis and predictive assessment of lung cancer. *Mol. Oncol.* **2020**, *14*, 2163–2175. [CrossRef] [PubMed]
9. Kawahara, A.; Fukumitsu, C.; Taira, T.; Abe, H.; Takase, Y.; Murata, K.; Yamaguchi, T.; Azuma, K.; Ishii, H.; Takamori, S.; et al. Epidermal growth factor receptor mutation status in cell-free DNA supernatant of bronchial washings and brushings. *Cancer Cytopathol.* **2015**, *123*, 620–628. [CrossRef]
10. Yanev, N.; Mekov, E.; Valev, D.; Yankov, G.; Milanov, V.; Bichev, S.; Gabrovska, N.; Kostadinov, D. EGFR mutation status yield from bronchoalveolar lavage in patients with primary pulmonary adenocarcinoma compared to a venous blood sample and tissue biopsy. *PeerJ* **2021**, *9*, e11448. [CrossRef]
11. Prim, N.; Quoix, E.; Beau-Faller, M. Tumor cell content for selection of molecular techniques for T790M EGFR mutation detection in non-small cell lung cancer. *J. Thorac. Oncol.* **2011**, *6*, 1615–1616. [CrossRef] [PubMed]
12. Park, S.; Hur, J.Y.; Lee, K.Y.; Lee, J.C.; Rho, J.K.; Shin, S.H.; Choi, C.-M. Assessment of EGFR mutation status using cell-free DNA from bronchoalveolar lavage fluid. *Clin. Chem. Lab. Med.* **2017**, *55*, 1489–1495. [CrossRef]
13. Kiura, K.; Yoh, K.; Katakami, N.; Nogami, N.; Kasahara, K.; Takahashi, T.; Okamoto, I.; Cantarini, M.; Hodge, R.; Uchida, H. Osimertinib in patients with epidermal growth factor receptor T790M advanced non-small cell lung cancer selected using cytology samples. *Cancer Sci.* **2018**, *109*, 1177–1184. [CrossRef] [PubMed]
14. Hur, J.Y.; Kim, H.J.; Lee, J.S.; Choi, C.-M.; Lee, J.C.; Jung, M.K.; Pack, C.G.; Lee, K.Y. Extracellular vesicle-derived DNA for performing EGFR genotyping of NSCLC patients. *Mol. Cancer* **2018**, *17*, 1–6. [CrossRef] [PubMed]
15. Gazdar, A.F.; Minna, J.D. Molecular detection of early lung cancer. *J. Natl. Cancer Inst.* **1999**, *91*, 299–301. [CrossRef] [PubMed]
16. Ahrendt, S.A.; Chow, J.T.; Xu, L.-H.; Yang, S.C.; Eisenberger, C.F.; Esteller, M.; Herman, J.G.; Wu, L.; Decker, P.A.; Jen, J.; et al. Molecular detection of tumor cells in bronchoalveolar lavage fluid from patients with early stage lung cancer. *J. Natl. Cancer Inst.* **1999**, *91*, 332–339. [CrossRef] [PubMed]
17. Oshita, F.; Nomura, I.; Yamada, K.; Kato, Y.; Tanaka, G.; Noda, K. Detection of K-ras mutations of bronchoalveolar lavage fluid cells aids the diagnosis of lung cancer in small pulmonary lesions. *Clin. Cancer Res.* **1999**, *5*, 617–620. [PubMed]
18. Li, J.; Hu, Y.-M.; Wang, Y.; Tang, X.-P.; Shi, W.-L.; Du, Y.-J. Gene mutation analysis in non-small cell lung cancer patients using bronchoalveolar lavage fluid and tumor tissue as diagnostic markers. *Int. J. Biol. Markers* **2014**, *29*, 328–336. [CrossRef] [PubMed]
19. Nakamichi, S.; Seike, M.; Miyanaga, A.; Chiba, M.; Matsuda, K.; Kobayashi, K.; Takahashi, A.; Takeuchi, S.; Minegishi, Y.; Kubota, K.; et al. RT-PCR for Detecting ALK Translocations in Cytology Samples from Lung Cancer Patients. *Anticancer Res.* **2017**, *37*, 3295. [PubMed]
20. Kim, H.; Kwon, Y.M.; Kim, J.S.; Lee, H.; Park, J.-H.; Shim, Y.M.; Han, J.; Park, J.; Kim, D.-H. Tumor-specific methylation in bronchial lavage for the early detection of non-small-cell lung cancer. *J. Clin. Oncol.* **2004**, *22*, 2363–2370. [CrossRef]
21. Destro, A.; Bianchi, P.; Alloisio, M.; Laghi, L.; di Gioia, S.; Malesci, A.; Cariboni, U.; Gribaudi, G.; Bulfamante, G.; Marchetti, A.; et al. K-ras and $p16^{INK4A}$ alterations in sputum of NSCLC patients and in heavy asymptomatic chronic smokers. *Lung Cancer* **2004**, *44*, 23–32. [CrossRef] [PubMed]
22. Konno, S.; Morishita, Y.; Fukasawa, M.; Shu, Y.; Wang, D.; Tanaka, R.; Minami, Y.; Iijima, T.; Noguchi, M. Anthracotic index and DNA methylation status of sputum contents can be used for identifying the population at risk of lung carcinoma. *Cancer Cytopathol.* **2004**, *102*, 348–354. [CrossRef] [PubMed]
23. Nikolaidis, G.; Raji, O.Y.; Markopoulou, S.; Gosney, J.R.; Bryan, J.; Warburton, C.; Walshaw, M.; Sheard, J.; Field, J.K.; Liloglou, T. DNA Methylation Biomarkers Offer Improved Diagnostic Efficiency in Lung Cancer. *Cancer Res.* **2012**, *72*, 5692–5701. [CrossRef]
24. Dietrich, D.; Kneip, C.; Raji, O.; Liloglou, T.; Seegebarth, A.; Schlegel, T.; Flemming, N.; Rausch, S.; Distler, J.; Fleischhacker, M.; et al. Performance evaluation of the DNA methylation biomarker SHOX2 for the aid in diagnosis of lung cancer based on the analysis of bronchial aspirates. *Int. J. Oncol.* **2012**, *40*, 825–832. [CrossRef] [PubMed]

25. Zhang, C.; Yu, W.; Wang, L.; Zhao, M.; Guo, Q.; Lv, S.; Hu, X.; Lou, J. DNA methylation analysis of the SHOX2 and RASSF1A panel in bronchoalveolar lavage fluid for lung cancer diagnosis. *J. Cancer* **2017**, *8*, 3585. [CrossRef] [PubMed]
26. Um, S.-W.; Kim, Y.; Lee, B.B.; Kim, D.; Lee, K.-J.; Kim, H.K.; Han, J.; Kim, H.; Shim, Y.M.; Kim, D.-H. Genome-wide analysis of DNA methylation in bronchial washings. *Clin. Epigenetics* **2018**, *10*, 1–9. [CrossRef] [PubMed]
27. Li, L.; Ye, Z.; Yang, S.; Yang, H.; Jin, J.; Zhu, Y.; Tao, J.; Chen, S.; Xu, J.; Liu, Y.; et al. Diagnosis of pulmonary nodules by DNA methylation analysis in bronchoalveolar lavage fluids. *Clin. Epigenetics* **2021**, *13*, 185. [CrossRef]
28. Tuo, L.; Sha, S.; Huayu, Z.; Du, K. P16^{INK4a} gene promoter methylation as a biomarker for the diagnosis of non-small cell lung cancer: An updated meta-analysis. *Thorac. Cancer* **2018**, *9*, 1032–1040. [CrossRef]
29. Molina-Pinelo, S.; Pastor, M.D.; Suarez, R.; Romero-Romero, B.; González De la Peña, M.; Salinas, A.; García-Carbonero, R.; De Miguel, M.J.; Rodríguez-Panadero, F.; Carnero, A.; et al. MicroRNA clusters: Dysregulation in lung adenocarcinoma and COPD. *Eur. Respir. J.* **2014**, *43*, 1740. [CrossRef] [PubMed]
30. Li, J.; Chen, P.; Li, X.-Q.; Bao, Q.-L.; Dai, C.-H.; Ge, L.-P. Elevated levels of survivin and livin mRNA in bronchial aspirates as markers to support the diagnosis of lung cancer. *Int. J. Cancer* **2013**, *132*, 1098–1104. [CrossRef]
31. Rehbein, G.; Schmidt, B.; Fleischhacker, M. Extracellular microRNAs in bronchoalveolar lavage samples from patients with lung diseases as predictors for lung cancer. *Clin. Chim. Acta* **2015**, *450*, 78–82. [CrossRef]
32. Kim, J.E.; Eom, J.S.; Kim, W.-Y.; Jo, E.J.; Mok, J.; Lee, K.; Kim, K.U.; Park, H.-K.; Lee, M.K.; Kim, M.-H. Diagnostic value of microRNAs derived from exosomes in bronchoalveolar lavage fluid of early-stage lung adenocarcinoma: A pilot study. *Thorac. Cancer* **2018**, *9*, 911–915. [CrossRef]
33. Kim, J.O.; Gazala, S.; Razzak, R.; Guo, L.; Ghosh, S.; Roa, W.H.; Bédard, E.L.R. Non-small Cell Lung Cancer Detection Using MicroRNA Expression Profiling of Bronchoalveolar Lavage Fluid and Sputum. *Anticancer Res.* **2015**, *35*, 1873. [PubMed]
34. Li, H.; Jiang, Z.; Leng, Q.; Bai, F.; Wang, J.; Ding, X.; Li, Y.; Zhang, X.; Fang, H.; Yfantis, H.G.; et al. A prediction model for distinguishing lung squamous cell carcinoma from adenocarcinoma. *Oncotarget* **2017**, *8*, 50704–50714. [CrossRef] [PubMed]
35. Mancuso, G.; Bovio, E.; Rena, O.; Rrapaj, E.; Mercalli, F.; Veggiani, C.; Paganotti, A.; Andorno, S.; Boldorini, R. Prognostic impact of a 3-MicroRNA signature in cytological samples of small cell lung cancer. *Cancer Cytopathol.* **2016**, *124*, 621–629. [CrossRef] [PubMed]
36. Rodríguez, M.; Silva, J.; López-Alfonso, A.; López-Muñiz, M.B.; Peña, C.; Domínguez, G.; García, J.M.; López-González, A.; Méndez, M.; Provencio, M.; et al. Different exosome cargo from plasma/bronchoalveolar lavage in non-small-cell lung cancer. *Genes Chromosomes Cancer* **2014**, *53*, 713–724. [CrossRef] [PubMed]
37. Kuo, C.-H.S.; Liu, C.-Y.; Pavlidis, S.; Lo, Y.-L.; Wang, Y.-W.; Chen, C.-H.; Ko, H.-W.; Chung, F.-T.; Lin, T.-Y.; Wang, T.-Y.; et al. Unique Immune Gene Expression Patterns in Bronchoalveolar Lavage and Tumor Adjacent Non-Neoplastic Lung Tissue in Non-Small Cell Lung Cancer. *Front. Immunol.* **2018**, *9*, 232. [CrossRef] [PubMed]
38. Macchia, V.; Mariano, A.; Cavalcanti, M.; Coppa, A.; Cecere, C.; Fraioli, G.; Elia, S.; Ferrante, G. Tumor markers and lung cancer: Correlation between serum and bronchial secretion levels of CEA, TPA, CanAg CA-50, NSE and ferritin. *Int. J. Biol. Markers* **1987**, *2*, 151–156. [CrossRef]
39. Edge, S.; Byrd, D.; Compton, C.; Fritz, A.; Greene, F.; Trotti, A. (Eds.) *AJCC Cancer Staging Manual*, 7th ed.; Springer: New York, NY, USA, 2010.
40. Naumnik, W.; Naumnik, B.; Niewiarowska, K.; Ossolinska, M.; Chyczewska, E. Novel cytokines: IL-27, IL-29, IL-31 and IL-33. Can they be useful in clinical practice at the time diagnosis of lung cancer? *Exp. Oncol.* **2012**, *34*, 348–353.
41. Naumnik, W.; Naumnik, B.; Niklińska, W.; Ossolińska, M.; Chyczewska, E. Clinical implications of hepatocyte growth factor, interleukin-20, and interleukin-22 in serum and bronchoalveolar fluid of patients with non-small cell lung cancer. In *Advancements in Clinical Research*; Pokorski, M., Ed.; Advances in Experimental Medicine and Biology; Springer: Cham, Switzerland, 2016; pp. 41–49.
42. Kontakiotis, T.; Katsoulis, K.; Hagizisi, O.; Kougioulis, M.; Gerou, S.; Papakosta, D. Bronchoalveolar lavage fluid alteration in antioxidant and inflammatory status in lung cancer patients. *Eur. J. Intern. Med.* **2011**, *22*, 522–526. [CrossRef] [PubMed]
43. Jakubowska, K.; Naumnik, W.; Niklińska, W.; Chyczewska, E. Clinical significance of HMGB-1 and TGF-β level in serum and BALF of advanced non-small cell lung cancer. In *Respiratory Carcinogenesis*; Pokorski, M., Ed.; Advances in Experimental Medicine and Biology; Springer: Cham, Switzerland, 2015; pp. 49–58.
44. Chen, Z.; Xu, Z.; Sun, S.; Yu, Y.; Lv, D.; Cao, C.; Deng, Z. TGF-β1, IL-6 and TNF-α in Bronchoalveolar Lavage Fluid: Useful Markers for Lung Cancer? *Sci. Rep.* **2014**, *4*, 5595. [CrossRef]
45. Xiong, W.; Ding, W.; Xu, M.; Pudasaini, B.; Sun, J.; Zhao, Y. The screening role of a biomarker panel in BALF among patients with cancer-suspected pulmonary nodules less than 8 mm. *Clin. Respir. J.* **2020**, *14*, 829–838. [CrossRef] [PubMed]
46. Charpidou, A.; Gkiozos, I.; Konstantinou, M.; Eleftheraki, A.; Demertzis, P.; Harrington, K.; Polyzos, A.; Syrigos, K.N. Bronchial washing levels of vascular endothelial growth factor receptor-2 (VEGFR2) correlate with overall survival in NSCLC patients. *Cancer Lett.* **2011**, *304*, 144–153. [CrossRef] [PubMed]
47. Cao, C.; Sun, S.F.; Lv, D.; Chen, Z.B.; Ding, Q.L.; Deng, Z.C. Utility of VEGF and sVEGFR-1 in bronchoalveolar lavage fluid for differential diagnosis of primary lung cancer. *Asian Pac. J. Cancer Prev.* **2013**, *14*, 2443–2446. [CrossRef] [PubMed]
48. Chen, L.; Li, Q.; Zhou, X.-D.; Shi, Y.; Yang, L.; Xu, S.-L.; Chen, C.; Cui, Y.-H.; Zhang, X.; Bian, X.-W. Increased pro-angiogenic factors, infiltrating neutrophils and CD163+ macrophages in bronchoalveolar lavage fluid from lung cancer patients. *Int. Immunopharmacol.* **2014**, *20*, 74–80. [CrossRef]

49. Pio, R.; Garcia, J.; Corrales, L.; Ajona, D.; Fleischhacker, M.; Pajares, M.J.; Cardenal, F.; Seijo, L.; Zulueta, J.J.; Nadal, E.; et al. Complement factor H is elevated in bronchoalveolar lavage fluid and sputum from patients with lung cancer. *Cancer Epidemiol. Biomark. Prev.* **2010**, *19*, 2665–2672. [CrossRef]
50. Ajona, D.; Pajares, M.J.; Corrales, L.; Perez-Gracia, J.L.; Agorreta, J.; Lozano, M.D.; Torre, W.; Massion, P.P.; de-Torres, J.P.; Jantus-Lewintre, E.; et al. Investigation of Complement Activation Product C4d as a Diagnostic and Prognostic Biomarker for Lung Cancer. *J. Natl. Cancer Inst.* **2013**, *105*, 1385–1393. [CrossRef] [PubMed]
51. Ajona, D.; Okrój, M.; Pajares, M.J.; Agorreta, J.; Lozano, M.D.; Zulueta, J.J.; Verri, C.; Roz, L.; Sozzi, G.; Pastorino, U.; et al. Complement C4d-specific antibodies for the diagnosis of lung cancer. *Oncotarget* **2018**, *9*, 6346–6355. [CrossRef]
52. Li, Q.K.; Shah, P.; Li, Y.; Aiyetan, P.O.; Chen, J.; Yung, R.; Molena, D.; Gabrielson, E.; Askin, F.; Chan, D.W.; et al. Glycoproteomic analysis of bronchoalveolar lavage (BAL) fluid identifies tumor-associated glycoproteins from lung adenocarcinoma. *J. Proteome Res.* **2013**, *12*, 3689–3696. [CrossRef] [PubMed]
53. Uribarri, M.; Hormaeche, I.; Zalacain, R.; Lopez-Vivanco, G.; Martinez, A.; Nagore, D.; Ruiz-Argüello, M.B. A new biomarker panel in bronchoalveolar lavage for an improved lung cancer diagnosis. *J. Thorac. Oncol.* **2014**, *9*, 1504–1512. [CrossRef]
54. Ortea, I.; Rodríguez-Ariza, A.; Chicano-Gálvez, E.; Arenas Vacas, M.S.; Jurado Gámez, B. Discovery of potential protein biomarkers of lung adenocarcinoma in bronchoalveolar lavage fluid by SWATH MS data-independent acquisition and targeted data extraction. *J. Proteom.* **2016**, *138*, 106–114. [CrossRef] [PubMed]
55. Almatroodi, S.A.; McDonald, C.F.; Collins, A.L.; Darby, I.A.; Pouniotis, D.S. Quantitative proteomics of bronchoalveolar lavage fluid in lung adenocarcinoma. *Cancer Genom. Proteom.* **2015**, *12*, 39–48.
56. Carvalho, A.S.; Cuco, C.M.; Lavareda, C.; Miguel, F.; Ventura, M.; Almeida, S.; Pinto, P.; de Abreu, T.T.; Rodrigues, L.V.; Seixas, S.; et al. Bronchoalveolar Lavage Proteomics in Patients with Suspected Lung Cancer. *Sci. Rep.* **2017**, *7*, 42190. [CrossRef]
57. Hmmier, A.; O'Brien, M.E.; Lynch, V.; Clynes, M.; Morgan, R.; Dowling, P. Proteomic analysis of bronchoalveolar lavage fluid (BALF) from lung cancer patients using label-free mass spectrometry. *BBA Clin.* **2017**, *7*, 97–104. [CrossRef] [PubMed]
58. Liu, L.; Li, D.; Shu, J.; Wang, L.; Zhang, F.; Zhang, C.; Yu, H.; Chen, M.; Li, Z.; Guo, X. Protein Glycopatterns in Bronchoalveolar Lavage Fluid as Novel Potential Biomarkers for Diagnosis of Lung Cancer. *Front. Oncol.* **2021**, *10*, 568433. [CrossRef]
59. Domagala-Kulawik, J.; Osinska, I.; Hoser, G.J. Mechanisms of immune response regulation in lung cancer. *Transl. Lung Cancer* **2014**, *3*, 15.
60. Kwiecien, I.; Stelmaszczyk-Emmel, A.; Polubiec-Kownacka, M.; Dziedzic, D.; Domagala-Kulawik, J. Elevated regulatory T cells, surface and intracellular CTLA-4 expression and interleukin-17 in the lung cancer microenvironment in humans. *Cancer Immunol. Immunother.* **2017**, *66*, 161–170. [CrossRef]
61. Zhang, B.; Yao, G.; Zhang, Y.; Gao, J.; Yang, B.; Rao, Z.; Gao, J. M2-polarized tumor-associated macrophages are associated with poor prognoses resulting from accelerated lymphangiogenesis in lung adenocarcinoma. *Clinics* **2011**, *66*, 1879–1886. [CrossRef]
62. Osińska, I.; Stelmaszczyk-Emmel, A.; Polubiec-Kownacka, M.; Dziedzic, D.; Domagała-Kulawik, J. CD4+/CD25high/FoxP3+/CD127−regulatory T cells in bronchoalveolar lavage fluid of lung cancer patients. *Hum. Immunol.* **2016**, *77*, 912–915. [CrossRef]
63. Hu, X.; Gu, Y.; Li, D.; Zhao, S.; Hua, S.; Jiang, Y. Analyzing the percentage of different PD-1+ T cell subsets in peripheral blood and bronchoalveolar lavage fluid of small cell lung cancer patients: A prospective study. *Clin. Exp. Pharmacol. Physiol.* **2019**, *46*, 1074–1083. [CrossRef] [PubMed]
64. Hu, X.; Gu, Y.; Zhao, S.; Hua, S.; Jiang, Y. Immunotherapy. Increased IL-10+ CD206+ CD14+ M2-like macrophages in alveolar lavage fluid of patients with small cell lung cancer. *Cancer Immunol.* **2020**, *69*, 2547–2560. [CrossRef] [PubMed]
65. Masuhiro, K.; Tamiya, M.; Fujimoto, K.; Koyama, S.; Naito, Y.; Osa, A.; Hirai, T.; Suzuki, H.; Okamoto, N.; Shiroyama, T.; et al. Bronchoalveolar lavage fluid reveals factors contributing to the efficacy of PD-1 blockade in lung cancer. *JCI Insight* **2022**, *7*, e157915. [CrossRef] [PubMed]
66. Krebs, M.G.; Hou, J.-M.; Ward, T.H.; Blackhall, F.H.; Dive, C. Circulating tumour cells: Their utility in cancer management and predicting outcomes. *Ther. Adv. Med. Oncol.* **2010**, *2*, 351–365. [CrossRef] [PubMed]
67. Chen, Q.; Ge, F.; Cui, W.; Wang, F.; Yang, Z.; Guo, Y.; Li, L.; Bremner, R.M.; Lin, P.P. Lung cancer circulating tumor cells isolated by the EpCAM-independent enrichment strategy correlate with Cytokeratin 19-derived CYFRA21-1 and pathological staging. *Clin. Chim. Acta* **2013**, *419*, 57–61. [CrossRef] [PubMed]
68. Wang, P.-P.; Liu, S.-H.; Chen, C.-T.; Lv, L.; Li, D.; Liu, Q.-Y.; Liu, G.-L.; Wu, Y. Circulating tumor cells as a new predictive and prognostic factor in patients with small cell lung cancer. *J. Cancer* **2020**, *11*, 2113. [CrossRef]
69. Yang, D.; Yang, X.; Li, Y.; Zhao, P.; Fu, R.; Ren, T.; Hu, P.; Wu, Y.; Yang, H.; Guo, N. Clinical significance of circulating tumor cells and metabolic signatures in lung cancer after surgical removal. *J. Transl. Med.* **2020**, *18*, 1–13. [CrossRef]
70. Zhong, M.; Zhang, Y.; Pan, Z.; Wang, W.; Zhang, Y.; Weng, Y.; Huang, H.; He, Y.; Liu, O. Clinical Utility of Circulating Tumor Cells in the Early Detection of Lung Cancer in Patients with a Solitary Pulmonary Nodule. *Technol. Cancer Res. Treat.* **2021**, *20*, 15330338211041465. [CrossRef]
71. Ciebiada, M.; Górski, P.; Antczak, A. Eicosanoids in Exhaled Breath Condensate and Bronchoalveolar Lavage Fluid of Patients with Primary Lung Cancer. *Dis. Markers* **2012**, *32*, 562862. [CrossRef]
72. Schmid, S.; Kübler, M.; Korcan Ayata, C.; Lazar, Z.; Haager, B.; Hoßfeld, M.; Meyer, A.; Cicko, S.; Elze, M.; Wiesemann, S.; et al. Altered purinergic signaling in the tumor associated immunologic microenvironment in metastasized non-small-cell lung cancer. *Lung Cancer* **2015**, *90*, 516–521. [CrossRef]

73. Callejón-Leblic, B.; García-Barrera, T.; Grávalos-Guzmán, J.; Pereira-Vega, A.; Gómez-Ariza, J.L. Metabolic profiling of potential lung cancer biomarkers using bronchoalveolar lavage fluid and the integrated direct infusion/ gas chromatography mass spectrometry platform. *J. Proteom.* **2016**, *145*, 197–206. [CrossRef]
74. Lui, G.Y.; Kovacevic, Z.; Richardson, V.; Merlot, A.M.; Kalinowski, D.S.; Richardson, D.R. Targeting cancer by binding iron: Dissecting cellular signaling pathways. *Oncotarget* **2015**, *6*, 18748. [CrossRef]
75. Callejon-Leblic, B.; Gómez-Ariza, J.L.; Pereira-Vega, A.; Garcia-Barrera, T. Metal dyshomeostasis based biomarkers of lung cancer using human biofluids. *Metallomics* **2018**, *10*, 1444–1451. [CrossRef] [PubMed]
76. Wang, K.; Huang, Y.; Zhang, Z.; Liao, J.; Ding, Y.; Fang, X.; Liu, L.; Luo, J.; Kong, J. A preliminary study of microbiota diversity in saliva and bronchoalveolar lavage fluid from patients with primary bronchogenic carcinoma. *Med. Sci. Monit.* **2019**, *25*, 2819. [CrossRef] [PubMed]
77. Cheng, C.; Wang, Z.; Wang, J.; Ding, C.; Sun, C.; Liu, P.; Xu, X.; Liu, Y.; Chen, B.; Gu, B. Characterization of the lung microbiome and exploration of potential bacterial biomarkers for lung cancer. *Transl. Lung Cancer Res.* **2020**, *9*, 693–704. [CrossRef] [PubMed]
78. Lee, S.H.; Sung, J.Y.; Yong, D.; Chun, J.; Kim, S.Y.; Song, J.H.; Chung, K.S.; Kim, E.Y.; Jung, J.Y.; Kang, Y.A.; et al. Characterization of microbiome in bronchoalveolar lavage fluid of patients with lung cancer comparing with benign mass like lesions. *Lung Cancer* **2016**, *102*, 89–95. [CrossRef]
79. Patnaik, S.K.; Cortes, E.G.; Kannisto, E.D.; Punnanitinont, A.; Dhillon, S.S.; Liu, S.; Yendamuri, S. Lower airway bacterial microbiome may influence recurrence after resection of early-stage non–small cell lung cancer. *J. Thorac. Cardiovasc. Surg.* **2021**, *161*, 419–429.e16. [CrossRef]
80. Zheng, L.; Sun, R.; Zhu, Y.; Li, Z.; She, X.; Jian, X.; Yu, F.; Deng, X.; Sai, B.; Wang, L.; et al. Lung microbiome alterations in NSCLC patients. *Sci. Rep.* **2021**, *11*, 11736. [CrossRef]
81. Jang, H.J.; Choi, J.Y.; Kim, K.; Yong, S.H.; Kim, Y.W.; Kim, S.Y.; Kim, E.Y.; Jung, J.Y.; Kang, Y.A.; Park, M.S.; et al. Relationship of the lung microbiome with PD-L1 expression and immunotherapy response in lung cancer. *Respir. Res.* **2021**, *22*, 322. [CrossRef]
82. Espuela-Ortiz, A.; Lorenzo-Diaz, F.; Baez-Ortega, A.; Eng, C.; Hernandez-Pacheco, N.; Oh, S.S.; Lenoir, M.; Burchard, E.G.; Flores, C.; Pino-Yanes, M. Bacterial salivary microbiome associates with asthma among African American children and young adults. *Pediatric Pulmonol.* **2019**, *54*, 1948–1956. [CrossRef]
83. Molyneaux, P.L.; Cox, M.J.; Wells, A.U.; Kim, H.C.; Ji, W.; Cookson, W.O.; Moffatt, M.F.; Kim, D.S.; Maher, T.M. Changes in the respiratory microbiome during acute exacerbations of idiopathic pulmonary fibrosis. *Respir. Res.* **2017**, *18*, 1–6. [CrossRef]
84. Yu, D.P.; Cheng, X.; Liu, Z.D.; Xu, S.F. Comparative beneficiary effects of immunotherapy against chemotherapy in patients with advanced NSCLC: Meta-analysis and systematic review. *Oncol. Lett.* **2017**, *14*, 1568–1580. [CrossRef]
85. Weber, J.S.; Kähler, K.C.; Hauschild, A. Management of immune-related adverse events and kinetics of response with ipilimumab. *J. Clin. Oncol.* **2012**, *30*, 2691–2697. [CrossRef] [PubMed]
86. Kyi, C.; Hellmann, M.D.; Wolchok, J.D.; Chapman, P.B.; Postow, M.A. Opportunistic infections in patients treated with immunotherapy for cancer. *J. Immunother. Cancer* **2014**, *2*, 19. [CrossRef]
87. Fujita, K.; Terashima, T.; Mio, T. Anti-PD1 antibody treatment and the development of acute pulmonary tuberculosis. *J. Thorac. Oncol.* **2016**, *11*, 2238–2240. [CrossRef] [PubMed]
88. Jain, P.; Sandur, S.; Meli, Y.; Arroliga, A.C.; Stoller, J.K.; Mehta, A.C. Role of flexible bronchoscopy in immunocompromised patients with lung infiltrates. *Chest* **2004**, *125*, 712–722. [CrossRef] [PubMed]
89. Hellmann, M.D.; Rizvi, N.A.; Goldman, J.W.; Gettinger, S.N.; Borghaei, H.; Brahmer, J.R.; Ready, N.E.; Gerber, D.E.; Chow, L.Q.; Juergens, R.A.; et al. Nivolumab plus ipilimumab as first-line treatment for advanced non-small-cell lung cancer (CheckMate 012): Results of an open-label, phase 1, multicohort study. *Lancet Oncol.* **2017**, *18*, 31–41. [CrossRef]
90. Suresh, K.; Naidoo, J.; Lin, C.T.; Danoff, S. Immune checkpoint immunotherapy for non-small cell lung cancer: Benefits and pulmonary toxicities. *Chest* **2018**, *154*, 1416–1423. [CrossRef]
91. Garon, E.B.; Rizvi, N.A.; Hui, R.; Leighl, N.; Balmanoukian, A.S.; Eder, J.P.; Patnaik, A.; Aggarwal, C.; Gubens, M.; Horn, L.; et al. Pembrolizumab for the treatment of non–small-cell lung cancer. *N. Engl. J. Med.* **2015**, *372*, 2018–2028. [CrossRef]
92. Gettinger, S.N.; Horn, L.; Gandhi, L.; Spigel, D.R.; Antonia, S.J.; Rizvi, N.A.; Powderly, J.D.; Heist, R.S.; Carvajal, R.D.; Jackman, D.M.; et al. Overall survival and long-term safety of nivolumab (anti–programmed death 1 antibody, BMS-936558, ONO-4538) in patients with previously treated advanced non–small-cell lung cancer. *J. Clin. Oncol.* **2015**, *33*, 2004. [CrossRef]
93. Delaunay, M.; Cadranel, J.; Lusque, A.; Meyer, N.; Gounant, V.; Moro-Sibilot, D.; Michot, J.-M.; Raimbourg, J.; Girard, N.; Guisier, F.; et al. Immune-checkpoint inhibitors associated with interstitial lung disease in cancer patients. *Eur. Respir. J.* **2017**, *50*, 1700050. [CrossRef]
94. Suresh, K.; Naidoo, J.; Zhong, Q.; Xiong, Y.; Mammen, J.; de Flores, M.V.; Cappelli, L.; Balaji, A.; Palmer, T.; Forde, P.M.; et al. The alveolar immune cell landscape is dysregulated in checkpoint inhibitor pneumonitis. *J. Clin. Investig.* **2019**, *129*, 4305–4315. [CrossRef]
95. Chopra, A.; Nautiyal, A.; Kalkanis, A.; Judson, M.A. Drug-induced sarcoidosis-like reactions. *Chest* **2018**, *154*, 664–677. [CrossRef] [PubMed]
96. Gkiozos, I.; Kopitopoulou, A.; Kalkanis, A.; Vamvakaris, I.N.; Judson, M.A.; Syrigos, K.N. Sarcoidosis-like reactions induced by checkpoint inhibitors. *J. Thorac. Oncol.* **2018**, *13*, 1076–1082. [CrossRef] [PubMed]

97. Costabel, U.; Hunninghake, G. ATS/ERS/WASOG statement on sarcoidosis. Sarcoidosis statement committee. American thoracic society. European respiratory society. World association for sarcoidosis and other granulomatous disorders. *Eur. Respir. J.* **1999**, *14*, 735–737. [CrossRef] [PubMed]
98. Montaudié, H.; Pradelli, J.; Passeron, T.; Lacour, J.P.; Leroy, S. Pulmonary sarcoid-like granulomatosis induced by nivolumab. *Br. J. Dermatol.* **2017**, *176*, 1060–1063. [CrossRef]
99. Zhao, L.; Wang, L.; Ji, W.; Wang, X.; Zhu, X.; Hayman, J.A.; Kalemkerian, G.P.; Yang, W.; Brenner, D.; Lawrence, T.S.; et al. Elevation of plasma TGF-β1 during radiation therapy predicts radiation-induced lung toxicity in patients with non-small-cell lung cancer: A combined analysis from Beijing and Michigan. *Int. J. Radiat. Oncol. Biol. Phys.* **2009**, *74*, 1385–1390. [CrossRef]
100. Chen, Y.; Rubin, P.; Williams, J.; Hernady, E.; Smudzin, T.; Okunieff, P. Circulating IL-6 as a predictor of radiation pneumonitis. *Int. J. Radiat. Oncol. Biol. Phys.* **2001**, *49*, 641–648. [CrossRef]
101. Park, K.-J.; Oh, Y.-T.; Kil, W.-J.; Park, W.; Kang, S.-H.; Chun, M. Bronchoalveolar lavage findings of radiation induced lung damage in rats. *J. Radiat. Res.* **2009**, *50*, 177–182. [CrossRef]
102. Hong, J.-H.; Jung, S.-M.; Tsao, T.C.Y.; Wu, C.-J.; Lee, C.-Y.; Chen, F.-H.; Hsu, C.-H.; Mcbride, W.H.; Chiang, C.-S. Bronchoalveolar lavage and interstitial cells have different roles in radiation-induced lung injury. *Int. J. Radiat. Biol.* **2003**, *79*, 159–167. [CrossRef]
103. Toma, C.L.; Serbescu, A.; Alexe, M.; Cervis, L.; Ionita, D.; Bogdan, M.A. The bronchoalveolar lavage pattern in radiation pneumonitis secondary to radiotherapy for breast cancer. *Maedica* **2010**, *5*, 250.
104. Crohns, M.; Saarelainen, S.; Laine, S.; Poussa, T.; Alho, H.; Kellokumpu-Lehtinen, P. Cytokines in bronchoalveolar lavage fluid and serum of lung cancer patients during radiotherapy—Association of interleukin-8 and VEGF with survival. *Cytokine* **2010**, *50*, 30–36. [CrossRef]
105. Yamagishi, T.; Kodaka, N.; Kurose, Y.; Watanabe, K.; Nakano, C.; Kishimoto, K.; Oshio, T.; Niitsuma, K.; Matsuse, H. Analysis of predictive parameters for the development of radiation-induced pneumonitis. *Ann. Thorac. Med.* **2017**, *12*, 252. [PubMed]
106. Aso, S.; Navarro-Martin, A.; Castillo, R.; Padrones, S.; Castillo, E.; Montes, A.; Martínez, J.I.; Cubero, N.; López, R.; Rodríguez, L.; et al. Severity of radiation pneumonitis, from clinical, dosimetric and biological features: A pilot study. *Radiat. Oncol.* **2020**, *15*, 1–10. [CrossRef] [PubMed]
107. Zarogoulidis, P.; Rapti, A.; Sardeli, C.; Chinelis, P.; Athanasiadou, A.; Paraskevaidou, K.; Kallianos, A.; Veletza, L.; Trakada, G.; Hohenforst-Schmidt, W.; et al. Re-biopsy after relapse of targeted therapy. T790M after epidermal growth factor mutation, where and why based on a case series. *Respir. Med. Case Rep.* **2017**, *21*, 171–175. [CrossRef] [PubMed]
108. Kawamura, T.; Kenmotsu, H.; Taira, T.; Omori, S.; Nakashima, K.; Wakuda, K.; Ono, A.; Naito, T.; Murakami, H.; Mori, K.; et al. Rebiopsy for patients with non-small-cell lung cancer after epidermal growth factor receptor-tyrosine kinase inhibitor failure. *Cancer Sci.* **2016**, *107*, 1001–1005. [CrossRef] [PubMed]

diagnostics

MDPI

Case Report

IDH Mutations Are Potentially the Intrinsic Genetic Link among the Multiple Neoplastic Lesions in Ollier Disease and Maffucci Syndrome: A Clinicopathologic Analysis from a Single Institute in Shanghai, China

Chunyan Chen [1,†], Jian Li [2,3,†], Ting Jiang [1], Juan Tang [1], Zhichang Zhang [4], Yanli Luo [1], Xinpei Wang [5], Keyang Sun [1], Zhiming Jiang [1], Juan Zhou [1,*] and Zhiyan Liu [1,*]

1 Department of Pathology, Shanghai Sixth People's Hospital Affiliated to Shanghai Jiao Tong University School of Medicine, Shanghai 200233, China
2 Department of Pathology, Peking University Shenzhen Hospital, Shenzhen 518036, China
3 State Key Laboratory of Chemical Oncogenomics, Peking University Shenzhen Graduate School, Shenzhen 518055, China
4 Department of Orthopaedics, Shanghai Sixth People's Hospital Affiliated to Shanghai Jiao Tong University School of Medicine, Shanghai 200233, China
5 Department of Pathology, School of Basic Medical Sciences, Cheeloo College of Medicine, Shandong University, Jinan 250012, China
* Correspondence: zjzycn@aliyun.com (J.Z.); zhiyanliu@shsmu.edu.cn (Z.L.)
† These authors contributed equally to this work.

Citation: Chen, C.; Li, J.; Jiang, T.; Tang, J.; Zhang, Z.; Luo, Y.; Wang, X.; Sun, K.; Jiang, Z.; Zhou, J.; et al. *IDH* Mutations Are Potentially the Intrinsic Genetic Link among the Multiple Neoplastic Lesions in Ollier Disease and Maffucci Syndrome: A Clinicopathologic Analysis from a Single Institute in Shanghai, China. *Diagnostics* **2022**, *12*, 2764. https://doi.org/10.3390/diagnostics12112764

Academic Editors: Yuli Huang, Yong Yuan and Peisong Chen

Received: 15 October 2022
Accepted: 10 November 2022
Published: 11 November 2022

Publisher's Note: MDPI stays neutral with regard to jurisdictional claims in published maps and institutional affiliations.

Abstract: Background: This study aims to investigate isocitrate dehydrogenase gene mutations in patients with the non-hereditary skeletal disorders of Ollier disease and Maffucci syndrome, particularly in the extraosseous tumours. Methods: A total of 16 tumours from three patients with Ollier disease and three patients with Maffucci syndrome were collected. Sanger sequencing was applied to determine the hotspot mutations of *IDH1* and *IDH2* genes in multiple neoplastic tissues. Results: A majority of the tumours displayed an *IDH1* mutation (p.R132C in 11 tumours including the paediatric ovarian tumour from one patient with Ollier disease, 4 cutaneous haemangiomas from three patients with Maffucci syndrome, 5 enchondromas and 1 chondrosarcoma; p.R132H in 2 cartilaginous tumours from one patient). Conclusions: *IDH1* mutations were demonstrated in multiple cartilaginous tumours and extraskeletal neoplasms in this case series. Specifically, identical *IDH1* mutations were confirmed in the separate lesions of each patient. These results are in concordance with findings that have been reported. However, here, we additionally reported the first case of Ollier disease with an ovarian tumour, which harboured the identical *IDH1* mutation with the corresponding cartilaginous tumour. We further provided evidence that *IDH* mutations are the potential genetic links among the multiple neoplastic lesions of Ollier disease and Maffucci syndrome.

Keywords: isocitrate dehydrogenase; Ollier disease; Maffucci syndrome; cartilage tumour; extraskeletal neoplasms; haemangioma; paediatric ovarian tumour

1. Introduction

Enchondromatosis is a syndrome hallmarked by a benign cartilage-forming tumour within the bones. The most common subtypes are non-hereditary Ollier disease and Maffucci syndrome [1,2], mainly involving the short tubular bone and long bone of the limbs asymmetrically, and the latter is accompanied by simultaneous haemangioma of the dermis or subcutis or internal organs. Patients with Ollier disease and Maffucci syndrome often manifest in early childhood, and surgical intervention is usually needed to correct deformities in the process of growth and development. However, the incidence of malignant transformation of enchondromas to chondrosarcoma is high; it ranges from 5% to 50% in

Ollier disease and could be as high as 53% in Maffucci syndrome [1,3,4]. In addition, both disorders have an increased risk of extraosseous malignant tumours, and particularly of ovarian juvenile granulosa cell tumours and central nervous system gliomas [2].

As is known, the hallmark genetic abnormality of enchondroma associated with Ollier disease and Maffucci syndrome is the somatic point mutations on isocitrate dehydrogenase (*IDH*) genes, *IDH1* and *IDH2* [5–8]. Tumours in separate sites of one individual display identical mutations. However, the genetic correlation among multiple tumours especially in extraosseous malignancies associated with the two disorders remains uncertain.

Herein, we retrospectively reviewed three cases of Ollier disease and three cases of Maffucci syndrome. *IDH1* and *IDH2* gene mutations were performed in a series of tumours from the 6 patients, including 1 ovarian tumour, 4 vascular tumours and 11 cartilage tumours. The specific *IDH* mutations within cartilage tumours and non-cartilage tumours of Ollier disease and Maffucci syndrome were compared.

2. Case Report

2.1. Case Selection

Three patients with Ollier disease and three patients with Maffucci syndrome were selected from Shanghai Sixth People's Hospital Affiliated to Shanghai Jiao Tong University School of Medicine during the period from 2016 to 2020. Ethical approval was given by the medical ethics committee of this hospital. The information of the patients was collected, including clinical symptoms, imaging reports (X-ray, CT or MRI) and follow-up data. All haematoxylin and eosin stain (H&E) sections were confirmed by C.Z. and J.Z. A total of 16 tumours and paired normal tissues were sampled from the six patients for the detection of *IDH1* and *IDH2* mutations by Sanger sequencing. Five enchondromas, one chondrosarcoma and one ovarian tumour were from the three patients with Ollier disease. Four enchondromas, one chondrosarcoma and four vascular tumours were from the three patients with Maffucci syndrome.

2.2. Clinical Data

All patients had no corresponding family history. The sites of the tumours ranged from one or several to dozens of lesions at most (Table 1). The cases of Ollier disease included two females and one male, and the age at diagnosis was mainly at young age (Table 1, case 1–3). In case one, the CT revealed lytic bone destructions in the right tibial metaphyseal, superior femur, ilium and pubis (Figure 1A), and the MRI showed a huge mass in the pelvis (Figure 1B,C). In case two, the patient presented with a nodular lesion of the right middle finger at 3-years-old. Eccentric low-density shadows were shown gradually in the proximal segment of the right middle finger and the right third metacarpal bone during the 13 years follow-up by X-rays. In case three, a swelling was found in the right finger at 7-years-old. The finger gradually enlarged in the following 48 years. Surgery was rendered because the nodule grew fast at the age of 55. Multiple nodular lesions with eccentric expansive growth and irregular calcification were demonstrated by imaging (Figure 1D,E), which locally penetrated the bone cortex and extended into the surrounding soft tissue.

Maffucci syndrome (cases 4–6) included two females and one male (Table 1). In case four, the patient presented with a swelling of the right finger and a left ulnar lesion at the age of three. The left ulnar nodule was proven to be enchondroma histologically by excision at the age of six. In the following 14 years, multiple intraosseous eccentric masses were presented gradually on his right fingers, accompanied by multiple subcutaneous soft tissue masses. Case five began to have nodules in her hands and feet when she was only two–three-years-old. In the following 33 years, extensive purple-blue skin masses appeared subsequently in the hands, feet, limbs, shoulders and back (Figure 1F,G). The patient showed rapid progression of swelling and pain in her right knee for about half a year before her consultation. The PET-CT scan showed multiple expansive bone destructions with irregular calcification and an accompanying soft tissue mass surrounding the right proximal tibia lesion (Figure 1H). In case six, a deformity of the left radius and ulna was identified at the age of six. An X-ray analysis showed a curved deformity of the left radius and ulna, and eccentric low-density shadow of multiple short tubular bones of the left hand and foot and the left distal tibia (Figure 1I,J). MRI showed multiple high-density nodules in the left plantar and medial subcutaneous.

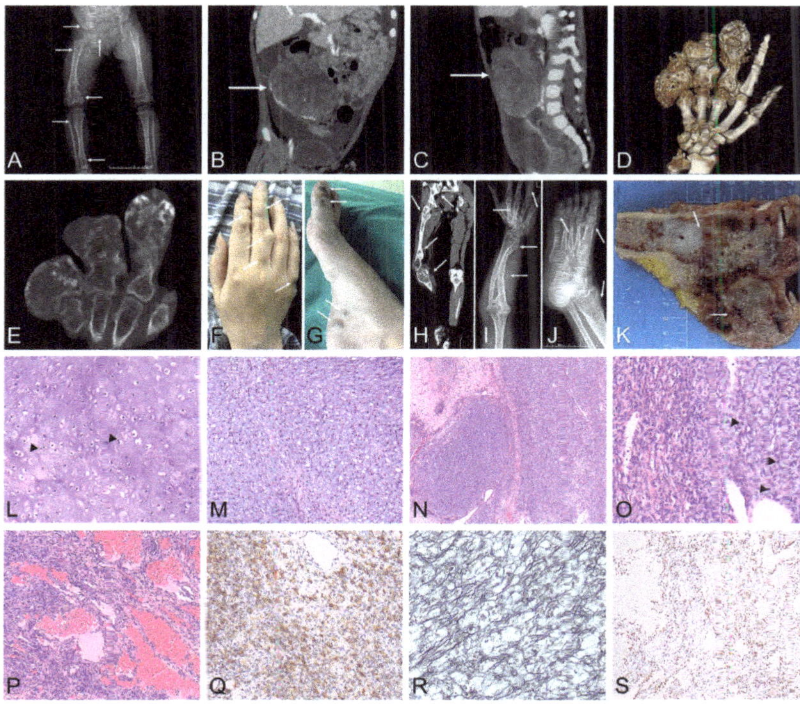

Figure 1. (**A**) Spotted bone destruction of the right upper femur, tibial metaphysis, iliac and pubis was shown by X-ray (arrow). (**B**,**C**) A huge soft tissue mass in pelvis could be observed from the MRI (arrow). (**D**,**E**) Multiple expansive bone destruction was shown in the right little finger; ring finger; middle finger; the third, fourth and fifth metacarpal bones; and the distal ulna by three-dimensional reconstruction and CT. (**F**,**G**) Multiple purplish blue subcutaneous nodules were found in the hands and feet, protruding on the surface of the skin (arrow). (**H**) Multiple bone destruction was seen in bilateral ilium, right acetabulum, left ischium, bilateral femur, right tibia, fibula and calcaneus by PET-CT (arrow, still image) (**I**,**J**) A curved deformity of the left radius and ulna, and eccentric low-density shadow of multiple short tubular bones of left hand and foot and the left distal tibia were presented by plain X-ray (arrow). (**K**) Well-differentiated hyaline cartilage (arrow, up) and dark red sticky area involving the surrounding soft tissue were seen in the cut section of the right proximal tibial tumour (arrow, down). (**L**) Active chondrocytes were distributed in clusters, and atypical binucleate cells (arrow) could occasionally be seen (Hematoxylin-eosin staining, $\times 200$). (**M**) Chondrocytes with severe atypia could be demonstrated in the sarcomatous area, no more cartilage lacunae could be found (Hematoxylin-eosin staining, $\times 100$). (**N**) Lobular-patterned ovarian sex cord-stromal tumour with mucinous degeneration was seen at low power (Hematoxylin-eosin staining, $\times 40$). (**O**) The tumour cells had obvious nucleoli and light eosinophilic cytoplasm, and mitotic figures (arrow) were easy to find at high power (Hematoxylin-eosin staining, $\times 200$). (**P**) Solid spindle cell and cavernous haemangioma-like areas of Maffucci syndrome (Hematoxylin-eosin staining, $\times 100$). (**Q**) Positive immunoreaction with α-inhibin was seen in the cytoplasm in ovarian tumour (EnVision, $\times 200$). (**R**) Reticular fibres were confirmed in the ovarian sex cord-stromal tumour (Reticular fibre staining, $\times 200$). (**S**) Positive immunoreaction with ERG was observed in the nucleus of the haemangioma (EnVision, $\times 100$).

Table 1. Clinicopathological data.

No	Sex	Onset Age	Age of Diagnosis	Locations	Treatment	Follow-Up (m)
1	F	2 yr	4 yr	Right upper femur, ilium, pubis, tibial metaphysis, right ovary right	Right upper femur biopsy and right ovarian mass resection	31
2	F	3 yr	18 yr	Right middle finger proximal segment, right third metacarpal bone, index finger proximal segment	Curettage of intraosseous lesions	40
3	M	7 yr	56 yr	Right little finger, ring finger, middle finger, right third, fourth, fifth metacarpal bone, right distal ulna	Resection of the fourth and fifth phalanges and half palms of the right hand	48
4	F	3 yr	20 yr	Left ulna, right index finger, little finger and ring finger; the right wrist and little finger subcutaneous soft tissue	Curettage of intraosseous lesions Resection of soft tissue mass	68
5	F	2 yr	36 yr	Bilateral scapula, sternum, left humerus, bilateral ilium, right acetabulum, left ischium, bilateral femur, right tibiofibula, right calcaneus; extensive subcutaneous nodules with calcification	Resection of right proximal tibia tumour Excision of subcutaneous mass in right leg	40
6	M	6 yr	33 yr	Multiple phalanges, metacarpals, ulna and radius in left hand, metatarsals, phalanges and distal tibia in left foot, and subcutaneous soft tissue nodules in the dorsum, sole and medial side of left foot	Curettage of short tubular lesions in left foot Resection of 2 subcutaneous nodules in the left foot	22

Benign cartilaginous lesions from six patients underwent curettage selectively. Resection was performed in 2 cartilaginous tumours which manifested local malignant transformation to chondrosarcoma from 2 patients, respectively. The ovarian tumour in case one was resected. Four subcutaneous nodules were resected from three cases of Maffucci syndrome.

2.3. Pathological Examination

The gross appearance of the chondrogenic tumour tissue was grey, translucent and cartilage-like. In case three: obvious deformity and swelling of the phalanx was observed in the half palmar resection specimen; multiple nodular cartilaginous lesions destroyed local bone cortex and invaded into the surrounding soft tissue. In case five: a translucent cartilaginous mass in the medullary cavity extending from the tibial bone end towards the metaphysis was demonstrated in the resected specimen of the right proximal tibial; a dark red jelly-like tumour with an irregular margin penetrated into the bone cortex and the surrounding soft tissue (Figure 1K).

All enchondromas showed a similar lobulated structure at low power, high cellularity at high power, accompanied by mild atypia and binuclear cells occasionally (Figure 1L). In the chondrosarcoma of 2 cases, atypical tumour cells with extensive mucinous degeneration of the cartilage matrix infiltrated into the cortex and soft tissue (1M). The ovarian tumour of case one consisted of primitive and immature tumour cells within the mucinous degeneration of interstitial components distributing in a flaky pattern with fibrous separation at low power, (Figure 1N). The tumour cells were ovate to elongated with light eosinophilic cytoplasm and had obvious nucleoli at high power (Figure 1O). Three cases with Maffucci syndrome presented a mixed spindle cell haemangioma and cavernous haemangioma (Figure 1P).

2.4. Immunohistochemical Features and Special Staining

The ovarian tumours in Ollier disease showed an immunoreaction with α-inhibin (Figure 1Q), and reticulin staining displayed tumour cells surrounded by reticular fibres (Figure 1R). The spindle cell haemangioma and cavernous haemangioma in Maffucci syndrome were positive for ERG (Figure 1S).

2.5. Molecular Features

IDH1 and *IDH2* mutations were detected by Sanger sequence. The molecular features were summarised in Table 2. *IDH1* mutations were discovered in most cartilage tumours except for three enchondromas. All 4 haemangiomas detected an *IDH1* mutation. The ovarian tumour from a patient with Ollier disease was shown to carry the R132C *IDH1* mutation. Multiple separate tumours in one individual shared identical *IDH1* mutations. The *IDH1* mutation variant of three patients with Ollier disease were p.R132C, p.R132C and p.R132H, respectively, and that of three patients with Maffucci syndrome was all p.R132C (Figure 2). The detection rate of *IDH1* mutations in cartilage tumours was 73% (8/11). All the non-chondrogenic tumours were detected to have identical hotspot mutation locations of *IDH1* with the corresponding cartilage tumours. There were no *IDH2* mutations detected.

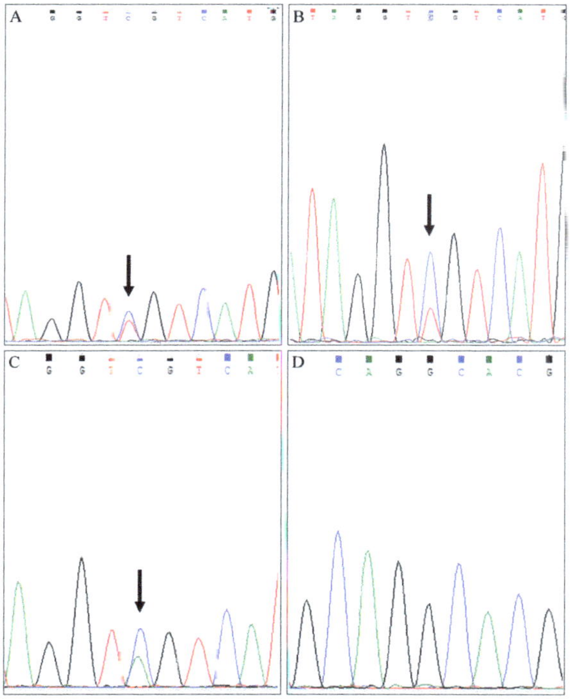

Figure 2. (**A**) *IDH1* mutation in ovarian tumour: p.R132C (c.394C > T) (indicated by arrow). (**B**,**C**) *IDH1* mutations in cartilage tumours: p.R132C (c.394C > T) (indicated by arrow) and p.R132H (c.395G > A) (indicated by arrow). (**D**) *IDH2* mutation: wild type.

2.6. Prognosis

Follow-up information as showed in Table 1 was available for all patients by radiographic examination and telephone interviews. In case two, no relapse or new lesions were demonstrated after surgery up to now. The remaining five patients were alive without recurrence and distant metastasis.

Table 2. Detection of *IDH1* mutation in 16 tumours.

No	Tumour Location	*IDH1* Mutation Results	*IDH2* Mutation Results
1	Right ovarian tumour	c.394C > T p.R132C	WT
	Endochondroma of right upper femur	c.394C > T p.R132C	WT
2	Endochondroma of proximal segment of right middle finger	c.394C > T p.R132C	WT
	Endochondroma of right third metacarpal bone	c.394C > T p.R132C	WT
	Endochondroma of proximal segment of index finger	WT	WT
3	Endochondroma of right little finger	c.395G > A p.R132H	WT
	Chondrosarcoma of right ring finger	c.395G > A p.R132H	WT
4	Subcutaneous haemangioma of right wrist	c.394C > T p.R132C	WT
	Subcutaneous haemangioma of right little finger	c.394C > T p.R132C	WT
	Endochondroma of right index finger	c.394C > T p.R132C	WT
	Endochondroma of right ring finger	WT	WT
5	Subcutaneous haemangioma of right leg	c.394C > T p.R132C	WT
	Chondrosarcoma of right proximal tibia	c.394C > T p.R132C	WT
6	Subcutaneous haemangioma of left medial malleolus	c.394C > T p.R132C	WT
	Endochondroma of left middle finger	c.394C > T p.R132C	WT
	Endochondroma of distal left tibia	WT	WT

3. Discussion

Ollier disease and Maffucci syndrome are characterized by multiple enchondromatosis, and the phenotypic difference between them is mainly the existence of vascular anomalies [1,9]. Maffucci syndrome may show more severe mesodermal dysplasia in that the extraskeletal areas are often involved. The cartilaginous skeletal dysplastic are easy to be identified; however, minimal soft tissue haemangiomas are hard to recognize initially. Therefore, Maffucci syndrome is usually mistaken for Ollier disease in the early stage. About 75% of those patients are diagnosed before the age of 20, and about 45% develop symptoms before the age of six [2,3]. All of our 6 patients presented symptoms before puberty, which were consistent with the previous studies. The incidence of secondary chondrosarcoma associated with Maffucci syndrome and Ollier disease is much higher than that in solitary enchondromas (3). As presented in our group, two cartilaginous tumours developed into chondrosarcomas, which were located in the short and long tubular bone, respectively. However, it is unnecessary to distinguish these two disorders from the clinical perspective; both of them need a lifetime follow-up by imaging after diagnosis. A timely surgical intervention is essential whenever the tumour has a tendency of malignant transformation.

IDH1 and *IDH2* genes are located on the long arms of chromosome 2 and chromosome 15, respectively [5–8]. It was reported that *IDH* mutations are identified in 81% of patients with Ollier disease and 77% of patients with Maffucci syndrome. The IDH mutation rate of benign cartilaginous tumours is about 87%, and that of vascular lesions is about 70%. In the current study, Sanger sequencing was used to detect the mutations of *IDH1* and *IDH2* in the 16 tumours of six patients. The mutation rate of *IDH1* in cartilaginous tumours was about 73% (8/11), which is slightly lower than reported. All 4 haemangiomas were confirmed to harbour a *IDH1* mutation. Since the mutant allele could be present as low as 1% (8), which may be below the detection level of the targeted mutation analysis, the true frequency of *IDH1/2* mutation in enchondroma could be underestimated. Saiji et al. reported the mutation rate of *IDH* was 100% in 13 enchondromas from eight patients with Ollier disease using next generation sequencing (NGS) [10]. The difference in methods may explain the lower occurrence (73%) of *IDH* mutations found in our current study, as NGS captures a broader spectrum of mutations than Sanger sequencing.

IDH mutations in enchondromatosis associated with Ollier disease most frequently involve Arginine 132 (R132) in exon 4 of *IDH1*, while the *IDH2* gene mutation (mainly p.R172S) appears to be rarely affected. In comparison with Ollier disease, enchondromas and spindle-cell haemangiomas in Maffucci syndrome are characterized by the exclusive predilection of *IDH1*. The mutation variants include mostly R132C, and rarely R132H, R132G. Here, we reported an R132C mutation of *IDH1* in two cases of Ollier disease and three cases of Maffucci syndrome, and R132H in the third case of Ollier disease. All the segregated extraskeletal non-cartilaginous tumours were found to possess the identical *IDH1* mutations with corresponding cartilage tumours. There were no *IDH2* mutations found.

IDH1/2 plays important roles in the tricarboxylic acid cycle by converting isocitrate to α-ketoglutarate (α-KG) [11]. The mutations of *IDH1/2* make the mutated enzyme catalyse α-ketoglutarate (α-KG) to (D)-2-hydroxyglutaric acid (D-2-HG) [12–15], which has already been confirmed that it could induce multiple enchondromas during bone formation through promoting chondrocyte differentiation and inhibiting osteogenic differentiation of mesenchymal stem cells [16,17]. As presented here, *IDH1* mutations were identified both in six enchondromas and two secondary chondrosarcomas. However, these mutations are merely an early driver of tumourigenesis in cartilage tumour formation, and further malignant transformation into chondrosarcoma may require additional genetic events [7,18]. Furthermore, the excessive intracellular accumulation of D-2-HG impairs histone demethylation, and leads to aberrant epigenetics changes of DNA methylation and abnormal cellular differentiation [7,19–23]. Therefore, *IDH1/2* mutations are believed to contribute to the development of a variety of tumours.

Ollier disease and Maffucci syndrome share the same phenotypic characteristics: (1) non-hereditary; (2) related tumours with asymmetrical polyostotic distribution, suggesting the genetic characteristic of somatic mosaic mutations. Multiple tumours in the same individual possess identical *IDH1* mutations, and a very low frequency of *IDH* mutant protein is observed in normal tissue in patients with Ollier disease and Maffucci syndrome [6–8]. Those reports further confirmed that *IDH* mutations occur in a somatic mosaic pattern. The identification of the same variant in multiple tumours available for analysis from the six patients in our group highlights the possibility of *IDH* somatic mosaicism; however, neither normal tissue nor blood were examined in the current report.

As is well-known, patients with Ollier disease and Maffucci syndrome have a marked risk of developing non-skeletal diverse malignancies, especially intracranial tumours of glial origin and ovarian sex cord-stromal tumours [2,24,25]. The genetic correlation among multiple tumours especially in extraosseous malignancies associated with the two disorders remains mysterious. We know that *IDH* mutations are well-detected in mostly enchondromas and spindle cell haemangiomas, and occasionally in other non-skeletal malignancies such as gliomas associated with Ollier disease, anaplastic astrocytoma and acute myeloid leukaemia in patients with Maffucci syndrome [26–28]. It seems that a mosaic pattern of *IDH* mutations could explain the pathogenesis of diverse tumours (multiple cartilaginous neoplasms, haemangiomas, gliomas or other uncommon extraosseous malignancies) in the same patient. However, juvenile granulosa cell tumours (JGCTs) are considered the most common reported extraskeletal tumours in patients with Ollier disease or Maffucci syndrome, which strongly suggests the possible connections between them [24,29]. However, there have been no previous reports on molecular features of JGCTs with Ollier disease or Maffucci syndrome. Other variants of ovarian sex cord-stromal tumour have rarely been reported. Only one case of ovarian cell-rich fibroma associated with Ollier disease has been reported to carry an *IDH1* mutation [30], but the detailed mutation position of it was not specified and the *IDH1* mutation status of the corresponding cartilage tumours is lacking.

In our group, the right ovarian tumour with Ollier disease in case one located ipsilaterally to the predominant skeletal dysplasia just as reported. Histologically, it was characterized by primitive undifferentiated ovate to elongated cells accompanied by stromal mucinous degeneration, and the mitotic figures were easy to find, but a pathological mitotic figure was absent. The tumour cells showed no immunoreaction with Desmin, MyoD1, Myogenin, SMA, S100, CD34 and BCOR, which were surrounded by reticular fibres. The only positive immunoreaction with α-inhibin suggested its ovarian sex cord-stroma origination, but this was insufficient for a conclusive diagnosis. Fortunately, an identical *IDH1 R132C* mutation was identified in both the ovarian tumour and the right femoral cartilage tumour. To the best of our knowledge, we reported, for the first time, that an identical *IDH1* mutation was represented in the ovarian tumour and corresponding cartilage tumour of Ollier disease. The demonstration of the *IDH1* mutation both in the ovarian tumour and cartilage tumour is of great significance. Theoretically, as the skeletal system and the gonads are both originated from the mesoderm, the involvement of the ovary is a manifestation of mesodermal dysplasia. Our result suggested that *IDH* mutation is pathognomonic for both the ovarian tumour and cartilage tumour in Ollier disease.

In conclusion, *IDH* mutation may not be merely the initial cause of cartilage tumours; they are probably the intrinsic genetic link among multiple neoplastic lesions in the setting of somatic mosaicism. However, a heterozygotic IDH mutation alone is insufficient to induce gliomas or ovarian tumours in patients with Ollier disease and Maffucci syndrome. It may at least increase the tissue susceptibility of non-cartilaginous tumours' formation. Future studies should demonstrate a causal effect and assess how these gene mutations lead to diverse neoplasms' formation.

Author Contributions: Writing—original draft preparation, C.C. and J.L.; investigation, T.J., J.T., Y.L., X.W., K.S. and Z.Z.; supervision and conception, Z.J., J.Z. and Z.L. All authors have read and agreed to the published version of the manuscript.

Funding: This research was funded by the National Nature Science Foundation of China, grant No. 81972500; clinical research funds from Shanghai Jiao Tong University Affiliated Sixth People's Hospital, grant No. ynhg201914; Shenzhen Science and Technology Program, grant No. JCYJ20220531094017039; and Open research fund of state key laboratory of chemical oncogenomics in 2021.

Institutional Review Board Statement: The study was conducted in accordance with the Declaration of Helsinki and approved by the Institutional Review Board of Shanghai Sixth People's Hospital Affiliated to Shanghai Jiao Tong University School of Medicine, with a protocol number of 2020-KY-041(K), (Approval date: 18 August 2020) and it conformed to the Declaration of Helsinki.

Informed Consent Statement: Informed written consent was obtained from the patients for publication of this case and any accompanying images.

Data Availability Statement: The data that support the findings of this study are available from the corresponding author, Zhiyan Liu, upon reasonable request.

Conflicts of Interest: The authors declare no conflict of interest.

References

1. Anderson, W.J.; Leona, A.D. Updates from the 2020 World Health Organization Classification of Soft Tissue and Bone Tumours. *Histopathology* **2021**, *78*, 644–657. [CrossRef]
2. Pansuriya, T.C.; Kroon, H.M.; Bovée, J.V. Enchondromatosis: Insights on the different subtypes. *Int. J. Clin. Exp. Pathol.* **2010**, *3*, 557–569. [PubMed]
3. Suzan, H.M.V.; Judith, V.M.G.B.; Twinkal, C.P.; Robert, J.G.; Harzem Ozger, P.C.J.; Mikel, S.J.; David, J.B.; Ingrid, C.M.G.; Andreas, L.; Arne, S.; et al. Incidence, predictive factors, and prognosis of chondrosarcoma in patients with Ollier disease and Maffucci syndrome: An international multicenter study of 161 patients. *Oncologist* **2011**, *16*, 1771–1779.
4. Herget, G.W.; Strohm, P.; Rottenburger, C.; Kontny, U.; Kraus, T.; Bohm, J.; Sudkamp, N.; Uhl, M. Insights into Enchondroma, Enchondromatosis and the risk of secondary Chondrosarcoma. Review of the literature with an emphasis on the clinical behaviour, radiology, malignant transformation and the follow up. *Neoplasma* **2014**, *61*, 365–378. [CrossRef]
5. Twinkal, P.; Jan, O.; Tibor, K.; Antonie, T.; Suzan, V.; Luca, S.; Raf, S.; Pancras, H.; Karoly, S.; Judith, B. Genome-wide analysis of Ollier disease: Is it all in the genes? *Orphanet J. Rare Dis.* **2011**, *6*, 2.
6. Fernanda, A.M.; Krisztian, B.; Francesca, M.; Stephen, D.; Dina, H.; Fitim, B.; Robin, P.; Paul, O.D.; Anita, G.; Tim, D.; et al. IDH1 and IDH2 mutations are frequent events in central chondrosarcoma and central and periosteal chondromas but not in other mesenchymal tumours. *J. Pathol.* **2011**, *224*, 334–343.
7. Twinkal, C.P.; Ronald, E.; van Ruler Maayke, A.J.H.; Marieke, L.K.; Jan, O.; Anne-Marie, C.-J.; van Oosterwijk, J.G.; Sofie, L.J.V.; Daniëlle, M.; van Wezel, T.; et al. Somatic mosaic IDH1 and IDH2 mutations are associated with enchondroma and spindle cell hemangioma in Ollier disease and Maffucci syndrome. *Nat. Genet.* **2011**, *43*, 1256–1261.
8. Fernanda, A.M.; Stephen, D.; Dina, H.; Malihe, E.; Fitim, B.; Fiona, B.; Stan, M.; Valeria, F.R.; Stanley, S.K.; Samira, L.; et al. Ollier disease and Maffucci syndrome are caused by somatic mosaic mutations of IDH1 and IDH2. *Nat. Genet.* **2011**, *43*, 1262–1265.

9. Fanburg, J.C.; Meis-Kindblom, J.M.; Rosenberg, A.E. Multiple enchondromas associated with spindle-cell hemangioendotheliomas. An overlooked variant of Maffucci's syndrome. *Am. J. Surg. Pathol.* **1995**, *19*, 1029–1038. [CrossRef]
10. Essia, S.; Gumy, P.F.; Pierre, L.; Christelle, C.B.; Lin, M.N.; Margaret, B.; Laura, M.; Anne-Laure, R. IDH1 immunohistochemistry reactivity and mosaic IDH1 or IDH2 somatic mutations in pediatric sporadic enchondroma and enchondromatosis. *Virchows Arch. Int. J. Pathol.* **2019**, *475*, 625–636.
11. Schaap, F.G.; French, P.J.; Bovée, J.V.M.G. Mutations in the isocitrate dehydrogenase genes IDH1 and IDH2 in tumors. *Adv. Anat. Pathol.* **2013**, *20*, 32–38. [CrossRef] [PubMed]
12. Lenny, D.; White, W.D.; Stefan, G.; Bennett, D.B.; Bittinger, A.M.; Driggers, M.E.; Fantin, R.V.; Gyung, J.H.; Shengfang, J.; Marie, K.C.; et al. Cancer-associated IDH1 mutations produce 2-hydroxyglutarate. *Nature* **2009**, *462*, 739–744.
13. Prensner, R.J.; Chinnaiyan, M.A. Metabolism unhinged: IDH mutations in cancer. *Nat. Med.* **2011**, *17*, 291–293. [CrossRef] [PubMed]
14. Reitman, J.Z.; Genglin, J.; Karoly, K.D.; Ivan, S.; Jian, Y.; Kinzler, K.W.; Yiping, H.; Darell, B.D.; Bert, V.; Hai, Y. Profiling the effects of isocitrate dehydrogenase 1 and 2 mutations on the cellular metabolome. *Proc. Natl. Acad. Sci. USA* **2011**, *108*, 3270–3275. [CrossRef] [PubMed]
15. Katarína, S.; Petr, J. The Role of Mitochondrial NADPH-Dependent Isocitrate Dehydrogenase in Cancer Cells. *Int. J. Cell Biol.* **2012**, *2012*, 273947.
16. Suijker, J.; Baelde, H.J.; Roelofs, H.; Cleton-Jansen, A.M.; Bovée, J.V. The oncometabolite D-2-hydroxyglutarate induced by mutant IDH1 or -2 blocks osteoblast differentiation in vitro and in vivo. *Oncotarget* **2015**, *6*, 14832–14842. [CrossRef]
17. Yonghui, J.; Hassan, E.; Makoto, W.; Sakura, T.; Sho, H.; Kazuhito, M.; Knut, W.; Yukiko, K.; Sanae, N.; Makoto, I.; et al. Mutant IDH1 Dysregulates the Differentiation of Mesenchymal Stem Cells in Association with Gene-Specific Histone Modifications to Cartilage- and Bone-Related Genes. *PLoS ONE* **2015**, *10*, e0131998.
18. Suijker, J.; Oosting, J.; Koornneef, A.; Struys, E.A.; Salomons, G.S.; Schaap, F.G.; Waaijer, C.J.; Wijers-Koster, P.M.; Briaire-de Bruijn, I.H.; Haazen, L.; et al. Inhibition of mutant IDH1 decreases D-2-HG levels without affecting tumorigenic properties of chondrosarcoma cell lines. *Oncotarget* **2015**, *6*, 12505–12519. [CrossRef]
19. War, P.S.; Cross, J.R.; Lu, C.; Weigert, O.; Abel-Wahab, O.; Levine, R.L.; Weinstock, D.M.; Sharp, K.A.; Thompson, C.B. Identification of additional IDH mutations associated with oncometabolite R(-)-2-hydroxyglutarate production. *Oncogene* **2012**, *31*, 2491–2498.
20. Dang, L.; Jin, S.; Su, S.M. IDH mutations in glioma and acute myeloid leukemia. *Trends Mol. Med.* **2010**, *16*, 387–397. [CrossRef]
21. Figueroa, M.E.; Abdel-Wahab, O.; Lu, C.; Ward, P.S.; Patel, J.; Shih, A.; Li, Y.; Bhagwat, N.; VasanthaKumar, A.; Fernandez, H.F.; et al. Leukemic IDH1 and IDH2 Mutations Result in a Hypermethylation Phenotype, Disrupt TET2 Function, and Impair Hematopoietic Differentiation. *Cancer Cell* **2010**, *18*, 553–567. [CrossRef] [PubMed]
22. Noushmehr, H.; Weisenberger, D.J.; Diefes, K.; Phillips, H.S.; Pujara, K.; Berman, B.P.; Pan, F.; Pelloski, C.E.; Sulman, E.P.; Bhat, K.P.; et al. Identification of a CpG Island Methylator Phenotype that Defines a Distinct Subgroup of Glioma. *Cancer Cell* **2010**, *17*, 510–522. [CrossRef] [PubMed]
23. Waitkus, M.S.; Diplas, B.H.; Yan, H. Biological Role and Therapeutic Potential of IDH Mutations in Cancer. *Cancer Cell* **2018**, *34*, 186–195. [CrossRef]
24. Burgetova, A.; Matejovsky, Z.; Zikan, M.; Slama, J.; Dundr, P.; Skapa, P.; Benkova, K.; Cibula, D.; Fischerova, D. The association of enchondromatosis with malignant transformed chondrosarcoma and ovarian juvenile granulosa cell tumor (Ollier disease). *Taiwan. J. Obstet. Gynecol.* **2017**, *56*, 253–257. [CrossRef] [PubMed]
25. Catherine, G.; Leora, W.; Stephanie, V.; Glenn, M.W.; Foulkes, D.W. Paediatric ovarian tumours and their associated cancer susceptibility syndromes. *J. Med. Genet.* **2018**, *55*, 1–10.
26. Kunihiko, M.; Kaneko, K.M.; Xing, L.; Masami, H.; Fumiyoshi, F.; Jun, S.; Satoshi, O.; Mika, W.; Yoji, S.; Shigeo, K.; et al. IDH2 and TP53 mutations are correlated with gliomagenesis in a patient with Maffucci syndrome. *Cancer Sci.* **2014**, *105*, 359–362.
27. Masaharu, A.; Masayoshi, Y.; Yoko, M.-T.; Wataru, O.; Kentaro, Y.; Yasuhiro, A.; Junko, T.; Hideaki, S.; Hisashi, Y. Somatic mosaic mutations of IDH1 and NPM1 associated with cup-like acute myeloid leukemia in a patient with Maffucci syndrome. *Int. J. Hematol.* **2015**, *102*, 723–728.
28. Loo, T.C.; Balamurugan, V.; Bingcheng, W.; Tsai, Y.T.; McLendon, E.R. Molecular profiling of different glioma specimens from an Ollier disease patient suggests a multifocal disease process in the setting of IDH mosaicism. *Brain Tumor Pathol.* **2018**, *35*, 202–208.
29. Xu, H.S.; Zhong, E.; Rotman, J. Juvenile granulosa cell tumor associated with Maffucci syndrome in pregnancy: A case report. *Clin. Imaging* **2019**, *56*, 77–80. [CrossRef]
30. Kenny, L.S.; Kaushik, P.; Anna, H.; Michael, D.; Flanagan, M.A.; Glenn, M.W. Ovarian cellular fibroma harbouring an isocitrate dehydrogenase 1 (1DH1) mutation in a patient with Ollier disease: Evidence for a causal relationship. *Histopathology* **2013**, *62*, 667–670. [CrossRef]

diagnostics

Systematic Review

Early Prediction and Monitoring of Treatment Response in Gastrointestinal Stromal Tumors by Means of Imaging: A Systematic Review

Ylva. A. Weeda [1,*], Gijsbert M. Kalisvaart [1], Floris H. P. van Velden [1], Hans Gelderblom [2], Aart. J. van der Molen [3], Judith V. M. G. Bovee [4], Jos A. van der Hage [5], Willem Grootjans [1] and Lioe-Fee de Geus-Oei [1,6,7]

1 Department of Radiology, Leiden University Medical Center, 2333 ZA Leiden, The Netherlands
2 Department of Medical Oncology, Leiden University Medical Center, 2333 ZA Leiden, The Netherlands
3 Department of Radiology, Section of Abdominal Radiology, Leiden University Medical Center, 2333 ZA Leiden, The Netherlands
4 Department of Pathology, Leiden University Medical Center, 2333 ZA Leiden, The Netherlands
5 Department of Surgical Oncology, Leiden University Medical Center, 2333 ZA Leiden, The Netherlands
6 Biomedical Photonic Imaging Group, University of Twente, 7522 NB Enschede, The Netherlands
7 Department of Radiation Science & Technology, Technical University of Delft, 2629 JB Delft, The Netherlands
* Correspondence: y.a.weeda@lumc.nl

Citation: Weeda, Y.A.; Kalisvaart, G.M.; van Velden, F.H.P.; Gelderblom, H.; van der Molen, A.J.; Bovee, J.V.M.G.; van der Hage, J.A.; Grootjans, W.; de Geus-Oei, L.-F. Early Prediction and Monitoring of Treatment Response in Gastrointestinal Stromal Tumors by Means of Imaging: A Systematic Review. *Diagnostics* **2022**, *12*, 2722. https://doi.org/10.3390/diagnostics12112722

Academic Editors: Yuli Huang, Yong Yuan and Peisong Chen

Received: 14 October 2022
Accepted: 4 November 2022
Published: 7 November 2022

Abstract: Gastrointestinal stromal tumors (GISTs) are rare mesenchymal neoplasms. Tyrosine kinase inhibitor (TKI) therapy is currently part of routine clinical practice for unresectable and metastatic disease. It is important to assess the efficacy of TKI treatment at an early stage to optimize therapy strategies and eliminate futile ineffective treatment, side effects and unnecessary costs. This systematic review provides an overview of the imaging features obtained from contrast-enhanced (CE)-CT and 2-deoxy-2-[^{18}F]fluoro-D-glucose ([^{18}F]FDG) PET/CT to predict and monitor TKI treatment response in GIST patients. PubMed, Web of Science, the Cochrane Library and Embase were systematically screened. Articles were considered eligible if quantitative outcome measures (area under the curve (AUC), correlations, sensitivity, specificity, accuracy) were used to evaluate the efficacy of imaging features for predicting and monitoring treatment response to various TKI treatments. The methodological quality of all articles was assessed using the Quality Assessment of Diagnostic Accuracy Studies, v2 (QUADAS-2) tool and modified versions of the Radiomics Quality Score (RQS). A total of 90 articles were included, of which 66 articles used baseline [^{18}F]FDG-PET and CE-CT imaging features for response prediction. Generally, the presence of heterogeneous enhancement on baseline CE-CT imaging was considered predictive for high-risk GISTs, related to underlying neovascularization and necrosis of the tumor. The remaining articles discussed therapy monitoring. Clinically established imaging features, including changes in tumor size and density, were considered unfavorable monitoring criteria, leading to under- and overestimation of response. Furthermore, changes in glucose metabolism, as reflected by [^{18}F]FDG-PET imaging features, preceded changes in tumor size and were more strongly correlated with tumor response. Although CE-CT and [^{18}F]FDG-PET can aid in the prediction and monitoring in GIST patients, further research on cost-effectiveness is recommended.

Keywords: gastrointestinal stromal tumor; prediction; response monitoring; FDG-PET; radiomics; tomography; X-ray computed; personalized medicine

1. Introduction

Gastrointestinal stromal tumors (GISTs) are rare mesenchymal neoplasms affecting the entire gastrointestinal tract and are presumed to originate from the interstitial cells of Cajal [1,2]. About 80–90% of GISTs harbor kinase-activating mutations in either receptor tyrosine kinase protein (KIT) or platelet-derived growth factor receptor α (PDGRF-α) [3,4]. Complete surgical excision remains the only curative treatment option for GIST patients. Since GISTs are generally insensitive to radio- and chemotherapy, non-surgical treatment is limited to tyrosine kinase inhibitor (TKI) therapy. This targeted molecular therapy is part of routine clinical practice for unresectable and metastatic disease [5,6].

Adjuvant TKI treatment is used in high-risk GISTs to improve survival [7]. Unfortunately, due to the varying aggressive nature of GISTs, about one-third of the patients will relapse within three years after surgery with curative-intent [8]. For localized disease, TKI treatment can be given to attain size reduction of the primary tumor and improve chances of complete resection while maintaining an acceptable risk of complications [9,10]. About 20–25% of patients do not benefit from the neoadjuvant TKI treatment, as no complete or partial response is observed [11,12]. The rarity and complex biological nature of this disease, makes it difficult to differentiate between good and poor responders. For example, GISTs harboring a KIT exon 11 mutation have a good response to TKI treatment, whereas the same treatment is less effective in tumors with KIT exon 9 mutations [13]. Additionally, progressive disease is common during long-term TKI treatment due to acquired treatment resistance [14,15].

In the era of personalized medicine, it is of utmost importance to evaluate the efficacy of TKI treatment at an early stage in order to optimize therapy strategies and protect patients from futile ineffective treatment, unnecessary side-effects and healthcare costs. Contrast-enhanced computed tomography (CE-CT) and 2-deoxy-2-[^{18}F]fluoro-D-glucose ([^{18}F]FDG) PET/CT are considered useful for diagnosis and response monitoring in GIST patients. The imaging modalities offer information on tumor morphology, perfusion characteristics, as well as tumor glucose metabolism [7]. However, optimal use of imaging for predicting and monitoring TKI treatment in patients with GIST is still a subject of debate. This systematic review aims to elucidate the added value of CE-CT and [^{18}F]FDG PET/CT imaging in the *prediction of response* and *early response monitoring* of TKI treatment in localized and advanced GISTs.

2. Methods

2.1. Search Strategy

From 29 April 2022 to 24 June 2022, the databases of PubMed, Web of Science, the Cochrane Library and Embase were systematically screened using predefined search queries (Supplementary Materials). The following terms and their corresponding synonyms were included: "gastrointestinal stromal tumors", "(neo)adjuvant", "TKI treatment" and "FDG-PET" and "Tomography, X-ray Computed" imaging. The search queries are wide-ranging and seek to cover the aspect of both response monitoring and prediction by including 'monitoring' and 'prediction models' as well as 'radiomics' and 'prognostics'. In addition to these search terms, other terms, such as 'patient selection' and 'personalized medicine', were also added, since these articles presumably covered the subject of TKI treatment evaluation and its efficacy in specific patient groups as well. The search strategy was implemented in consultation with an experienced research directorate, and access to the databases was granted by the Leiden University Medical Center.

2.2. Article Selection

Articles were screened and considered eligible for full-text assessment if the title or abstract mentioned (i) quantitative outcome measures to evaluate the efficacy of imaging features (ii) retrieved from CE-CT and/or [^{18}F]FDG PET/CT imaging (iii) in predicting or monitoring (neo)adjuvant TKI treatment response (iv) in localized and advanced GISTs. Response monitoring is defined as the evaluation of disease over the course of treatment using multiple medical imaging time points. Predicting response, however, solely involves the use of baseline scans made prior to TKI treatment administration. Articles assessing the prognostic value of different clinical and imaging parameters (e.g., risk of recurrence and metastatic potential) that can guide TKI treatment duration or timing for specific patient groups were also included, since these findings may improve patient selection in the future. Exclusion criteria comprised non-English and non-human studies, reviews, guidelines, recommendations, editorials, conference papers and abstracts. Case reports and studies analyzing less than ten patients were also excluded. If the title and abstract did not contain sufficient information, full-text evaluation was used for judgement of relevance.

Subsequently, the articles were screened on full-text and excluded if they did not meet the previously mentioned inclusion criteria or if full-texts were not available. During this assessment, the focus was primarily on quantitative outcome measure(s) of studies. Outcome measures that were included in this analysis were correlations, associations, area under the curve (AUC), sensitivity, specificity and accuracy.

Finally, the reference lists from included articles were screened to find additional articles on this topic. The articles were independently assessed by the first two authors (Y.A.W., G.M.K.) and in cases of discrepancy, consensus reading was performed to make a final decision that led to either inclusion or exclusion.

2.3. Quality Assessment

Articles using a radiomics pipeline were assessed through the radiomics quality score (RQS). The RQS is a scoring system that assigns points to a radiomics study based on specific criteria, where a maximum score of 36 points can be awarded. In this paper, the RQS is modified to focus on the methodological aspects of the included studies. The following criteria were omitted from the RQS, yielding a modified RQS (RQS_m); 'imaging at multiple time points', 'trial database registry' and 'multivariable analysis on non-radiomics features', since they were considered less relevant for the quality of the obtained models [16]. The criteria from the RQS_m were also used to create a quality assessment tool to assess studies on non-radiomics prediction models and correlational research. Modifying the RQS_m for non-radiomics studies yielded the $RQS_{m,nonrad}$. This $RQS_{m,nonrad}$ had a maximum score of ten points (Supplementary Materials). Articles were considered high quality if they reached a score above 50%. To assess applicability concerns and the risk of bias in articles covering the topic of monitoring, the Quality Assessment of Diagnostic Accuracy Studies Tool, Version 2 (QUADAS-2) was applied [17]. Articles on response monitoring were considered to have a high risk of bias or applicability concerns if two or more of the domains were scored as 'high' or 'unclear'. Subsequently these articles were scored as low-quality.

Quality assessment was performed by the first author (Y.A.W.). The quality score was not considered as an exclusion criterion, as the authors considered it important to review all relevant evidence [17–19].

2.4. Data Analysis

The eligible studies were categorized based on their topic concerning either response prediction or therapy monitoring. From each study, detailed information on the publication year, first author, patient groups, type of TKI treatment and imaging technique(s), was obtained. The specific CE-CT and [^{18}F]FDG-PET imaging features and their corresponding conclusions on efficacy, along with the attributed quality score, were briefly summarized. In the results section, only studies that were considered to be high-quality, were analyzed in depth by clarifying conclusions on clinical relevance, discrepancies in results and insights on biological correlates.

2.4.1. Response Prediction

In response prediction, imaging features from baseline/diagnostic CE-CT and [^{18}F]FDG-PET/CT are retrieved to predict responder status, prior to TKI administration. Articles on this topic were divided into five categories: mutational status, proliferative activity, risk stratification, radiological response and prognosis. These categories were considered important, as they all influence treatment strategies. Clinical genotyping is essential for clinical decision making regarding neoadjuvant therapy, since the sensitivity and resistance towards TKI treatment in GISTs is dependent on the mutational status. In addition, patients with a high-risk GIST (and thus high proliferative activity) receive adjuvant TKI treatment for a period of three years to eliminate remaining disease and reduce chances of relapse [7]. Predicting whether patients will have a radiological response or a good prognosis at baseline could also aid the development of a more personalized TKI treatment.

2.4.2. Therapy Monitoring

In therapy monitoring, one uses the visual and quantitative differences between baseline and follow-up scans to determine treatment response. The efficacy of CE-CT and [^{18}F]FDG-PET are first discussed separately, followed by a qualitative comparison between both imaging modalities.

3. Results

3.1. Search Strategy and Article Selection

The search query identified a total of 599 articles from the databases of PubMed, Web of Science, the Cochrane library and Embase. The study selection process led to a total of 90 articles eligible for analysis (Figure 1). Articles that were excluded based on imaging criteria included, for example, the use of radiotracers other than [^{18}F]FDG [20]. Additionally, some articles discussed the use of molecular genotyping and DNA sequencing to predict or determine response and therefore did not involve the use of any imaging modality [21]. Other excluded articles discussed the efficacy of a specific TKI treatment but did not quantify the efficacy of imaging features in predicting or monitoring response [22,23]. Of the 90 eligible articles, 67 were concerning response prediction [24–90] and 23 discussed response monitoring [91–113].

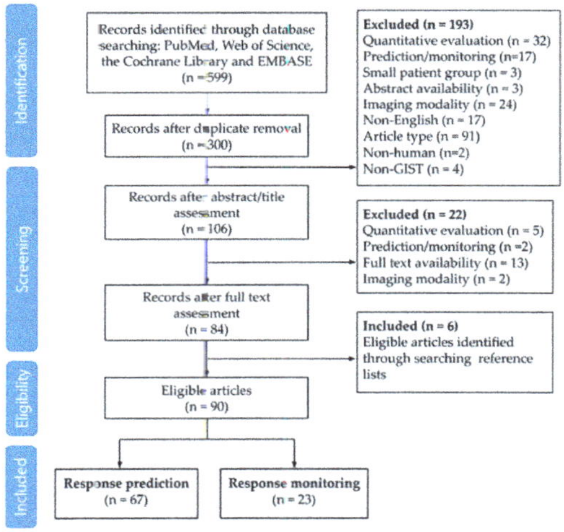

Figure 1. Preferred Reporting Items for Systematic Reviews and Meta-analysis (PRISMA) flowchart showing all the exclusion criteria. A total of 90 articles were included for this systematic review. Sixty-seven articles covered the topic of response prediction, and 23 articles covered the topic of response monitoring.

3.2. Quality Assessment

Twenty-two articles discussed the use of radiomic models, and six out of 22 studies were of low quality (score < 50%). The mean RQS_m of the included articles was 13.5 (SD ± 2.60) out of 26 points. Low scores were mainly caused by a lack of transparency, biological correlates and gold standard comparison. Two articles received a score of 18 points (69.2%), which was the highest attributed score [70,88]. The forty-five studies on non-radiomic prediction models and correlational research scored an average $RQS_{m,nonrad}$ of 3.91 (SD ± 1.23) out of ten points, where eighteen articles scored above 50.0%. This was mainly caused by the fact that only a few articles used gold standard comparison [31,35,46] or an undescribed test set to validate their results [44,48,68]. The results of the QUADAS-2 tool are graphically displayed in Figure 2. Eight articles on response monitoring had high risk of bias or concerns for applicability and were therefore scored as low-quality. High risk of bias was often introduced by using reference standards involving follow-up (e.g., progression free survival, overall survival, time-to-treatment failure). Concerns for applicability were mainly caused by a lack of reporting on patient characteristics. In this way, judgement on whether the included patients matched the review question was unclear.

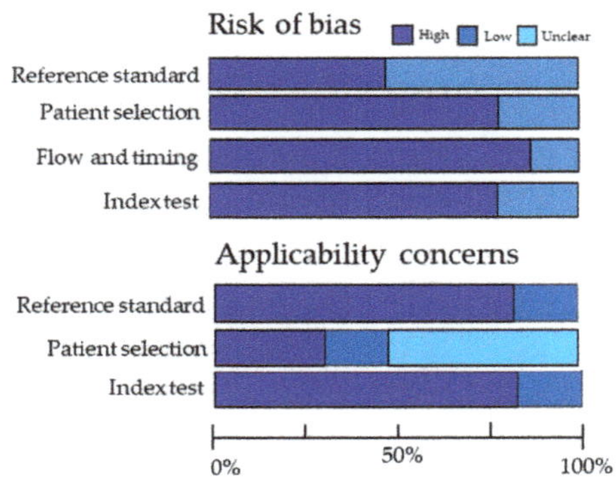

Figure 2. Summary of Methodological Quality Scored According to Quality Assessment of Diagnostic Accuracy Studies Tool, Version 2 (QUADAS-2) for 23 articles discussing the topic of response monitoring.

3.3. *Response Prediction*

All articles on response prediction have been summarized in the Supplementary Materials. Studies that were considered high-quality will be discussed in more detail.

3.3.1. Mutational Status

The radiomic model of Starmans et al. was validated on unseen data and achieved an AUC, sensitivity and specificity of 0.51, 96.0% and 3.00% for predicting KIT mutation presence [81]. The model, based on portal venous radiomic features, requires further improvement in order to be clinically applicable.

Three studies developed a model or nomogram based on radiomic features obtained from CE-CT imaging (arterial, venous and delayed phase) to predict the presence of KIT exon 11 mutations, which resulted in varying AUC outcomes, namely 0.57, 0.72 and 0.81 [75,76,81]. Deletions in exon 11 may indicate more aggressive tumor behavior, and for this reason, Liu et al. also assessed the efficacy of their model in predicting exon 11 deletion affecting codons 557–558 and achieved an AUC of 0.85 [76].

In clinical practice, patients with KIT exon 9 mutations often receive a high-dose imatinib regimen (800 mg) to improve progression-free survival (PFS). Yin et al. showed significantly greater tumor sizes and higher enhancement ratios (Hounsfield units (HU) for tumor parenchyma divided by the HUs of the erector spinae muscle) on portal venous CE-CT imaging compared to KIT exon 11 mutations. Using a 1.60 cut-off point, KIT exon 9 mutated small intestine tumors could be differentiated with an AUC, sensitivity and specificity of 0.76 and 86.7% and 98.5%, respectively. This threshold has, however, not been validated on independent validation data [67].

3.3.2. Proliferative Activity

Since high-risk GISTs have a high proliferation rate, several studies attempted to link the mitotic index and Ki-67 proliferation index to imaging features in order to make a non-invasive assessment of expected tumor behavior. On CE-CT imaging, intralesional hypodensity and concurrent heterogeneous enhancement patterns were significantly more common in high-mitotic tumors (Figures 3 and 4) [29,46]. Hypodensity was, in this case, defined as an area of low attenuation on portal venous phase CE-CT with HUs between 0 and 30 and when no HU increase (max 5 HUs) was observed between unenhanced and post-contrast images [46]. The changes in enhancement patterns were attributed to the principle of neovascularization. Tumors with high proliferative activity can induce the formation of hyperpermeable disorganized blood vessels and consequent development of necrosis [29,61]. Therefore, the supply and washout of contrast agent is affected, which has a direct impact on tumor enhancement patterns.

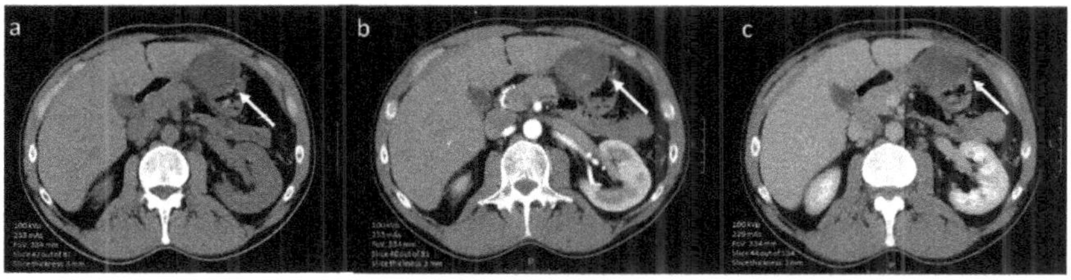

Figure 3. (**a**–**c**). Axial contrast-enhanced (iodinated contrast media) CT image of a 45-year-old male diagnosed with a (histopathologically confirmed) low mitotic gastric GIST. The lesion (arrow) shows a round tumor shape and homogeneous enhancement in (**a**) nonenhanced phase, (**b**) arterial phase and (**c**) portal venous phase (scale bars 5 cm).

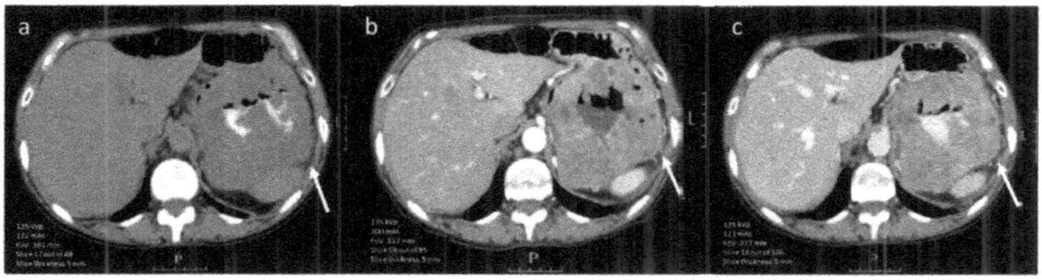

Figure 4. (**a**–**c**). Axial contrast-enhanced (iodinated contrast media) CT images of a 60-year-old male diagnosed with a (histopathologically confirmed) high-mitotic gastric GIST. The lesion (arrow) shows a lobulated tumor shape, heterogeneous enhancement in (**a**) nonenhanced phase, (**b**) arterial phase and (**c**) portal venous phase (scale bars 5 cm).

A radiomic model using 42 quantitative and semantic imaging features (tumor location, first-order and texture radiomic features) retrieved from portal venous CE-CT imaging, differentiated high- from low-mitotic tumors with an AUC, sensitivity and specificity of 0.54, 27.0% and 75.0%, respectively [81]. Although on theoretical grounds CE-CT should be able to visualize poor neo-vasculature due to rapid tumor growth, no radiomic study has been able to establish this correlation. However, radiomic models predicting high Ki-67 proliferation index in localized and advanced GISTs achieved AUC values above 0.75 [77,88,89].

Comparison of studies investigating the relation between imaging and Ki-67 indices is complicated by the fact that different thresholds (e.g., 4%, 5%, 8% and 10%) for Ki-67 expression were used. Due to the small study sizes and heterogeneous outcomes with respect to Ki-67 indices, the true relationship between CE-CT imaging and proliferation has yet to be established.

3.3.3. Risk Stratification

Research on the use of [^{18}F]FDG-PET imaging features for risk stratification in GISTs is limited. In two studies, high metabolic tumor volume (MTV) and total lesion glycolysis

(TLG) were predictive for high risk GISTs [25,35]. The use of quantitative imaging features showed improved predictive accuracy during follow-up when compared to a clinical reference standard (NIH modified criteria) [35]. Although these results suggest the added role of [^{18}F]FDG-PET for risk stratification, there are only a few studies that investigated [^{18}F]FDG-PET for this purpose.

Larger tumor sizes, mixed or extra-luminal growth patterns, ill-defined tumor shape, presence of vessels feeding or vessels draining the tumor mass, necrosis and ulceration on CE-CT imaging were all associated with high-risk GISTs [44,53,58,60,63,64,68]. Of note, Wei et al. used the angle between the longest and shortest tumor diameter to quantify tumor shape. This parameter was able distinguish intermediate- and high-risk from low-risk GISTs more accurately when compared to using solely the longest diameter [58]. Heterogeneous enhancement patterns on portal venous phase CE-CT proved to be predictive for high-risk GISTs as well (Figure 5) [53]. Incomplete enhancement of the overlying gastric mucosa on arterial phase, was also significantly more common in high-risk gastric GISTs [51]. In a study by Tang et al., HUs of the arterial phase CE-CT were subtracted from the attenuation coefficients in the portal venous phase to derive quantitative features describing contrast enhancement. Using the subtraction CT, small regions of interest (ROIs) of 30–50 mm^2, were placed in the most enhancing solid components of the tumors. The difference in HUs was significantly lower in high-risk gastric GISTs [53]. Additionally, the peak value of enhancement on CE-CT (arterial and portal venous phase) imaging was strongly correlated with risk [45]. Both articles suggest a rapid inflow of iodinated contrast agent in high-risk GISTs and thus the presence of permeable and leaky tumor vessels. The mean of the positive pixels (HU > 0) of the entire tumor volume on portal venous CT imaging was lower in high-risk GISTs [31]. This observation can be attributed to the presence of tumor necrosis, which was more commonly found in the high-risk group.

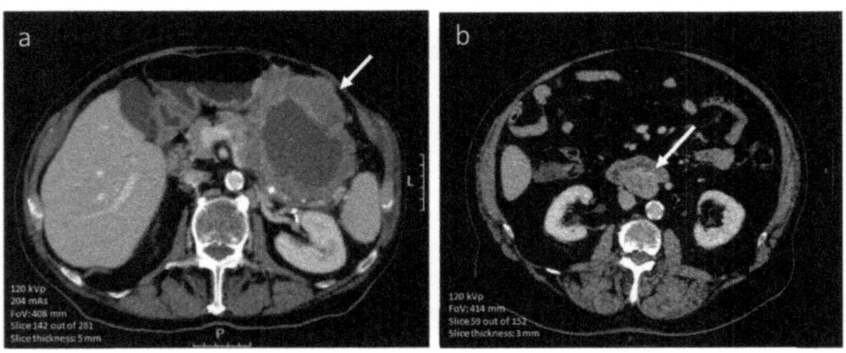

Figure 5. (**a**,**b**). Axial portal venous phase (iodinated contrast media) CT image of an 83-year-old male diagnosed with a high-risk (Miettinen AFIP classification) gastric GIST (scale bar 5 cm). The large lesion is lobulated and has central necrosis (arrow). (**b**) Axial portal venous phase (iodinated contrast media) CT slice of a 72-year-old male diagnosed with a low-risk GIST affecting the small intestine (scale bar 5 cm). It shows a well-defined and rounded lesion with a homogeneous enhancement pattern (arrow).

By contrast, Li et al. included gastric, intestinal and extra gastrointestinal tumors and did not find a significant difference in enhancement patterns between risk groups [43]. Although tumor enhancement has been established as a relevant factor in the risk stratification of GISTs, there are discrepancies in the results.

Machine learning used for the prediction of risk is extensively investigated with a total of twelve articles covering this topic [71,72,79,83,86,87]. All models achieved an AUC above 0.83 for predicting high-risk GISTs, with an average AUC of 0.87. In many of the models, texture radiomic features (gray level co-occurrence matrix (GLCM), neighboring gray-tone difference matrix (NGTDM) and gray run-length matrix (GRLM) and gray level size zoned matrix (GLSZM)) were used to develop the model. These texture features reflect enhancement patterns and inter-pixel relationships in a three-dimensional tumor volume.

3.3.4. Prediction of Radiological Response

There was one article attempting to predict radiological response using baseline imaging. Disease progression was in this case defined by the modified Choi criteria, which is currently one of the reference standards used for GIST response evaluation [114]. In this case, four textural portal venous features (features retrieved from GLCM, GLRLM and NGTDM) predicted disease progression with an AUC of 0.83 [32].

3.3.5. Prognosis

Of the selected articles, two articles discussed the use of imaging features obtained from [^{18}F]FDG-PET/CT imaging to predict PFS through detection of disease recurrence (locally and or development of distant metastases). They found significantly higher MTV and TLG values in patients with a lower PFS. In addition to quantitative [^{18}F]FDG-PET imaging features, larger tumor sizes were also a significant factor contributing to lower PFS [25,35].

On CE-CT imaging, one study with a relatively large patient group (n = 143) observed that tumor sizes greater than 10 cm, ill-defined tumor outline and enhancing solid components contributed to a poor patient prognosis, as reflected by their overall survival (OS) [48]. The study by Jung et al. combined relevant predictive parameters (tumor location, ill-defined tumor outline and presence of feeding vessels) to create a nomogram. The nomogram was internally validated and achieved an AUC of 0.86 [37]. In addition to semantic CT features, Ekert et al. assessed the efficacy of four quantitative textural features from portal venous phase CT imaging to predict prognosis of GIST patients. This study found that high values for these texture features were all associated with poor PFS [32].

In another study, three-year recurrence-free survival (RFS) was predicted by a deep learning ResNet model based on features retrieved from arterial phase CT images. Results show that, using an internal validation cohort, a predictive model with an AUC of 0.91 was obtained [70]. Furthermore, Zheng et al. investigated whether the occurrence of liver metastasis in high risk GISTs could be predicted. They found that a model based on portal venous CT radiomic features reached an AUC and accuracy of 0.87 and 84.9% [90].

3.4. *Therapy Monitoring*

All articles on therapy monitoring have been summarized in the Supplementary Materials. Studies that were considered high-quality will be discussed in more detail.

3.4.1. CE-CT Imaging

Many articles discussed the use of the well-established Response Evaluation Criteria in Solid Tumors (RECIST 1.1) to assess tumor response. RECIST 1.1 is a method in which the sum of the longest diameter of (a maximum of 5) target lesions is used to evaluate treatment response. The RECIST scoring system categorizes patients into four types of response, namely complete response (disappearance of all lesions), partial response ($\geq$30% reduction of the sum of the target lesions (SLD)), progressive disease ($\geq$20% increase of the SLD compared to the smallest SLD ever measured) and stable disease (neither progressive disease nor partial response) [115]. Nonetheless, substantial tumor shrinkage is often not observed during effective TKI treatment. Subtle and moderate changes in tumor size may be more accurately quantified by means of volumetric measurements. This is shown by Schiavon et al., who showed that size changes in GIST liver metastases larger than 20% were more frequently detected by volumetric measurements compared to the RECIST 1.1 criteria [110]. By using solely one-dimensional measurements, one presumes tumors remain spherical and that response occurs equally along three orthogonal axes during TKI treatment. However, liver metastasis in GIST patients showed significant changes in morphology over the course of imatinib treatment, which was better reflected by an ellipsoid volumetric approach [109].

In addition to RECIST 1.1, Choi et al. proposed a new method (Choi criteria) by including treatment-related changes in portal venous CT tumor densities [95]. Suppression of vascular endothelial growth factor expression can be induced by TKI treatment [116,117]. Therefore, treatment leads to changes in tumor vascularity and can lead to a reduction in tumor density, as reflected by the value of the HUs measured on CT (Figure 6). Using RECIST1.1 and Choi, comparable results were obtained for predicting PFS for patients treated with second line sunitinib assessed during an early follow-up of about 2–3 months [96,105]. Nonetheless, the Choi criteria gradually overestimated the number of patients with a partial response to sunitinib and regorafenib during longer follow-up periods (up to a year), leading to poorer PFS [105,106]. It was speculated that a drop in tumor density could also be caused by tumor necrosis, which is often a sign of progressive disease. So, instead of measuring a reduction in tumor vascularization, one may be measuring progressive disease over longer follow-up periods [105].

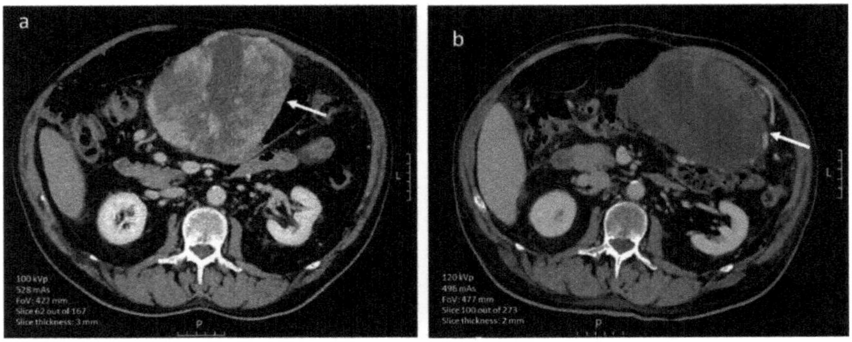

Figure 6. (**a**,**b**). Axial portal venous phase CT images (iodinated contrast media) of a 67-year-old male diagnosed with a primary GIST (arrows) of the stomach. (**a**) Pre-treatment imaging shows a large gastric mass with heterogeneous enhancement. (**b**) After 1.5 months of avapritinib treatment, the lesion has become hypodense (scale bars 5 cm).

3.4.2. [18F]FDG-PET Imaging

In [18F]FDG-PET imaging, the European Organization for Research and Treatment in Cancer (EORTC) PET criteria are most commonly used, in which a metabolic response is determined by a reduction in SUVmax of 25% or more [118]. Metabolic response was significantly associated with prolonged PFS and could be detected as early as seven days, after the induction of TKI treatment (imatinib and sunitinib) [100,102]. On the contrary, the prospective study of Chacón et al. did not find a significant association between PFS and metabolic response determined by the EORTC PET criteria.

Additionally, two retrospective studies by Farag et al. evaluated the impact of [18F]FDG-PET/CT on clinical decision making in the treatment of localized and advanced GIST patients. Changes in surgical management, systemic treatment and treatment objective were all included in the evaluation [111,112]. In 27.1% of GIST patients treated with neoadjuvant intent, management was changed because of [18F]FDG-PET/CT findings at an interval of eight weeks. The lack of metabolic response was correlated with therapeutic changes in management, especially in non-KIT exon 11 mutations [111]. In the advanced disease setting, specifically late [18F]FDG-PET response findings (median of 293 days) proved to have an impact on therapeutic decision [112].

3.4.3. CE-CT vs. [18F]FDG-PET Imaging

When comparing the aforementioned response evaluation criteria on CE-CT imaging with the EORTC PET criteria on [18F]FDG-PET imaging, articles reported high agreement and RECIST responders also showed significant reductions in SUVmax [91,98,100,108]. Choi et al. showed greater sensitivity and specificity (97.0% and 100%) when compared to the EORTC PET criteria [95]. Metabolic response could, however, be observed within a week and preceded changes in tumor size and volume in localized and advanced GIST patients treated with imatinib (Figure 7) [92,97,100,107]. By using the RECIST criteria, the early effect of TKI treatment may be underestimated. For example, Choi et al. showed that 70% of the stable disease RECIST patients had an SUVmax reduction between 61 and 100% at the two-month follow-up [94].

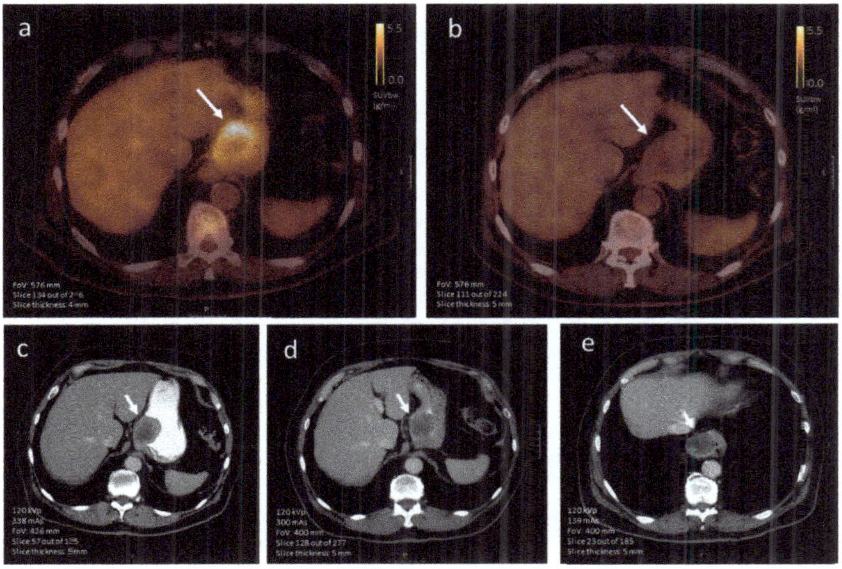

Figure 7. (**a**–**e**). Axial [18F]FDG-PET images of a 71-year-old male diagnosed with a primary gastric GIST (**a**) before treatment induction and (**b**) after about three months of TKI treatment, where the standardized uptake value (SUV) is normalized (scale bar 5 cm). Corresponding contrast-enhanced CT imaging (iodinated contrast media) visualizing the same lesion (arrow) (**c**) at diagnosis and after (**d**) 2.5 months and (**e**) 6.5 months of imatinib treatment, showing minimal to no change in tumor size (scale bar 5 cm). In the last image, the intrathoracic tumor location is caused by a sliding hernia.

4. Discussion

The aim of this review was to provide an overview of the value of CE-CT and [18F]FDG PET/CT imaging to predict and monitor TKI treatment response in GIST patients.

There is limited literature available on the use of baseline [18F]FDG-PET findings to predict tumor response. Although there are only a few studies available, generally imaging features, such as MTV and TLG, were correlated with more aggressive tumor behavior. On the contrary, there is more data available on the potential of CE-CT imaging features to predict treatment response. Results indicate that larger tumor sizes (>5 cm), ill-defined or lobulated tumor outline, mixed or exophytic growth patterns, the presence of (enlarged) and feeding vessels are associated with patient outcome. The presence of heterogeneous enhancement patterns was a recurring observation in high-risk GISTs. The hypodensities observed on CE-CT imaging were devoted to the biological phenomena of neovascularization and necrosis. It should be noted that the correlation between hypodensities on radiological imaging and actual pathological necrosis and neovascularization in GIST tumors is still disputable.

Many articles discussed the use of radiomic and deep learning models for response prediction on baseline CE-CT imaging. High performance scores were stated for models predicting RFS and risk stratifications, while mutational status remained difficult to predict with variable AUC values. Radiomics offers the possibility to identify clinically relevant imaging features that would normally be imperceptible to the naked eye. For example, it has proven to be difficult to obtain a sufficient amount of tissue samples from biopsy material, which makes it difficult to determine the mutational status or a reliable mitotic count. Additionally, if the mitotic count is determined on postoperative surgical specimens, the results can be inaccurate due to the occasional administration of neoadjuvant TKI treatment. It would, therefore, be very helpful if imaging could provide additional information, other than tumor size. Nonetheless, the biological explanation behind the efficacy of radiomic features was often missing in the included articles. Before advanced and objective learning techniques can be introduced in clinical practice, they should be clinically relevant and biologically meaningful. It is recommended to further explore the prediction of actual radiological response using semantic or quantitative imaging features selected based upon tumor biology.

The three evaluation methods currently used to monitor response in GIST patients, are the RECIST, Choi and EORTC PET criteria. The main disadvantage of the RECIST criteria is the one-dimensional nature of its measurements, presuming a spherical tumor shape throughout the entire course of TKI treatment. To overcome this limitation, an additional set of criteria was developed by Choi et al. involving CT densities. The Choi criteria are occasionally applied in clinical practice. However, its efficacy and prognostic value in determining response in GIST patients remains unclear. Supposedly, the antiangiogenic effect of TKI treatment would lead to a consequent reduction in HU values. As previously stated, necrosis and heterogeneous enhancement patterns at baseline were considered predictive for more aggressively behaving tumors. Using reductions in CT densities as a criterion for response monitoring may, therefore, be misleading, since it can reflect a decrease in angiogenesis induced by TKI treatment, as well as necrosis induced by aggressive tumor behavior. This hypothesis was supported by literature, since response evaluation using Choi criteria led to an overestimation in the number of partial responders at longer follow-up periods.

[^{18}F]FDG-PET proved to be useful in the early monitoring of GISTs, since significant reductions in SUVmax could be observed within a week of TKI treatment and metabolic changes preceded morphological changes in size. However, this imaging technique is often not considered for early response monitoring in clinical practice because of higher costs. Since some of the targeted treatments are more expensive than PET-CT scans, further research should, therefore, be focused on the cost-effectiveness of [^{18}F]FDG-PET imaging in the treatment of GISTs.

Particularly, the combined use of different imaging modalities, also known as multimodality imaging, might provide more detailed information that can assist in making early image-guided treatment decisions. The use of such a multimodality imaging approach might be useful to gather as much information as possible on the biological behavior of GIST. However, currently, no literature is available on the specific use of combining these different imaging modalities for response prediction or monitoring.

5. Conclusions

In conclusion, imaging features obtained from CE-CT and [^{18}F]FDG PET/CT imaging can aid in the development of a more personalized treatment of GIST patients by enabling early prediction and monitoring of TKI therapy response. Heterogeneous enhancement patterns on baseline CE-CT imaging were predictive for high-risk GISTs, reflecting neovascularization and necrosis.

For the purpose of response monitoring, current RECIST and Choi criteria are still lacking sensitivity and are prone to errors when predicting or monitoring treatment response. [18F]FDG-PET is a promising imaging technique that visualizes functional metabolic changes in GISTs, which precedes measurable changes in tumor size. Although promising, the true added value of [18F]FDG-PET remains elusive, and research on cost-effectiveness is warranted.

Radiomics is an emerging topic in medicine and shows potential for the prediction of RFS and risk stratifications in GISTs. However, future research should mainly focus on clinical utility, explainability and correlation with actual tumor biology.

Supplementary Materials: The following supporting information can be downloaded at: https://www.mdpi.com/article/10.3390/diagnostics12112722/s1.

Author Contributions: Conceptualization, Y.A.W., G.M.K., J.A.v.d.H., W.G. and L.-F.d.G.-O. methodology, Y.A.W., G.M.K., W.G., J.A.v.d.H. and L.-F.d.G.-O.; software; Y.A.W. and G.M.K.; validation, Y.A.W., G.M.K. and W.G.; formal analysis, Y.A.W., G.M.K., W.G. and L.-F.d.G.-O.; investigation, Y.A.W., G.M.K., W.G. and L.-F.d.G.-O.; resources, Y.A.W., G.M.K., W.G., J.A.v.d.H. and L.-F.d.G.-O.; data curation, Y.A.W., G.M.K., W.G. and L.-F.d.G.-O.; writing—Original draft preparation, Y.A.W., G.M.K., W.G. and L.-F.d.G.-O.; writing—Review and editing, G.M.K., W.G., J.A.v.d.H., F.H.P.v.V., H.G., J.V.M.G.B., A.J.v.d.M. and L.-F.d.G.-O.; visualization, Y.A.W., A.J.v.d.M. and L.-F.d.G.-O.; supervision, G.M.K., W.G. and L.-F.d.G.-O.; project administration, Y.A.W. All authors have read and agreed to the published version of the manuscript.

Funding: We declare the following financial interests/personal relationships which may be considered as potential competing interests: G.M. Kalisvaart was the recipient of an educational grant (LEI-05) from Philips Electronics Nederland B.V, Eindhoven and supported by a public grant (LSHM18089) from Health Holland TKI Life Sciences & Health. The funders had no role in the design of the study, in the collection, analyses, or interpretation of data, in the writing of the manuscript or in the decision to publish the results.

Institutional Review board Statement: The study was conducted in accordance with the Declaration of Helsinki, and approved by the Ethics Committee Leiden Den Haag Delft (METC LDD) (protocol code: B19.050, date of approval: 14 January 2020).

Informed Consent Statement: Patient consent was waived due to the retrospective nature of the study. Patients who objected to the use of their data were excluded.

Data Availability Statement: The datasets generated during and/or analyzed during the current study are available from the corresponding author upon reasonable request.

Acknowledgments: The authors would like to thank Jan W. Schoones for his contribution to the search strategy.

Conflicts of Interest: The other authors declare no conflict of interest.

References

1. Balachandran, V.P.; DeMatteo, R.P. Gastrointestinal stromal tumors: Who should get imatinib and for how long? *Adv. Surg.* **2014**, *48*, 165–183. [CrossRef] [PubMed]
2. Zappacosta, R.; Zappacosta, B.; Capanna, S.; D'Angelo, C.; Gatta, D.; Rosini, S. *GISTs: From the History to the Tailored Therapy, Gastrointestinal Stromal Tumor*; IntechOpen: London, UK, 2012. [CrossRef]
3. Joensuu, H.; Hohenberger, P.; Corless, C.L. Gastrointestinal stromal tumour. *Lancet* **2013**, *382*, 973–983. [CrossRef]
4. Wu, C.E.; Tzen, C.Y.; Wang, S.Y.; Yeh, C.N. Clinical Diagnosis of Gastrointestinal Stromal Tumor (GIST): From the Molecular Genetic Point of View. *Cancers* **2019**, *11*, 679. [CrossRef] [PubMed]
5. Verweij, J.; Casali, P.G.; Zalcberg, J.; LeCesne, A.; Reichardt, P.; Blay, J.Y.; Issels, R.; van Oosterom, A.; Hogendoorn, P.C.; Van Glabbeke, M.; et al. Progression-free survival in gastrointestinal stromal tumours with high-dose imatinib: Randomised trial. *Lancet* **2004**, *364*, 1127–1134. [CrossRef]
6. Reichardt, P. The Story of Imatinib in GIST—A Journey through the Development of a Targeted Therapy. *Oncol. Res. Treat.* **2018**, *41*, 472–477. [CrossRef]
7. Casali, P.G.; Blay, J.Y.; Abecassis, N.; Bajpai, J.; Bauer, S.; Biagini, R.; Bielack, S.; Bonvalot, S.; Boukovinas, I.; Bovee, J.; et al. Gastrointestinal stromal tumours: ESMO-EURACAN-GENTURIS Clinical Practice Guidelines for diagnosis, treatment and follow-up. *Ann. Oncol.* **2022**, *33*, 20–33. [CrossRef]

8. Casali, P.G.; Le Cesne, A.; Poveda Velasco, A.; Kotasek, D.; Rutkowski, P.; Hohenberger, P.; Fumagalli, E.; Judson, I.R.; Italiano, A.; Gelderblom, H.; et al. Time to Definitive Failure to the First Tyrosine Kinase Inhibitor in Localized GI Stromal Tumors Treated With Imatinib As an Adjuvant: A European Organisation for Research and Treatment of Cancer Soft Tissue and Bone Sarcoma Group Intergroup Randomized Trial in Collaboration with the Australasian Gastro-Intestinal Trials Group, UNICANCER, French Sarcoma Group, Italian Sarcoma Group, and Spanish Group for Research on Sarcomas. *J. Clin. Oncol.* **2015**, *33*, 4276–4283. [CrossRef]
9. Lopes, L.F.; Bacchi, C.E. Imatinib treatment for gastrointestinal stromal tumour (GIST). *J. Cell Mol. Med.* **2010**, *14*, 42–50. [CrossRef]
10. Reynoso, D.; Trent, J.C. Neoadjuvant and adjuvant imatinib treatment in gastrointestinal stromal tumor: Current status and recent developments. *Curr. Opin. Oncol.* **2010**, *22*, 330–335. [CrossRef]
11. Eisenberg, B.L.; Trent, J.C. Adjuvant and neoadjuvant imatinib therapy: Current role in the management of gastrointestinal stromal tumors. *Int. J. Cancer* **2011**, *129*, 2533–2542. [CrossRef]
12. Yang, W.; Liu, Q.; Lin, G.; Zhang, B.; Cao, H.; Zhao, Y.; Xia, L.; Feng, F.; Xiong, Z.; Hu, J.; et al. The effect of neoadjuvant imatinib therapy on outcome and survival in rectal gastrointestinal stromal tumors: A multiinstitutional study. *J. Surg. Oncol.* **2021**, *124*, 1128–1135. [CrossRef] [PubMed]
13. Miettinen, M.; Lasota, J. Gastrointestinal stromal tumors: Review on morphology, molecular pathology, prognosis, and differential diagnosis. *Arch. Pathol. Lab. Med.* **2006**, *130*, 1466–1478. [CrossRef] [PubMed]
14. Betz, M.; Kopp, H.G.; Spira, D.; Claussen, C.D.; Horger, M. The benefit of using CT-perfusion imaging for reliable response monitoring in patients with gastrointestinal stromal tumor (GIST) undergoing treatment with novel targeted agents. *Acta Radiol.* **2013**, *54*, 711–721. [CrossRef] [PubMed]
15. Cassier, P.A.; Fumagalli, E.; Rutkowski, P.; Schöffski, P.; Van Glabbeke, M.; Debiec-Rychter, M.; Emile, J.F.; Duffaud, F.; Martin-Broto, J.; Landi, B.; et al. Outcome of patients with platelet-derived growth factor receptor alpha-mutated gastrointestinal stromal tumors in the tyrosine kinase inhibitor era. *Clin. Cancer Res.* **2012**, *18*, 4458–4464. [CrossRef] [PubMed]
16. Lambin, P.; Leijenaar, R.T.H.; Deist, T.M.; Peerlings, J.; de Jong, E.E.C.; van Timmeren, J.; Sanduleanu, S.; Larue, R.; Even, A.J.G.; Jochems, A.; et al. Radiomics: The bridge between medical imaging and personalized medicine. *Nat. Rev. Clin. Oncol.* **2017**, *14*, 749–762. [CrossRef] [PubMed]
17. Whiting, P.F.; Rutjes, A.W.; Westwood, M.E.; Mallett, S.; Deeks, J.J.; Reitsma, J.B.; Leeflang, M.M.; Sterne, J.A.; Bossuyt, P.M. QUADAS-2: A revised tool for the quality assessment of diagnostic accuracy studies. *Ann. Intern. Med.* **2011**, *155*, 529–536. [CrossRef] [PubMed]
18. Jüni, P.; Witschi, A.; Bloch, R.; Egger, M. The Hazards of Scoring the Quality of Clinical Trials for Meta-analysis. *JAMA* **1999**, *282*, 1054–1060. [CrossRef]
19. Whiting, P.; Harbord, R.; Kleijnen, J. No role for quality scores in systematic reviews of diagnostic accuracy studies. *BMC Med. Res. Methodol.* **2005**, *5*, 19. [CrossRef]
20. Pretze, M.; Reffert, L.; Diehl, S.; Schönberg, S.O.; Wängler, C.; Hohenberger, P.; Wängler, B. GMP-compliant production of [(68)Ga]Ga-NeoB for positron emission tomography imaging of patients with gastrointestinal stromal tumor. *EJNMMI Radiopharm. Chem.* **2021**, *6*, 22. [CrossRef]
21. Hedenström, P.; Andersson, C.; Sjövall, H.; Enlund, F.; Nilsson, O.; Nilsson, B.; Sadik, R. Pretreatment Tumor DNA Sequencing of KIT and PDGFRA in Endosonography-Guided Biopsies Optimizes the Preoperative Management of Gastrointestinal Stromal Tumors. *Mol. Diagn. Ther.* **2020**, *24*, 201–214. [CrossRef]
22. Cai, P.Q.; Lv, X.F.; Tian, L.; Luo, Z.P.; Mitteer, R.A., Jr.; Fan, Y.; Wu, Y.P. CT Characterization of Duodenal Gastrointestinal Stromal Tumors. *AJR Am. J. Roentgenol.* **2015**, *204*, 988–993. [CrossRef] [PubMed]
23. Nannini, M.; Pantaleo, M.A.; Maleddu, A.; Saponara, M.; Mandrioli, A.; Lolli, C.; Pallotti, M.C.; Gatto, L.; Santini, D.; Paterini, P.; et al. Duration of adjuvant treatment following radical resection of metastases from gastrointestinal stromal tumours. *Oncol. Lett.* **2012**, *3*, 677–681. [CrossRef] [PubMed]
24. Al-Balas, H.A.; Shaib, Y.H. Gastrointestinal stromal tumours: Role of computed tomography in predicting tumour behaviour. *Hong Kong J. Radiol.* **2012**, *15*, 155–161.
25. Albano, D.; Bosio, G.; Tomasini, D.; Bonù, M.; Giubbini, R.; Bertagna, F. Metabolic behavior and prognostic role of pretreatment 18F-FDG PET/CT in gist. *Asia Pac. J. Clin. Oncol.* **2020**, *16*, e207–e215. [CrossRef] [PubMed]
26. Cannella, R.; Tabone, E.; Porrello, G.; Cappello, G.; Gozzo, C.; Incorvaia, L.; Grignani, G.; Merlini, A.; D'Ambrosio, L.; Badalamenti, G.; et al. Assessment of morphological CT imaging features for the prediction of risk stratification, mutations, and prognosis of gastrointestinal stromal tumors. *Eur. Radiol.* **2021**, *31*, 8554–8564. [CrossRef] [PubMed]
27. Chen, T.; Xu, L.; Dong, X.; Li, Y.; Yu, J.; Xiong, W.; Li, G. The roles of CT and EUS in the preoperative evaluation of gastric gastrointestinal stromal tumors larger than 2 cm. *Eur. Radiol.* **2019**, *29*, 2481–2489. [CrossRef] [PubMed]
28. Chen, X.S.; Shan, Y.C.; Dong, S.Y.; Wang, W.T.; Yang, Y.T.; Liu, L.H.; Xu, Z.H.; Zeng, M.S.; Rao, S.X. Utility of preoperative computed tomography features in predicting the Ki-67 labeling index of gastric gastrointestinal stromal tumors. *Eur. J. Radiol.* **2021**, *142*, 109840. [CrossRef]
29. Chen, Z.; Yang, J.; Sun, J.; Wang, P. Gastric gastrointestinal stromal tumours (2–5 cm): Correlation of CT features with malignancy and differential diagnosis. *Eur. J. Radiol.* **2020**, *123*, 108783. [CrossRef]

30. Cho, M.H.; Park, C.K.; Park, M.; Kim, W.K.; Cho, A.; Kim, H. Clinicopathologic Features and Molecular Characteristics of Glucose Metabolism Contributing to ^{18}F-fluorodeoxyglucose Uptake in Gastrointestinal Stromal Tumors. *PLoS ONE* **2015**, *10*, e0141413. [CrossRef]
31. Choi, I.Y.; Yeom, S.K.; Cha, J.; Cha, S.H.; Lee, S.H.; Chung, H.H.; Lee, C.M.; Choi, J. Feasibility of using computed tomography texture analysis parameters as imaging biomarkers for predicting risk grade of gastrointestinal stromal tumors: Comparison with visual inspection. *Abdom. Radiol.* **2019**, *44*, 2346–2356. [CrossRef]
32. Ekert, K.; Hinterleitner, C.; Horger, M. Prognosis assessment in metastatic gastrointestinal stromal tumors treated with tyrosine kinase inhibitors based on CT-texture analysis. *Eur. J. Radiol.* **2019**, *116*, 98–105. [CrossRef] [PubMed]
33. Fuster, D.; Ayuso, J.R.; Poveda, A.; Cubedo, R.; Casado, A.; Martinez-Trufero, J.; Lopez-Pousa, A.; Del Muro, X.G.; Lomena, F.; Maurel, J.; et al. Value of FDG-PET for monitoring treatment response in patients with advanced GIST refractory to high-dose imatinib. A multicenter GEIS study. *Q. J. Nucl. Med. Mol. Imaging* **2011**, *55*, 680–687.
34. Grazzini, G.; Guerri, S.; Cozzi, D.; Danti, G.; Gasperoni, S.; Pradella, S.; Miele, V. Gastrointestinal stromal tumors: Relationship between preoperative CT features and pathologic risk stratification. *Tumori* **2021**, *107*, 556–563. [CrossRef] [PubMed]
35. Hwang, S.H.; Jung, M.; Jeong, Y.H.; Jo, K.; Kim, S.; Wang, J.; Cho, A. Prognostic value of metabolic tumor volume and total lesion glycolysis on preoperative (18)F-FDG PET/CT in patients with localized primary gastrointestinal stromal tumors. *Cancer Metab.* **2021**, *9*, 8. [CrossRef] [PubMed]
36. Iannicelli, E.; Carbonetti, F.; Federici, G.F.; Martini, I.; Caterino, S.; Pilozzi, E.; Panzuto, F.; Briani, C.; David, V. Evaluation of the Relationships between Computed Tomography Features, Pathological Findings, and Prognostic Risk Assessment in Gastrointestinal Stromal Tumors. *J. Comput. Assist. Tomogr.* **2017**, *41*, 271–278. [CrossRef]
37. Jung, H.; Lee, S.M.; Kim, Y.C.; Byun, J.; Park, J.Y; Oh, B.Y.; Kwon, M.J.; Kim, J. Gastrointestinal stromal tumours: Preoperative imaging features to predict recurrence after curative resection. *Eur. J. Radiol.* **2022**, *149*, 110193. [CrossRef]
38. Kim, H.C.; Lee, J.M.; Kim, K.W.; Park, S.H.; Kim, S.H.; Lee, J.Y.; Han, J.K.; Choi, B.I. Gastrointestinal stromal tumors of the stomach: CT findings and prediction of malignancy. *AJR Am. J. Roentgenol.* **2004**, *183*, 893–898. [CrossRef]
39. Kamiyama, Y.; Aihara, R.; Nakabayashi, T.; Mochiki, E.; Asao, T.; Kuwano, H.; Oriuchi, N.; Endo, K. 18F-fluorodeoxyglucose positron emission tomography: Useful technique for predicting malignant potential of gastrointestinal stromal tumors. *World J. Surg.* **2005**, *29*, 1429–1435. [CrossRef]
40. Kim, H.C.; Lee, J.M.; Kim, S.H.; Park, S.H.; Lee, J.W.; Lee, M.; Han, J.K.; Choi, B.I. Small gastrointestinal stromal tumours with focal areas of low attenuation on CT: Pathological correlation. *Clin. Radiol.* **2005**, *60*, 384–388. [CrossRef]
41. Kurata, Y.; Hayano, K.; Ohira, G.; Narushima, K.; Aoyagi, T.; Matsubara, H. Fractal analysis of contrast-enhanced CT images for preoperative prediction of malignant potential of gastrointestinal stromal tumor. *Abdom. Radiol.* **2018**, *43*, 2659–2664. [CrossRef]
42. Kwon, Y.; Park, E.; Pahk, K.; Kim, S.; Kim, M.J.; Graf, D.; Park, S. Preoperative assessment of malignant potential of gastrointestinal stromal tumor by dual-time-point 18F-fluorodeoxyglucose positron emission tomography imaging: Usefulness of standardized uptake value and retention index. *J. Cancer Res. Ther.* **2019**, *15*, 142–147. [CrossRef] [PubMed]
43. Li, H.; Ren, G.; Cai, R.; Chen, J.; Wu, X.; Zhao, J. A correlation research of Ki67 index, CT features, and risk stratification in gastrointestinal stromal tumor. *Cancer Med.* **2018**, *7*, 4467–4474. [CrossRef] [PubMed]
44. Li, C.; Fu, W.; Huang, L.; Chen, Y.; Xiang, P.; Guan, J.; Sun, C. A CT-based nomogram for predicting the malignant potential of primary gastric gastrointestinal stromal tumors preoperatively. *Abdom. Radiol.* **2021**, *46*, 3075–3085. [CrossRef] [PubMed]
45. Liu, S.; Pan, X.; Liu, R.; Zheng, H.; Chen, L.; Guan, W.; Wang, H.; Sun, Y.; Tang, L.; Guan, Y; et al. Texture analysis of CT images in predicting malignancy risk of gastrointestinal stromal tumours. *Clin. Radiol.* **2018**, *73*, 266–274. [CrossRef]
46. Mazzei, M.A.; Cioffi Squitieri, N.; Vindigni, C.; Guerrini, S.; Gentili, F.; Sadotti, G.; Mercuri, P.; Righi, L.; Lucii, G.; Mazzei, F.G.; et al. Gastrointestinal stromal tumors (GIST): A proposal of a "CT-based predictive model of Miettinen index" in predicting the risk of malignancy. *Abdom. Radiol.* **2020**, *45*, 2989–2996. [CrossRef]
47. Miyake, K.K.; Nakamoto, Y.; Mikami, Y.; Tanaka, S.; Higashi, T.; Tadamura, E.; Saga, T.; Minami, S.; Togashi, K. The predictive value of preoperative (18)F-fluorodeoxyglucose PET for postoperative recurrence in patients with localized primary gastrointestinal stromal tumour. *Eur. Radiol.* **2016**, *26*, 4664–4674. [CrossRef]
48. O'Neill, A.C.; Shinagare, A.B.; Kurra, V.; Tirumani, S.H.; Jagannathan, J.P.; Baheti, A.D.; Hornick, J.L.; George, S.; Ramaiya, N.H. Assessment of metastatic risk of gastric GIST based on treatment-naïve CT features. *Eur. J. Surg. Oncol.* **2016**, *42*, 1222–1228. [CrossRef]
49. Pelandre, G.L.; Djahjah, M.C.; Gasparetto, E.L.; Nacif, M.S.; Marchiori, E.; De Mello, E.L.R. To mographic findings of gastric gastrointestinal stromal tumor and correlation with the mitotic index. *Arq. Gastroenterol.* **2013**, *50*, 244–250. [CrossRef]
50. Palatresi, D.; Fedeli, F.; Danti, G.; Pasqualini, E.; Castiglione, F.; Messerini, L.; Massi, D.; Bettarini, S.; Tortoli, P.; Busoni, S.; et al. Correlation of CT radiomic features for GISTs with pathological classification and molecular subtypes: Preliminary and monocentric experience. *Radiol. Med.* **2022**, *127*, 117–128. [CrossRef]
51. Peng, G.; Huang, B.; Yang, X.; Pang, M.; Li, N. Preoperative CT feature of incomplete overlying enhancing mucosa as a high-risk predictor in gastrointestinal stromal tumors of the stomach. *Eur. Radiol.* **2021**, *31*, 3276–3285. [CrossRef]
52. Pinaikul, S.; Woodtichartpreecha, P.; Kanngurn, S.; Leelakiatpaiboon, S. 1189 Gastrointestinal stromal tumor (GIST): Computed tomographic features and correlation of CT findings with histologic grade. *J. Med. Assoc. Thai.* **2014**, *97*, 1189–1198. [PubMed]
53. Tang, B.; Feng, Q.X.; Liu, X.S. Comparison of Computed Tomography Features of Gastric and Small Bowel Gastrointestinal Stromal Tumors with Different Risk Grades. *J. Comput. Assist. Tomogr.* **2022**, *46*, 175–182. [CrossRef] [PubMed]

54. Tateishi, U.; Hasegawa, T.; Satake, M.; Moriyama, N. Gastrointestinal stromal tumor. Correlation of computed tomography findings with tumor grade and mortality. *J. Comput. Assist. Tomogr.* **2003**, *27*, 792–798. [CrossRef] [PubMed]
55. Tokumoto, N.; Tanabe, K.; Misumi, T.; Fujikuni, N.; Suzuki, T.; Ohdan, H. The usefulness of preoperative 18FDG positron-emission tomography and computed tomography for predicting the malignant potential of gastrointestinal stromal tumors. *Dig. Surg.* **2014**, *31*, 79–86. [CrossRef] [PubMed]
56. Ulusan, S.; Koc, Z.; Kayaselcuk, F. Gastrointestinal stromal tumours: CT findings. *Br. J. Radiol.* **2008**, *81*, 618–623. [CrossRef] [PubMed]
57. Verde, F.; Hruban, R.H.; Fishman, E.K. Small Bowel Gastrointestinal Stromal Tumors: Multidetector Computed Tomography Enhancement Pattern and Risk of Progression. *J. Comput. Assist. Tomogr.* **2017**, *41*, 407–411. [CrossRef] [PubMed]
58. Wei, S.C.; Xu, L.; Li, W.H.; Li, Y.; Guo, S.F.; Sun, X.R.; Li, W.W. Risk stratification in GIST: Shape quantification with CT is a predictive factor. *Eur. Radiol.* **2020**, *30*, 1856–1865. [CrossRef]
59. Xu, D.; Si, G.Y.; He, Q.Z. Correlation analysis of multi-slice computed tomography (MSCT) findings, clinicopathological factors, and prognosis of gastric gastrointestinal stromal tumors. *Transl. Cancer Res.* **2020**, *9*, 1787–1794. [CrossRef]
60. Yang, T.H.; Hwang, J.I.; Yang, M.S.; Hung, S.W.; Chan, S.W.; Wang, J.; Tyan, Y.S. Gastrointestinal stromal tumors: Computed tomographic features and prediction of malignant risk from computed tomographic imaging. *J. Chin. Med. Assoc.* **2007**, *70*, 367–373. [CrossRef]
61. Yang, C.W.; Liu, X.J.; Zhao, L.; Che, F.; Yin, Y.; Chen, H.J.; Zhang, B.; Wu, M.; Song, B. Preoperative prediction of gastrointestinal stromal tumors with high Ki-67 proliferation index based on CT features. *Ann. Transl. Med.* **2021**, *9*, 1556. [CrossRef]
62. Yoshikawa, K.; Shimada, M.; Kurita, N.; Sato, H.; Iwata, T.; Morimoto, S.; Miyatani, T.; Kashihara, H.; Takasu, C.; Matsumoto, N. Efficacy of PET-CT for predicting the malignant potential of gastrointestinal stromal tumors. *Surg. Today* **2013**, *43*, 1162–1167. [CrossRef] [PubMed]
63. Zhou, C.; Duan, X.; Zhang, X.; Hu, H.; Wang, D.; Shen, J. Predictive features of CT for risk stratifications in patients with primary gastrointestinal stromal tumour. *Eur. Radiol.* **2016**, *26*, 3086–3093. [CrossRef] [PubMed]
64. Zhu, M.P.; Ding, Q.L.; Xu, J.X.; Jiang, C.Y.; Wang, J.; Wang, C.; Yu, R.S. Building contrast-enhanced CT-based models for preoperatively predicting malignant potential and Ki67 expression of small intestine gastrointestinal stromal tumors (GISTs). *Abdom. Radiol.* **2022**, *47*, 3161–3173. [CrossRef] [PubMed]
65. Otomi, Y.; Otsuka, H.; Morita, N.; Terazawa, K.; Furutani, K.; Harada, M.; Nishitani, H. Relationship between FDG uptake and the pathological risk category in gastrointestinal stromal tumors. *J. Med. Investig.* **2010**, *57*, 270–274. [CrossRef]
66. Park, J.-W.; Cho, C.-H.; Jeong, D.-S.; Chae, H.-D. Role of F-fluoro-2-deoxyglucose Positron Emission Tomography in Gastric GIST: Predicting Malignant Potential Pre-operatively. *J. Gastric Cancer* **2011**, *11*, 173–179. [CrossRef]
67. Yin, Y.Q.; Liu, C.J.; Zhang, B.; Wen, Y.; Yin, Y. Association between CT imaging features and KIT mutations in small intestinal gastrointestinal stromal tumors. *Sci. Rep.* **2019**, *9*, 7257. [CrossRef]
68. Wang, J.K. Predictive value and modeling analysis of MSCT signs in gastrointestinal stromal tumors (GISTs) to pathological risk degree. *Eur. Rev. Med. Pharmacol. Sci.* **2017**, *21*, 999–1005.
69. Ao, W.; Cheng, G.; Lin, B.; Yang, R.; Liu, X.; Zhou, C.; Wang, W.; Fang, Z.; Tian, F.; Yang, G.; et al. A novel CT-based radiomic nomogram for predicting the recurrence and metastasis of gastric stromal tumors. *Am. J. Cancer Res.* **2021**, *11*, 3123–3134.
70. Chen, T.; Liu, S.; Li, Y.; Feng, X.; Xiong, W.; Zhao, X.; Yang, Y.; Zhang, C.; Hu, Y.; Chen, H.; et al. Developed and validated a prognostic nomogram for recurrence-free survival after complete surgical resection of local primary gastrointestinal stromal tumors based on deep learning. *EBioMedicine* **2019**, *39*, 272–279. [CrossRef]
71. Chen, T.; Ning, Z.; Xu, L.; Feng, X.; Han, S.; Roth, H.R.; Xiong, W.; Zhao, X.; Hu, Y.; Liu, H.; et al. Radiomics nomogram for predicting the malignant potential of gastrointestinal stromal tumours preoperatively. *Eur. Radiol.* **2019**, *29*, 1074–1082. [CrossRef]
72. Chen, Z.H.; Xu, L.Y.; Zhang, C.M.; Huang, C.C.; Wang, M.H.; Feng, Z.; Xiong, Y. CT Radiomics Model for Discriminating the Risk Stratification of Gastrointestinal Stromal Tumors: A Multi-Class Classification and Multi-Center Study. *Front. Oncol.* **2021**, *11*, 654114. [PubMed]
73. Chu, H.; Pang, P.; He, J.; Zhang, D.; Zhang, M.; Qiu, Y.; Li, X.; Lei, P.; Fan, B.; Xu, R. Value of radiomics model based on enhanced computed tomography in risk grade prediction of gastrointestinal stromal tumors. *Sci. Rep.* **2021**, *11*, 12009. [CrossRef] [PubMed]
74. Kang, B.; Yuan, X.; Wang, H.; Qin, S.; Song, X.; Yu, X.; Zhang, S.; Sun, C.; Zhou, Q.; Wei, Y.; et al. Preoperative CT-Based Deep Learning Model for Predicting Risk Stratification in Patients With Gastrointestinal Stromal Tumors. *Front. Oncol.* **2021**, *11*, 750875. [CrossRef] [PubMed]
75. Liu, B.; Liu, H.; Zhang, L.; Song, Y.; Yang, S.; Zheng, Z.; Zhao, J.; Hou, F.; Zhang, J. Value of contrast-enhanced CT based radiomic machine learning algorithm in differentiating gastrointestinal stromal tumors with KIT exon 11 mutation: A two-center study. *Diagn. Interv. Radiol.* **2022**, *28*, 29–38. [CrossRef]
76. Liu, X.; Yin, Y.; Wang, X.; Yang, C.; Wan, S.; Yin, X.; Wu, T.; Chen, H.; Xu, Z.; Li, X.; et al. Gastrointestinal stromal tumors: Associations between contrast-enhanced CT images and KIT exon 11 gene mutation. *Ann. Transl. Med.* **2021**, *9*, 1496. [CrossRef]
77. Feng, Q.; Tang, B.; Zhang, Y.; Liu, X. Prediction of the Ki-67 expression level and prognosis of gastrointestinal stromal tumors based on CT radiomics nomogram. *Int. J. Comput. Assist. Radiol. Surg.* **2022**, *17*, 1167–1175. [CrossRef]
78. Shao, M.; Niu, Z.; He, L.; Fang, Z.; He, J.; Xie, Z.; Cheng, G.; Wang, J. Building Radiomics Models Based on Triple-Phase CT Images Combining Clinical Features for Discriminating the Risk Rating in Gastrointestinal Stromal Tumors. *Front. Oncol.* **2021**, *11*, 737302. [CrossRef]

79. Ren, C.; Wang, S.; Zhang, S. Development and validation of a nomogram based on CT images and 3D texture analysis for preoperative prediction of the malignant potential in gastrointestinal stromal tumors. *Cancer Imaging* **2020**, *20*, 5. [CrossRef]
80. Ren, C.; Wang, S.; Zhang, S.; Jiang, Z. Value of CT-Based Texture Analysis in Preoperative Prediction of the Grade of Gastrointestinal Stromal Tumors Compared to Conventional CT Imaging. *Iran. J. Radiol.* **2019**, *16*, e85703. [CrossRef]
81. Starmans, M.P.A.; Timbergen, M.J.M.; Vos, M.; Renckens, M.; Grünhagen, D.J.; van Leenders, G.; Dwarkasing, R.S.; Willemssen, F.; Niessen, W.J.; Verhoef, C.; et al. Differential Diagnosis and Molecular Stratification of Gastrointestinal Stromal Tumors on CT Images Using a Radiomics Approach. *J. Digit. Imaging* **2022**, *35*, 127–136. [CrossRef]
82. Wang, C.; Li, H.; Jiaerken, Y.; Huang, P.; Sun, L.; Dong, F.; Huang, Y.; Dong, D.; Tian, J.; Zhang, M. Building CT Radiomics-Based Models for Preoperatively Predicting Malignant Potential and Mitotic Count of Gastrointestinal Stromal Tumors. *Transl. Oncol.* **2019**, *12*, 1229–1236. [CrossRef] [PubMed]
83. Wang, M.; Feng, Z.; Zhou, L.; Zhang, L.; Hao, X.; Zhai, J. Computed-Tomography-Based Radiomics Model for Predicting the Malignant Potential of Gastrointestinal Stromal Tumors Preoperatively: A Multi-Classifier and Multicenter Study. *Front. Oncol.* **2021**, *11*, 582847. [CrossRef] [PubMed]
84. Xu, F.; Ma, X.; Wang, Y.; Tian, Y.; Tang, W.; Wang, M.; Wei, R.; Zhao, X. CT texture analysis can be a potential tool to differentiate gastrointestinal stromal tumors without KIT exon 11 mutation. *Eur. J. Radiol.* **2018**, *107*, 90–97. [CrossRef]
85. Xu, J.; Zhou, J.; Wang, X.; Fan, S.; Huang, X.; Xie, X.; Yu, R. A multi-class scoring system based on CT features for preoperative prediction in gastric gastrointestinal stromal tumors. *Am. J. Cancer Res.* **2020**, *10*, 3867–3881. [PubMed]
86. Zhang, L.; Kang, L.; Li, G.; Zhang, X.; Ren, J.; Shi, Z.; Li, J.; Yu, S. Computed tomography-based radiomics model for discriminating the risk stratification of gastrointestinal stromal tumors. *Radiol. Med.* **2020**, *125*, 465–473. [CrossRef]
87. Zhang, Q.W.; Zhou, X.X.; Zhang, R.Y.; Chen, S.L.; Liu, Q.; Wang, J.; Zhang, Y.; Lin, J.; Xu, J.R.; Gao, Y.J.; et al. Comparison of malignancy-prediction efficiency between contrast and non-contract CT-based radiomics features in gastrointestinal stromal tumors: A multicenter study. *Clin. Transl. Med.* **2020**, *10*, e291. [CrossRef]
88. Zhang, Q.W.; Gao, Y.J.; Zhang, R.Y.; Zhou, X.X.; Chen, S.L.; Zhang, Y.; Liu, Q.; Xu, J.R.; Ge, Z.Z. Personalized CT-based radiomics nomogram preoperative predicting Ki-67 expression in gastrointestinal stromal tumors: A multicenter development and validation cohort. *Clin. Transl. Med.* **2020**, *9*, 12. [CrossRef]
89. Zhao, Y.; Feng, M.; Wang, M.; Zhang, L.; Li, M.; Huang, C. CT Radiomics for the Preoperative Prediction of Ki67 Index in Gastrointestinal Stromal Tumors: A Multi-Center Study. *Front. Oncol.* **2021**, *11*, 689136. [CrossRef]
90. Zheng, J.; Xia, Y.; Xu, A.; Weng, X.; Wang, X.; Jiang, H.; Li, Q.; Li, F. Combined model based on enhanced CT texture features in liver metastasis prediction of high-risk gastrointestinal stromal tumors. *Abdom. Radiol.* **2022**, *47*, 85–93. [CrossRef]
91. Antoch, G.; Kanja, J.; Bauer, S.; Kuehl, H.; Renzing-Koehler, K.; Schuette, J.; Bockisch, A.; Debatin, J.F.; Freudenberg, L.S. Comparison of PET, CT, and dual-modality PET/CT imaging for monitoring of imatinib (STI571) therapy in patients with gastrointestinal stromal tumors. *J. Nucl. Med.* **2004**, *45*, 357–365.
92. Beheshti, M.; Li, S.R.; Vali, R.; Schima, W.; Dudczak, R.; Langsteger, W. The Potential Value of F-18 FDG PET in Comparison to CT in Early Prediction of Response to Imatinib (STI571) Therapy in Patients with Gastrointestinal Stromal Tumors. *Iran. J. Nucl. Med.* **2007**, *15*, 34–42.
93. Chacón, M.; Eleta, M.; Espindola, A.R.; Roca, E.; Méndez, G.; Rojo, S.; Pupareli, C. Assessment of early response to imatinib 800 mg after 400 mg progression by ^{18}F-fluorodeoxyglucose PET in patients with metastatic gastrointestinal stromal tumors. *Future Oncol.* **2015**, *11*, 953–964. [CrossRef] [PubMed]
94. Choi, H.; Charnsangavej, C.; de Castro Faria, S.; Tamm, E.P.; Benjamin, R.S.; Johnson, M.M.; Macapinlac, H.A.; Podoloff, D.A. CT evaluation of the response of gastrointestinal stromal tumors after imatinib mesylate treatment: A quantitative analysis correlated with FDG PET findings. *AJR Am. J. Roentgenol.* **2004**, *183*, 1619–1628. [CrossRef] [PubMed]
95. Choi, H.; Charnsangavej, C.; Faria, S.C.; Macapinlac, H.A.; Burgess, M.A.; Patel, S.R.; Chen, L.L.; Podoloff, D.A.; Benjamin, R.S. Correlation of computed tomography and positron emission tomography in patients with metastatic gastrointestinal stromal tumor treated at a single institution with imatinib mesylate: Proposal of new computed tomography response criteria. *J. Clin. Oncol.* **2007**, *25*, 1753–1759. [CrossRef]
96. Dudeck, O.; Zeile, M.; Reichardt, P.; Pink, D. Comparison of RECIST and Choi criteria for computed tomographic response evaluation in patients with advanced gastrointestinal stromal tumor treated with sunitinib. *Ann. Oncol.* **2011**, *22*, 1828–1833. [CrossRef]
97. Gayed, I.; Vu, T.; Iyer, R.; Johnson, M.; Macapinlac, H.; Swanston, N.; Podoloff, D. The role of 18F-FDG PET in staging and early prediction of response to therapy of recurrent gastrointestinal stromal tumors. *J. Nucl. Med.* **2004**, *45*, 17–21.
98. Goerres, G.W.; Stupp, R.; Barghouth, G.; Hany, T.F.; Pestalozzi, B.; Dizendorf, E.; Schnyder, P.; Luthi, F.; von Schulthess, G.K.; Leyvraz, S. The value of PET, CT and in-line PET/CT in patients with gastrointestinal stromal tumours: Long-term outcome of treatment with imatinib mesylate. *Eur. J. Nucl. Med. Mol. Imaging* **2005**, *32*, 153–162. [CrossRef]
99. Holdsworth, C.H.; Badawi, R.D.; Manola, J.B.; Kijewski, M.F.; Israel, D.A.; Demetri, G.D.; Van den Abbeele, A.D. CT and PET: Early prognostic indicators of response to imatinib mesylate in patients with gastrointestinal stromal tumor. *AJR Am. J. Roentgenol.* **2007**, *189*, W324–W330. [CrossRef]
100. Jager, P.L.; Gietema, J.A.; van der Graaf, W.T. Imatinib mesylate for the treatment of gastrointestinal stromal tumours: Best monitored with FDG PET. *Nucl. Med. Commun.* **2004**, *25*, 433–438. [CrossRef]

101. Phongkitkarun, S.; Phaisanphrukkun, C.; Jatchavala, J.; Sirachainan, E. Assessment of gastrointestinal stromal tumors with computed tomography following treatment with imatinib mesylate. *World J. Gastroenterol.* **2008**, *14*, 892–898. [CrossRef]
102. Prior, J.O.; Montemurro, M.; Orcurto, M.V.; Michielin, O.; Luthi, F.; Benhattar, J.; Guillou, L.; Elsig, V.; Stupp, R.; Delaloye, A.B.; et al. Early prediction of response to sunitinib after imatinib failure by 18F-fluorodeoxyglucose positron emission tomography in patients with gastrointestinal stromal tumor. *J. Clin. Oncol.* **2009**, *27*, 439–445. [CrossRef] [PubMed]
103. Ryu, M.H.; Lee, J.L.; Chang, H.M.; Kim, T.W.; Kang, H.J.; Sohn, H.J.; Lee, J.S.; Kang, Y.K. Patterns of progression in gastrointestinal stromal tumor treated with imatinib mesylate. *Jpn. J. Clin. Oncol.* **2006**, *36*, 17–24. [CrossRef] [PubMed]
104. Schindler, E.; Amantea, M.A.; Karlsson, M.O.; Friberg, L.E. PK-PD modeling of individual lesion FDG-PET response to predict overall survival in patients with sunitinib-treated gastrointestinal stromal tumor. *CPT Pharmacomet. Syst. Pharmacol.* **2016**, *5*, 173–181. [CrossRef] [PubMed]
105. Schramm, N.; Englhart, E.; Schlemmer, M.; Hittinger, M.; Übleis, C.; Becker, C.R.; Reiser, M.F.; Berger, F. Tumor response and clinical outcome in metastatic gastrointestinal stromal tumors under sunitinib therapy: Comparison of RECIST, Choi and volumetric criteria. *Eur. J. Radiol.* **2013**, *82*, 951–958. [CrossRef]
106. Shinagare, A.B.; Barysauskas, C.M.; Braschi-Amirfarzan, M.; O'Neill, A.C.; Catalano, P.J.; George, S.; Ramaiya, N.H. Comparison of performance of various tumor response criteria in assessment of sunitinib activity in advanced gastrointestinal stromal tumors. *Clin. Imaging* **2016**, *40*, 880–884. [CrossRef]
107. Van den Abbeele, A.D.; Gatsonis, C.; de Vries, D.J.; Melenevsky, Y.; Szot-Barnes, A.; Yap, J.T.; Godwin, A.K.; Rink, L.; Huang, M.; Blevins, M.; et al. ACRIN 6665/RTOG 0132 phase II trial of neoadjuvant imatinib mesylate for operable malignant gastrointestinal stromal tumor: Monitoring with 18F-FDG PET and correlation with genotype and GLUT4 expression. *J. Nucl. Med.* **2012**, *53*, 567–574. [CrossRef]
108. Stroobants, S.; Goeminne, J.; Seegers, M.; Dimitrijevic, S.; Dupont, P.; Nuyts, J.; Martens, M.; van den Borne, B.; Cole, P.; Sciot, R.; et al. 18FDG-Positron emission tomography for the early prediction of response in advanced soft tissue sarcoma treated with imatinib mesylate (Glivec). *Eur. J. Cancer* **2003**, *39*, 2012–2020. [CrossRef]
109. Schiavon, G.; Ruggiero, A.; Bekers, D.J.; Barry, P.A.; Sleijfer, S.; Kloth, J.; Krestin, G.P.; Schöffski, P.; Verweij, J.; Mathijssen, R.H. The effect of baseline morphology and its change during treatment on the accuracy of Response Evaluation Criteria in Solid Tumours in assessment of liver metastases. *Eur. J. Cancer* **2014**, *50*, 972–980. [CrossRef]
110. Schiavon, G.; Ruggiero, A.; Schöffski, P.; van der Holt, B.; Bekers, D.J.; Eechoute, K.; Vandecaveye, V.; Krestin, G.P.; Verweij, J.; Sleijfer, S.; et al. Tumor volume as an alternative response measurement for imatinib treated GIST patients. *PLoS ONE* **2012**, *7*, e48372. [CrossRef]
111. Farag, S.; Geus-Oei, L.F.; van der Graaf, W.T.; van Coevorden, F.; Grunhagen, D.; Reyners, A.K.L.; Boonstra, P.A.; Desar, I.; Gelderblom, H.; Steeghs, N. Early Evaluation of Response Using (18)F-FDG PET Influences Management in Gastrointestinal Stromal Tumor Patients Treated with Neoadjuvant Imatinib. *J. Nucl. Med.* **2018**, *59*, 194–196. [CrossRef]
112. Farag, S.; NS, I.J.; Houdijk, M.P.M.; Reyners, A.K.L.; Arens, A.I.; Grünhagen, D.J.; Desar, I.M.E.; Gelderblom, H.; Steeghs, N.; de Geus-Oei, L.F. Early response evaluation using 18F-FDG-PET/CT does not influence management of patients with metastatic gastrointestinal stromal tumors (GIST) treated with palliative intent. *Nuklearmedizin* **2021**, *60*, 411–416. [CrossRef] [PubMed]
113. Goh, B.K.; Chow, P.K.; Chuah, K.L.; Yap, W.M.; Wong, W.K. Pathologic, radiologic and PET scan response of gastrointestinal stromal tumors after neoadjuvant treatment with imatinib mesylate. *Eur. J. Surg. Oncol.* **2006**, *32*, 961–963. [CrossRef] [PubMed]
114. Arshad, J.; Ahmed, J.; Subhawong, T.; Trent, J.C. Progress in determining response to treatment in gastrointestinal stromal tumor. *Expert Rev. Anticancer Ther.* **2020**, *20*, 279–288. [CrossRef] [PubMed]
115. Padhani, A.R.; Ollivier, L. The RECIST (Response Evaluation Criteria in Solid Tumors) criteria: Implications for diagnostic radiologists. *Br. J. Radiol.* **2001**, *74*, 983–986. [CrossRef] [PubMed]
116. Faivre, S.; Demetri, G.; Sargent, W.; Raymond, E. Molecular basis for sunitinib efficacy and future clinical development. *Nat. Rev. Drug Discov.* **2007**, *6*, 734–745. [CrossRef] [PubMed]
117. Jin, T.; Nakatani, H.; Taguchi, T.; Nakano, T.; Okabayashi, T.; Sugimoto, T.; Kobayashi, M.; Araki, K. STI571 (Glivec) suppresses the expression of vascular endothelial growth factor in the gastrointestinal stromal tumor cell line, GIST-T1. *World J. Gastroenterol.* **2006**, *12*, 703–708. [CrossRef] [PubMed]
118. Young, H.; Baum, R.; Cremerius, U.; Herholz, K.; Hoekstra, O.; Lammertsma, A.A.; Pruim, J.; Price, P. Measurement of clinical and subclinical tumour response using [^{18}F]-fluorodeoxyglucose and positron emission tomography: Review and 1999 EORTC recommendations. European Organization for Research and Treatment of Cancer (EORTC) PET Study Group. *Eur. J. Cancer* **1999**, *35*, 1773–1782. [CrossRef]

diagnostics

MDPI

Protocol

Early Diagnosis of Chemotherapy-Linked Cardiotoxicity in Breast Cancer Patients Using Conventional Biomarker Panel: A Prospective Study Protocol †

Saule Balmagambetova [1,*], Zhenisgul Tlegenova [2], Bekbolat Zholdin [2], Gulnara Kurmanalina [2], Iliada Talipova [2], Arip Koyshybaev [1], Dinara Nurmanova [2], Gulmira Sultanbekova [3], Mira Baspayeva [4], Saule Madinova [4], Kulparshan Kubenova [5] and Ainel Urazova [4]

1 Department of Oncology, West Kazakhstan Marat Ospanov Medical University, 68 Maresyev Street, Aktobe 030019, Kazakhstan
2 Department of Internal Diseases No. 2, West Kazakhstan Marat Ospanov Medical University, 68 Maresyev Street, Aktobe 030019, Kazakhstan
3 Cardiology Division at University Medical Center, Building 8G, Zhanakonys, Aktobe 030017, Kazakhstan
4 Chemotherapy Division at University Medical Center, Building 8G, Zhanakonys, Aktobe 030017, Kazakhstan
5 Clinical Laboratory at University Medical Center, Building 8G, Zhanakonys, Aktobe 030017, Kazakhstan
* Correspondence: saule.balmagambetova@zkmu.kz; Tel.: +7-705-579-46-37
† The study is registered in ISRCTN (https://www.isrctn.com/ISRCTN12628444) 21 July 2022.

Citation: Balmagambetova, S.; Tlegenova, Z.; Zholdin, B.; Kurmanalina, G.; Talipova, I.; Koyshybaev, A.; Nurmanova, D.; Sultanbekova, G.; Baspayeva, M.; Madinova, S.; et al. Early Diagnosis of Chemotherapy-Linked Cardiotoxicity in Breast Cancer Patients Using Conventional Biomarker Panel: A Prospective Study Protocol. *Diagnostics* **2022**, *12*, 2714. https://doi.org/10.3390/diagnostics12112714

Academic Editors: Yuli Huang, Yong Yuan and Peisong Chen

Received: 25 September 2022
Accepted: 2 November 2022
Published: 6 November 2022

Abstract: The prognosis of cancer treatment depends on, among other aspects, the cardiotoxicity of chemotherapy. This research aims to create a feasible algorithm for the early diagnosis of antitumor therapy cardiotoxicity in breast cancer patients. The paper represents a protocol for a prospective cohort study with N 120 eligible participants admitted for treatment with anthracyclines and/or trastuzumab. These patients will be allocated into four risk groups regarding potential cardiotoxic complications. Patients will be examined five times every three months for six biomarkers: cardiac troponin I (cTnI), brain natriuretic peptide (BNP), C-reactive protein (CRP), myeloperoxidase (MPO), galectin-3 (Gal-3), and D-dimer, simultaneously with echocardiographic methods, including speckle tracking. The adjusted relative risk (aOR) of interrupting an entire course of chemotherapy due to cardiotoxic events will be assessed using multiple analyses of proportional Cox risks. The Cox model will also assess associations between baseline biomarker values and time to cardiotoxic events. Moreover, partly conditional survival models will be applied to determine associations between repeated assessments of changes in biomarkers from baseline and time to cancer therapy-related cardiac dysfunction. All models will be adjusted for cancer therapy regimen, baseline LVEF, groups at risk, baseline biomarker values, and age. The decision-tree and principal component analysis (PCA) methods will also be applied. Thus, feasible patterns will be detected.

Keywords: cardiotoxicity; chemotherapy; breast cancer; speckle tracking; biomarkers; kazakhstan

1. Introduction

Breast cancer (BC) was the most commonly diagnosed cancer in 2020 and remains as such as of 2021, accounting for 12% of all new annual cancer cases worldwide [1,2]. In Kazakhstan, the leading country in Central Asia, BC ranks second following lung cancer and accounts for 12.4% of all new cases as of 2021 [3]. Nevertheless, owing to cancer treatment achievements, the overall 5-year relative survival rate for breast cancer has reached 90% [4]. The prognosis of BC treatment depends on the tumor tissue's histochemical properties, the tumor's aggressiveness, the cancer process staging, and the cardiotoxicity of chemotherapy, now defined in the literature as cancer therapy-related cardiac dysfunction (CTRCD) [5]. At present, the CTRCD definition includes all possible cardiovascular complications occurring during anticancer treatment, such as left ventricular dysfunction, heart failure, myocardial infarction, arrhythmias or conduction disorders, acute myocarditis or pericarditis, hypertension arterial or hypotonia, and, in case of radiotherapy, coronary heart disease [6]. Currently, researchers distinguish two main typesof CTRCD: the anthracycline-associated, irreversible type with a cumulative effect, and the trastuzumab-induced, reversible, dose-independent type. Moreover, clinicians deal with its early-onset and late-onset varieties, as well as acute, subacute, and chronic forms [7,8]. As CTRCD can affect cancer survivors for years, decreasing their quality of life, the scope of the problem increases yearly. Only in the USA, 9.7 million females out of more than 18 million Americans with a history of cancer were alive on 1 January 2022, and the most prevalent tumor type was BC (4,055,770) [9]. A retrospective analysis of outcomes on all sarcoma patients in Denmark revealed that cardiotoxicity was observed in 31% of these patients. Of them, 70% experienced early-onset CTRCD [10].

The challenge of heart consequences after chemotherapy appeared so large-scale that it gave rise to a new subspecialty at the intersection of oncology and cardiology: cardio-oncology [11]. Cardio-oncologists primarily focused on echocardiographic methods in searching for methodsto detect CTRCD as early as possible. Transthoracic echocardiography (TTE) was recognized as the standard method to evaluate cardiotoxicity [12]. In the 2000s, the novel speckle tracking echocardiography (STE) method was widely introduced. STE allows for tracking of the displacement of "speckles" in two-dimensional (2D) echocardiographic images in an angle-independent way and assessment of their movement (strain) during the cardiac cycle. Its high feasibility, reproducibility, and accuracy have been demonstrated [13–15]. CTRCD is defined by either a decrease in left ventricular ejection fraction (LVEF) above 10% from baseline to a value of LVEF under 53%, or a decrease in global longitudinal strain (GLS) deformation below 15% from the baseline value [5,12,16]. Current echocardiography recommendations set the low normal value of 2D LVEF as 54% for women and 52% for men; hence, in the previous European Association of Cardiovascular Imaging (EACVI) position statement, a reduction in LVEF below 53% was classified as abnormal [17]. To date, researchers mainly focus on applying cardiac markers as they aresimple and relatively cost-effective compared to echocardiographic methods, particularly GLS assessment. Moreover, including biomarkers in the algorithms of CTRCD detection allows a comprehensive evaluation of cardiac function. Among biomarkers, cardiac troponins and B-type natriuretic peptide (BNP) seem to display the most proven efficacy [18,19].

Common practices include serial echocardiography and mostly troponin measurements. Nevertheless, although the pathophysiology of CTRCD has been characterized, there is currently no evidence-based approach for monitoring and managing patients who develop CTRCD [20]. Primarily, the task is to detect CTRCD at a preclinical stage, before any echocardiographic evidence of dysfunction or clinical symptoms are present, which would minimize cardiac complications. If the GLS assessment technique and three-dimensional (3D) echocardiography can cope with this task, the more affordable TTE cannot.

Therefore, searching for reliable biomarkers remains the research target regarding CTRCD's feasible monitoring. Regrettably, to date, the cardio-oncologic integrative approach and development of appropriate tools to diagnose CTRCD conditions have not been implemented in the practice of Kazakhstani cardiologists and oncologists. Cardiac dysfunctions provoked by cancer treatment remain largely unexplored. The attitude to and awareness of CTRCD within involved specialties in Kazakhstan are similar to the state in many other countries. According to J. Peng et al. [21], the majority of cardiologists felt that cardiotoxicity should be monitored, even in asymptomatic cancer patients (55.8% vs. 12% of oncologists); the majority of cardiologists stated that cardio-oncology clinics would significantly improve cancer patients' prognosis (88.3% vs. 45.8% of oncologists).

Thus, the current study is a first step toward forming a new subspecialty, cardio-oncology, in the country, by developing a program for the early diagnosis and timely correction of cardiotoxic complications of breast cancer antitumor therapy.

The rationale for this study is to use currently available methods to elaborate an algorithm to reveal chemotherapy-linked cardiotoxic complications early, to minimize their negative consequences. It is known that assessing the global longitudinal myocardial strain is the most effective technique in evaluating heart activity, although it is quite expensive. On the contrary, cardiac markers are considered the most cost-effective tools. Determining the predictive value of myocardial damage, inflammation, and oxidative stress biomarkers by assessing GLS in chemotherapy patients would allow the further application of these tests without combination with GLS, which is not always accessible. Detecting the incipient stages of cardiotoxicity using relatively cheap and the earliest-responding biomarkers would enable the development of a cost-effective algorithm for CTRCD's timely correction. Generally, this research aims to identify steady patterns between the decrease in systolic myocardial function recorded by EchoCG, including speckle tracking, and the values of selected biomarkers.

In this context, we designed a multimarker panel focused chiefly on the tests' affordability and relative comprehensiveness. The biomarkers' values will be matched with EchoCG data (GLS) in definite periods. As such, we present an opportunity for the synchronous tracing of the changes and an evaluation of the comparative efficiency of all selected tests. The article presents a protocol for a prospective cohort study focused on biomarker research.

2. Materials and Methods

2.1. Protocol for Prospective Cohort Study

2.1.1. Study Design

In this observational single-center study, a mixed design was used, with retrospective and prospective stages. The main points of the prospective cohort study are presented in Figure 1 (stage II). During the first, retrospective stage (from April to September 2021), we extracted all available information from the Aktobe Regional Oncologic Registry on BC patients who received chemotherapy with anthracyclines and/or targeted therapy in 2018–2019. The findings of this database research were published in 2022 [22]. Subsequently, survival rates from both samples (retrospective and prospective) will be matched within the framework of a quasi-experimental design.

2.1.2. Study Population

At this stage, we consecutively recruited a single-center cohort of eligible women newly diagnosed with BC on admission to the chemotherapy division of the University's Medical Center for treatment with anthracyclines an d/or trastuzumab. Given the single-center design, we chose the BC diagnosis due to its relatively high incidence, facilitating the recruitment of participants. The recruitment started on September 2021 and lasted until late August 2022. Criteria for participants' eligibility are shown in Figure 1. Notably, the inclusion criteria do not imply participants who are administered targeted therapy with Kinase Inhibitors (KIs) or Immune Checkpoint Inhibitors (ICIs), only human epidermal growth factor receptor 2 (HER2)-positive individuals with indications of trastuzumab treatment. We also include patients who received radiation treatment before chemotherapy, as many schemes imply combined treatment. These patients will be allocated to a separate subgroup during analysis. The form of chemotherapy is also of significance. We select patients of any age (18+) and at any stage of BC but eligible for adjuvant or neoadjuvant chemotherapy, not palliative. Exclusion criteria are commonly accepted: patients showing LVEF $\leq$ 40% by Simpson or having established cardiac dysfunction or severe comorbidities are believed to be not eligible.

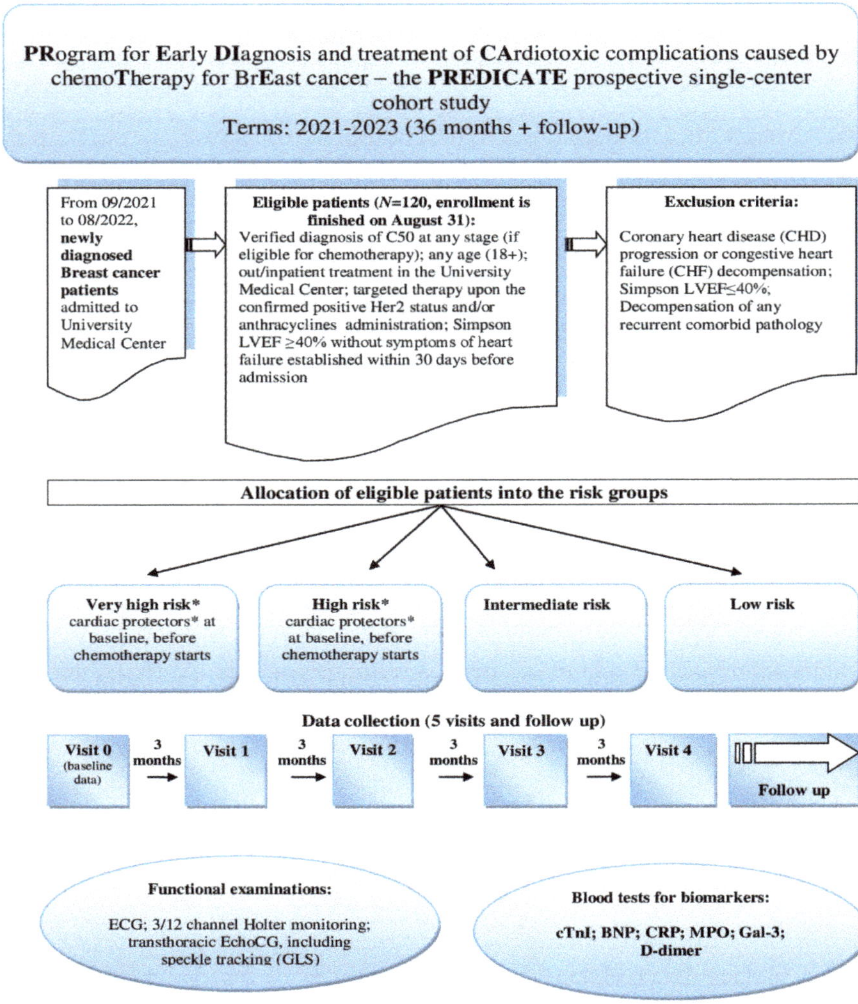

Figure 1. Design of the prospective cohort study.*"Very high" risk means presence of existing chronic heart failure or dilated cardiomyopathy, according to the current position papers on CTRCD [17,23–25]; "high" risk includes previous severe valvular heart disease, past myocardial infarction and/or revascularization, baseline left ventricular ejection fraction <50%, stable angina, or prior treatment with anthracyclines or radiation therapy (see Section 2.1.7). * Cardiac protectors: ACE inhibitors/ARBs; beta-blockers; statins; trimetazidine/analogues; others (see Section 2.1.5).

2.1.3. Sample Size Estimation

To calculate the sample size for the prospective cohort, we analyzed data from the regional oncologic registry for ten years, 2010–2019, and literature sources on the incidence of LV dysfunction associated with chemotherapy [5,7,20]. The sample size was calculated using the Epi Info TM v.7 software (CDC USA) (https://www.cdc.gov/epiinfo (accessed on 2 September 2020)). We entered the following information: population size—150 (the average annual number of BC patients admitted to chemotherapy treatment at the University Medical center), the expected incidence of cardiovascular disease in women in the menopause transition aged 50+—9.7% [26], and the assessment accuracy—5%. The calculated sample size appeared to be equal to 100. Given the possible losses, the sample might be increased to 120 patients. As is known, the predictive model requires at least ten observations per predictor. We believe that the calculated sample size will be sufficient for a clinically relevant predictive model, as the resulting sample size allows for the detection of small-to-medium-sized group differences.

2.1.4. Patient and Public Involvement

There is no patient or public involvement. There is no potential risk for patients due to the observational nature of this research.

2.1.5. Data Collection and Timing

The research implies five participants' visits, including an initial visit and one every three months (see Figure 1) Thus, the recruitment lasts twelve months, and the overall data collection will last until late August 2023. Given the follow-up, the prospective stage duration is thirty months. Data to be collected during visits are displayed in Table 1.

These data are listed in the patient's individual registration card (IRC, Appendix No. 5 to the Protocol) and are entered into the dataset. The patient's IRC example and other supplemental materials of the study are placed in a publicly available repository, osf.io [https://osf.io/nykmw/ (accessed on 10 June 2022)]. According to the Individual Participant Data Sharing Statement, Excel datasets would also be placed in this repository upon their completion (https://www.isrctn.com/ISRCTN12628444 (accessed on 21 July 2022).

2.1.6. Definitions and Outcomes of the Study

The definition applied in the present study is as follows: CTRCD is defined by either a decrease in left ventricular ejection fraction (LVEF) above 10% from baseline to a value of LVEF under 53%, or a decrease in global longitudinal strain (GLS) deformation below 15% from the baseline value. Early cardiotoxicity is defined as <one year and late cardiotoxicity as ≥one year [5].

Accordingly, we apply the following definitions of measured outcomes.

- Primary outcomes:

1. The number of patients who develop cardiotoxic complications, including subclinical dysfunction during chemotherapy, measured using LVEF monitoring (<50% or >10% decline from baseline) and GLS assessment (decrease > 15% from baseline) at 12 months after the chemotherapy course started, through all groups at risk.

2. One-year survival without cardiotoxic complications, measured using LVEF monitoring and GLS assessment (as defined above) at 12 months after chemotherapy completion through all groups at risk.

In the primary outcomes, we included not only manifested CTRCD but also subclinical cardiac dysfunction during chemotherapy.

- Secondary outcomes:

Measured using biomarker measurement units on immunofluorescence analyzers at 3, 6, 9, and 12 months for all below:

1. Presence of increased values of the tests during chemotherapy treatment: cardiac troponin I (cTnI)—≥0.3 ng/mL; brain natriuretic peptide (BNP)—>100 pg/mL; C-reactive protein (CRP)—>5 mg/l; antibodies to myeloperoxidase (MPO)—>5 U/mL; galectin-3 (Gal-3)—>28.7 ng/mL; D-dimer—≥0.5 mg/L (presented values are taken from the analyzers' manuals).

2. Time trends in the onset of increasing biomarker values in the LVEF and GLS data.

3. The predictive value of a positive result (PPV) and the predictive value of a negative result (PVN) for all enlisted biomarkers.

4. The number of all identified cardiovascular events during observation.

Table 1. Type of data subject to collection during visits.

Domain	Measures (Itemized)
General information	Patient's code (individual identification No.); age; ethnicity; main occupation for the last 12 months; menopause; BMI*
Breast cancer features	Tumor clinical staging; tumor histotype; tumor clinical classification; IHC* data
Presence of risk factors	BC hereditary factor; smoking; physical activity (hour, min); history of stroke; IHD*, CHF* (functional class); Charlson comorbidity index (disease, scores); arterial hypertension; diabetes mellitus (presence/absence)
Clinical examination at admission	Heart rate; SBP*; DBP*; 6-min walk test (functional class, meters); selected clinical and biochemical lab tests; ECG* and Holter monitoring (HM every 6 months)
Breast cancer treatment	Medications prescribed (dose, regimen) Previous treatment, before admission to chemotherapy division: radiation (dose, regimen), surgical (radical, sectoral)
Cardiac protectors prescribed	ACE* inhibitors/ARBs*; beta-blockers; statins; trimetazidine/analogues; others
Echocardiographic data and biomarkers	See Figure 1 (including time periods of change onset)
Complications of Chemotherapy CTRCD conditions	 Code; type; form; time of emergence
Other complications	Hematopoietic system complications; thromboembolic complications; pericarditis; gastrointestinal tract; respiratory system; urinary system; allergic reactions; neurotoxicity; toxic effects on the skin and appendages; toxic hyperthermic reactions; toxic phlebitis
Chemotherapy outcomes	Without complications; CT* correction due to CTRCD; CT interruption due to CTRCD; CT correction due to non-CTRCD complications; CT interruption due to non-CTRCD complications
Completeness of chemotherapy	(%)
Death	(Cause; date)
Dropped out of the study	Reason

*ACE—angiotensin-converting enzyme inhibitors; *ARBs—angiotensin II receptor blockers; *BMI—body mass index; *CHF—chronic heart failure; *CT—chemotherapy; *ECG—electrocardiogram; *IHC—immunohistochemical test; *IHD—ischemic heart disease; *SBP, *DBP—systolic, diastolic blood pressure.

2.1.7. The Risk Stratification Strategy

During baseline clinical examination (visit 0), the potential risk of CTRCD emergence is calculated. We use the strategy proposed in recent studies and adopted by current position papers on CTRCD [17,23–25], except for additional measurements after three–four medication cycles. According to the current strategy, the risk scores are calculated considering all possible risk factors—existing (recorded) cardiovascular diseases, the toxicity of chemotherapy prescribed, and lifestyle risk factors. If there is one intermediate risk factor or an absence of risks at the moment of examination, the patient is allocated to the low-risk group. If two to four intermediate risk factors are present, they are allocated to the intermediate risk group. If more than five intermediate risk factors or at least one high-risk factor are revealed, they are allocated to the high-risk group. Respectively, patients with one very high-risk factor are allocated to the very-high-risk group. Thus, all patients are divided into four risk groups, and patients from the groups of very high and high risk are provided with cardiac protectors from the pharmaceutical classes pointed out in Table 1. For all enrolled participants, a particular individual form, "Cardiovascular risk stratification for upcoming chemotherapy treatment," is filled out (Appendix No. 6, [https://osf.io/nykmw/ (accessed on 10 June 2022)].) The patient's risk scores will be calculated and stated in this form based on the listed possible risk factors and their defined hazard level and individual risk level.

2.1.8. Outcome Measurements

General blood and biochemical tests for this study are performed according to the "Breast Cancer" in-country protocol dated 1 March 2019, No. 56, at the University Medical Center's clinical lab. Some biomarkers' detection (cTnI, BNP, CRP, D-dimer) is also performed in this lab using immunoassay analyzers and Finecare rapid quantitative tests, in line with Good Laboratory Practice (GLP). Galectin-3 and MPO assays were directed to a third-party collaborator, the "Olymp" labs network. The types of tests are the ELISA Galectin-3 S and ELiA MPO Well tests.

- Functional outcomes:

ECG machines and 24-record 12-channel wearable devices for outpatient Holter monitoring are used to detect rhythm and conduction disturbances, including the QT interval duration, in each enrolled participant during scheduled visits (*HM every six months, not three.) If cardiac complaints have emerged, out-of-schedule invitations are provided. Two project staffers are responsible for TTE and speckle tracking. Cardiac imaging is performed using "automated function imaging (AFI)—automatic non-Doppler assessment of longitudinal strain of the left ventricular myocardium" software and M5-SD transducer, 1.5–4.5 MHz. Left ventricular end-diastolic and end-systolic volumes are calculated using Simpson's approach to derive LVEF. The same specialists also analyze GLS according to currently accepted protocols [17,27]. The study area is corrected to cover the whole thickness of the myocardial wall. Measurements are carried out from the apical 3-chamber (3C), 4-chamber (4C), and 2-chamber positions. When deriving apical positions, care is taken to ensure that the long axis of the ventricle is perpendicular to the plane of the mitral annulus in the LV apical views. Intra- and inter-observer coefficients of variation are not set for LVEF and GLS. CTRCD is defined as a $\geq$10% absolute decline in LVEF to a value of <53%. Subclinical LV myocardial dysfunction is considered to be a drop in GLS below (-) 18%, those in the range (0% to 17.9%), or a decrease in this indicator >15% from baseline. Of note, although we scheduled functional and laboratory measurements in terms of 0, 3, 6, 9, and 12 months, if patients receive the treatment on these dates, we postpone the tests until the chemotherapy cycle is completed or until the treatment interval. Usually, it takes up to 1–2 weeks.

2.1.9. Adherence Control

It is known that poor adherence has been characterized as the most common cause of an unsatisfactory response to medication [28]. As the overwhelming majority of participants are not present in the inpatient division, we cannot monitor the number of pills (cardiac protectors) prescribed to patients from the two high-risk groups or use other vigorous methods of compliance control. Instead, we practice monthly encouraging calls to patients under observation to ensure retention and adherence to protocol activities. Moreover, during the baseline examination, we assist willing participants who are prescribed cardiac protectors in installing appropriate apps on their telephones to remind them to take their pills. There are two project staffers responsible for these activities. According to research on experience in the discussed issue, 89% adherence (medication taken) under long-term observation is recognized as an excellent result [29].

2.2. Statistical Analysis

In this study, baseline characteristics will be summarized using proportions for categorical variables (cancer treatment, revealed cardiotoxic complications, etc.), and medians (interquartile range) be presented for continuous variables (biomarker values, etc.). The Pearson χ^2 criterion will be applied to identify intergroup differences for categorical variables. Quantitative variables will be compared using the nonparametric U Mann–Whitney test in terms of rhythm disturbances (ECG/HM), dynamics in the level of biomarkers (0, 3, 6, 9, 12 months), the degree of myocardial dysfunction identified by LVEF and GLS (0, 3, 6, 9, 12 months), etc., for two unrelated groups (with and without the development of cardiotoxicity.) A comparison of three or more independent groups (e.g., four risk groups) will be carried out using parametric analysis of variance (one-way ANOVA) or a nonparametric Kruskal–Wallis H-test. For repeated measurements (related samples), analysis of variance with Greenhouse–Geisser corrections will be applied.

To characterize the changes in biomarker levels according to treatment groups, mean estimated changes from baseline will be plotted over time. Considering radiation treatment received immediately before the chemotherapy as a source of potential bias, we arrange not only anthracycline, trastuzumab, and combined treatment groups but also a separate subgroup after radiation treatment. Mean changes will be determined using repeated-measures linear regression estimated via generalized equations. Each model will be adjusted for the baseline values of the biomarker under consideration and the time since the treatment started. Contemporaneous associations between changes in biomarkers from baseline and changes in LVEF will also be determined using repeated-measures linear regression.

To assess the influence of independent factors on the binary variable of response (cardiotoxicity), multiple logistic regression analysis (LRA) will be used by the sequential exclusion of variables (age, biomarkers, arrhythmias, LVEF according to Simpson and GLS data, etc.). The presence of a statistically significant relationship with the predicted event in the one-dimensional analysis will be the criterion for inclusion in the multivariate analysis. The results will be presented as unadjusted (uOR), adjusted odds ratios (aOR), and 95% CI. The incidence of cardiotoxic complications of chemotherapy will also be presented with the 95% CI, calculated according to the Wilson method. Sensitivity analysis (the Cornfield method) will also be applied for the binary variable of response (cardiotoxicity).

Indices of sensitivity (Sn), specificity (Sp), PPV, and PVN will be calculated for each diagnostic test under consideration (biomarkers, TTE, rhythm disturbances, etc.), as well as for predictive models in general.

The risk of interrupting a course of chemotherapy due to cardiotoxicity in the analyzed groups will be assessed through Kaplan–Meier survival curves. Given the influence of potential confounders, the relative risk of interrupting an entire course of chemotherapy due to cardiotoxic events will be assessed using multiple analyses of proportional Cox risks. The results will be presented as aOR with a 95% CI. Associations between baseline biomarker values and time to CTRCD will also be assessed using Cox proportional hazards model. Moreover, partly conditional survival models will be applied to determine associations between repeated assessments of changes in biomarkers from baseline and time to CTRCD. All models will be adjusted for cancer therapy regimen, baseline LVEF and group at risk (and cardiac protection if administered), baseline biomarker values, age, comorbidities index, and body mass index. Associations between changes from baseline in biomarker levels over time and subsequent changes in LVEF and longitudinal strain will be assessed similarly. Differences in the associations between changes in biomarkers and LVEF/GLS across the different treatment groups will be evaluated by including biomarker–treatment interaction terms, as shown in the recent paper by Demissei et al. [30].

Due to the number of variables subjected to analysis, and the anticipated presence of multiple outcomes, the included tests may not be sufficient to solve the task of building a clinically relevant decision algorithm, and so the decision-tree method will be applied. Decision trees break down complex data into more manageable parts, which is beneficial in prediction analysis, data classification, and regression. A dimensionalityreduction method, principal component analysis (PCA), will also be applied.

Two-sided levels <0.05 are assumed to be statistically significant. For statistical processing, software packages SPSS (IBM, Armonk, NY, USA, v.25), Statistica (StatSoft, Inc., Tulsa, OK, USA, v. 10), and R 3.3.2 (R Foundation for Statistical Computing, Vienna, Austria).

2.3. Study Ethics and Dissemination Plan

As mentioned above, this study poses no risk to the participating individuals. Study participation does not imply restrictions in receiving any clinical care determined by oncologists (additional CT, MRI, etc.). All study procedures are conducted according to the principles of the Declaration of Helsinki (2013), and patient rights are observed. Participation in the study is voluntary, and participants can leave the study at any time for any reason. This provision was stated in the informed consent form at enrollment in the study (https://osf.io/nykmw/ (accessed on 10 June 2022). Subsequent publications will include the interim research results and final results with developed algorithms (in peer-reviewed journals). Moreover, the dissemination plan permits the preparation of country-scale clinical recommendations on the management of CTRCD for cardiologists and oncologists.

3. Preliminary Results of the Study

Previous retrospective research for 2018–2019, the first stage of the current project, aimed to clarify the proportion and structure of complications that led to chemotherapy interruption in BC patients. This registry study included all available characteristics of the oncologic process: tumor staging and histological type, medications used, duration of chemotherapy, occurring complications, treatment outcomes, and survival rates

Overall, 305 BC cases were analyzed. Chemotherapy was completed successfully in 65.9% of patients and interrupted in 10.5% due to adverse events. Cardiovascular disorders were identified in 6.2%. There were significant differences in the number of detected cardiovascular complications between the two groups of patients—those under EchoCG monitoring and those without ($p < 0.001$)—but no difference in survival was found in the two groups (p 0.814). Survival in patients with cardiovascular complications was 28.1 months, compared to 34.3 months in the group without CV events ($p < 0.005$). Those who completed a full course of chemotherapy had a survival rate of 34.9 months, compared to 17.6 months in persons whose treatment was interrupted due to complications ($p < 0.001$). We identified four fatal cases of cardiotoxicity that were not recorded in the oncologic registry as CTRCD.

4. Discussion

In the mentioned preface to the cohort study (retrospective research for 2018–2019), we obtained an idea of CTRCD's scope in the region, establishing a 6.2% CTRCD prevalence and reporting four fatal cases [22]. Analysis of these cases revealed that only one of the women was administered cardiac protectors, and three women received adjuvant radiation treatment before admission to chemotherapy. Breast cancer patients who have undergone even small-dose radiation, particularly left-sided exposure, are known to carry an increased risk of ischemic heart disease, which is proportional to the radiation dose [31,32]. In the present cohort study, we cannot exclude those who underwent radiation treatment before chemotherapy due to the relatively small number of eligible patients in a single-center setting. This circumstance can lead to bias, and we will manage these patients as a separate subgroup during analysis, as stated above. Recent research reported the cardiotoxicity risk of relatively new classes of targeted therapies such as KIs or ICIs [33–36]. We exclude those who are administered the mentioned treatments to minimize the subgroups under analysis.

A great deal of research has proposed and tested different panels, combining various markers. Emerging biomarkers include myeloperoxidase (MPO), placental growth factor (PIGF), growth differentiation factor 15 (GDF-15), C-reactive protein (CRP), Galectin-3 (Gal-3), microRNAs, etc., having potential value in predicting CTRCD before any signs of overt cardiotoxicity are apparent [37,38]. Some of these biomarkers (CRP, Gal-3, MPO) reflect such aspects of heart pathophysiology as oxidative stress, inflammation, and fibrosis [39]. A high serum level of CRP (>10 mg/L) is known to predict mortality in patients with acute decompensated heart failure (HF) one year after discharge [40]. CRP has recently been shown to have important predictive value in tumor immunotherapy. CRP is a prognostic biomarker for Immune Checkpoint Inhibitor (ICI)treatment [41,42]. Gal-3, a member of the beta-galactoside-binding lectin family, is involved in the occurrence and development of cardiac fibrosis, HF, and atherosclerosis [43]. In addition, Gal-3 inhibitors can effectively block lung adenocarcinoma growth and metastasis and increase the efficacy of PD-L1 ICIs [44]. Moreover, Gal-3 concentrations appear to increase before heart failure manifests, making it a potential screening tool for patients at risk of heart failure [45]. MPO is known to be an independent predictive factor of 1-year mortality in acute HF patients [46]. D-dimer is a sensitive biomarker for cancer-associated thrombosis, but little is known about its significance in CTRCD. Japanese researchers recently established that the occurrence of CTRCD was higher in the high D-dimer group compared to the low group (16.2 vs. 4.5%, $p = 0.0146$, n 169) [47].

Undoubtedly, high-sensitivity cardiac troponin T (hs-cTnT) and NT-proBNP (N-terminal pro-B-type natriuretic peptide) are currently recognized as the most effective predictors of CTRCD [18,19,30,48]. According to a recent review, only NPs (BNP and NT-proBNP) come close to the characteristics of "ideal" HF biomarkers. They are often regarded as the reference standard against which other potential biomarkers must be evaluated, as they may allow the identification of patients with subclinical LV dysfunction [49]. Compared with BNP, NT-proBNP has a much broader "gray zone". Notwithstanding, these hormones are clinically equivalent in diagnosing CHF, and recalculation is possible [50–52]. Presumably, BNP appears more convenient in many actual situations. Moreover, a definite correlation between D-dimer and BNP is found in patients with various pathologies. However, this can lead to an incorrect diagnosis regarding suspected cases of hypercoagulable states in heart failure. More clinical data are needed to rely on such a correlation [53,54].

Troponin test elevation occurs even in chronic heart failure patients as an indicator of myocardial stress. Within the panel of troponins, hsTn assays are believed to be the most sensitive tests. Nevertheless, there is a loss of specificity unless specific protocols, particularly those involving "deltas", are used [55,56]. Moreover, researchers found that when using hsTnI, a sex-specific threshold for MI diagnosis doubled the diagnosis of MI in female participants [57]. Meanwhile, recently, the longitudinal LVEF trajectory, but not hs-cTnT or NT-proBNP, was reported as allowing for a dynamic assessment of cardiotoxicity risk in early BC [58]. Thus, an evidence-based approach for managing CTRCD patients is yet to be developed.

Strengths and Limitations of This Study

- This study is the first of its kind across Kazakhstan and Central Asia, thus creating a springboard for future developments in cardio-oncology.

- All biomarkers selected for the panel are relatively affordable and can provide a comprehensive assessment of cardiovascular disorders at definite periods over two years of CTRCD monitoring in cancer patients, irrespective of cancer nosology.
- The single-center design may be considered the study's main limitation, resulting in a relatively small sample size compared to other researchers' data and potentially affecting the results' generalizability.
- One of the limitations potentially leading to bias is the impossibility of excluding those who receive radiation treatment before chemotherapy.
- The absence of set intra- and inter-observer coefficients of variation for LVEF and GLS may also be referred to as a limitation of this study.

5. Conclusions

The PREDICATE research aims to study the effect of chemotherapy on myocardial function and structure, presenting groups of BC patients for whom GLS assessment and a proposed panel of biomarkers (cTnI, BNP, D-dimer, CRP, MPO, and Gal-3) would be the most appropriate for the timely diagnosis and treatment of cardiotoxicity. The selected tests were not combined in previous research by other authors. Some of these biomarkers are included in routine examinations and do not require additional financing. We believe that it would be helpful to establish further patterns in their relationship at certain time intervals. In the future, the involvement of other in-country clinics to enlarge the number of participants under CTRCD monitoring is needed.

Supplementary Materials: As mentioned in the text, all supplementary materials relevant to the presented study protocol are placed in a publicly available repository: [https://osf.io/nykmw/ (accessed on 10 June 2022)]. The STROBE checklist for cohort studies is added as a supplementary file.

Author Contributions: Conceptualization, B.Z., Z.T. and S.B.; methodology, B.Z. and Z.T.; software, A.U.; validation, A.K., D.N., G.S., M.B. and K.K.; formal analysis, Z.T., S.B., G.K., M.B., I.T. and S.M.; investigation, G.S., D.N., M.B., S.M. and K.K.; resources, A.K.; data curation, S.M., G.S. and K.K.; writing—original draft preparation, S.B.; writing—review and editing, S.B., Z.T. and B.Z.; visualization, I.T., Z.T.; ECG and HM performance, D.N. and G.K.; supervision, B.Z.; project administration, Z.T. and G.K.; funding acquisition, D.N. All authors have read and agreed to the published version of the manuscript.

Funding: This research was funded by the Science Committee of the Ministry of Education and Science of the Republic of Kazakhstan (grant IRN AP09259524, state reg. No. 0121PK00565). The funders had no role in the design of the study; in the collection, analyses, or interpretation of data; in the writing of the manuscript, or in the decision to publish the results.

Institutional Review Board Statement: The study was conducted in accordance with the Declaration of Helsinki, and approved by the Bioethics Committee of the West Kazakhstan Marat Ospanov Medical University (Ref. No. 7, 09/09/2020).

Informed Consent Statement: Informed consent was obtained from all subjects involved in the study. Written informed consent has been obtained from the patients to publish this paper and all subsequent papers under the following conditions: their data must be placed anonymously in the research database without mentioning personal details, i.e., under coded numbers. These provisions were stated in the patient's informed consent document at enrollment in the study.

Data Availability Statement: The data obtained in this study will be openly available at [https://www.isrctn.com/ISRCTN12628444 (accessed on 21 July 2022)] and [https://osf.io/nykmw/ (accessed on 10 June 2022)].

Acknowledgments: The research team acknowledges the West Kazakhstan University Medical Center authorities for their administrative and technical support in implementing this project.

Conflicts of Interest: The authors declare no conflict of interest. The funders had no role in the design of the study; in the collection, analyses, or interpretation of data; in the writing of the manuscript; or in the decision to publish the results.

References

1. Ferlay, J.; Colombet, M.; Soerjomataram, I.; Parkin, D.M.; Pineros, M.; Znaor, A.; Bray, F. Cancer statistics for the year 2020: An overview. *Int. J. Cancer* **2021**. [CrossRef] [PubMed]
2. Sung, H.; Ferlay, J.; Siegel, R.L.; Laversanne, M.; Soerjomataram, I.; Jemal, A.; Bray, F. Global Cancer Statistics 2020: GLOBOCAN Estimates of Incidence and Mortality Worldwide for 36 Cancers in 185 Countries. *CA Cancer J. Clin.* **2021**, *71*, 209–249. [CrossRef]
3. The Global Cancer Observatory. Kazakhstan. Available online: https://gco.iarc.fr/today/data/factsheets/populations/398-kazakhstan-fact-sheets.pdf (accessed on 10 March 2021).
4. Stump-Sutliff, K.A. Breast Cancer Survival Rates. Available online: https://www.webmd.com/breast-cancer/guide/breast-cancer-survival-rates (accessed on 12 June 2022).
5. Zamorano, J.L.; Lancellotti, P.; Rodriques, M.D.; Aboyans, V.; Asteggiano, R.; Galderisi, M.; Habib, G.; Lenihan, D.J.; Lip, G.Y.H.; Lyon, A.R.; et al. 2016 ESC Position Paper on cancer treatments and cardiovascular toxicity developed under the auspices of the ESC Committee for Practice Guldelines: The Task Force for cancer treatment and cardiovascular toxicity of the European Society of Cardioloqy (ESC). *Eur. Heart J.* **2016**, *37*, 2768–2801. [CrossRef] [PubMed]
6. Herrmann, J.; Lenihan, D.; Armenian, S.; Barac, A.; Blaes, A.; Cardinale, D.; Carver, J.; Dent, S.; Ky, B.; Lyon, A.R.; et al. Defining cardiovascular toxicities of cancer therapies: An International Cardio-Oncology Society (IC-OS) consensus statement. *Eur. Heart J.* **2022**, *43*, 280–299. [CrossRef]
7. Cardinale, D.; Iacopo, F.; Cipolla, C.M. Cardiotoxicity of Anthracyclines. *Front. Cardiovasc. Med.* **2020**, *7*, 26. [CrossRef]
8. Tocchetti, C.G.; Ameri, P.; de Boer, R.A.; D'Alessandra, Y.; Russo, M.; Sorriento, D.; Ciccarelli, M.; Kiss, B.; Bertrand, L.; Dawson, D.; et al. Cardiac dysfunction in cancer patients: Beyond direct cardiomyocyte damage of anticancer drugs: Novel cardio-oncology insights from the joint 2019 meeting of the ESC Working Groups of Myocardial Function and Cellular Biology of the Heart. *Cardiovasc. Res.* **2020**, *116*, 1820–1834. [CrossRef]
9. Miller, K.D.; Nogueira, L.; Devasia, T.; Mariotto, A.B.; Yabroff, K.R.; Jemal, A.; Kramer, J.; Siegel, R.L. Cancer treatment and survivorship statistics, 2022. *CA Cancer J. Clin.* **2022**. [CrossRef]
10. Vitfell-Rasmussen, J.; Krarup-Hansen, A.; Vaage-Nilsen, M.; Kümler, T.; Zerahn, B. Real-life incidence of cardiotoxicity and associated risk factors in sarcoma patients receiving doxorubicin. *Acta Oncol.* **2022**, *61*, 801–808. [CrossRef]
11. Barish, R.; Lynce, F.; Unger, K.; Barac, A. Management of Cardiovascular Disease in Women With Breast Cancer. *Circulation* **2019**, *139*, 1110–1120. [CrossRef]
12. Plana, J.C.; Galderisi, M.; Barac, A.; Ewer, M.S.; Ky, B.; Scherrer-Crosbie, M.; Ganame, J.; Sebag, I.A.; Agler, D.A.; Badano, L.P.; et al. Expert consensus for multimodality imaging evaluation of adult patients during and after cancer therapy: A report from the American Society of Echocardiography and the European Association of Cardiovascular Imaging. *Eur. Heart J. Cardiovasc. Imaging* **2014**, *15*, 1063–1093. [CrossRef]
13. Cameli, M.; Mandoli, G.E.; Sciaccaluga, C.; Mondillo, S. More than 10 years of speckle tracking echocardiography: Still a novel technique or a definite tool for clinical practice? *Echocardiography* **2019**, *36*, 958–970. [CrossRef] [PubMed]
14. 14. Cameli, M.; Mondillo, S.; Galderisi, M.; Mandoli, G.E.; Ballo, P.; Nistri, S.; Capo, V.; D'Ascenzi, F.; D'Andrea, A.; Esposito, R.; et al. Speckle tracking echocardiography: A practical guide. *G. Ital. Di Cardiol.* **2017**, *18*, 253–269. [CrossRef]
15. Ye, L.; Yang, Z.G.; Selvanayagam, J.B.; Luo, H.; Yang, T.; Perry, R.; Diao, K.; Huang, S.; Yang, M.; Yang, P.; et al. Myocardial strain imaging by echocardiography for the prediction of cardiotoxicity in chemotherapy-treated patients: A meta-analysis. *JACC Cardiovasc. Imaging* **2020**, *13*, 881–882. [CrossRef] [PubMed]
16. Thavendiranathan, P.; Poulin, F.; Lim, K.D.; Plana, J.C.; Woo, A.; Marwick, T.H. Use of myocardial strain imaging by echocardiography for the early detection of cardiotoxicity in patients during and after cancer chemotherapy: A systematic review. *J. Am. Coll. Cardiol.* **2014**, *63*, 2751–2768. [CrossRef]
17. Čelutkienė, J.; Pudil, R.; López-Fernández, T.; Grapsa, J.; Nihoyannopoulos, P.; Bergler-Klein, J.; Cohen-Solal, A.; Farmakis, D.; Tocchetti, C.G.; von Haehling, S.; et al. Role of cardiovascular imaging in cancer patients receiving cardiotoxic therapies: A position statement on behalf of the Heart Failure Association (HFA), the European Association of Cardiovascular Imaging (EACVI) and the Cardio-Oncology Council of the European Society of Cardiology (ESC). *Eur. J. Heart Fail.* **2020**, *22*, 1504–1524, Erratum in *Eur. J. Heart Fail* **2021**, *23*, 345.
18. Semeraro, G.C.; Cipolla, C.M.; Cardinale, D.M. Role of Cardiac Biomarkers in Cancer Patients. *Cancers* **2021**, *13*, 5426. [CrossRef]
19. Pudil, R.; Mueller, C.; Čelutkienė, J.; Henriksen, P.A.; Lenihan, D.; Dent, S.; Barac, A.; Stanway, S.; Moslehi, J.; Suter, T.M.; et al. Role of serum biomarkers in cancer patients receiving cardiotoxic cancer therapies: A position statement from the Cardio-Oncology Study Group of the Heart Failure Association and the Cardio-Oncology Council of the European Society of Cardiology. *Eur. J. Heart Fail.* **2020**, *22*, 1966–1983. [CrossRef]
20. Hamo, C.E.; Bloom, M.W.; Cardinale, D.; Ky, B.; Nohria, A.; Baer, L.; Skopicki, H.; Lenihan, D.J.; Gheorghiade, M.; Lyon, A.R.; et al. Cancer Therapy-Related Cardiac Dysfunction and Heart Failure: Part 2: Prevention, Treatment, Guidelines, and Future Directions. *Circ. Heart Fail.* **2016**, *9*, e002843. [CrossRef]
21. Peng, J.; Rushton, M.; Johnson, C.; Brezden-Masley, C.; Sulpher, J.; Chiu, M.G.; Graham, I.D.; Dent, S. An international survey of healthcare providers' knowledge of cardiac complications of cancer treatments. *Cardio-Oncol.* **2019**, *5*, 12. [CrossRef]
22. Tlegenova, Z.; Balmagambetova, S.; Zholdin, B.; Kurmanalina, G.; Talipova, I.; Koyshybaev, A.; Urazova, A.; Nurmanova, D.; Urazayev, O.; Sultanbekova, G.; et al. A first approach to identifying cardiotoxic effects of Breast cancer chemotherapeutic treatment in Kazakhstan. *J.Clin. Med. Kaz.* **2022**, *19*, 28–35. [CrossRef]

23. Martel, S.; Maurer, C.; Lambertini, M.; Ponde, N.; De Azambuya, E. Breast cancer treatment induced cardiotoxicity. *Expert Opin. Drug Saf.* **2017**, *16*, 1021–1038. [CrossRef] [PubMed]
24. Larsen, C.M.; Mulvagh, S.L. Cardio-oncology: What you need to know now for clinical practice and echocardiography. *Echo Res. Pract.* **2017**, *4*, R33–R41. [CrossRef]
25. Lyon, A.R.; Dent, S.; Stanway, S.; Earl, H.; Brezden-Masley, C.; Cohen-Solal, A.; Tocchetti, C.G.; Moslehi, J.J.; Groarke, J.D.; Bergler-Klein, J.; et al. Baseline cardiovascular risk assessment in cancer patients scheduled to receive cardiotoxic cancer therapies: A position statement and new risk assessment tools from the Cardio-Oncology Study Group of the Heart Failure Association of the European Society of Cardiology in collaboration with the International Cardio-Oncology Society. *Eur. J. Heart Fail.* **2020**, *22*, 1945–1960. [CrossRef] [PubMed]
26. El Khoudary, S.R.; Aggarwal, B.; Beckie, T.M.; Hodis, H.N.; Johnson, A.E.; Langer, R.D.; Limacher, M.C.; Manson, J.E.; Stefanick, M.L.; Allison, M.A.; et al. Menopause Transition and Cardiovascular Disease Risk: Implications for Timing of Early Prevention: A Scientific Statement From the American Heart Association. *Circulation* **2020**, *142*, e506–e5320. [CrossRef] [PubMed]
27. Duncan, A.E.; Alfirevic, A.; Sessler, D.I.; Popovic, Z.B.; Thomas, J.D. Perioperative assessment of myocardial deformation. *Anesth Analg.* **2014**, *118*, 525–544. [CrossRef] [PubMed]
28. Boudes, P. Drug compliance in therapeutic trials. A review. *Control. Clin. Trials* **1998**, *19*, 257–268. [CrossRef]
29. Hamstra, M.S.; Pemberton, V.L.; Dagincourt, N.; Hollenbeck-Pringle, D.; Trachtenberg, F.L.; Cnota, J.F.; Atz, A.M.; Cappella, E.; De Nobele, S.; Grima, J.; et al. Pediatric Heart Network Investigators. Recruitment, retention, and adherence in a clinical trial: The Pediatric Heart Network's Marfan Trial experience. *Clin. Trials* **2020**, *17*, 684–695. [CrossRef]
30. Demissei, B.G.; Hubbard, R.A.; Zhang, L.; Smith, A.M.; Sheline, K.; McDonald, K.; Narayan, V.; Domchek, S.M.; De Michele, A.; Shah, P.; et al. Changes in Cardiovascular Biomarkers With Breast Cancer Therapy and Associations With Cardiac Dysfunction. *J. Am. Heart Assoc.* **2020**, *9*, e014708. [CrossRef]
31. Flachskampf, F.A. Aortic Stenosis in Cancer Survivors AfterChest Radiation. *J. Am. Coll. Cardiol. Img.* **2018**, *11*, 1081–1083. [CrossRef]
32. Suchorska, V.M. Radiobiological models in prediction of radiation cardiotoxicity. *Rep. Pract. Oncol. Radiother.* **2020**, *25*, 46–49. [CrossRef]
33. Grela-Wojewoda, A.; Pacholczak-Madej, R.; Adamczyk, A.; Korman, M.; Püsküllüoğlu, M. Cardiotoxicity Induced by Protein Kinase Inhibitors in Patients with Cancer. *Int. J. Mol. Sci.* **2022**, *23*, 2815. [CrossRef] [PubMed]
34. Chen, C.; Chen, T.; Liang, J.; Guo, X.; Xu, J.; Zheng, Y.; Guo, Z.; Chi, L.; Wei, L.; Chen, X.; et al. Cardiotoxicity Induced by Immune Checkpoint Inhibitors: A Pharmacovigilance Study From 2014 to 2019 Based on FAERS. *Front. Pharmacol.* **2021**, *12*, 616505. [CrossRef]
35. Zhou, Y.-W.; Zhu, Y.-J.; Wang, M.-N.; Xie, Y.; Chen, C.; Zhang, T.; Xia, F.; Ding, Z.; Liu, J. Immune Checkpoint Inhibitor-Associated Cardiotoxicity: Current Understanding on Its Mechanism, Diagnosis and Management. *Front. Pharmacol.* **2019**, *10*, 1350. [CrossRef] [PubMed]
36. Xavier, C.B.; Lopes, C.D.H.; Harada, G.; Peres, E.D.B.; Katz, A.; Jardim, D.L. Cardiovascular toxicity following immune checkpoint inhibitors: A systematic review and meta-analysis. *Transl. Oncol.* **2022**, *19*, 101383. [CrossRef] [PubMed]
37. Xiao, H.; Wang, X.; Li, S.; Liu, Y.; Cui, Y.; Deng, X. Advances in Biomarkers for Detecting Early Cancer Treatment-Related Cardiac Dysfunction. *Front. Cardiovasc. Med.* **2021**, *8*, 753313. [CrossRef] [PubMed]
38. Srikanthan, K.; Klug, R.; Tirona, M.; Thompson, E.; Visweshwar, H.; Puri, N.; Shapiro, J.; Sodhi, K. Creating a Biomarker Panel for Early Detection of Chemotherapy Related Cardiac Dysfunction in Breast Cancer Patients. *J. Clin. Exp. Cardiolog.* **2017**, *8*, 507. [CrossRef]
39. Pauklin, P.; Zilmer, M.; Eha, J.; Tootsi, K.; Kals, M.; Kampus, P. Markers of Inflammation, Oxidative Stress, and Fibrosis in Patients with Atrial Fibrillation. *Oxid. Med. Cell. Longev.* **2022**, *2022*, 4556671. [CrossRef]
40. Suzuki, K.; Terakawa, T.; Furukawa, J.; Harada, K.; Hinata, N.; Nakano, Y.; Fujisawa, M. C-reactive protein and the neutro-phil-to-lymphocyte ratio are prognostic biomarkers in metastatic renal cell carcinoma patients treated with nivolumab. *Int. J. Clin. Oncol.* **2020**, *25*, 135–144. [CrossRef]
41. Nishimoto, Y.; Kato, T.; Morimoto, T.; Yaku, H.; Inuzuka, Y.; Tamaki, Y.; Yamamoto, E.; Yoshikawa, Y.; Kitai, T.; Taniguchi, R.; et al. C-reactive protein at discharge and 1-year mortality in hospitalised patients with acute decompensated heart failure: An observational study. *BMJ Open* **2020**, *10*, e041068. [CrossRef]
42. Kuster, N.; Huet, F.; Dupuy, A.-M.; Akodad, M.; Battistella, P.; Agullo, A.; Leclercq, F.; Kalmanovich, E.; Meilhac, A.; Aguilhon, S.; et al. Multimarker approach including CRP, sST2 and GDF-15 for prognostic stratification in stable heart failure. *ESC Heart Fail.* **2020**, *7*, 2230–2239. [CrossRef]
43. Suthahar, N.; Meijers, W.C.; Silljé, H.H.W.; Ho, J.E.; Liu, F.-T.; de Boer, R.A. Galectin-3 activation and inhibition in heart failure and cardiovascular disease: An update. *Theranostics* **2018**, *8*, 593–609. [CrossRef] [PubMed]
44. Chen, H.; Chen, C.; Fang, J.; Wang, R.; Nie, W. Circulating galectin-3 on admission and prognosis in acute heart failure patients: A meta-analysis. *Heart Fail. Rev.* **2020**, *25*, 331–341. [CrossRef] [PubMed]
45. Iqbal, N.; Wentworth, B.; Choudhary, R.; Landa Ade, L.; Kipper, B.; Fard, A.; Maisel, A.S. Cardiac biomarkers: New tools for heart failure management. *Cardiovasc. Diagn. Ther.* **2012**, *2*, 147–164. [CrossRef] [PubMed]
46. Ndrepepa, G. Myeloperoxidase—A bridge linking inflammation and oxidative stress with cardiovascular disease. *Clin. Chim. Acta* **2019**, *493*, 36–51. [CrossRef]

47. Oikawa, M.; Yaegashi, D.; Yokokawa, T.; Misaka, T.; Sato, T.; Kaneshiro, T.; Kobayashi, A.; Yoshihisa, A.; Nakazato, K.; Ishida, T.; et al. D-Dimer Is a Predictive Factor of Cancer Therapeutics-Related Cardiac Dysfunction in Patients Treated With Cardiotoxic Chemotherapy. *Front. Cardiovasc. Med.* **2022**, *8*, 807754. [CrossRef]
48. Sawaya, H.; Sebag, I.A.; Plana, J.C.; Januzzi, J.L.; Ky, B.; Tan, T.C.; Cohen, V.; Banchs, J.; Carver, J.R.; Wiegers, S.E.; et al. Assessment of echocardiography and biomarkers for the extended prediction of cardiotoxicity in patients treated with anthracyclines, taxanes, and trastuzumab. *Circ. Cardiovasc. Imaging* **2012**, *5*, 596–603. [CrossRef]
49. Castiglione, V.; Aimo, A.; Vergaro, G.; Saccaro, L.; Passino, C.; Emdin, M. Biomarkers for the diagnosis and management of heart failure. *Heart Fail. Rev.* **2022**, *27*, 625–643. [CrossRef]
50. Ontario Health (Quality). Use of B-Type Natriuretic Peptide (BNP) and N-Terminal proBNP (NT-proBNP) as Diagnostic Tests in Adults With Suspected Heart Failure: A Health Technology Assessment. *Ont. Health Technol. Assess. Ser.* **2021**, *21*, 1–125. eCollection 2021.
51. Maisel, A.S.; Duran, J.M.; Wettersten, N. Natriuretic Peptides in Heart Failure: Atrial and B-type Natriuretic Peptides. *Heart Fail. Clin.* **2018**, *14*, 13–25. [CrossRef]
52. Kasahara, S.; Sakata, Y.; Nochioka, K.; Miura, M.; Abe, R.; Sato, M.; Aoyanagi, H.; Fujihashi, T.; Yamanaka, S.; Shiroto, T. Conversion formula from B-type natriuretic peptide to N-terminal proBNP values in patients with cardiovascular diseases. *Int. J. Cardiol.* **2019**, *280*, 184–189. [CrossRef]
53. Okazaki, T.; Yamamoto, Y.; Yoda, K.; Nagahiro, S. The ratio of D-dimer to brain natriuretic peptide may help to differentiate between cerebral infarction with and without acute aortic dissection. *J. Neurol. Sci.* **2014**, *340*, 133–138. [CrossRef] [PubMed]
54. Tayyab, H.; Tayyab, A.; Haider Zaidi, S.M.; Javaid, A.; Naeem, S.; Furquan uddin Kasi, A.; Saleem, S.; Burhanuddin Kasi, A. Correlation between BNP and D-dimer levels in Heart Failure: A short report. *J. Clin. Images Med. Case Rep.* **2022**, *3*, 1949. [CrossRef]
55. Gaggin, H.K.; Januzzi, J.L.; Cardiac Biomarkers and Heart Failure. American College of Cardiology. Available online: http://www.acc.org/latest-in-cardiology/articles/2015/02/09/13/00/cardiac-biomarkers-and-heart-failure (accessed on 9 October 2017).
56. Jarolim, P. High-sensitivity cardiac troponin assays in the clinical laboratories. *Clin. Chem. Lab. Med.* **2015**, *53*, 635–652. [CrossRef]
57. Shah, A.S.V.; Griffiths, M.; Lee, K.K.; McAllister, D.A.; Hunter, A.L.; Ferry, A.V.; Cruikshank, A.; Reid, A.; Stoddart, M.; Strachan, F. High sensitivity cardiac troponin and the under-diagnosis of myocardial infarction in women: Prospective cohort study. *BMJ* **2015**, *350*, g7873, Erratum in *BMJ* **2016**, *354*, i4840. [CrossRef] [PubMed]
58. Posch, F.; Niedrist, T.; Glantschnig, T.; Firla, S.; Moik, F.; Kolesnik, E.; Wallner, M.; Verheyen, N.; Jost, P.J.; Zirlik, A. Left ventricular ejection fraction and cardiac biomarkers for dynamic prediction of cardiotoxicity in early breast cancer. *Front. Cardiovasc. Med.* **2022**, *9*, 933428. [CrossRef]

diagnostics

MDPI

Article

miR-146a, miR-221, and miR-155 are Involved in Inflammatory Immune Response in Severe COVID-19 Patients

Noemí Gaytán-Pacheco [1], Alejandro Ibáñez-Salazar [1], Ana Sofía Herrera-Van Oostdam [2], Juan José Oropeza-Valdez [3], Martín Magaña-Aquino [4], Jesús Adrián López [5], Joel Monárrez-Espino [6] and Yamilé López-Hernández [7,*]

1 Clinical Analysis Laboratory UAZ-Siglo-XXI, Academic Unit of Chemical Sciences, Autonomous University of Zacatecas, Zacatecas 98000, Mexico
2 Faculty of Medicine, Autonomous University of San Luis Potosí, San Luis Potosí 78210, Mexico
3 Metabolomics and Proteomics Laboratory, Academic Unit of Biological Sciences, Autonomous Universityof Zacatecas, Zacatecas 98600, Mexico
4 Central Hospital Dr. Ignacio Morones Prieto, San Luis Potosí 78210, Mexico
5 MicroRNAs and Cancer Laboratory, Academic Unit of Biological Sciences, Autonomous University of Zacatecas, Zacatecas 98000, Mexico
6 Department of Health Research, Christus Muguerza del Parque Hospital Chihuahua, University of Monterrey, San Pedro Garza García 66238, Mexico
7 CONACyT-Metabolomics and Proteomics Laboratory, Autonomous University of Zacatecas, Zacatecas 98000, Mexico
* Correspondence: ylopezher@conacyt.mx

Abstract: COVID-19 infection triggered a global public health crisis during the 2020–2022 period, and it is still evolving. This highly transmissible respiratory disease can cause mild symptoms up to severe pneumonia with potentially fatal respiratory failure. In this cross-sectional study, 41 PCR-positive patients for SARS-CoV-2 and 42 healthy controls were recruited during the first wave of the pandemic in Mexico. The plasmatic expression of five circulating miRNAs involved in inflammatory and pathological host immune responses was assessed using RT-qPCR (Reverse Transcription quantitative Polymerase Chain Reaction). Compared with controls, a significant upregulation of miR-146a, miR-155, and miR-221 was observed; miR-146a had a positive correlation with absolute neutrophil count and levels of brain natriuretic propeptide (proBNP), and miR-221 had a positive correlation with ferritin and a negative correlation with total cholesterol. We found here that CDKN1B gen is a shared target of miR-146a, miR-221-3p, and miR-155-5p, paving the way for therapeutic interventions in severe COVID-19 patients. The ROC curve built with adjusted variables (miR-146a, miR-221-3p, miR-155-5p, age, and male sex) to differentiate individuals with severe COVID-19 showed an AUC of 0.95. The dysregulation of circulating miRNAs provides new insights into the underlying immunological mechanisms, and their possible use as biomarkers to discriminate against patients with severe COVID-19. Functional analysis showed that most enriched pathways were significantly associated with processes related to cell proliferation and immune responses (innate and adaptive). Twelve of the predicted gene targets have been validated in plasma/serum, reflecting their potential use as predictive prognosis biomarkers.

Keywords: miRNAs; SARS-CoV-2; immune response

Citation: Gaytán-Pacheco, N.; Ibáñez-Salazar, A.; Herrera-Van Oostdam, A.S.; Oropeza-Valdez, J.J.; Magaña-Aquino, M.; Adrián López, J.; Monárrez-Espino, J.; López-Hernández, Y. miR-146a, miR-221, and miR-155 are Involved in Inflammatory Immune Response in Severe COVID-19 Patients. *Diagnostics* **2023**, *13*, 133. https://doi.org/10.3390/diagnostics13010133

Academic Editors: Yuli Huang, Yong Yuan and Peisong Chen

Received: 16 November 2022
Revised: 19 December 2022
Accepted: 22 December 2022
Published: 30 December 2022

1. Introduction

SARS-CoV-2 virus is transmitted through aerosols and droplets. So far, the COVID-19 pandemic has resulted in more than 530 million cases and 6.3 million deaths [1].

Upon transmission, the virus particles bind the lung epithelial cells through interactions between the spike (S) protein and the host cellular entry receptor angiotensin-converting enzyme 2 (ACE2) [2], causing a negative regulation of ACE-2 with the consequent loss of its catalytic capacity to degrade angiotensin II. During the infectivity phase, the virus can also produce its own miRNAs, which could further regulate host miRNAs and their targets [3]. These host miRNAs may influence different phases of the viral life cycle, including translation, by attaching itself to the viral RNA or mRNA.

Viruses can induce the up-/downregulation of certain host miRNAs to evade the host's immune system by suppressing antiviral factors, such as interferon (IFN) [4–6]. On the other hand, several miRNAs could have an antiviral effect, enabling the defense mechanisms to fight the infection [5,7].

Previous data on the mechanisms regulated by miRNAs suggest a possible role in COVID-19; miRNAs for which significant alterations have been reported are implicated in the regulation of the immune and/or inflammatory pathways at different levels: cytokine and chemokine synthesis; T-cell development, differentiation, and activation; or B-cell development, differentiation, and activation, among others [8]. Well-described associations have been reported in ventilated patients between miR-155 and inflammation; miR-208a/miR-499 and myocardial/cardiomyocyte damage; and miR-21/miR-126 in cardiac fibroblast and endothelial-cell dysfunction [9].

In this context, the use of molecular principles of gene regulation mechanisms could represent an innovative way to identify potential biomarkers of infection for developing antiviral therapeutic agents for certain diseases that do not have yet an effective treatment, such as COVID-19. Despite there are validated and approved methods available for COVID-19 diagnosis, prognosis is still complicated by only assuming the positive results of rapid antigen tests, serological tests, or RT-qPCR. For this reason, we are continuously looking for reproducible methods that are able to predict poor prognosis even at early points after infection. Most COVID-19 studies have focused on proteomic, metabolomic, and cellular biomarkers [8]. In the last decade, noncoding RNAs (ncRNAs) and miRNAs, have emerged as novel tools to aid in medical decision-making and can be easily measured through standard techniques already employed in clinical laboratories, such as RT-qPCR. miRNAs are sensitive, robust, and cost-effective biomarkers that offer additional information to already established clinical variables and clinical indicators [10]. They are stable in various body fluids and offer advantages as biomarkers because they are highly conserved between species, and their expression patterns are tissue and life-stage specific. The advantage of miRNAs is their consistent detectability in patients with lower COVID-19 severity and a lesser dependence on sampling time. The integration of miRNAs into biomarker signatures may improve the performance of established biomarkers, as demonstrated for binary and triplet combinations with D-dimer, troponin T, SARS-CoV-2, RNAemia, age, and BMI [11]. Target genes for these miRNAs play an important role in the immune dysregulation observed in COVID-19 patients, and they may also be evaluated as potential biomarkers in the context of long-COVID, because cumulative evidence suggest immune alterations in patients with persistent symptoms or post-COVID-19 symptoms.

For the aims of the present work, we chose five miRNAs with previously validated involvement in inflammatory processes related to immune system activation [12–15] to examine their expression in plasma from patients with severe COVID-19 and to explore their role as potential biomarkers of disease severity. It is well acknowledged that hyperinflammation and massive cytokine dysregulation are mechanisms leading to poor outcomes in severe COVID-19 patients [16]. Due to the heterogeneity of the factors that could affect the course of the disease, we propose here an age- and gender-adjusted model to differentiate controls from severe COVID-19 patients.

2. Materials and Methods

2.1. Study Population and Sampling

This is a cross-sectional study with 41 patients suffering from severe COVID-19 and 42 negative controls attending hospitals for COVID-19 diagnosis or treatment. SARS-CoV-2 diagnosis was conducted using reverse transcription polymerase chain reaction (RT-PCR) from a nasopharyngeal specimen using standard methods [17,18].

This study was revised and approved by the Ethics and Research Committees of the Christus Muguerza del Parque Hospital (folio: CEI-HCMP-15042020-3) and Health Secretary Services of San Luis Potosí (folio: CEI-003-20161034). The study was conducted in accordance with the Declaration of Helsinki. Informed consent was obtained from all participants prior to the collection of the blood samples. Figure 1 shows the workflow for the experimental design and data analysis.

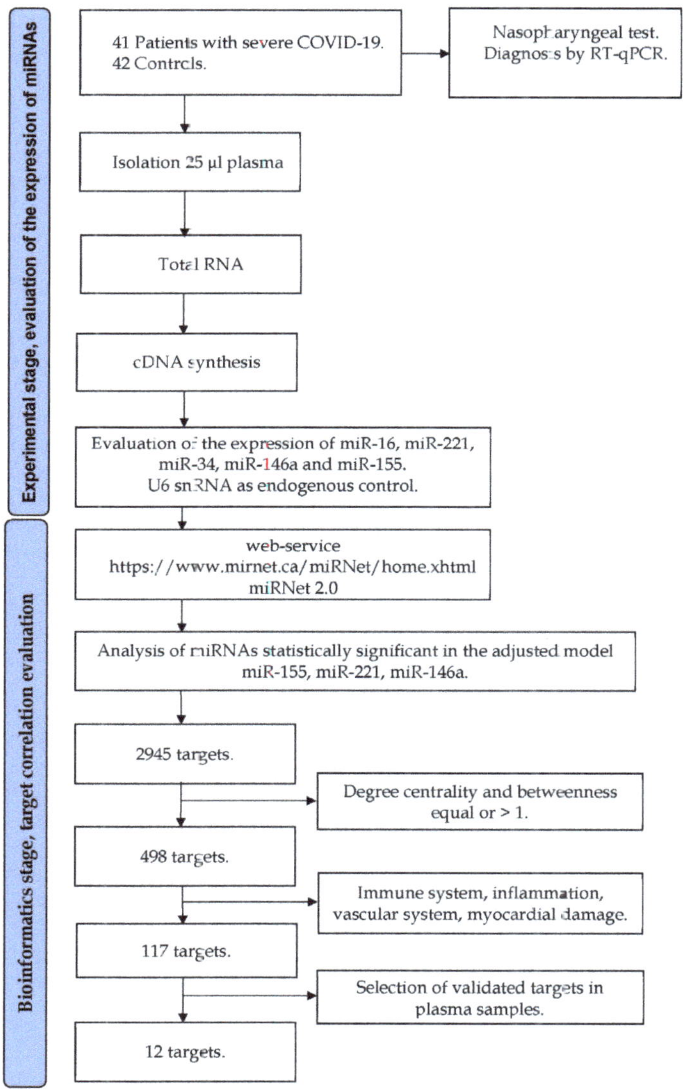

Figure 1. Flowchart diagram of the experimental design and data analysis.

2.2. *miRNAs Selection*

One of the preferred methods to study miRNAs is microarrays, which requires a large amount of RNA sample (usually more than 1 µg). Since COVID-19 has been widely acknowledged as a hyperinflammatory diseases with a strong dysregulation in the immune system, it is remarkable to find the factors that control this inflammatory unbalance. For this reason, more than to perform a massive screening of all miRNAs, we selected some miRNAs that have been previously related to inflammation, immune response, vascular complications, metabolic signaling, and organ damage. Based on recent bioinformatic prediction studies [11,19], we selected five miRNAs associated with some of the processes previously mentioned and ranked within the most important miRNAs. Our goal was to validate previous bioinformatic predictions for these miRNAs, this time for a Mexican population, which remains poorly explored until now. Supplementary Table S1 shows the miRNAs selected for validation and the justification for their inclusion in our study based on previous findings.

2.3. Circulating miRNA Isolation and Relative Expression Determination by RT-qPCR

Total RNA from each sample was isolated from 25 μL plasma using TRI-Reagent (Sigma-Aldrich, Germany), according to manufacturer's instructions. The isolated RNA was resuspended in DEPC-treated and RNAse inhibitors. Then, the RNA quality was measured at 260/280 nm using a spectrophotometer (Q3000 Quawell Technology, Inc., San José, CA; USA). A value greater than 1.8 was considered acceptable. The cDNAs for the mature miRNAs (U6 snRNA, miR-16-5p, miR-221-3p, miR-34-5p, miR-146a, and miR-155) were synthesized from 300 ng of total RNA by one-step RT-qPCR using the GoldBio's Probe-Based One Step RT-qPCR Kit (Gold Biotechnology® St Louis, MO; USA); TaqMan probes were used for each of the miRNAs to avoid unspecific amplification. The miRNA amplification by qRT-PCR was carried out using TaqMan MicroRNA Assay specific primers (Applied Biosystems, Foster City, CA; USA) in a thermocycler (qTOWER^3; AnalytikJena, Gottingen, Germany) with the following amplification conditions: first strand cDNA synthesis at 42 °C for 30 min, initial denaturation/RT inactivation at 95 °C for 3 min, followed by 40 cycles of denaturation at 95 °C for 5 s, annealing/extension at 60 °C for 30 s.

All RT-qPCR reactions were performed in duplicate. Cq values were averaged and the ΔΔCq method [20] was used to obtain the relative expression, where ΔΔCq was calculated by subtracting the ΔCq value from the mean of the control group with the ΔCq value of the COVID-19 patients. NormFinder [21] was used to estimate the best candidate reference gene between U6 snRNA and RNU48; miRNA levels were normalized with the use of the average of reference Cq value as the housekeeping gene, according to NormFinder parameters.

2.4. Bioinformatic Analysis

Bioinformatic analysis of targets was performed through the page miRNet 2.0 and its web service https://www.mirnet.ca/miRNet/home.xhtml (accessed on 14 July 2022). This analysis identified targets that were shared by 2 or more miRNAs. We assessed hsa-miR-155-5p, hsa-miR-221-3p, and hsa-miR-146a-3p, which were the significant miRNAs. The selection of most relevant genes was made, setting degree centrality and betweenness to equal or >1. Using the same web service, a miRNA-gene network with selected targets was constructed. Subsequently, all genes related to immune system functions were selected. Functional association analysis using the TAM 2.0 http://www.lirmed.com/tam2/ (accessed on 17 July 2022) [22] was also conducted to identify the functional terms for differentially expressed miRNAs. To control for multiple comparisons, a false-discovery rate (FDR) < 0.05 was used.

2.5. Statistical Analysis

Frequencies and proportions were used to describe sociodemographic characteristics, main comorbidities, symptomatology, and epidemiological data for the study participants.

Statistical analyses were carried out using GraphPad (version 5.0) software (GraphPad, La Jolla, CA, USA). The mean ± SD or median ± IQR were used to represent continuous data with parametric or nonparametric distribution, respectively. For clinical data, ANOVA with Tukey's post hoc tests for continuous variables were used to identify differences across categories, and Fisher's exact tests were used for nominal data. Nonparametric

Kruskal–Wallis with Dunn's post hoc tests were employed to identify differences in miRNA data. Statistical significance was set at $p < 0.05$.

Spearman's correlation coefficients and plots between miRNAs and laboratory values were computed using R studio (4.1.2).

Crude and adjusted logistic regression models were built to predict COVID-19 severity. Odds ratios (OR) with 95% confidence intervals (CI) were computed. The full model included all variables, with a $p \leq 0.10$ in crude analyses, but only independent variables with a $p < 0.05$ in at least one category in the comparisons remained in the final model. A receiving operating characteristics (ROC) curve was produced from the final logistic model, and the area under the curve (AUC) was reported. The Nagelkerke pseudo-R^2 statistic, ranging from 0 to 1, was used to provide an indication of the amount of variation in the dependent variable explained by the model.

3. Results

3.1. Sociodemographic Characteristics

The sociodemographic characteristics of the patients recruited in the present study are summarized in Table 1. The mean age of the patients with severe COVID-19 was higher than that of healthy controls (52.4 vs. 39.1 years, $p < 0.01$). Male patients with severe COVID-19 accounted for 82.9%, compared with 45.2% in the control group ($p < 0.01$). In terms of comorbidities, the proportion of patients with diabetes (26.8% vs. 2.3%; $p < 0.01$) and hypertension (41.4% vs. 9.5%; $p < 0.01$) was higher among those with severe COVID-19, but no statistical differences were seen for the other conditions assessed. All general and respiratory symptoms measured were clearly more frequent ($p < 0.01$) in patients with severe COVID-19 than among controls.

Table 1. Selected sociodemographic characteristics, main comorbidities, and clinical symptoms for patients with severe COVID-19 (n = 41) and healthy controls (n = 42).

Variable	Category	Group, % (n)		p-Value *
		Controls	COVID-19	
Sex	Male	45.2 (19)	82.9 (34)	<0.01
	Female	54.7 (23)	14.6 (6)	<0.01
Age (years)	20–45	73.8 (31)	31.7 (13)	<0.01
	46–65	23.8 (10)	53.6 (22)	<0.01
	66–85	2.38 (1)	14.6 (6)	0.04
Current smoking		11.9 (5)	7.3 (3)	0.71
Type 2 diabetes (DM-II)		2.3 (1)	26.2 (11)	<0.01
Hypertension		9.5 (4)	41.4 (17)	<0.01
COPD or asthma		4.7 (2)	4.8 (2)	0.98
Immunosuppressed		0 (0)	2.4 (1)	0.49
Chronic kidney disease		0 (0)	4.8 (2)	0.24
Obesity, BMI ≥ 30 kg/m^2		71.4 (30)	63.4 (26)	0.48
General symptomatology	Fever	2.3 (1)	73.1 (30)	<0.01
	Headache	16.6 (7)	58.5 (24)	<0.01
	Myalgia	16.6 (7)	63.4 (26)	<0.01
	Arthralgia	7.1 (3)	58.5 (24)	<0.01
Respiratory symptomatology	Cough	19.0 (8)	80.4 (33)	<0.01
	Odynophagia	9.5 (4)	31.7 (13)	0.01
	Dyspnea	7.1 (3)	85.3 (35)	<0.01
	Chest pain	11.9 (5)	36.5 (15)	<0.01
Other signs/symptoms	Anosmia/dysgeusia	2.3 (1)	4.8 (2)	0.61
	Diarrhea	7.1 (3)	24.3 (10)	0.03

COPD: chronic obstructive pulmonary disease; BMI: body mass index. Chi^2 and Fisher's exact tests were used. * $p < 0.05$ was considered statistically significant.

3.2. Differential Expression of miRNAs in the Plasma of Patients with COVID-19

Figure 2 shows the relative expression (normalized against U6 snRNA) in controls and COVID-19 patients for each studied miRNA. Significant upregulation of miR-16, miR-155, and miR-221 was observed for COVID-19 patients (Figure 2). miR-146a was found to be marginally significant in the crude analysis. NormFinder analysis identified U6 snRNA as the single most stable gene, with a stability of 0.589 and a standard error of 0.33.

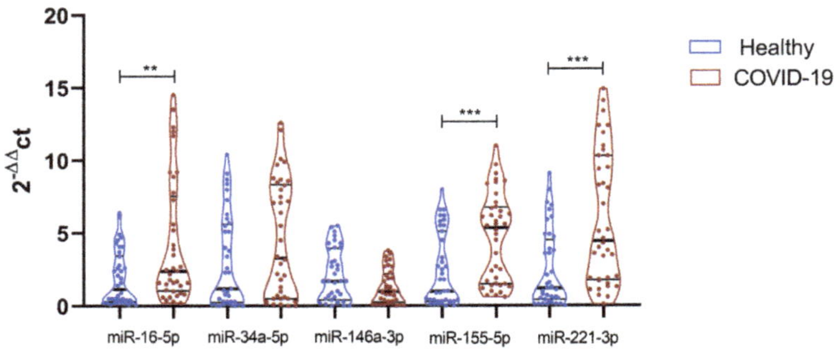

Figure 2. Violin plot shows the relative expression of miRNAs in COVID-19 patient vs. healthy controls. The violins represent medians and interquartile ranges. Data were analyzed using the Kruskal–Wallis test and Dunn's post hoc test when significant values were obtained (** and *** $p < 0.05$).

3.3. Logistic Regression Model Based on miRNAs for Classification of COVID-19 Patients

Crude and adjusted regression models (OR; 95% CI) to differentiate patients with COVID-19 from healthy controls are presented in Table 2. The results showed a statistically significant upregulated expression of miR-155, miR-16, and miR-221 in patients with severe COVID-19 compared with healthy controls in crude analyses. However, when adjusting by age and sex, only miR-155, miR-146a, and miR-221 remained significant for identifying COVID-19 patients.

Table 2. Computed crude and adjusted odds ratios (OR) with 95% confidence intervals (CI) from logistic regression for the probability to identify severe COVID-19.

Variables	OR (95% CI)	
	Crude	**Adjusted ***
miR-16	1.28 (1.06–1.54)	–
miR-34a	1.12 (0.99–1.28)	–
miR-146a	0.70 (0.51–0.97)	0.24 (0.09–0.62)
miR-155	1.34 (1.13–1.60)	1.68 (1.19–2.37)
miR-221	1.29 (1.11–1.50)	1.36 (1.04–1.78)
Sex (male)	6.02 (2.16–16.7)	7.75 (1.40–42.7)
Age (years)	1.09 (1.04–1.14)	1.10 (1.04–1.18)
DM-II	15.1 (1.85–124.1)	-
HTN	6.83 (2.04–22.8)	-

DM-II: diabetes mellitus type II; HTN: hypertension; *Only significant variables ($p < 0.05$) remained in the final adjusted model; Nagelkerke $R^2 = 0.71$, Hosmer–Lemeshow Chi^2, * p-value = 0.53.

3.4. Model Performance

Figure 3A represents the ROC curve for all the individual miRNAs, without adjustment. However, when adjusted by sex and age, the combination of miR-155, miR-146a, and miR-221 showed an adequate performance (AUC: 0.95, 95%CI 0.89–0.98) (Figure 3B) for discrimination of severe patients.

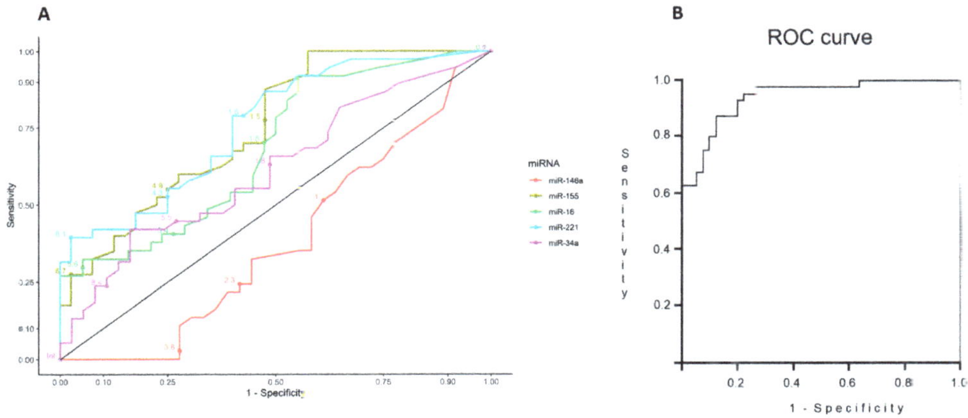

Figure 3. (**A**) ROC curve with individual miRNAs. (**B**) ROC curve built with the variables included in the logistic regression model. (AUC: 0.95, 95%CI 0.89–0.98). Nagelkerke pseudo-R^2 statistic, ranging from 0 to 1.

3.5. Correlation of Significant miRNAs with Clinical Variables in Severe COVID-19 Patients

To assess if differences in miRNA expression were related to other variables, we performed a correlation analysis. There was a significant positive association between miR-146a and absolute neutrophil count (r = 0.57, p = 0.007), and proBNP (r = 0.40, p = 0.0001), also between miR-221 and ferritin (r = 0.35, p = 0.03). On the other side, we observed a significant negative correlation (r = −0.61, p = 0.01) between miR-221 and total cholesterol (Figure 4).

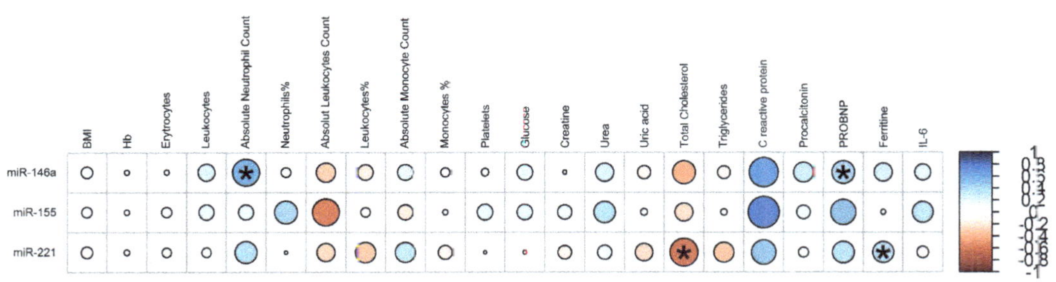

Figure 4. Correlations between significant miRNAs and clinical characteristics. Blue circles represent positive correlations, while red circles represent negative correlations. Significance of correlation is, accordingly, the circle diameter. Spearman's correlation (* $p < 0.05$).

3.6. Evaluation of the Interaction Networks of Common Target Genes across the Studied miRNAs

Only validated target genes with a correlation ≥1 were entered in the analysis to identify genes regulated by miR-155-5p, miR-221-3p, and miR-146a-3p, which were the miRNAs that contributed to the adjusted logistic regression model. A total of 498 targets were related to the analyzed miRNAs (Figure 5A). The Venn diagram (Figure 5B) shows the number of genes shared by the miRNAs under study. These genes are listed in Table 3.

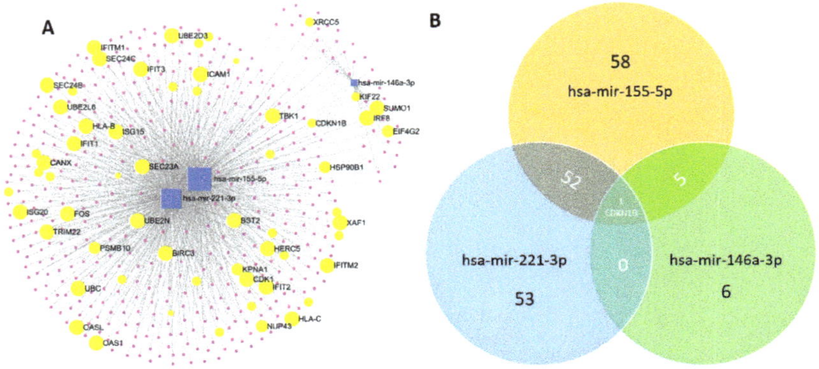

Figure 5. miRNA-target gene network related to immune system. (**A**) Results of pathway and network analyses for the 3 miRNAs, selected according to the logistic regression model. Blue squares indicate miRNAs, pink dots indicate corresponding targets, and yellow circles represent genes involved in immune system functions that can be targeted by two or more of the selected miRNAs. (**B**) Venn diagram shows the miRNAs 155-5p, 146a-3p, and 221-3p and the number of their immune system targets. CDKN1B gene is shared between miRNAs.

Table 3. Target genes validated in plasma/serum for the significant miRNAs.

Target Official Symbol	Official Full Name	Sequence Accession ID (Gene)	miRNA Associated with Regulation	Predicted/Validated	Gene Function	Reference
INPP5D (SHIP1)	Inositol polyphosphate-5-phosphatase D	NC_000002	hsa-miR-155-5p	Validated (qPCR, assay, luciferase reporter assay, and Western blot)	Tumor suppressor recognized to inhibit cell proliferation in many types of tumor cells.	[23,24]
CDKN1B	Cyclin-dependent kinase inhibitor 1B	NC_000012	hsa-miR-155-5p	Validated (qPCR, assay, luciferase reporter assay, and Western blot)	Plays a critical role in controlling cell growth and division. Macrophage proliferation.	[23]
SOCS1	Suppressor of cytokine signaling 1	NC_000016	hsa-miR-155-5p	Validated (qPCR, assay, luciferase reporter assay, and Western blot)	Acts as a negative feedback regulator to inhibit JAK2/STAT3 signaling. Control of systemic inflammation and promotes the proliferation and inflammation of macrophages through downregulating SHIP1.	[25]
FOXO3	Forkhead box O3	NC_000006	hsa-miR-155-5p	Validated (qPCR, dual-luciferase reporter system, and Western blot)	Regulatory effects on cell proliferation, apoptosis, metabolism, and oxidative stress. Plays an important role in both inflammation and regulation of cell proliferation. Regulates inflammation by NF-κB, T cells, and autoinflammation.	[26]
ICAM-1	Intercellular adhesion molecule 1	NC_000019	hsa-miR-155-5p	Validated (reporter gene assay, qPCR, and Western Blot)	Indicator of vascular inflammation.	[27]
PTEN	Phosphatase and tensin homolog	NC_000010	hsa-miR-221-3p	Validated (RT-qPCR)	Tumor suppressor by negatively regulating the AKT/PKB signaling pathway.	[28]
ICAM1	Intercellular adhesion molecule 1	NC_000019	hsa-miR-221-3p	Validated (RT-qPCR)	It binds to integrins of type CD11a / CD18, or CD11b / CD18, and is also exploited by rhinovirus as a receptor.	[29]
FOS	Fos proto-oncogene, AP-1 transcription factor subunit	NC_000014	hsa-miR-221-3p	Validated (RT-qPCR)	Implicated as regulators of cell proliferation, differentiation, and transformation.	[30]

Table 3. *Cont.*

Target Official Symbol	Official Full Name	Sequence Accession ID (Gene)	miRNA Associated with Regulation	Predicted/Validated	Gene Function	Reference
NFKB1	Nuclear factor kappa B subunit 1	NC_000004	hsa-miR-221-3p	Validated (qRT-PCR)	Transcription regulator that is activated by various intra- and extracellular stimuli such as cytokines.	[29]
P27KIP1	Cyclin-dependent kinase inhibitor 1B	NC_000012	hsa-miR-221-3p	Validated (luciferase reporter assays, Western blotting, and qPCR)	p27kip1 is an inhibitor of cell-cycle progression. Plays a decisive role in nonproliferating cell types, such as eosinophils and dendritic cells (DCs).	[24]
EGFR	Epidermal growth factor receptor	NC_000007	hsa-miR-146a-3p	Validated (qPCR)	It is a receptor for members of the epidermal growth factor family. EGFR is a component of the cytokine storm, which contributes to a severe form of COVID-19 resulting from infection with severe acute respiratory syndrome coronavirus-2 (SARS-CoV-2).	[31]
SUMO1	Small ubiquitin-like modifier 1	NC_000002	hsa-miR-146a-3p	Validated (dual-luciferase assay, Western blot)	SUMOylation of SERCA2a, Ca^{2+} handling in cardiomyocytes.	[32]

From the 498 validated targets, 12 genes have been validated for miRNAs regulation in plasma/serum. Table 3 shows the validation method, as well as the function associated with each target gene. These genes are the most important target genes to be considered as potential biomarkers since previous studies have validated them in the same matrix in which miRNAs were measured in our study.

Finally, in the functional association analysis (Figure 6), miRNAs statistically different in COVID-19 patients showed enriched functional terms, including aging, apoptosis, T-cell differentiation, hematopoiesis, and immune response, among others.

Figure 6. Significantly over-represented functional terms for miRNAs up-reulated amongst predicted targets in severe cases. The heat map shows that T-cell activation, inflammation, hematopoiesis, cell death and apoptosis and aging were the most enriched cell activities. False discovery rate (FDR) ($p < 0.05$).

4. Discussion

This study looked at the differential expression of five circulating miRNAs in plasma from healthy controls and patients with severe COVID-19 to explore their potential role in the inflammatory and host immune response against infection, and to assess their value as biomarkers of disease severity. Our group has been focused in the role of miRNAs as regulators for certain diseases [33–35]. Evidence suggests that these molecules are important predictors of disease complications and an important source of diagnosis biomarkers or therapeutic targets. The Mexican population is complex to study, since multiple factors, including ethnicity, comorbidities, genetic factors, and lifestyle, could modify disease outcomes when comparing with other populations. We have previously seen these miRNAs modulated in other inflammatory diseases [36], and we aimed to find a link between COVID-19 and all the inflammatory mediators that influence the hyperinflammatory status observed in severe cases. Therefore, with the present study, we validated not only our previous findings about the involvement of the studied miRNAs in inflammation and immune system, but also, we validated other author contributions, demonstrating that, independently of population differences, these miRNAs regulate important processes related to COVID-19 complications. Analysis of the identified miRNAs showed regulatory functions associated with inflammation, immune response, and vascular and metabolic diseases, indicating that the infection caused by the SARS-CoV-2 virus has multiple effects that alter the homeostasis of different organs and tissues.

Previous studies have shown that viral infections affect host homeostasis by regulating miRNA expression [11,37–41]. The altered expression of miRNAs causes other genes to regulate the host immune response to viral infection. Each miRNA can then target multiple genes, making them important regulators of numerous cellular functions. Eyileten et al. (2022) reported a bioinformatic miRNA prediction and subsequent validation in Poland patients. The authors looked for miRNAs regulating the highest number of top network-medicine-based integrative approach (NERI) nodes and top NERI targets associated with coagulation. In their study, top miRNAs were identified based on their regulation of the highest number of the top differentially expressed genes associated with coagulation and involved in the coagulation process. The miRNAs selected in our study are ranked in the top 30 miRNAs having a role in ACE2-related thrombosis in coronavirus infection [19].

The results presented here showed a statistically significant dysregulated expression of miR-155, miR-16, and miR-221 in patients with severe COVID-19 compared with healthy controls in crude analyses. However, for the establishment of an adequate predictive model, adjustment for cofounders must be accomplished. Due to the presence of different factors influencing COVID-19 outcomes, such as sex, age, and comorbidities, we propose here an adjusted model. When adjusting by age and sex, only miR-155, miR-146a, and miR-221 remained significant for identifying severe COVID-19 patients. ROC built with the adjusted model showed an adequate performance (AUC: 0.95, 95% CI 0.89–0.98). However, interpretation of these results must be done with caution. A low sensitivity was achieved in our study. Sensitivity refers to the proportion of subjects who have the target condition (reference standard positive) and give positive test results. This parameter is highly dependent on the simple size. For the sample size calculation, the prevalence of the target population must be considered to obtain a representative sample. With a low or high prevalence, the study may be overpowered in one subpopulation. Since COVID-19 was an emergent disease, these estimations were difficult to calculate at the beginning of the pandemic. Sammut-Powell et al. [42] reported a simulation study to evaluate the effect of sample size calculations in the sensitivity and specificity of COVID-19 diagnostic tests in practice. Under the current emergency guidelines from the Medicines and Healthcare Products Regulatory Agency, companies are required to evaluate diagnostic tests in 30 positive and 30 negative cases. The authors demonstrated that, in practice, in a test performed with 80% sensitivity and 93% specificity in 30 positive and negative samples, respectively, their real-world sensitivity and specificity could be as low as 57.7% and 83.2%, respectively.

Our results are consistent with previous studies in which expression of miR-155 has been increased in patients with COVID-19 [43–45]. A study by Garg et al., found overexpressed values of miR-155 in critically ill and mechanically-ventilated COVID-19 patients, suggesting that this miRNA could be useful for evaluating the severity of the disease [9]. This overexpression seems to contribute to the overall exacerbated proinflammatory state widely described in patients with COVID-19. miR-155 has its origin in leukocyte cells, and there is increasing evidence that macrophage activation contributes to the initiation of inflammatory responses resulting in tissue damage [23]; miR-155 responds to many inflammatory stimuli, such as TNF-α, IL-1β, pathogen-associated molecular patterns (PAMPs), and damage-associated molecular patterns (DAMPs) that act by potentiating the inflammatory response [15]. Additionally, miR-155 has an important role in innate immunity and differentiation and activation of T and NK cells [15,23,45,46]. miR-155-5p also has an important role in inflammatory and immunological processes, endothelial dysfunction, and cardiometabolic diseases. Moreover, it has been shown that miR-155 upregulates IL-1, IL-6, TNF-α, and IL-12 signaling pathways, as well as the NF-kB and JAK/STAT3 pathways. Therefore, over-regulation of miR-155 in SARS-CoV-2 infection could be a good predictor of inflammatory status and immune disorders [9,43,44,47–49]. Studies have found that this miRNA is also differentially expressed in other conditions, including acute lung damage, viral infections (influenza), pulmonary fibrosis, and asthma [44,50–52].

On the other hand, miR-221-3p is able to target molecules belonging to pathways with key roles in inflammatory responses, including toll-like receptors (TLRs), transcription factors (NF-kB), and cytokines/chemokines (TNF-α, IL-6, and IL- 8) [53–55]. miR-221 has been differentially expressed downwards in patients with COVID-19 compared to those with community-acquired pneumonia [39]. Molinero et al. found an overexpression in the ratio of miR-221-3p in bronchial pulmonary aspirate samples in patients seriously ill with COVID-19, compared with healthy controls (in which decreased levels of miR-221-3p in non survivors were reported), suggesting that the signature of miRNAs may change depending on the degree of severity, the type of tissue studied, and the outcome of the disease [44]. The proposed miR-221-3p mechanisms may be severe endothelial injury and coagulopathy (features observed in lung samples of fatal COVID-19 cases) [56] to the targeting core receptor protein ADAM17 (a disintegrin and metalloproteinase 17) involved in ACE2-dependent shedding (associated with lung pathogenesis) [57], and to the suppression of the innate immune response and promotion of the viral infection via the TBK1 (TANK binding kinase 1) gene [58,59].

Previous experimental studies have described downregulation of miR-146a during SARS-CoV-2 infection [50,53,60]. miR-146a is known to be an anti-inflammatory miRNA. It regulates inflammation by targeting the factor 6-associated TNF receptor (TRAF6), thereby reducing the expression of NF-kB (nuclear factor kappa B) [12,61,62]. miR-146a depletion leads to IL-1, IL-6, and TNFα overproduction [63]. It is also related to the toll-like receptor signaling pathway and is a regulator of IL-1B and TGFB1 transcription factors [64,65].

miR-146a expression also correlated with biochemical parameters such as proBNP and absolute neutrophil count. Natriuretic peptides are sensitive indicators of cardiac and hemodynamic stress, which may be due to left ventricular systolic/diastolic dysfunction, ischemic or inflammatory dysfunction, and right cardiac overload secondary to pulmonary consequences of the disease (i.e., pulmonary embolism, pulmonary hypertension, hypoxic vasoconstriction, or acute respiratory distress syndrome) [66]. In fact, the potential usefulness of this cardiac parameter as a prognostic biomarker associated with the severity of COVID-19 has been previously suggested [67,68]. Neutrophils are at the intersection of innate immune responses, including pathogen destruction, thrombosis, and activation of the adaptive immune system [69,70].

Lastly, miR-146a directly regulates thromboinflammatory processes by inhibiting several proinflammatory elements of the NF-κB pathway [13,71]. This miRNA is predominantly expressed in cells that promote thrombosis (i.e., macrophages, platelets, neutrophils, and endothelial cells) [72]. In addition, it has been shown that decreased levels of miR-146a in patients with pneumonia are associated with an increased risk of adverse cardiovascular events by exacerbating the inflammatory and prothrombotic responses associated with severe COVID-19 [70].

We also account for a positive correlation between ferritin and miR-221. It has been reported that ferritin is a nonspecific marker of inflammation and a key mediator of immune dysregulation through direct immunosuppressive and proinflammatory effects that contribute to the cytokine storm [73]. Clinically, cytokine release storms are a common phenomenon in patients with SARS-CoV-2. This process results in multiple deleterious effects on both the innate and acquired immunity, potentially related to the activation and differentiation of the T-cellular process wherein miRNAs have essential functions in various immune-related diseases and could therefore modulate the response during viral infections.

The negative correlation between total cholesterol and miR-221 ($r = -0.61$) is also relevant, as lipid metabolism plays an essential role in the COVID-19 disease. Cholesterol has been shown in several studies to interact with the S protein of SARS-CoV-2 [74]. Decreased serum total cholesterol levels have been associated with poor prognosis in patients with COVID-19. Ressaire et al., reported that low total cholesterol levels could result from SARS-CoV-2-induced vasculopathy; the authors also observed a positive correlation between total blood cholesterol levels and COVID-19 severity, which was evaluated using the Kirby index [75].

Functional analysis showed that most enriched pathways were significantly associated with processes related to cell proliferation and immune responses (innate and adaptive). CDKN1B was found to be the unique shared target between the significant miRNAs. This gene has an important role controlling the cellular cycle and apoptosis. A recent study found that p27Kip1, encoded by CDKN1B, was positively regulated by the innate immune signaling activated by the Influenza A virus. The authors suggested that increased expression of p27Kip1 could limit the viral replication, constituting a potential therapeutic approach [76].

In conclusion, SARS-CoV-2 infection appears to induce an important response in the host's miRNA profile, suggesting that the severity of the symptoms is associated with epigenetic factors, which in turn regulate a large number of functions in different tissues that can modulate the host response. The differential expression of miRNAs involved in inflammatory processes, including miR-155, miR-221, and miR-146a found in patients with severe COVID-19 in this study, point to their potential role as regulators of cellular processes in SARS-CoV-2 infections. Therefore, the expression levels of these dysregulated miRNAs could be of diagnostic and prognostic value as biomarkers to predict the severity of the disease and to develop therapeutic strategies against the virus through their regulation. To our knowledge, this is the first study reflecting miRNA dysregulation in severe Mexican COVID-19 patients and their association with inflammation, immune system, and vascular complications leading to organ failure. Since several studies about miRNA dysregulation are based on bioinformatic prediction, the validation of these findings, such as we are presenting here, will contribute to the inclusion of these molecules in multiomics prognostic panels. However, we need to acknowledge study limitations. Sample size was the major limitation, which makes it conflictive to establish highly accurate models of prediction. Moreover, the sample amount (volume) limited the validation of targets by traditional methods such as Western blot.

For future studies, the sample size needs to be increased, including patients with mild COVID-19. miRNAs could also be evaluated in survivors and non survivors, as well as in other types of samples that could be less invasive, such as urine, sputum, saliva, etc.

Supplementary Materials: The following supporting information can be downloaded at: https://www.mdpi.com/article/10.3390/diagnostics13010133/s1, Supplementary Table S1: miRNAs selected for validation and the justification for their inclusion in our study based on previous findings. References [77–90] are cited in the supplementary materials.

Author Contributions: Conceptualization, Y.L.-H. and J.M.-E.; methodology, N.G.-P., A.I.-S., A.S.H.-V.O., J.J.O.-V. and J.M.-E.; software, N.G.-P., A.I.-S., J.J.O.-V. and J.M.-E.; validation, Y.L.-H., J.M.-E., N.G.-P., A.I.-S., A.S.H.-V.O. and J.J.O.-V.; formal analysis, N.G.-P., A.I.-S., A.S.H.-V.O., J.J.O.-V., Y.L.-H. and J.M.-E.; investigation, N.G.-P. and Y.L.-H.; resources, Y.L.-H., M.M.-A. and N.G.-P.; data curation, M.M.-A., A.S.H.-V.O., J.J.O.-V. and N.G.-P.; writing—original draft preparation, N.G.-P., Y.L.-H. and J.M.-E.; writing—review and editing, J.A.L. and J.M.-E.; visualization, Y.L.-H. and N.G.-P.; supervision, Y.L.-H.; project administration, Y.L.-H.; funding acquisition, Y.L.-H. All authors have read and agreed to the published version of the manuscript.

Funding: This study was support by grants from CONACyT (numbers 311880, 216258, and 319503).

Institutional Review Board Statement: The study was conducted in accordance with the Declaration of Helsinki and was revised and approved by the Ethics and Research Committees of the Christus Muguerza del Parque Hospital (folio: CEI-HCMP-15042020-3) and Health Secretary Services of San Luis Potosí (folio: CEI-003-20161034).

Informed Consent Statement: Informed consent was obtained from all participants prior to the collection of the blood samples.

Data Availability Statement: Data is contained within the article or Supplementary material.

Acknowledgments: The authors want to thank Mariana Salgado-Bustamante for supporting some experimental approaches.

Conflicts of Interest: There are no conflicts of interest for any author.

References

1. WHO Coronavirus (COVID-19) Dashboard, (n.d.). Available online: https://covid19.who.int/ (accessed on 20 April 2022).
2. Mulay, A.; Konda, B.; Garcia, G.; Yao, C.; Beil, S.; Villalba, J.; Koziol, C.; Sen, C.; Purkayastha, A.; Kolls, J.; et al. SARS-CoV-2 infection of primary human lung epithelium for COVID-19 modeling and drug discovery. *Cell Rep.* **2021**, *35*, 1–16. [CrossRef] [PubMed]
3. Roberts, A.; Lewis, A.; Jopling, C. The Role of MicroRNAs in Viral Infection. *Prog. Mol. Biol. Transl. Sci.* **2011**, *102*, 101–139. [CrossRef] [PubMed]
4. Scheel, T.; Luna, J.; Liniger, M.; Nishiuchi, E.; Rozen-Gagnon, K.; Shlomai, A.; Auray, G.; Gerber, M.; Fak, J.; Keller, I.; et al. A Broad RNA Virus Survey Reveals Both miRNA Dependence and Functional Sequestration. *Cell Host Microbe* **2016**, *19*, 409–423. [CrossRef] [PubMed]
5. Girardi, E.; López, P.; Pfeffer, S. On the Importance of Host MicroRNAs During Viral Infection. *Front. Genet.* **2018**, *9*, 1–17. [CrossRef] [PubMed]
6. Trobaugh, D.; Gardner, C.; Sun, C.; Haddow, A.; Wang, E.; Chapnik, E.; Mildner, A.; Weaver S.; Ryman, K.; Klimstra, W. RNA viruses can hijack vertebrate microRNAs to suppress innate immunity. *Nature* **2014**, *506*, 245–248. [CrossRef] [PubMed]
7. Guo, X.; Zhang, Q.; Gao, L.; Li, N.; Chen, X.; Feng, W. Increasing Expression of MicroRNA 181 Inhibits Porcine Reproductive and Respiratory Syndrome Virus Replication and Has Implications for Controlling Virus Infection. *J. Virol.* **2013**, *87*, 1159–1171. [CrossRef]
8. de Gonzalo-Calvo, D.; Benítez, I.; Pinilla, L.; Carratalá, A.; Moncusí-Moix, A.; Gort-Paniello, C. Molinero, M.; González, J.; Torres, G.; Bernal, M.; et al. Circulating microRNA profiles predict the severity of COVID-19 in hospitalized patients. *Transl. Res.* **2021**, *236*, 147–159. [CrossRef]
9. Garg, A.; Seeliger, B.; Derda, A.; Xiao, K.; Gietz, A., Scherf, K.; Sonnenschein, K.; Pink, I.; Hoeper, M.; Welte, T.; et al. Circulating cardiovascular microRNAs in critically ill COVID-19 patients. *Eur. J. Heart Fail.* **2021**, *23*, 468–475. [CrossRef]
10. Kang, K.; Peng, X.; Luo, J.; Gou, D. Identification of circulating miRNA biomarkers based on global quantitative real-time PCR profiling. *J. Anim. Sci. Biotechnol.* **2012**, *3*, 4. [CrossRef]
11. Bautista-Becerril, B.; Pérez-Dimas, G.; Sommerhalder-Nava, P.; Hanono, A.; Martínez-Cisneros, J.; Zarate-Maldonado, B.; Muñoz-Soria, E.; Aquino-Gálvez, A.; Castillejos-López, M.; Juárez-Cisneros, A.; et al. miRNAs, from Evolutionary Junk to Possible Prognostic Markers and Therapeutic Targets in COVID-19. *Viruses* **2021**, *14*, 41. [CrossRef]
12. Balasubramanyam, M.; Aravind, S.; Gokulakrishnan, K.; Prabu, P.; Sathishkumar, C.; Ranjani, H.; Mohan, V. Impaired miR-146a expression links subclinical inflammation and insulin resistance in Type 2 diabetes. *Mol. Cell. Biochem.* **2011**, *351*, 197–205. [CrossRef] [PubMed]
13. Taganov, K.; Boldin, M.; Chang, K.-J.; Baltimore, D. NF-κB-dependent induction of microRNA miR-146, an inhibitor targeted to signaling proteins of innate immune responses. *Proc. Natl. Acad. Sci. USA* **2006**, *103*, 12481–12486. [CrossRef] [PubMed]
14. Meerson, A.; Traurig, M.; Ossowski, V.; Fleming, J.; Mullins, M.; Baier, L. Human adipose microRNA-221 is upregulated in obesity and affects fat metabolism downstream of leptin and TNF-α. *Diabetologia* **2013**, *56*, 1971–1979. [CrossRef]
15. Mahesh, G.; Biswas, R. MicroRNA-155: A Master Regulator of Inflammation. *J. Interf. Cytokine Res.* **2019**, *39*, 321–330. [CrossRef] [PubMed]
16. Montazersaheb, S.; Khatibi, S.H.; Hejazi, M.; Tarhriz, V.; Farjami, A.; Sorbeni, F.G.; Farahzadi, R.; Ghasemnejad, T. COVID-19 infection: An overview on cytokine storm and related interventions. *Virol. J.* **2022**, *19*, 92. [CrossRef]
17. Tali, S.S.; LeBlanc, J.; Sadiq, Z.; Oyewunmi, O.; Camargo, C.; Nikpour, B.; Armanfard, N.; Sagan, S.; Jahanshahi-Anbuhi, S. Tools and Techniques for Severe Acute Respiratory Syndrome Coronavirus 2 (SARS-CoV-2)/COVID-19 Detection. *Clin. Microbiol. Rev.* **2021**, *34*, 1–63. [CrossRef]
18. Böger, B.; Fachi, M.; Vilhena, R.; Cobre, A.; Tonin, F.; Pontarolo, R. Systematic review with meta-analysis of the accuracy of diagnostic tests for COVID-19. *Am. J. Infect. Control* **2021**, *49*, 21–29. [CrossRef] [PubMed]

19. Eyileten, C.; Wicik, Z.; Simões, S.; Martins-Jr, D.; Klos, K.; Wlodarczyk, W.; Assinger, A.; Soldacki, D.; Chcialowski, A.; Siller-Matula, J.; et al. Thrombosis-related circulating miR-16-5p is associated with disease severity in patients hospitalised for COVID-19. *RNA Biol.* **2022**, *19*, 963–979. [CrossRef]
20. Livak, K.; Schmittgen, T. Analysis of Relative Gene Expression Data Using Real-Time Quantitative PCR and the $2-\Delta\Delta$CT Method. *Methods* **2001**, *25*, 402–408. [CrossRef]
21. Andersen, C.; Jensen, J.; Ørntoft, T. Normalization of Real-Time Quantitative Reverse Transcription-PCR Data: A Model-Based Variance Estimation Approach to Identify Genes Suited for Normalization, Applied to Bladder and Colon Cancer Data Sets. *Cancer Res.* **2004**, *64*, 5245–5250. [CrossRef]
22. Li, J.; Han, X.; Wan, Y.; Zhang, S.; Zhao, Y.; Fan, R.; Cui, Q.; Zhou, Y. TAM 2.0: Tool for MicroRNA set analysis. *Nucleic Acids Res.* **2018**, *46*, W180–W185. [CrossRef] [PubMed]
23. Jiang, K.; Yang, J.; Guo, S.; Zhao, G.; Wu, H.; Deng, G. Peripheral Circulating Exosome-Mediated Delivery of miR-155 as a Novel Mechanism for Acute Lung Inflammation. *Mol. Ther.* **2019**, *27*, 1758–1771. [CrossRef] [PubMed]
24. Lu, C.; Huang, X.; Zhang, X.; Roensch, K.; Cao, Q.; Nakayama, K.; Blazar, B.; Zeng, Y.; Zhou, X. miR-221 and miR-155 regulate human dendritic cell development, apoptosis, and IL-12 production through targeting of p27kip1, KPC1, and SOCS-1. *Blood* **2011**, *117*, 4293–4303. [CrossRef] [PubMed]
25. Wang, C.; Zhang, C.; Liu, L.; Xi, A.; Chen, B.; Li, Y.; Du, J. Macrophage-Derived mir-155-Containing Exosomes Suppress Fibroblast Proliferation and Promote Fibroblast Inflammation during Cardiac Injury. *Mol. Ther.* **2017**, *25*, 192–204. [CrossRef]
26. Wang, Y.; Feng, T.; Duan, S.; Shi, Y.; Li, S.; Zhang, X.; Zhang, L. miR-155 promotes fibroblast-like synoviocyte proliferation and inflammatory cytokine secretion in rheumatoid arthritis by targeting FOXO3a. *Exp. Ther. Med.* **2019**, *19*, 1288–1296. [CrossRef]
27. Choi, S.; Kim, J.; Kim, J.-H.; Lee, D.-K.; Park, W.; Park, M.; Kim, S.; Hwang, J.; Won, M.-H.; Choi, Y.; et al. Carbon monoxide prevents TNF-α-induced eNOS downregulation by inhibiting NF-κB-responsive miR-155-5p biogenesis. *Exp. Mol. Med.* **2017**, *49*, e403. [CrossRef]
28. Gill, P.; Dweep, H.; Rose, S.; Wickramasinghe, P.; Vyas, K.; McCullough, S.; Porter-Gill, P.; Frye, R. Integrated microRNA–mRNA Expression Profiling Identifies Novel Targets and Networks Associated with Autism. *J. Pers. Med.* **2022**, *12*, 920. [CrossRef]
29. Liu, F.; Dong, J.; Zhou, D.; Zhang, Q. Identification of Key Candidate Genes Related to Inflammatory Osteolysis Associated with Vitamin E-Blended UHMWPE Debris of Orthopedic Implants by Integrated Bioinformatics Analysis and Experimental Confirmation. *J. Inflamm. Res.* **2021**, *14*, 3537–3554. [CrossRef]
30. Tu, Y.; Wu, W.; Guo, Y.; Lu, F.; Li, X.; Xu, D.; Zou, D.; Tu, Y.; Chai, Y.; He, L. Up-regulation of hsa-miR-221-3p induced by UVB affects proliferation and apoptosis of keratinocytes via Bcl-xL/Bax pathway. *Photodermatol. Photoimmunol. Photomed.* **2021**, *37*, 269–277. [CrossRef]
31. Bohatá, J.; Horváthová, V.; Pavlíková, M.; Stibůrková, B. Circulating microRNA alternations in primary hyperuricemia and gout. *Arthritis Res. Ther.* **2021**, *23*, 186. [CrossRef]
32. Yan, F.; Meng, W.; Ye, S.; Zhang, X.; Mo, X.; Liu, J.; Chen, D.; Lin, Y. MicroRNA-146a as a potential regulator involved in the pathogenesis of atopic dermatitis. *Mol. Med. Rep.* **2019**, *20*, 4645–4653. [CrossRef]
33. Oostdam, A.H.-V.; Salgado-Bustamante, M.; López, J.; Oostdam, D.H.-V.; López-Hernández, Y. Placental exosomes viewed from an 'omics' perspective: Implications for gestational diabetes biomarkers identification. *Biomark. Med.* **2019**, *13*, 675–684. [CrossRef]
34. Córdova-Rivas, S.; Fraire-Soto, I.; Torres, A.M.-C.; Servín-González, L.; Granados-López, A.; López-Hernández, Y.; Reyes-Estrada, C.; Gutiérrez-Hernández, R.; Castañeda-Delgado, J.; Ramírez-Hernández, L.; et al. 5p and 3p Strands of miR-34 Family Members Have Differential Effects in Cell Proliferation, Migration, and Invasion in Cervical Cancer Cells. *Int. J. Mol. Sci.* **2019**, *20*, 545. [CrossRef] [PubMed]
35. Oostdam, A.H.-V.; Toro, J.; López, J.; Noyola, D.; García, D.; Durán, N.; Martínez, E.; Portales, D.; Salgado, M.; López, Y. Placental exosomes isolated from urine of patients with gestational diabetes exhibit a differential profile expression of microRNAs across gestation. *Int. J. Mol. Med.* **2020**, *46*, 546–560. [CrossRef]
36. Gaytán-Pacheco, N.; Lima-Rogel, V.; Méndez-Mancilla, A.; Escalante-Padrón, F.; Toro-Ortíz, J.; Jiménez-Capdeville, M.; Zaga-Clavellina, V.; Portales-Pérez, D.; Noyola, D.; Salgado-Bustamante, M. Changes in PPAR-γ Expression Are Associated with microRNA Profiles during Fetal Programming due to Maternal Overweight and Obesity. *Gynecol. Obstet. Investig.* **2021**, *86*, 415–426. [CrossRef]
37. Wilson, J.; Kealy, D.; James, S.; Plowman, T.; Newling, K.; Jagger, C.; Filbey, K.; Mann, E.; Konkel, J.; Menon, M.; et al. Integrated miRNA/cytokine/chemokine profiling reveals severity-associated step changes and principal correlates of fatality in COVID-19. *IScience* **2022**, *25*, 1036–1072. [CrossRef]
38. Narożna, M.; Rubiś, B. Anti-SARS-CoV-2 Strategies and the Potential Role of miRNA in the Assessment of COVID-19 Morbidity, Recurrence, and Therapy. *Int. J. Mol. Sci.* **2021**, *22*, 8663. [CrossRef]
39. Martínez-Fleta, P.; Vera-Tomé, P.; Jiménez-Fernández, M.; Requena, S.; Roy-Vallejo, E.; Sanz-García, A.; Lozano-Prieto, M.; López-Sanz, C.; Vara, A.; Lancho-Sánchez, Á.; et al. A Differential Signature of Circulating miRNAs and Cytokines Between COVID-19 and Community-Acquired Pneumonia Uncovers Novel Physiopathological Mechanisms of COVID-19. *Front. Immunol.* **2022**, *12*, 1–11. [CrossRef]
40. Li, C.; Hu, X.; Li, L.; Li, J. Differential microRNA expression in the peripheral blood from human patients with COVID-19. *J. Clin. Lab. Anal.* **2020**, *34*, e23590. [CrossRef]

41. Li, C.; Wu, A.; Song, K.; Gao, J.; Huang, E.; Bai, Y.; Liu, X. Identifying Putative Causal Links between MicroRNAs and Severe COVID-19 Using Mendelian Randomization. *Cells* **2021**, *10*, 3504. [CrossRef]
42. Sammut-Powell, C.; Reynard, C.; Allen, J.; McDermott, J.; Braybrook, J.; Parisi, R.; Lasserson, D.; Body, R.; Body, R.; Hayward, G.; et al. Examining the effect of evaluation sample size on the sensitivity and specificity of COVID-19 diagnostic tests in practice: A simulation study. *Diagnostic Progn. Res.* **2022**, *6*, 12. [CrossRef] [PubMed]
43. Haroun, R.-H.; Osman, W.; Amin, R.; Hassan, A.; Abo-Shanab, W.; Eessa, A. Circulating plasma miR-155 is a potential biomarker for the detection of SARS-CoV-2 infection. *Pathology* **2022**, *54*, 104–110. [CrossRef] [PubMed]
44. Molinero, M.; Benítez, I.D.; González, J.; Gort-Paniello, C.; Moncusí-Moix, A.; Rodríguez-Jara, F.; García-Hidalgo, M.; Torres, G.; Vengoechea, J.; Gómez, S.; et al. Bronchial Aspirate-Based Profiling Identifies MicroRNA Signatures Associated With COVID-19 and Fatal Disease in Critically Ill Patients. *Front. Med.* **2022**, *8*, 1–15. [CrossRef]
45. Wyler, E.; Mösbauer, K.; Franke, V.; Diag, A.; Gottula, L.; Arsiè, R.; Klironomos, F.; Koppstein, D.; Hönzke, K.; Ayoub, S.; et al. Transcriptomic profiling of SARS-CoV-2 infected human cell lines identifies HSP90 as target for COVID-19 therapy. *IScience* **2021**, *24*, 102151. [CrossRef] [PubMed]
46. Lindsay, M. microRNAs and the immune response. *Trends Immunol.* **2008**, *29*, 343–351. [CrossRef] [PubMed]
47. Jablonski, K.; Gaudet, A.; Amici, S.; Popovich, P.; Guerau-de-Arellano, M. Control of the Inflammatory Macrophage Transcriptional Signature by miR-155. *PLoS ONE* **2016**, *11*, e0159724. [CrossRef] [PubMed]
48. Bala, S.; Csak, T.; Saha, B.; Zatsiorsky, J.; Kodys, K.; Catalano, D.; Satishchandran, A.; Szabo, G. The pro-inflammatory effects of miR-155 promote liver fibrosis and alcohol-induced steatohepatitis. *J. Hepatol.* **2016**, *64*, 1378–1387. [CrossRef]
49. Liu, Y.; Pan, Q.; Zhao, Y.; He, C.; Bi, K.; Chen, Y.; Zhao, B.; Chen, Y.; Ma, X.; Production, R.R.; et al. Apoptosis and Multiple Functions of Human Brain Microvessel Endothelial Cells Under Physiological and Pathological Conditions. *J. Cell. Biochem.* **2015**, *116*, 2870–2881. [CrossRef]
50. Wang, W.; Liu, Z.; Su, J.; Chen, W.-S.; Wang, X.-W.; Bai, S.-X.; Zhang, J.-Z.; Yu, S.-Q. Macrophage micro-RNA-155 promotes lipopolysaccharide-induced acute lung injury in mice and rats. *Am. J. Physiol. Cell. Mol. Physiol.* **2016**, *311*, L494–L506. [CrossRef]
51. Inchley, C.; Sonerud, T.; Fjærli, H.; Nakstad, B. Nasal mucosal microRNA expression in children with respiratory syncytial virus infection. *BMC Infect. Dis.* **2015**, *15*, 150. [CrossRef]
52. Malmhäll, C.; Johansson, K.; Winkler, C.; Alawieh, S.; Ekerljung, L.; Rådinger, M. Altered miR-155 Expression in Allergic Asthmatic Airways. *Scand. J. Immunol.* **2017**, *85*, 300–307. [CrossRef] [PubMed]
53. Tahamtan, A.; Teymoori-Rad, M.; Nakstad, B.; Salimi, V. Anti-Inflammatory MicroRNAs and Their Potential for Inflammatory Diseases Treatment. *Front. Immunol.* **2018**, *9*, 1–14. [CrossRef] [PubMed]
54. Chakraborty, C.; Sharma, A.; Sharma, G.; Lee, S.-S. The Interplay among miRNAs, Major Cytokines, and Cancer-Related Inflammation. *Mol. Ther. Nucleic Acids.* **2020**, *20*, 606–620. [CrossRef] [PubMed]
55. O'Neill, L.; Sheedy, F.; McCoy, C. MicroRNAs: The fine-tuners of Toll-like receptor signalling. *Nat. Rev. Immunol.* **2011**, *11*, 163–175. [CrossRef]
56. Ackermann, M.; Verleden, S.; Kuehnel, M.; Haverich, A.; Welte, T.; Laenger, F.; Vanstapel, A.; Werlein, C.; Stark, H.; Tzankov, A.; et al. Pulmonary Vascular Endothelialitis, Thrombosis, and Angiogenesis in Covid-19. *N Engl. J. Med.* **2020**, *383*, 120–128. [CrossRef]
57. Kim, W.; Park, E.; Kang, K.-W.; Lee, S.-M.; Kim, B.; Kim, H.-S. Expression Analyses of MicroRNAs in Hamster Lung Tissues Infected by SARS-CoV-2. *Mol. Cells.* **2020**, *43*, 953–963. [CrossRef]
58. Pineau, P.; Volinia, S.; McJunkin, K.; Marchio, A.; Battiston, C.; Terris, B.; Mazzaferro, V.; Lowe, S.; Croce, C.; Dejean, A. miR-221 overexpression contributes to liver tumorigenesis. *Proc. Natl. Acad. Sci. USA* **2010**, *107*, 264–269. [CrossRef]
59. Chen, W.-X.; Hu, Q.; Qiu, M.-T.; Zhong, S.-L.; Xu, J.-J.; Tang, J.-H.; Zhao, J.-H. miR-221/222: Promising biomarkers for breast cancer. *Tumor Biol.* **2013**, *34*, 1361–1370. [CrossRef]
60. Sabbatinelli, J.; Giuliani, A.; Matacchione, G.; Latini, S.; Laprovitera, N.; Pomponio, G.; Ferrarini, A.; Baroni, S.S.; Pavani, M.; Moretti, M.; et al. Decreased serum levels of the inflammaging marker miR-146a are associated with clinical non-response to tocilizumab in COVID-19 patients. *Mech. Ageing Dev.* **2021**, *193*, 111413. [CrossRef]
61. Deng, Y.; Yan, Y.; Tan, K.S.; Liu, J.; Chow, V.; Tao, Z.-Z.; Wang, D.-Y. MicroRNA-146a induction during influenza H3N2 virus infection targets and regulates TRAF6 levels in human nasal epithelial cells (hNECs). *Exp. Cell Res.* **2017**, *352*, 184–192. [CrossRef]
62. Su, Y.-L.; Wang, X.; Mann, M.; Adamus, T.; Wang, D.; Moreira, D.; Zhang, Z.; Ouyang, C.; He, X.; Zhang, B.; et al. Myeloid cell–targeted miR-146a mimic inhibits NF-κB–driven inflammation and leukemia progression in vivo. *Blood* **2020**, *135*, 167–180. [CrossRef]
63. Arghiani, N.; Nissan, T.; Matin, M. Role of microRNAs in COVID-19 with implications for therapeutics. *Biomed. Pharmacother* **2021**, *144*, 112247. [CrossRef]
64. Yamasaki, K.; Nakasa, T.; Miyaki, S.; Ishikawa, M.; Deie, M.; Adachi, N.; Yasunaga, Y.; Asahara, H.; Ochi, M. Expression of MicroRNA-146a in osteoarthritis cartilage. *Arthritis Rheum.* **2009**, *60*, 1035–1041. [CrossRef]
65. Jurkin, J.; Schichl, Y.; Koeffel, R.; Bauer, T.; Richter, S.; Konradi, S.; Gesslbauer, B.; Strobl, H. miR-146a Is Differentially Expressed by Myeloid Dendritic Cell Subsets and Desensitizes Cells to TLR2-Dependent Activation. *J. Immunol.* **2010**, *184*, 4955–4965. [CrossRef]

66. Mueller, C.; McDonald, K.; de Boer, R.; Maisel, A.; Cleland, J.; Kozhuharov, N.; Coats, A.; Metra, M.; Mebazaa, A.; Ruschitzka, F.; et al. Heart Failure Association of the European Society of Cardiology practical guidance on the use of natriuretic peptide concentrations. *Eur. J. Heart Fail.* **2019**, *21*, 715–731. [CrossRef]
67. Calvo-Fernández, A.; Izquierdo, A.; Subirana, I.; Farré, N.; Vila, J.; Durán, X.; García-Guimaraes, M.; Valdivielso, S.; Cabero, P.; Soler, C.; et al. Marcadores de daño miocárdico en la predicción del pronóstico a corto plazo de los pacientes con COVID-19. *Rev. Española Cardiol.* **2021**, *74*, 576–583. [CrossRef]
68. Cannata, F.; Bombace, S.; Stefanini, G. Marcadores cardiacos en pacientes con COVID-19: Un instrumento práctico en tiempos difíciles. *Rev. Española Cardiol.* **2021**, *74*, 566–568. [CrossRef]
69. Widiasta, A.; Sribudiani, Y.; Nugrahapraja, H.; Hilmanto, D.; Sekarwana, N.; Rachmadi, D. Potential role of ACE2-related microRNAs in COVID-19-associated nephropathy. *Non-Coding RNA Res.* **2020**, *5*, 153–166. [CrossRef]
70. Arroyo, A.B.; Fernández-Pérez, M.P.; Del Monte, A.; Águila, S.; Méndez, R.; Hernández-Antolín, R.; García-Barberá, N.; Ascensión, M.; González-Jiménez, P.; Arcas, M.I.; et al. miR-146a is a pivotal regulator of neutrophil extracellular trap formation promoting thrombosis. *Haematologica* **2020**, *106*, 1636–1646. [CrossRef]
71. Saba, R.; Sorensen, D.; Booth, S. MicroRNA-146a: A Dominant, Negative Regulator of the Innate Immune Response. *Front. Immunol.* **2014**, *5*, 1–11. [CrossRef]
72. Boldin, M.; Taganov, K.; Rao, D.; Yang, L.; Zhao, J.; Kalwani, M.; Garcia-Flores, Y.; Luong, M.; Devrekanli, A.; Xu, J.; et al. miR-146a is a significant brake on autoimmunity, myeloproliferation, and cancer in mice, *J. Exp. Med.* **2011**, *208*, 1189–1201. [CrossRef]
73. Vargas-Vargas, M.; Cortés-Rojo, C. Ferritin levels and COVID-19. *Rev. Panam. Salud Pública.* **2020**, *44*, 1. [CrossRef]
74. Kočar, E.; Režen, T.; Rozman, D. Cholesterol, lipoproteins, and COVID-19: Basic concepts and clinical applications. *Biochim. Biophys. Acta Mol. Cell Biol. Lipids.* **2021**, *1866*, 158849. [CrossRef]
75. Ressaire, Q.; Dudoignon, E.; Moreno, N.; Coutrot, M.; Dépret, F. Low total cholesterol blood level is correlated with pulmonary severity in COVID-19 critical ill patients. *Anaesth. Crit. Care Pain Med.* **2020**, *39*, 733–735. [CrossRef]
76. Rai, K.; Chen, B.; Zhao, Z.; Chen, Y.; Hu, J.; Liu, S.; Maarouf, M.; Li, Y.; Xiao, M.; Liao, Y.; et al. Robust expression of p27Kip1 induced by viral infection is critical for antiviral innate immunity. *Cell. Microbiol.* **2020**, *22*, 132–142. [CrossRef]
77. Wang, M.; Li, J.; Cai, J.; Cheng, L.; Wang, X.; Xu, P.; Li, G.; Liang, X. Overexpression of MicroRNA-16 Alleviates Atherosclerosis by Inhibition of Inflammatory Pathways. *Bio. Med. Res. Int.* **2020**, *2020*, 1–12. [CrossRef]
78. Yang, Y.; Yang, F.; Yu, X.; Wang, B.; Yang, Y.; Zhou, X.; Cheng, R.; Xia, S.; Zhou, X. miR-16 inhibits NLRP3 inflammasome activation by directly targeting TLR4 in acute lung injury. *Biomed. Pharmacother.* **2019**, *112*, 108664. [CrossRef]
79. Möhnle, P.; Hirschberger, S.; Hinske, L.C.; Briegel, J.; Hübner, M.; Weis, S.; Dimopoulos, G.; Bauer, M.; Giamarellos-Bourboulis, E.J.; Kreth, S. MicroRNAs 143 and 150 in whole blood enable detection of T-cell immunoparalysis in sepsis. *Mol. Med.* **2018**, *24*, 54. [CrossRef]
80. Jafarinejad-Farsangi, S.; Jazi, M.M.; Rostamzadeh, F.; Hadizadeh, M. High affinity of host human microRNAs to SARS-CoV-2 genome: An in silico analysis. *Non-coding RNA Res.* **2020**, *5*, 222–231. [CrossRef]
81. Wicik, Z.; Eyileten, C.; Jakubik, D.; Simões, S.N.; Martins, J.D.C.; Pavão, R.; Siller-Matula, J.M.; Postula, M. ACE2 Interaction Networks in COVID-19: A Physiological Framework for Prediction of Outcome in Patients with Cardiovascular Risk Factors. *J. Clin. Med.* **2020**, *9*, 3743. [CrossRef]
82. Wang, J.; Li, K.; Zhang, X.; Li, G.; Liu, T.; Wu, X.; Brown, S.L.; Zhou, L.; Mi, Q.-S. MicroRNA-155 Controls iNKT Cell Development and Lineage Differentiation by Coordinating Multiple Regulating Pathways. *Front. Cell Dev. Biol.* **2021**, *8*, 619220. [CrossRef]
83. Katze, M.G.; He, Y.; Gale, M., Jr. Viruses and interferon: A fight for supremacy. *Nat. Rev. Immunol.* **2002**, *2*, 675–687. [CrossRef]
84. Mitash, N.; Donovan, J.E.; Swiatecka-Urban, A. The Role of MicroRNA in the Airway Surface Liquid Homeostasis. *Int. J. Mol. Sci.* **2020**, *21*, 3848. [CrossRef]
85. Gracias, D.T.; Stelekati, E.; Hope, J.L.; Boesteanu, A.C.; Doering, T.A.; Norton, J.; Mueller, Y.M.; Fraietta, J.A.; Wherry, E.J.; Turner, M.; et al. The microRNA miR-155 controls CD8+ T cell responses by regulating interferon signaling. *Nat. Immunol.* **2013**, *14*, 593–602. [CrossRef]
86. Centa, A.; Fonseca, A.S.; Ferreira, S.G.D.S.; Azevedo, M.L.V.; de Paula, C.B.V.; Nagashima, S.; Machado-Souza, C.; Miggiolaro, A.F.R.D.S.; Baena, C.P.; de Noronha, L.; et al. Deregulated miRNA expression is associated with endothelial dysfunction in post-mortem lung biopsies of COVID-19 patients. *Am. J. Physiol. Cell. Mol. Physiol.* **2021**, *320*, L405–L412. [CrossRef]
87. Bartoszewski, R.; Dabrowski, M.; Jakiela, B.; Matalon, S.; Harrod, K.S.; Sanak, M.; Collawn, J.F. SARS-CoV-2 may regulate cellular responses through depletion of specific host miRNAs. *Am. J. Physiol. Cell. Mol. Physiol.* **2020**, *319*, L444–L455. [CrossRef]
88. Comer, B.; Camoretti-Mercado, B.; Kogut, P.C.; Halayko, A.J.; Solway, J.; Gerthoffer, W.T. MicroRNA-146a and microRNA-146b expression and anti-inflammatory function in human airway smooth muscle. *Am. J. Physiol. Cell. Mol. Physiol.* **2014**, *307*, L727–L734. [CrossRef]

89. Anzola, A.; González, R.; Gámez-Belmonte, R.; Ocón, B.; Aranda, C.J.; Martínez-Moya, P.; López-Posadas, R.; Hernández-Chirlaque, C.; de Medina, F.S.; Martínez-Augustin, O. miR-146a regulates the crosstalk between intestinal epithelial cells, microbial components and inflammatory stimuli. *Sci. Rep.* **2018**, *8*, 1–12. [CrossRef]
90. Bátkai, S.; Thum, T. MicroRNAs in Hypertension: Mechanisms and Therapeutic Targets. *Curr. Hypertens. Rep.* **2011**, *14*, 79–87. [CrossRef]

diagnostics

MDPI

Article

COVID-19 Patients in the COVID-19 Recovery and Engagement (CORE) Clinics in the Bronx

Anna Eligulashvili [1,†], Megan Darrell [1,†], Carolyn Miller [1], Jeylin Lee [1], Seth Congdon [2], Jimmy S. Lee [1], Kevin Hsu [1], Judy Yee [1], Wei Hou [3], Marjan Islam [2] and Tim Q. Duong [1,*]

1 Department of Radiology, Montefiore Medical Center, Albert Einstein College of Medicine, Bronx, NY 10461, USA
2 Department of Medicine, Montefiore Medical Center, Albert Einstein College of Medicine, Bronx, NY 10461, USA
3 Department of Family, Population and Preventive Medicine, Renaissance School of Medicine at Stony Brook University, Stony Brook, NY 11794, USA
* Correspondence: tim.duong@einsteinmed.edu
† These authors contributed equally to this work.

Abstract: Background: Early in the pandemic, we established COVID-19 Recovery and Engagement (CORE) Clinics in the Bronx and implemented a detailed evaluation protocol to assess physical, emotional, and cognitive function, pulmonary function tests, and imaging for COVID-19 survivors. Here, we report our findings up to five months post-acute COVID-19. Methods: Main outcomes and measures included pulmonary function tests, imaging tests, and a battery of symptom, physical, emotional, and cognitive assessments 5 months post-acute COVID-19. Findings: Dyspnea, fatigue, decreased exercise tolerance, brain fog, and shortness of breath were the most common symptoms but there were generally no significant differences between hospitalized and non-hospitalized cohorts ($p > 0.05$). Many patients had abnormal physical, emotional, and cognitive scores, but most functioned independently; there were no significant differences between hospitalized and non-hospitalized cohorts ($p > 0.05$). Six-minute walk tests, lung ultrasound, and diaphragm excursion were abnormal but only in the hospitalized cohort. Pulmonary function tests showed moderately restrictive pulmonary function only in the hospitalized cohort but no obstructive pulmonary function. Newly detected major neurological events, microvascular disease, atrophy, and white-matter changes were rare, but lung opacity and fibrosis-like findings were common after acute COVID-19. Interpretation: Many COVID-19 survivors experienced moderately restrictive pulmonary function, and significant symptoms across the physical, emotional, and cognitive health domains. Newly detected brain imaging abnormalities were rare, but lung imaging abnormalities were common. This study provides insights into post-acute sequelae following SARS-CoV-2 infection in neurological and pulmonary systems which may be used to support at-risk patients and develop effective screening methods and interventions.

Keywords: PASC; long COVID; pulmonary function tests; brain imaging; chest imaging; COVID symptoms; fatigue; shortness of breath

Citation: Eligulashvili, A.; Darrell, M.; Miller, C.; Lee, J.; Congdon, S.; Lee, J.S.; Hsu, K.; Yee, J.; Hou, W.; Islam, M.; et al. COVID-19 Patients in the COVID-19 Recovery and Engagement (CORE) Clinics in the Bronx. *Diagnostics* **2023**, *13*, 119. https://doi.org/10.3390/diagnostics13010119

Academic Editors: Yuli Huang, Yong Yuan and Peisong Chen

Received: 4 November 2022
Revised: 22 December 2022
Accepted: 26 December 2022
Published: 30 December 2022

1. Introduction

Many survivors of coronavirus disease 2019 (COVID-19) experience lingering neurological and pulmonary symptoms that persist long after severe acute respiratory syndrome coronavirus 2 (SARS-CoV-2) infection has resolved [1–4]. These symptoms are commonly referred to as post-acute sequelae of SARS-CoV-2 infection (PASC). Neuro-PASC include, but are not limited to, altered mental status, anxiety, depression, dizziness, headaches, memory loss, and post-traumatic stress disorder [5–8]. Pulmonary PASC include shortness of breath, persistent cough, and fatigue. Neuro-PASC could be caused by direct infection of the central nervous system, cytokine storm, systematic illness secondary to the initial viral infection, or psychological stressors such as social isolation, stigma, and future uncertainty [9]. Pulmonary-PASC could be caused by direct infection of the pulmonary system and/or systematic illness secondary to the initial viral infection, including pneumonia, acute respiratory distress, hypoxia, sepsis, pro-inflammatory mediators, cytokine storm and other host-mediated immunological responses [10–12]. Individuals with mild symptoms from SARS-CoV-2 infection (i.e., not requiring hospitalization) may also be susceptible to neurological and pulmonary PASC [13,14].

New York City was hit hard by the first wave of COVID-19 and by multiple subsequent surges of infection from different variants [15]. The Montefiore Health System consists of 15 hospitals located in the New York Metropolitan area in the Bronx and its environs, serving a large and diverse patient population, including many patients with lower socioeconomic status. Early in the pandemic, we established two COVID-19 Recovery and Engagement (CORE) Clinics for COVID-19 survivors with protracted symptoms. A detailed evaluation protocol was implemented to assess physical, emotional, and cognitive function, including pulmonary function tests and post-COVID imaging findings. In this study, we report a detailed analysis of a subset of this CORE patient cohort that had pulmonary function tests, imaging tests, and a battery of symptom, physical, emotional, and cognitive assessments.

2. Methods

2.1. CORE Patients

This study was approved by the Einstein IRB (IRB# 2021-13658) with a waiver of informed consent and followed all relevant regulatory guidelines. This is a prospective observational study of COVID-19 adult patients with protracted symptoms who were referred to Montefiore Medical Center's COVID-19 Recovery and Engagement (CORE) clinics between 26 June 2020 and 7 January 2022. Eligible patients were adults ($\geq$18 years old) who had probable or confirmed COVID-19 and were experiencing new or continued symptoms 4 or more weeks later. Patients who were terminally ill, referred to hospice or interested in comfort measures only were not eligible for referral to the CORE. On average, patients visited the CORE Clinics 133 $\pm$ 108 days after COVID-19 diagnosis. There were 11 patients who returned time over 300 days. If these patients were to be removed, the standard deviation was 59.93 days.

Patients who had less than 70 answers to survey questions and clinical variables (out of 100) were excluded. This threshold albeit arbitrary was chosen to avoid using subjects with many missing data. The final cohort consisted of 97 CORE patients.

2.2. Demographics and Laboratory Data

Age, sex, race, ethnicity, hospitalization status, and critical illness information was collected through patient questionnaires. Data from electronic medical records (EMR) were extracted automatically as described previously [16–18]. Preexisting comorbidities included body mass index (BMI), congestive heart failure (CHF), chronic kidney disease (CKD), hypertension, chronic obstructive pulmonary disease (COPD) and asthma that were designated by ICD10 codes at admission or prior. Hospitalization status and intensive-care-unit (ICU) admission were also extracted.

2.3. Symptom, Physical, Emotional, and Cognitive Assessments

All patients were surveyed for the presence of various post-infection symptoms. They were additionally administered a Modified Edmonton Physical Symptom Assessment (MEPSA) [19]. The original Edmonton Physical Symptom Assessment, initially validated in advanced cancer patients, included a list of common symptoms and allowed for repeated quantitative measurement of symptom intensity with minimal patient burden [20]. In our MEPSA, we asked patients to quantify their level of 14 symptoms over the past week prior to presentation to the clinic using a numerical rating scale between 0 (no symptom or best overall wellbeing) and 10 (the highest level of symptom imaginable or worst overall wellbeing).

Assessments of physical, emotional, and cognitive functions were performed. Disability and functional status were assessed using a modified Katz Activities of Daily Living (ADL) [21], Instrumental Activities of Daily Living (IADL) [22], and markers of frailty [23]. We assessed for depression, anxiety and post-traumatic symptoms using the Patient Health Questionnaire (PHQ-9) [24], Generalized Anxiety Disorder (GAD-7) [25], and Post-Traumatic Stress List for DSM-5 (PCL-5) questionnaires, respectively. We used previously validated cut-points for depression (PHQ $\geq$ 10) [26,27], anxiety (GAD-7 $\geq$ 8) [25,28] and post-traumatic symptoms (PCL-5 > 33) [29]. Social determinants of health (SDOH) information was collected through patient questionnaires.

2.4. Pulmonary Assessments

During the clinic visit (which could be remote or in-person), patients were evaluated by either general internal medicine or a pulmonary/critical care physician. The visits included a debrief of the acute illness experience, a summary of the questionnaire scores, an exploration of the patients' persistent symptoms, a physical examination (for in-person visits), medication reconciliation, and interventions based on the clinical assessment. We used chart review to collect additional information on the patient's history and demographics and to summarize relevant clinical interventions. Clinical measures of physical pulmonary health included the Borg Dyspnea Scale at rest and exertion, Lung Ultrasound (LUS) Score, and triplicate Diaphragm Excursion measurements with quiet and deep inspiration for survivors of critically ill COVID-19 pneumonia, defined as hypoxemia requiring high-flow nasal cannula or invasive or non-invasive positive pressure ventilation. All LUS videos were independently reviewed and scored by a blinded reviewer trained in scoring Lung Ultrasound Scores. Patients also underwent spirometry testing, including the Six Minute Walk Test (6MWT) and Pulmonary Function Test (PFT), which provided measures of distance, maximum heart rate (HR), and percent oxygen saturation (SpO_2), as well as forced vital capacity (FVC % Predicted), forced expiratory volume (FEV1 % Predicted), the FEV1/FVC ratio, the respiratory volume to total lung capacity (RV/TLC % Predicted) ratio, and the diffusing capacity of lung carbon monoxide (DLCO % Predicted), respectively [30].

2.5. Lung Ultrasound Protocol

For the LUS, subjects were scanned in the upright position. Twelve fields were assessed (six zones per hemithorax). Each zone was scanned based on predetermined anatomical landmarks to assess anterior, medial, and posterior lung areas per published protocols [31,32].

LUS was scored by identification of four sonographic patterns of lung ultrasound [31,32]: normal lung by the presence of lung sliding with A-lines (score 0); the presence of significant B-lines (score 1); confluent B-lines with or without subpleural consolidations (score 2); and extensive subpleural consolidations with B-lines (score 3). The total lung ultrasound score was calculated by a composite of the sum of all individual zone scores, ranging between 0 and 36 [33].

2.6. Diaphragm Ultrasound Protocol

Subjects underwent scanning in a supine position of the right hemidiaphragm via a subcostal approach, using two-dimensional B-mode and M-mode ultrasound views [34]. Diaphragm excursion was measured during resting respiration (quiet breathing) and deep inspiration (deep breathing) [34]. We obtained (3) consecutive measurements which were averaged into a final composite distance recording in centimeters (cm).

2.7. Imaging Assessments

Images and radiology reports of chest x-ray, chest computed tomography (CT) images, head CT, and brain magnetic resonance imaging (MRI) were evaluated at three time points: pre-COVID-19, during COVID-19, and post-COVID-19, if available. Pre-COVID-19 images were at least one month prior to COVID-19 diagnosis. During-COVID images included the first image taken after COVID-19 diagnosis. Post-COVID images included the first images taken at least one month after discharge from COVID-19 hospitalization. For patients who were not hospitalized for acute COVID-19 infection, there were generally no during-COVID-19 images.

For brain imaging data, two board-certified neuroradiologists with 16 and 11 years of experience evaluated the brain images and radiology reports in a single setting and consensus was reached for all assessments. The primary findings on brain imaging included new and prior stroke and hemorrhage. Secondary findings included microvascular disease (MVD), atrophy, and white matter (WM) changes. The presence or changes (for patients with multiple time point data) in stroke and hemorrhage. MVD, volume loss, and WM (i.e., hyperintensity or lesion) were noted with qualitative adjustment for age (i.e., whether brain volume loss is appropriate for patient's age). Mass effect (from tumor or infarction) if any was also noted.

For chest imaging data, radiology reports were evaluated for the presence of unilateral or bilateral lung opacity or consolidation/infiltrate and the presence of fibrosis-like changes in the lung at the three time points, if available, were tabulated.

2.8. Statistical Analysis

Analysis of group differences for categorical variables employed χ^2 tests and for categorical variables student *t*-tests via the statistical library from SciPy Python package in JupyterLab (©2018, Project Jupyter, https://jupyter.org, accessed on 25 December 2022). Sensitivity power analysis was performed. Cramer's V and Cohen's D values for size effect investigation were computed. $p < 0.05$ was considered statistically significant unless specified otherwise.

3. Results

3.1. Patient Demographics

Table 1 shows the profiles of the CORE patients. The entire cohort was 58.61 ± 14.43 years old, 49.48% female, 49.48% Hispanic, and 26.80% Black Non-Hispanic, with a BMI of 32.61 ± 7.63. Prevalence of diabetes was 32.0%, hypertension 56.7%, COPD/asthma 38.1%, CHF 9.3% and CKD 9.3%. Most patients were hospitalized (77/97) and about half (50) were critically ill. Effect size measures across all variables in Table 1 were small. Age was the only variable that was significantly different between hospitalized and non-hospitalized cohort ($p < 0.05$). For the sample size of 20 non-hospitalized and 77 hospitalized patients, the minimum detectable effect sizes for the 80% and 90% statistical power were 0.71 and 0.82, respectively.

3.2. Symptom, Physical, Emotional, and Cognitive Assessments

Table 2 summarizes the symptom, MEPSA, ADL, IADL, frailty, PHQ-9, GAD-7 and PCL-5 assessments. For all patients, dyspnea (77.32%), fatigue (70.1%), decreased exercise tolerance (55.67%), and brain fog/cognitive issues (40.21%) were the most reported symptoms. Many CORE patients reported fatigue (3.87 ± 3.23), overall well-being score (3.45 ± 3.14), and shortness of breath (3.32 ± 2.96). The average ADL score was 0.57 ± 1.14 (out of 6) and IADL was 2.11 ± 2.53 (out of 8), indicative of high function but some dependency. Challenges with bathing (16.84%) and dressing (15.54%) were the most reported ADLs and help with shopping (41.24%) and laundry (36.84%) were the most reported IADLs. The average frailty score was 1.66 ± 1.4 (out of 5), indicating pre-frailty. Of the behavioral health surveys, the GAD-7 anxiety score was 3.8 ± 5.43, PCL-5 PTSD score was 13.12 ± 14.9, and PHQ-9 depression score was 5.64 ± 6.08. Few patients reported SDOH challenges (0.40 ± 0.79 out of 8) and safety concerns (0.1 ± 0.31 out of 2).

These test scores were not significantly different between hospitalized and non-hospitalized cohort after multiple comparison correction ($p > 0.05$), except GAD-7 score which was higher in hospitalized compared to the non-hospitalized group (4.23 ± 5.94

vs. 2.15 ± 2.06, $p = 0.039$). The difference between the hospitalized and non-hospitalized groups are in the range of small to medium effect sizes in Table 2.

3.3. Pulmonary Assessments

Table 3 summarizes the 6MWT, Borg dyspnea scale, LUS, diaphragm excursion, and PFT results for the hospitalized and non-hospitalized cohorts. The 6MWT distance and 6MWT maximum heart rate were below normative values but were not significantly different between hospitalized and non-hospitalized cohorts ($p > 0.05$). 6MWT SpO_2% was abnormal for the hospitalized cohort (89 ± 7%) but normal (95 ± 1.8%) for the non-hospitalized cohort with a significant difference between groups ($p < 0.001$).

The Borg dyspnea data indicated patients had slight shortness of breath at rest and moderate difficulty breathing during exertion, but there were no significant differences between groups ($p > 0.05$). The mean LUS score was abnormal for hospitalized patients but normal for non-hospitalized patients, with a significant difference between groups ($p < 0.001$). The average diaphragm excursion measurement with quiet inspiration was significantly higher in hospitalized patients ($p < 0.05$); however, the average diaphragm excursion measurement with deep inspiration was not significantly different between groups ($p > 0.05$).

FVC% Predicted, FEV1% Predicted, RV/TLC% Predicted, and DLCO% Predicted were abnormal for the hospitalized cohort, but normal for the non-hospitalized cohort, with significant differences between groups ($p < 0.031$, 0.04, 0.010, and 0.018, respectively). FEV1/FVC was normal for both groups and was not significantly different between groups ($p > 0.05$). Across all variables, effect size measures ranged from small to large in Table 3.

3.4. Imaging Assessments

Table 4 summarizes the brain imaging findings pre-, during and post-COVID-19. The average follow-up imaging was 5 months post-COVID-19 diagnosis. Many but not all patients (N = 31, 6, and 13 out of 97, for pre-, during and post-COVID-19, respectively) had clinically indicated brain MRI or head CT.

Four patients had prior stroke, one patient had newly detected hemorrhage during COVID-19 hospitalization, and 2 patients had newly detected hemorrhage post-COVID. Overall, newly detected major neurological events post-COVID-19 were rare.

Table 1. Demographics of CORE patients. Mean ± SD or (%), N = 97. * $p < 0.05$ between hospitalized and non-hospitalized cohort.

	All (N = 97)	Hospitalized (N = 77)	Non-Hospitalized (N = 20)	Cohen's D	Cramer's V
Age (years)	58.61 ± 14.43	60.36 ± 13.53	50.8 ± 14.76 *	0.71	
Female	48 (49.48%)	32 (41.56%)	16 (80.00%)		0.311
Race/Ethnicity					
Hispanic	48 (49.48%)	42 (54.55%)	6 (30.00%)		0.199
Black Non-Hispanic	26 (26.8%)	16 (20.78%)	9 (45.00%)		0.224
White Non-Hispanic	13 (13.4%)	10 (12.99%)	3 (15.00%)		0.024
Other/Unknown	11 (11.34%)	9 (11.70%)	2 (10.00%)		0.022
BMI	32.61 ± 7.63	32.77 ± 7.79	32.01 ± 6.74	0.10	
Comorbidities					
CHF	9 (9.3%)	8 (10.39%)	1 (5.00%)		0.075
CKD	9 (9.3%)	8 (10.39%)	1 (5.00%)		0.075
Hypertension	55 (56.7%)	45 (58.44%)	10 (50.00%)		0.069
COPD/Asthma	37 (38.1%)	33 (42.86%)	8 (40.00%)		0.197
Diabetes	31 (32.0%)	29 (37.66%)	2 (10.00%)		0.240
Hospitalized	77 (79.38%)	na	na		
Critically Ill (IMV/ICU)	50 (51.55%)	50 (64.94%)	na		

Table 2. Symptoms, MEPSA, ADL, IADL, frailty and neuropsychiatric test scores of CORE patients (mean ± SD or N (%), N = 97). Continuous variables: *t*-test; Categorical variables: chi-squared (>10), Fisher's test (<10). *p* (Bonf corr) = Bonferroni correction for multiple comparison. * indicates statistical difference.

	Variables	N	All Patients (N = 97)	Hospitalized (N = 77, 79%)	Non-Hospitalized (N = 20, 21%)	*p*	*p* (Bonf. Corr)	Cohen's D	Cramer's V
Presence of Symptoms	Dyspnea	97	75 (77.32%)	60 (77.9%)	15 (75%)	1.000	1		0.028
	Fatigue	97	68 (70.1%)	53 (68.8%)	15 (75%)	0.977	1		0.055
	Decreased Exercise Tolerance	97	54 (55.67%)	43 (55.8%)	11 (55%)	1.000	1		0.007
	Brain Fog/Cognitive Issues	97	39 (40.21%)	30 (39%)	9 (45%)	0.818	1		0.050
	Cough	97	23 (23.71%)	16 (20.8%)	7 (35%)	0.404	1		0.135
	Palpitations	97	15 (15.46%)	11 (14.3%)	4 (20%)	0.735	1		0.064
	Chest Pain	97	10 (10.31%)	7 (9.1%)	3 (15%)	0.445	1		0.079
	Abnormal Smell/Taste	97	13 (13.4%)	9 (11.7%)	4 (20%)	0.475	1		0.099
	Joint Pain	97	16 (16.49%)	13 (16.9%)	3 (15%)	1.000	1		0.021
	Lightheadedness/Dizziness	97	8 (8.25%)	6 (7.8%)	2 (10%)	0.672	1		0.032
	Symptoms Resolved	97	6 (6.19%)	6 (7.8%)	0 (0%)	0.594	1		0.131
	Post-Exertional Malaise	97	4 (4.12%)	1 (1.3%) *	3 (15%) *	0.036	0.504		0.279
	Headache	97	4 (4.12%)	4 (5.2%)	0 (0%)	0.582	1		0.106
	Back Pain	97	2 (2.06%)	2 (2.6%)	0 (0%)	1.000	1		0.074
MEPSA Questions Out of 10	Q2: Fatigue	97	3.87 ± 3.23	3.68 ± 3.25	4.6 ± 3.1	0.249	1	0.29	
	Q14: Overall Well-Being	97	3.45 ± 3.14	3.29 ± 3.22	4.1 ± 2.83	0.273	1	0.26	
	Q8: Shortness of Breath	97	3.32 ± 2.96	3.25 ± 2.85	3.6 ± 3.41	0.673	1	0.12	
	Q1: Pain	97	2.43 ± 2.90	2.55 ± 2.89	2 ± 2.97	0.468	1	0.19	
	Q3: Drowsiness	96	2.20 ± 2.76	2.25 ± 2.87	2 ± 2.34	0.688	1	0.09	
	Q12: Anxiety	97	2.18 ± 3.08	2.29 ± 3.17	1.75 ± 2.73	0.455	1	0.17	
	Q11: Depression	97	1.82 ± 2.89	1.97 ± 3.08	1.25 ± 1.94	0.201	1	0.25	
	Q6: Lack of Appetite	97	0.86 ± 1.95	0.86 ± 1.98	0.85 ± 1.84	0.988	1	0.00	
	Q4: Nausea/vomiting	97	0.43 ± 1.34	0.44 ± 1.4	0.4 ± 1.1	0.888	1	0.00	
ADL (out of 6) = 0.57 (1.14) High function, independent	Q4: Requires help bathing/getting in or out of shower	95	16 (16.84%)	16 (20.8%)	0 (0%)	0.071	0.426		0.225
	Q2: Requires help getting completely dressed	96	13 (13.54%)	13 (16.9%)	0 (0%)	0.120	0.72		0.203
	Q1: Requires help walking across room	96	11 (11.46%)	11 (14.3%)	0 (0%)	0.211	1		0.185
	Q3: Requires help transferring from bed to chair	97	9 (9.28%)	9 (11.7%)	0 (0%)	0.203	1		0.163
	Q5: Requires help because of problems controlling bladder or bowel	96	5 (5.21%)	4 (5.2%)	1 (5%)	1.000	1		0.005
	Q6: Requires help feeding self	96	1 (1.04%)	1 (1.3%)	0 (0%)	1.000	1		0.053

Table 2. *Cont.*

	Variables	N	All Patients (N = 97)	Hospitalized (N = 77, 79%)	Non-Hospitalized (N = 20, 21%)	*p*	*p* (Bonf. Corr)	Cohen's D	Cramer's V
IADL (out of 8) = 2.11 (2.53) High function, independent	Q6: Requires help taking care of shopping needs	97	40 (41.24%)	36 (46.8%)	4 (20%)	0.215	1		0.220
	Q4: Requires help with doing laundry	95	35 (36.84%)	30 (39%)	5 (25%)	0.464	1		0.127
	Q3: Requires help with household tasks like cleaning, doing the dishes, or making the bed	96	34 (35.42%)	28 (36.4%)	6 (30%)	0.807	1		0.058
	Q8: Requires help traveling outside of home	97	31 (31.96%)	30 (39%) *	1 (5%) *	0.025	0.2		0.295
	Q5: Requires help handling purchases or other financial matters	96	28 (29.17%)	24 (31.2%)	4 (20%)	0.590	1		0.104
	Q7: Requires help planning, preparing or serving meals	94	27 (28.72%)	26 (33.8%) *	1 (5%) *	0.043	0.344		0.273
	Q2: Requires help taking medicine at correct time and dosage	97	7 (7.22%)	7 (9.1%)	0 (0%)	0.341	1		0.142
	Q1: Requires help using telephone	97	3 (3.09%)	3 (3.9%)	0 (0%)	1.000	1		0.091
Frailty Markers (out of 5) = 1.66 (1.4) Pre-Frail	Q1: Cannot stand up from chair without using arms	97	62 (63.92%)	46 (59.7%)	16 (80%)	0.570	1		0.171
	Q3: Cannot do moderate activities	97	49 (50.52%)	40 (51.9%)	9 (45%)	0.829	1		0.056
	Q2: Cannot climb one flight of stairs	96	48 (50%)	40 (51.9%)	8 (40%)	0.659	1		0.078
	Q5: Poor vision	97	19 (19.59%)	17 (22.1%)	2 (10%)	0.522	1		0.123
	Q4: Hard of hearing	97	10 (10.31%)	8 (10.4%)	2 (10%)	1.000	1		0.005
Behavioral Score	PHQ-9 Score (Depression)	97	5.64 ± 6.08	5.96 ± 6.67	4.4 ± 2.58	0.106	0.318	0.26	
	GAD-7 Score (Anxiety)	97	3.8 ± 5.43	4.23 ± 5.94 *	2.15 ± 2.06 *	0.013	0.039	0.39	
	PCL-5 Score (PTSD)	97	13.12 ± 14.9	14.05 ± 16.12	9.55 ± 8.04	0.085	0.255	0.30	
SDOH	SDOH Challenges Present (out of 8)	67	0.40 ± 0.79	0.48 ± 0.85	0.13 ± 0.52	0.058	0.117	0.44	
	SDOH Safety Concerns Present (out of 2)	67	0.1 ± 0.31	0.1 ± 0.3	0.13 ± 0.35	0.713	1	0.12	

Table 3. 6MWT, Borg Dyspnea scale, LUS Score, Diaphragm Excursion Measurements and PFT results (Mean ± SD). * $p < 0.05$, ** $p < 0.01$, *** $p < 0.001$ (t-test).

Variable	N	Hospitalized (N = 77, 79%)	Non-Hospitalized (N = 20, 21%)	p-Value	Normal Values or Ranges	Cohen's D
6MWT (Distance, feet)	59	985 ± 429	773 ± 258	0.091	1873 ± 295 [35]	0.51
6MWT (Maximum heart rate, bpm)	55	111 ± 15	101 ± 24	0.341	128 ± 18 [36]	0.58
6MWT (SpO_2 % Nadir)	56	89 ± 7 ***	95 ± 1.8 ***	<0.001	95 ± 2 [35]	0.74
Borg Dyspnea Scale (Rest)	76	0.96 ± 1.47	0.82 + 1.23	0.734	0 (normal) to 10 (worst).	0.10
Borg Dyspnea Scale (Exertion)	62	2.93 ± 2.23	2.86 ± 2.19	0.939	0 (normal) to 10	0.03
LUS Score	53	8.43 ± 6.73 ***	0.5 ± 0.71 ***	<0.001	0 (normal) to 36	1.19
Diaphragm Excursion, quiet inspiration (cm)	47	2.02 ± 0.72 *	1.74 ± 0.02 *	0.014	Range: 1.9–9	0.40
Diaphragm Excursion, deep inspiration (cm)	47	4.72 ± 1.82	5.18 ± 0.39	0.307	Range: 1.9–9	0.25
Pulmonary Function Tests						
FVC%	42	66.34 ± 18.93 *	90.6 ± 17.29 *	0.031	>80	1.29
FEV1%	42	68.73 ± 19 **	92 ± 10 **	0.004	>80	1.24
FEV1/FVC	42	81.02 ± 13	81.8 ± 7.6	0.852	>70	0.06
RV/TLC	19	122.65 ± 26.63 **	105 ± 2 **	0.010	<120	0.94
DLCO%	20	62.33 ± 23.90 *	94 ± 9 *	0.018	75–140%	1.92

6MWT—6 min walk test; HR—heart rate; bpm—beats per minute; SpO_2—oxygen saturation, LUS—lung ultrasound score; FVC—forced vital capacity; FEV1—forced expiratory volume in the first second. For diaphragm excursion measurements, lower values are more abnormal. Please note that the last columns that indicate typical normal ranges are for general reference only. They are not matched controls (i.e., for age, BMI, sex, etc.). Reference [35]—normative cohort comparison: 46% female, mean age of 58, BMI of 27, 5% non-white. Reference [36]—normative cohort comparison: 20–50 years old, 100% non-white.

Table 4. Major brain imaging findings pre-, during and post-COVID-19 diagnosis. Post-COVID-19 average 182 days after diagnosis. Note that each patient could have multiple radiological findings and thus the counts are larger than unique patient counts.

	Pre-COVID-19	During-COVID-19	Post-COVID-19
Total N with Imaging	31/97	6/97	13/97
No remarkable findings	20	2	4
Positive findings	11	4	9
Primary Outcomes			
Acute Stroke	1	0	0
Prior Stroke	3	1	1
Newly Detected Hemorrhage	0	1	2
Secondary Outcomes			
MVD	6	2	8
Atrophy	0	0	2
WM HI/Lesions	3	0	0

Although the presence of MVD was frequently noted, it was consistent with patient's age. Newly detected MVD was rare post COVID-19. Similarly, some patients had brain volume loss (atrophy) consistent with their age and newly detected atrophy was also rare post-COVID-19. For a patient subgroup who had both pre- and post-COVID-19 brain imaging (N = 6), 3 patients exhibited longitudinal changes (3 MVD, 1 atrophy, and 3 WM changes). The average time difference between the two time points was 7.06 years in this subgroup.

Table 5 summarizes the pulmonary imaging findings pre-, during- and post- COVID-19. The average follow-up scan was also 5 months post-COVID-19 diagnosis. Most patients had clinically indicated chest imaging performed (64, 75 and 70 out of 97 patients for pre, during and post-COVID-19, respectively). Opacity findings were rare pre-COVID-19 (N = 4), but very common during COVID-19 (N = 60), most were bilateral (N = 52). Most patients with post-COVID-19 lung imaging had opacity findings (N = 40), and most were also bilateral (N = 31). Fibrosis-like changes were rare pre- and during- COVID-19 (N = 5 and 1, respectively) but common post-COVID-19 (N = 14).

Table 5. Major lung imaging findings pre-, during and post-COVID-19 diagnosis. N reflects the number of unique patients. Post-COVID-19 average 5 months after diagnosis (N = 97).

	Pre-COVID-19	During-COVID-19	Post-COVID-19
Total N with imaging	64	75	70
No opacity	59	15	30
Opacity	4	60	40
Unilateral	2	8	9
Bilateral	2	52	31
No fibrosis	59	74	56
Fibrosis	5	1	14

With respect to hospitalization status, lung opacity was more common in the hospitalized compared to the non-hospitalized group both during- and post-COVID-19 ($p < 0.05$, Supplementary Table S1). Fibrosis-like changes were more common in the hospitalized cohort post-COVID-19 ($p < 0.05$).

4. Discussion

This study characterized a subset of the COVID-19 CORE clinic patients. The major findings are: (i) dyspnea, fatigue, decreased exercise tolerance, brain fog, and shortness of breath were the most common symptoms, (ii) some patients have abnormal physical, emotional, and cognitive scores but most are able to live and function independently, (iii) some variables of the 6MWT, LUS, and diaphragm excursion are abnormal in the hospitalized cohort, suggestive of interstitial lung disease, (iv) pulmonary function tests suggest moderately restrictive pulmonary function in the hospitalized cohort but no obstructive pulmonary function nor air trapping in the lung, (v) newly detected major neurological events, microvascular disease, atrophy, and WM changes are rare, and (vi) many patients have persistent lung opacity and fibrosis-like findings post-COVID-19.

4.1. Symptom, Physical, Emotional, and Cognitive Assessments

The high incidence of dyspnea, fatigue, brain fog, decreased exercise tolerance, and shortness of breath are consistent with the literature. Goertz et al. found that fatigue, dyspnea, headache, and chest tightness were the four most common persistent symptoms across both hospitalized and non-hospitalized COVID-19 patients 3 months after infection [37]. Huang et al. reported that fatigue or muscle weakness was by far the most common symptom in hospitalized COVID-19 patients at 6 months followed up [38]. Carfi et al. found fatigue, dyspnea, and joint pains to be the three most prevalent symptoms in COVID survivors about 2 months after their COVID-19 hospitalization [39]. These symptomologies are not surprising given many COVID-19 patients were often discharged with major medical referrals [40,41].

Collectively, ADL, IADL, and frailty marker scores indicate that CORE patients have a mild dependency and prefrailty. Although CORE patients showed signs of anxiety, depression, and PTSD, they were below the clinical cutoff in both hospitalized and non-hospitalized patients. Other studies reported clinically significant neurological and psychiatric symptoms after COVID-19 diagnosis [42].

Surprisingly, there were few differences in symptom, physical, emotional, and cognitive scores between hospitalized and non-hospitalized cohorts. A likely explanation is that non-hospitalized COVID-19 patients who visited our CORE clinics likely had more symptomatic COVID-19 disease. Previously studies have reported individuals with mild symptoms from SARS-CoV-2 infection (i.e., not requiring hospitalization) may also be susceptible to neuro-PASC [13,14].

4.2. Pulmonary Assessments

For most pulmonary measures, only the hospitalized cohort exhibited abnormalities. Both 6MWT distance and maximum heart rates were below normal (compared to some normative means) because patients as a cohort were not able to accomplish the task. 6MWT and SpO_2 saturation were clearly abnormal in the hospitalized cohort. Note that the normative data used for reference were from a very different population that were not matched (i.e., for age or sex, etc.) and thus comparisons need to be interpreted with caution.

LUS has been previously used to measure the degree of loss of aeration in patients with acute respiratory distress syndrome, with higher scores indicating more loss of normal lung aeration. B-lines in lung ultrasound denote artifacts reflecting a widening of the interlobular septa of the secondary pulmonary lobule. As a result, this finding is non-specific, and acute respiratory failure may represent cardiogenic or non-cardiogenic pulmonary edema as in ARDS, or other diagnoses [43]. In the post-COVID-19 population, however, high LUS likely denotes the presence of interstitial lung disease and lung injury following severe viral pneumonia. The abnormalities in spirometry measurements are suggestive of residual lung disease likely from COVID-19 pneumonia. Our spirometry analysis did not account for lung volume differences among patients thus such comparisons need to be interpreted with caution.

A combination of moderately reduced FVC%, FEV1%, RV/TLC ratio and DLCO% were abnormal in the hospitalized cohort, suggesting restrictive pulmonary function and air-trapping. FEV1/FVC in the hospitalized cohort was however normal, suggesting there was no significant obstructive pulmonary function. A normal FEV1/FVC ratio with a decreased FVC indicates a restrictive lung condition, which includes pulmonary fibrosis and infections such as pneumonia. A decreased FEV1/FVC ratio indicates an obstructive condition, such as asthma or COPD. In contrast, all these PFT variables were normal in the non-hospitalized cohort.

4.3. Imaging Findings

Although the sample size for brain imaging is small and both brain MRI and CT were included (which provide different sensitivity to detection of abnormalities), the main findings will likely hold, namely, that newly detected major neurological events post-COVID-19 are rare, and newly detected MVD, atrophy, and WM changes are also rare 5 months post -COVID-19 in our cohort. We thus concluded that there is no evidence of widespread changes in routine clinical brain imaging in our CORE patients 5 months post-COVID-19. However, it is possible that subtle brain changes existed, but were not detectable by routine clinical imaging. Imaging studies using more advanced imaging methods (such as diffusion tensor imaging, quantitative susceptibility mapping, and functional MRI) are warranted. It is surprising that given the broad spectrum of neuro-PASC symptoms reported in this cohort, there were comparatively few observable radiological abnormalities post-COVID-19. It is possible that these symptoms and neurological abnormalities have not yet manifested into structural changes in the brain on routine clinical imaging methods, and thus longer follow-up and more sophisticated imaging tools are necessary.

A few studies have reported brain imaging findings in post-COVID-19 patients. A cohort study found a greater reduction in grey matter thickness and tissue contrast in the orbitofrontal cortex and parahippocampal gyrus, greater changes in markers of tissue damage in regions that are functionally connected to the primary olfactory cortex and a greater reduction in global brain size in the COVID-19 cases compared to negative controls [44]. Abnormalities and cerebral microstructural changes in the brain of COVID-19 survivors both with and without neurological manifestations were noted [45,46], and persistent WM changes and ischemic stroke were associated with COVID-19 [47,48]. Note that many published studies to date were case reports or did not have pre-COVID-19 imaging data or controls (including the current study), which makes it difficult to definitively discern whether imaging abnormalities were pre-existing or a consequence of COVID-19 disease. Thus, there is likely reporting bias of positive clinical imaging findings associated with COVID-19. Brain imaging studies with proper controls with a correlation of neurological function at longer follow-up intervals are needed.

In contrast to brain imaging findings, there are clear anatomical abnormalities in the lung 5 months post-COVID-19. It is concerning that lung opacity in many patients has not completely resolved and that many patients developed pulmonary fibrosis-like changes 5 months after COVID-19. Pre-COVID-19 abnormalities were very rare and thus essentially all new lung findings were due to COVID-19. Hospitalized patients had more opacity and fibrosis-like changes post-COVID-19, consistent with disease severity. The incidence of fibrosis-like changes is likely underestimated in our study because a chest radiograph was included which has lower sensitivity for fibrosis detection compared to CT.

A few studies have previously reported persistent fibrosis-like lung changes post-COVID-19. CT abnormalities were common at 3 months after COVID-19 but with signs of fibrosis in a minority. More severe acute disease was linked with CT abnormalities at 3 months [49]. One study reported that although COVID-19 survivors showed continuous improvement in chest CT, residual lesions could still be observed and correlated with lung volume parameters one-year post-COVID-19 and the risk of developing residual CT opacities increases with age [50]. Another study found a significant percentage of individuals develop pulmonary sequelae after COVID-19 pneumonia, regardless of the severity of the acute process [51]. Six-month follow-up CT showed fibrotic-like changes in the lung in more than one-third of patients who survived severe COVID-19 pneumonia. These changes were associated with older age, acute respiratory distress syndrome, longer hospital stays, tachycardia, noninvasive mechanical ventilation, and higher initial chest CT score [52]. Given the numerous reports of persistent post-COVID-19 pulmonary sequela in many COVID-19 patients, longitudinal monitoring by chest imaging and pulmonary function in at-risk patients is warranted. Moreover, SARS-CoV-2 infection could also worsen existing pulmonary diseases. Many COVID-19 survivors are already being treated with pulmonary medications and pulmonary rehabilitation for pulmonary sequela.

4.4. Limitations

This study has several limitations. Our findings were limited to COVID-19 survivors who came to our CORE clinics and who were more likely to have more severe COVID-19 symptoms, and thus our cohort was not representative of the general population. It was not possible to definitively distinguish abnormalities that were due to COVID-19, pre-existing or worsened by COVID-19 disease although attempts were made to evaluate patients' pre-pandemic data. SARS-CoV-2 infection often resulted in multi-organ injury and future studies should also investigate long COVID regarding multi-organ injury [17,18,40,41,53].

Sample sizes of imaging data were small and consisted of a mixture of imaging modalities with different sensitivities to pathology, and thus results must be interpreted with caution. The small sample size also precluded quantitative statistical parametric analysis of imaging data. Large multicenter longitudinal imaging studies with proper controls are needed. We hope to be able to report longer follow-up findings on this cohort in the future.

5. Conclusions

We established early in the pandemic the CORE Clinics for COVID-19 survivors. Many CORE patients experienced significant symptoms across the physical, emotional, and cognitive health domains. Pulmonary function tests suggest moderately restrictive pulmonary function in patients hospitalized with COVID-19. Newly detected major neurological events, microvascular disease, atrophy, and white-matter changes were rare, but persistent lung opacity and COVID-19-related lung fibrosis-like findings were common. Our study provides insights into neurological and pulmonary COVID-19 sequela which may be used to support at-risk patients and develop effective screening methods and interventions to address the potentially high burden of care needed among COVID-19 survivors.

Supplementary Materials: The following supporting information can be downloaded at: https://www.mdpi.com/article/10.3390/diagnostics13010119/s1. Table S1: Major lung imaging findings pre-, during and post-COVID-19 diagnosis broken down by hospitalization status. N reflects the number of unique patients. Post covid average 5 months after diagnosis (N = 97). * $p < 0.05$, *** $p < 0.001$.

Author Contributions: Conceptualization, M.D., A.E., S.C., M.I. and T.Q.D.; Methodology, M.D., A.E., S.C., J.S.L., K.H., M.I., W.H. and T.Q.D.; Software, J.L., W.H; Validation, M.D., A.E., C.M., J.L., S.C., T.Q.D., K.H. and M.I.; Formal analysis, M.D., A.E., W.H. and J.L.; Investigation, M.D., A.E., S.C., T.Q.D., K.H. and M.I.; Resources, S.C., T.Q.D., K.H., J.Y., M.I. and T.Q.D.; Data curation, M.D., A.E., C.M., J.L., S.C., T.Q.D., K.H. and M.I.; Writing—original draft, M.D., A.E. and T.Q.D.; Writing—review & editing, M.D., A.E., S.C., T.Q.D., K.H., J.Y., M.I. and T.Q.D.; Visualization, M.D., A.E. and T.Q.D.; Supervision, S.C., T.Q.D., K.H., J.Y., M.I. and T.Q.D. All authors have read and agreed to the published version of the manuscript.

Funding: This research received no external funding.

Institutional Review Board Statement: This study was approved by the Einstein IRB (IRB# 2021-13658) with a waiver of informed consent and followed all relevant regulatory guidelines. Approval date: 1 December 2021 and reapproved 1 December 2022.

Informed Consent Statement: Informed consent was waived the IRB due to retrospective nature of the study.

Data Availability Statement: Upon reasonable request via corresponding author.

Conflicts of Interest: The authors declare no conflict of interest.

Abbreviations

6MWT	Six-Minute Walk Test
ADL	Activities of Daily Living
ALT	Alanine aminotransferase
AST	Aspartate transaminase
BMI	Body mass index
BNP	Brain natriuretic peptide
BUN	Blood urea nitrogen
CHF	Congestive heart failure
CKD	Chronic kidney disease
COPD	Chronic obstructive pulmonary disease
CORE	COVID-19 Recovery and Engagement
COVID-19	Coronavirus disease 2019
Cr	Creatinine
CRP	C-reactive protein
CT	Computed tomography

DDIM	D-dimer
DLCO% Predicted	Diffusing capacity of lung carbon monoxide
EMR	Electronic medical records
FERR	Ferritin
FEV1% Predicted	Forced expiratory volume
FVC% Predicted	Forced vital capacity
GAD-7	Generalized Anxiety Disorder Questionnaire
HI	Hyperintensity
HR	Heart rate
IADL	Instrumental Activities of Daily Living
ICU	Intensive care unit
IMV	Intermittent mandatory ventilation
INR	International normalized ratio
LDH	Lactate dehydrogenase
LUS	Lung ultrasound
Lymph	Lymphocyte count
MEPSA	Modified Edmonton Physical Symptom Assessment
MRI	Magnetic resonance imaging
MVD	Microvascular disease
PASC	Post-acute sequelae of SARS-CoV-2
PCL-5	Post-Traumatic Stress List Questionnaire
PFT	Pulmonary Function Test
PHQ-9	Patient Health Questionnaire
RV	Respiratory volume
SARS-CoV-2	Severe acute respiratory syndrome coronavirus 2
SBP	Systolic blood pressure
SDOH	Social determinants of health
Sodium	Glucose
SpO_2	Percent oxygen saturation
TLC	Total lung capacity
TNT	Troponin-T
WBC	White blood cell count
WM	White matter

References

1. Davis, H.E.; Assaf, G.S.; McCorkell, L.; Wei, H.; Low, R.J.; Re'em, Y.; Redfield, S.; Austin, J.P.; Akrami, A. Characterizing long COVID in an international cohort: 7 months of symptoms and their impact. *EClinicalMedicine* **2021**, *38*, 101019. [CrossRef] [PubMed]
2. Rubin, R. As Their Numbers Grow, COVID-19 "Long Haulers" Stump Experts. *JAMA* **2020**, *324*, 1381–1383. [CrossRef] [PubMed]
3. Zhao, Y.M.; Shang, Y.M.; Song, W.B.; Li, Q.Q.; Xie, H.; Xu, Q.F.; Jia, J.L.; Li, L.M.; Mao, H.L.; Zhou, X.M.; et al. Follow-up study of the pulmonary function and related physiological characteristics of COVID-19 survivors three months after recovery. *EClinicalMedicine* **2020**, *25*, 100463. [CrossRef]
4. Iosifescu, A.L.; Hoogenboom, W.S.; Buczek, A.J.; Fleysher, R.; Duong, T.Q. New-onset and persistent neurological and psychiatric sequelae of COVID-19 compared to influenza: A retrospective cohort study in a large New York City healthcare network. *Int. J. Methods Psychiatr. Res.* **2022**, *31*, e1914. [CrossRef]
5. Moghimi, N.; Di Napoli, M.; Biller, J.; Siegler, J.E.; Shekhar, R.; McCullough, L.D.; Harkins, M.S.; Hong, E.; Alaouieh, D.A.; Mansueto, G.; et al. The Neurological Manifestations of Post-Acute Sequelae of SARS-CoV-2 infection. *Curr. Neurol. Neurosci. Rep.* **2021**, *21*, 44. [CrossRef]
6. Collantes, M.E.V.; Espiritu, A.I.; Sy, M.C.C.; Anlacan, V.M.M.; Jamora, R.D.G. Neurological Manifestations in COVID-19 Infection: A Systematic Review and Meta-Analysis. *Can. J. Neurol. Sci.* **2021**, *48*, 66–76. [CrossRef]
7. Graham, E.L.; Clark, J.R.; Orban, Z.S.; Lim, P.H.; Szymanski, A.L.; Taylor, C.; DiBiase, R.M.; Jia, D.T.; Balabanov, R.; Ho, S.U.; et al. Persistent neurologic symptoms and cognitive dysfunction in non-hospitalized Covid-19 "long haulers". *Ann. Clin. Transl. Neurol.* **2021**, *8*, 1073–1085. [CrossRef]
8. Wang, F.; Kream, R.M.; Stefano, G.B. Long-Term Respiratory and Neurological Sequelae of COVID-19. *Med. Sci. Monit.* **2020**, *26*, e928996. [CrossRef]
9. Mazza, M.G.; De Lorenzo, R.; Conte, C.; Poletti, S.; Vai, B.; Bollettini, I.; Melloni, E.M.T.; Furlan, R.; Ciceri, F.; Rovere-Querini, P.; et al. Anxiety and depression in COVID-19 survivors: Role of inflammatory and clinical predictors. *Brain Behav. Immun.* **2020**, *89*, 594–600. [CrossRef]

10. Khan, S.; Chen, L.; Yang, C.R.; Raghuram, V.; Khundmiri, S.J.; Knepper, M.A. Does SARS-CoV-2 Infect the Kidney? *J. Am. Soc. Nephrol.* **2020**, *31*, 2746–2748. [CrossRef]
11. Ahmadian, E.; Hosseiniyan Khatibi, S.M.; Razi Soofiyani, S.; Abediazar, S.; Shoja, M.M.; Ardalan, M.; Zununi Vahed, S. Covid-19 and kidney injury: Pathophysiology and molecular mechanisms. *Rev. Med. Virol.* **2020**, *31*, e2176. [CrossRef] [PubMed]
12. Adapa, S.; Chenna, A.; Balla, M.; Merugu, G.P.; Koduri, N.M.; Daggubati, S.R.; Gayam, V.; Naramala, S.; Konala, V.M. COVID-19 Pandemic Causing Acute Kidney Injury and Impact on Patients With Chronic Kidney Disease and Renal Transplantation. *J. Clin. Med. Res.* **2020**, *12*, 352–361. [CrossRef]
13. Townsend, L.; Dyer, A.H.; Jones, K.; Dunne, J.; Mooney, A.; Gaffney, F.; O'Connor, L.; Leavy, D.; O'Brien, K.; Dowds, J.; et al. Persistent fatigue following SARS-CoV-2 infection is common and independent of severity of initial infection. *PLoS ONE* **2020**, *15*, e0240784. [CrossRef] [PubMed]
14. van den Borst, B.; Peters, J.B.; Brink, M.; Schoon, Y.; Bleeker-Rovers, C.P.; Schers, H.; van Hees, H.W.H.; van Helvoort, H.; van den Boogaard, M.; van der Hoeven, H.; et al. Comprehensive Health Assessment 3 Months After Recovery From Acute Coronavirus Disease 2019 (COVID-19). *Clin. Infect. Dis.* **2021**, *73*, e1089–e1098. [CrossRef] [PubMed]
15. Hoogenboom, W.S.; Pham, A.; Anand, H.; Fleysher, R.; Buczek, A.; Soby, S.; Mirhaji, P.; Yee, J.; Duong, T.Q. Clinical characteristics of the first and second COVID-19 waves in the Bronx, New York: A retrospective cohort study. *Lancet Reg. Health Am.* **2021**, *3*, 100041. [CrossRef] [PubMed]
16. Hoogenboom, W.S.; Fleysher, R.; Soby, S.; Mirhaji, P.; Mitchell, W.B.; Morrone, K.A.; Manwani, D.; Duong, T.Q. Individuals with sickle cell disease and sickle cell trait demonstrate no increase in mortality or critical illness from COVID-19—A fifteen hospital observational study in the Bronx, New York. *Haematologica* **2021**, *106*, 3014–3016. [CrossRef]
17. Lu, J.Q.; Lu, J.Y.; Wang, W.; Liu, Y.; Buczek, A.; Fleysher, R.; Hoogenboom, W.S.; Zhu, W.; Hou, W.; Rodriguez, C.J.; et al. Clinical predictors of acute cardiac injury and normalization of troponin after hospital discharge from COVID-19. *EBioMedicine* **2022**, *76*, 103821. [CrossRef]
18. Lu, J.Y.; Hou, W.; Duong, T.Q. Longitudinal prediction of hospital-acquired acute kidney injury in COVID-19: A two-center study. *Infection* **2022**, *50*, 109–119. [CrossRef]
19. Watanabe, S.M.; Nekolaichuk, C.L.; Beaumont, C. The Edmonton Symptom Assessment System, a proposed tool for distress screening in cancer patients: Development and refinement. *Psychooncology* **2012**, *21*, 977–985. [CrossRef]
20. Bruera, E.; Kuehn, N.; Miller, M.J.; Selmser, P.; Macmillan, K. The Edmonton Symptom Assessment System (ESAS): A simple method for the assessment of palliative care patients. *J. Palliat. Care* **1991**, *7*, 6–9. [CrossRef]
21. Katz, S.; Downs, T.D.; Cash, H.R.; Grotz, R.C. Progress in development of the index of ADL. *Gerontologist* **1970**, *10*, 20–30. [CrossRef] [PubMed]
22. Lawton, M.P.; Brody, E.M. Assessment of older people: Self-maintaining and instrumental activities of daily living. *Gerontologist* **1969**, *9*, 179–186. [CrossRef] [PubMed]
23. Hope, A.A.; Hsieh, S.J.; Petti, A.; Hurtado-Sbordoni, M.; Verghese, J.; Gong, M.N. Assessing the Utility and Validity of Frailty Markers in Critically Ill Adults. *Ann. Am. Thorac. Soc.* **2017**, *14*, 952–959. [CrossRef] [PubMed]
24. Kroenke, K.; Spitzer, R.L.; Williams, J.B. The PHQ-9: Validity of a brief depression severity measure. *Gen. Intern. Med* **2001**, *16*, 606–613. [CrossRef] [PubMed]
25. Spitzer, R.L.; Kroenke, K.; Williams, J.B.; Löwe, B. A brief measure for assessing generalized anxiety disorder: The GAD-7. *Arch. Intern. Med.* **2006**, *166*, 1092–1097. [CrossRef]
26. Manea, L.; Gilbody, S.; McMillan, D. Optimal cut-off score for diagnosing depression with the Patient Health Questionnaire (PHQ-9): A meta-analysis. *CMAJ* **2012**, *184*, E191–E196. [CrossRef]
27. Moriarty, A.S.; Gilbody, S.; McMillan, D.; Manea, L. Screening and case finding for major depressive disorder using the Patient Health Questionnaire (PHQ-9): A meta-analysis. *Gen. Hosp. Psychiatry* **2015**, *37*, 567–576. [CrossRef]
28. Kroenke, K.; Spitzer, R.L.; Williams, J.B.; Monahan, P.O.; Löwe, B. Anxiety disorders in primary care: Prevalence, impairment, comorbidity, and detection. *Ann. Intern. Med.* **2007**, *146*, 317–325. [CrossRef]
29. Johnson, S.U.; Ebrahimi, O.V.; Hoffart, A. PTSD symptoms among health workers and public service providers during the COVID-19 outbreak. *PLoS ONE* **2020**, *15*, e0241032. [CrossRef]
30. Trakada, G.; Kastritis, E.; Gavriatopoulou, M.; Velentza, L.; Fotiou, D.; Ziogas, D.C.; Panagiotidis, I.; Eleutherakis-Papaiakovou, E.; Roussou, M.; Migkou, M.; et al. Pulmonary function abnormalities are common in patients with multiple myeloma and are independently associated with worse outcome. *Ann. Hematol.* **2019**, *98*, 1427–1434. [CrossRef]
31. Baciarello, M.; Bonetti, A.; Vetrugno, L.; Saturno, F.; Nouvenne, A.; Bellini, V.; Meschi, T.; Bignami, E. Is lung ultrasound score a useful tool to monitoring and handling moderate and severe COVID-19 patients in the general ward? An observational pilot study. *J. Clin. Monit. Comput.* **2022**, *36*, 785–793. [CrossRef] [PubMed]
32. Dargent, A.; Chatelain, E.; Kreitmann, L.; Quenot, J.P.; Cour, M.; Argaud, L.; COVID-LUS Study Group. Lung ultrasound score to monitor COVID-19 pneumonia progression in patients with ARDS. *PLoS ONE* **2020**, *15*, e0236312. [CrossRef] [PubMed]
33. Soldati, G.; Smargiassi, A.; Inchingolo, R.; Buonsenso, D.; Perrone, T.; Briganti, D.F.; Perlini, S.; Torri, E.; Mariani, A.; Mossolani, E.E.; et al. Proposal for International Standardization of the Use of Lung Ultrasound for Patients With COVID-19: A Simple, Quantitative, Reproducible Method. *J. Ultrasound Med.* **2020**, *39*, 1413–1419. [CrossRef]
34. Boussuges, A.; Rives, S.; Finance, J.; Bregeon, F. Assessment of diaphragmatic function by ultrasonography: Current approach and perspectives. *World J. Clin. Cases* **2020**, *8*, 2408–2424. [CrossRef]

35. Casanova, C.; Celli, B.R.; Barria, P.; Casas, A.; Cote, C.; de Torres, J.P.; Jardim, J.; Lopez, M.V.; Marin, J.M.; Montes de Oca, M.; et al. The 6-min walk distance in healthy subjects: Reference standards from seven countries. *Eur. Respir. J.* **2011**, *37*, 150–156. [CrossRef]
36. Chetta, A.; Zanini, A.; Pisi, G.; Aiello, M.; Tzani, P.; Neri, M.; Olivieri, D. Reference values for the 6-min walk test in healthy subjects 20-50 years old. *Respir. Med.* **2006**, *100*, 1573–1578. [CrossRef]
37. Goertz, Y.M.J.; Van Herck, M.; Delbressine, J.M.; Vaes, A.W.; Meys, R.; Machado, F.V.C.; Houben-Wilke, S.; Burtin, C.; Posthuma, R.; Franssen, F.M.E.; et al. Persistent symptoms 3 months after a SARS-CoV-2 infection: The post-COVID-19 syndrome? *ERJ Open Res.* **2020**, *6*, 00542–02020. [CrossRef]
38. Huang, C.; Huang, L.; Wang, Y.; Li, X.; Ren, L.; Gu, X.; Kang, L.; Guo, L.; Liu, M.; Zhou, X.; et al. 6-month consequences of COVID-19 in patients discharged from hospital: A cohort study. *Lancet* **2021**, *397*, 220–232. [CrossRef]
39. Carfi, A.; Bernabei, R.; Landi, F.; for the Gemelli Against COVID-19 Post-Acute Care Study Group. Persistent Symptoms in Patients After Acute COVID-19. *JAMA* **2020**, *324*, 603–605. [CrossRef]
40. Musheyev, B.; Borg, L.; Janowicz, R.; Matarlo, M.; Boyle, H.; Singh, G.; Ende, V.; Babatsikos, I.; Hou, W.; Duong, T.Q. Functional status of mechanically ventilated COVID-19 survivors at ICU and hospital discharge. *J. Intensive Care* **2021**, *9*, 31. [CrossRef]
41. Musheyev, B.; Janowicz, R.; Borg, L.; Matarlo, M.; Boyle, H.; Hou, W.; Duong, T.Q. Characterizing non-critically ill COVID-19 survivors with and without in-hospital rehabilitation. *Sci. Rep.* **2021**, *11*, 21039. [CrossRef] [PubMed]
42. Taquet, M.; Luciano, S.; Geddes, J.R.; Harrison, P.J. Bidirectional associations between COVID-19 and psychiatric disorder: Retrospective cohort studies of 62354 COVID-19 cases in the USA. *Lancet Psychiatry* **2021**, *8*, 130–140. [CrossRef] [PubMed]
43. Lichtenstein, D.A.; Meziere, G.A. Relevance of lung ultrasound in the diagnosis of acute respiratory failure: The BLUE protocol. *Chest* **2008**, *134*, 117–125. [CrossRef] [PubMed]
44. Douaud, G.; Lee, S.; Alfaro-Almagro, F.; Arthofer, C.; Wang, C.; McCarthy, P.; Lange, F.; Andersson, J.L.R.; Griffanti, L.; Duff, E.; et al. SARS-CoV-2 is associated with changes in brain structure in UK Biobank. *Nature* **2022**, *604*, 697–707. [CrossRef]
45. Lu, Y.; Li, X.; Geng, D.; Mei, N.; Wu, P.Y.; Huang, C.C.; Jia, T.; Zhao, Y.; Wang, D.; Xiao, A.; et al. Cerebral Micro-Structural Changes in COVID-19 Patients—An MRI-based 3-month Follow-up Study. *EClinicalMedicine* **2020**, *25*, 100484. [CrossRef]
46. Hellgren, L.; Birberg Thornberg, U.; Samuelsson, K.; Levi, R.; Divanoglou, A.; Blystad, I. Brain MRI and neuropsychological findings at long-term follow-up after COVID-19 hospitalisation: An observational cohort study. *BMJ Open* **2021**, *11*, e055164. [CrossRef]
47. Huang, S.; Zhou, Z.; Yang, D.; Zhao, W.; Zeng, M.; Xie, X.; Du, Y.; Jiang, Y.; Zhou, X.; Yang, W.; et al. Persistent white matter changes in recovered COVID-19 patients at the 1-year follow-up. *Brain* **2022**, *145*, 1830–1838. [CrossRef]
48. Kamal, M.; Abo Omirah, M.; Hussein, A.; Saeed, H. Assessment and characterisation of post-COVID-19 manifestations. *Int. J. Clin. Pract.* **2021**, *75*, e13746. [CrossRef] [PubMed]
49. Vijayakumar, B.; Tonkin, J.; Devaraj, A.; Philip, K.E.J.; Orton, C.M.; Desai, S.R.; Shah, P.L. CT Lung Abnormalities after COVID-19 at 3 Months and 1 Year after Hospital Discharge. *Radiology* **2022**, *303*, 444–454. [CrossRef]
50. Chen, Y.; Ding, C.; Yu, L.; Guo, W.; Feng, X.; Yu, L.; Su, J.; Xu, T.; Ren, C.; Shi, D.; et al. One-year follow-up of chest CT findings in patients after SARS-CoV-2 infection. *BMC Med.* **2021**, *19*, 191. [CrossRef]
51. Tarraso, J.; Safont, B.; Carbonell-Asins, J.A.; Fernandez-Fabrellas, E.; Sancho-Chust, J.N.; Naval, E.; Amat, B.; Herrera, S.; Ros, J.A.; Soler-Cataluna, J.J.; et al. Lung function and radiological findings 1 year after COVID-19: A prospective follow-up. *Respir. Res.* **2022**, *23*, 242. [CrossRef] [PubMed]
52. Han, X.; Fan, Y.; Alwalid, O.; Li, N.; Jia, X.; Yuan, M.; Li, Y.; Cao, Y.; Gu, J.; Wu, H.; et al. Six-month Follow-up Chest CT Findings after Severe COVID-19 Pneumonia. *Radiology* **2021**, *299*, E177–E186. [CrossRef] [PubMed]
53. Lu, J.Y.; Ho, S.L.; Buczek, A.; Fleysher, R.; Hou, W.; Chacko, K.; Duong, T.Q. Clinical predictors of recovery of COVID-19 associated-abnormal liver function test 2 months after hospital discharge. *Sci. Rep.* **2022**, *12*, 17972. [CrossRef] [PubMed]

diagnostics

MDPI

Article

Predictive Value of MR-proADM in the Risk Stratification and in the Adequate Care Setting of COVID-19 Patients Assessed at the Triage of the Emergency Department

Marilena Minieri [1,2,*], Vito N. Di Lecce [3], Maria Stella Lia [2], Massimo Maurici [4], Francesca Leonardis [5,6], Susanna Longo [7], Luca Colangeli [7], Carla Paganelli [3], Stefania Levantesi [3], Alessandro Terrinoni [1,2], Vincenzo Malagnino [8], Domenico J. Brunetti [9], Alfredo Giovannelli [2], Massimo Pieri [1,2], Marco Ciotti [10], Cartesio D'Agostini [1,11], Mariachiara Gabriele [1,12], Sergio Bernardini [1,2] and Jacopo M. Legramante [7,8]

1 Department of Experimental Medicine, University of Rome Tor Vergata, 00133 Rome, Italy
2 Laboratory Medicine Unit, Tor Vergata University Hospital, 00133 Rome, Italy
3 Emergency Department, Tor Vergata University Hospital, 00133 Rome, Italy
4 Department of Biomedicine and Prevention, University of Rome Tor Vergata, 00133 Rome, Italy
5 Department of Surgical Sciences, University of Rome Tor Vergata, 00133 Rome, Italy
6 Intensive Care Unit, Emergency Department, Tor Vergata University Hospital, 00133 Rome, Italy
7 Department of Systems Medicine, University of Rome Tor Vergata, 00133 Rome, Italy
8 Infectious Disease Unit, Tor Vergata University Hospital, 00133 Rome, Italy
9 Anaesthesia and Intensive Care Unit, Tor Vergata University Hospital, 00133 Rome, Italy
10 Virology Unit, Tor Vergata University Hospital, 00133 Rome, Italy
11 Clinical Microbiology Unit, Tor Vergata University Hospital, 00133 Rome, Italy
12 Respiratory Medicine Unit, Tor Vergata University Hospital, 00133 Rome, Italy
* Correspondence: minieri@uniroma2.it; Tel.: +39-06-20902365

Citation: Minieri, M.; Di Lecce, V.N.; Lia, M.S.; Maurici, M.; Leonardis, F.; Longo, S.; Colangeli, L.; Paganelli, C.; Levantesi, S.; Terrinoni, A.; et al. Predictive Value of MR-proADM in the Risk Stratification and in the Adequate Care Setting of COVID-19 Patients Assessed at the Triage of the Emergency Department. *Diagnostics* **2022**, *12*, 1971. https://doi.org/10.3390/diagnostics12081971

Academic Editors: Yuli Huang, Yong Yuan and Peisong Chen

Received: 28 June 2022
Accepted: 8 August 2022
Published: 15 August 2022

Abstract: In the past two pandemic years, Emergency Departments (ED) have been overrun with COVID-19-suspicious patients. Some data on the role played by laboratory biomarkers in the early risk stratification of COVID-19 patients have been recently published. The aim of this study is to assess the potential role of the new biomarker mid-regional proadrenomedullin (MR-proADM) in stratifying the in-hospital mortality risk of COVID-19 patients at the triage. A further goal of the present study is to evaluate whether MR-proADM together with other biochemical markers could play a key role in assessing the correct care level of these patients. Data from 321 consecutive patients admitted to the triage of the ED with a COVID-19 infection were analyzed. Epidemiological; demographic; clinical; laboratory; and outcome data were assessed. All the biomarkers analyzed showed an important role in predicting mortality. In particular, an increase of MR-proADM level at ED admission was independently associated with a threefold higher risk of IMV. MR-proADM showed greater ROC curves and AUC when compared to other laboratory biomarkers for the primary endpoint such as in-hospital mortality, except for CRP. This study shows that MR-proADM seems to be particularly effective for early predicting mortality and the need of ventilation in COVID-19 patients admitted to the ED.

Keywords: emergency department; triage; COVID-19 biomarkers; mid-regional proadrenomedullin

1. Introduction

The severe acute respiratory syndrome coronavirus 2 (SARS-CoV-2) pandemic has spread worldwide and reached catastrophic proportions in the past two years. The SARS-CoV-2 infection has been named coronavirus disease 2019 (COVID-19). The World Health Organization (WHO) declared the SARS-CoV-2 infection a 'Public Health Emergency of International Concern' due to its rapid transmission among humans. To date, there have been more than 539 million cases worldwide and more than 6 million deaths SARS-CoV-2 infection presents a wide clinical spectrum ranging from asymptomatic infection to mild upper respiratory tract illness or severe interstitial pneumonia with respiratory failure [1].

During the COVID-19 pandemic, the high number of hospital admissions complicated patient management, highlighting the weakness of our health system, resulting in an increase of the mortality rate. Furthermore, the lack of specific clinical features of COVID-19 pneumonia complicated the differential diagnosis from other forms of severe pneumonia [1].

The lack of immediate results from the current microbiological tests to confirm COVID-19, coupled with the reported suboptimal sensitivity of swab tests by real-time reverse transcription-polymerase chain reaction (RT-PCR) assays, made the situation worse [2]. Consequently, Emergency Departments have been overwhelmed by COVID-19-suspicious patients, inducing the need to stratify the patient risk immediately upon entering the triage. Since predicting the course of this disease at symptom onset is difficult and very often clinical conditions tends to worsen abruptly, prognostic tools and/or biochemical markers have been fundamental to address patients through the right clinical pathway in the ED. However, although several laboratory biomarkers have been so far identified to diagnose COVID-19 pneumonia more rapidly, to date, there are no data on biomarkers with high specificity and sensibility able to early stratify the mortality risk of patients affected by viral pneumonia [3].

C-reactive protein (CRP) has been one of the most used biomarkers to assess the evolution of COVID-19 inflammatory processes, even though its use is limited by a low sensitivity for community-acquired pneumonia (CAP). While a high CRP value (>100 mg/L) can indicate a severe bacterial infection, lower values are common in both viral infections and noninfectious diseases [4].

Another biomarker evaluated in COVID-19 patients has been procalcitonin (PCT), since it has an important role in detecting a bacterial superinfection to manage antibiotic therapy [5]. As is known, PCT can predict microbial etiology in pneumonia [6]. On the other hand, in patients with a high pneumonia severity index (PSI classes III-V), PCT has proven to be a good prognostic marker rather than a diagnostic marker [7].

Procalcitonin, CRP and white blood cell count have been shown to be significantly higher in CAP patients with a typical bacterial etiology as compared to cases in which the pathogen was represented by an atypical bacterium or by a virus [8].

The mid-regional proadrenomedullin (MR-proADM) is the precursor molecule of adrenomedullin (ADM). Unlike ADM, which is unstable and characterized by a short half-life, MR-proADM is more stable and appears promising as biomarker for detecting endothelial dysfunction, therefore, predicting severity and long-term adverse outcomes in CAP. Christ-Crain et al. [9] showed that the level of MR-proADM increased with the severity of CAP, in contrast to CRP levels and leukocytes. In the study by Valenzuela Sanchez et al. [10], MR-proADM predicted unfavorable outcomes in patients with pneumonia caused by influenza virus. In addition, MR-proADM can be used to stratify the clinical risk in patients affected by CAP [11]. Determination of MR-proADM level within 6 h of arrival to the hospital has prognostic value. It has also been reported that MR-proADM obtained within 6 h from arrival at the hospital has considerable prognostic value, irrespective of the causal agent of CAP, and in association with PSI and CURB-65 scores, improves prognostic accuracy [11,12].

In the presence of an alteration of the microcirculatory integrity due to an endothelium damage with consequent capillary leak, as observed in sepsis, MR-proADM plasma concentrations tend to increase [13].

Accordingly, Hupf et al. [14] have recently reported significantly higher adrenomedullin RNA blood expression in patients with severe COVID-19 vs. patients with a mild disease.

It has been hypothesized by Li et al. [15] that the integrity of the epithelial-endothelial barrier is severely damaged in critical ill patients with COVID-19 pneumonia, introducing the concept of "viral sepsis". Considering this pathogenetic mechanism, several recent studies reported an increased level of proinflammatory cytokines and chemokines such as tumor necrosis factor-α (TNF-α) and interleukin-6 (IL-6) in COVID-19 patients [16,17] suggesting that this cytokine storm might have a critical role in the evolution of SARS-CoV-2 infection [18].

The predictive value of MR-proADM in patients with COVID-19 pneumonia has been recently reported [19,20]. Our group showed that MR-proADM might have a predictive value in the early risk stratification of patients with COVID-19 infection at the triage in the Emergency Department [21].

Therefore, the aim of this study is to evaluate more in detail if the laboratory biomarkers can play a key role in predicting the correct care level of these patients, contributing to optimizing the hospital resources and helping the emergency physician in the decision about the adequate setting of care of patients at the entry to the Emergency Department.

2. Materials and Methods

2.1. Study Design

The present study has an observational, retrospective single-center design. Data from 321 consecutive patients admitted to the Emergency Department of the University Hospital Tor Vergata (Rome, Italy) from April to December 2020 with a confirmed COVID-19 infection were analyzed. A diagnosis of COVID-19 was made by a positive real-time reverse transcription polymerase chain reaction (RT-PCR) taken from nasopharyngeal swabs and through radiological imaging, where indicated, in accordance with WHO interim guidelines. Adult patients aged >18 years with a positive swab test were enrolled.

The demographic, epidemiological and clinical data were obtained from the electronic clinical records.

Patients underwent chest X-rays or computed tomography (CT) scans based on the physician's clinical assessment, and these data were further revised by the Emergency Department's radiologist. Blood culture, sputum, urine, bronchial aspirate and/or bronchoalveolar samples were analyzed when deemed necessary.

The final diagnosis was considered as that provided by the ED physician. A patient follow-up was performed up to 45 days.

The design of the study was evaluated and approved by the local Ethics Committee at Tor Vergata University Hospital (approval number 87/20) and was carried out according to the Declaration of Helsinki. Written informed consent was waived because of the rapid spread of this infectious disease.

2.2. Blood Sample Collection

Blood samples were collected at the triage admission. Upon arrival at the laboratory, blood samples were centrifuged at 4500× g for 5 min to obtain serum or plasma samples.

Blood examinations were for mid-regional proadrenomedullin (MR-proADM), C-reactive protein (CRP), procalcitonin (PCT), D-dimer, lactate dehydrogenase (LDH).

CRP (normality range 0.01–5.0 mg/L) and LDH (normality range 125–220 IU/L) levels were measured in serum using an Abbott ARCHITECT c16000 (Abbott, Chicago, IL, USA) clinical chemistry analyzer. PCT (normality range 0.01–0.50 ng/mL; cut-off for suspected infection 0.50 ng/mL) was detected in serum with a BRAHMS PCT chemiluminescent microparticle immunoassay (CMIA) by an Abbott ARCHITECT i2000SR instrument. MR-proADM (normality range 0.05–0.55 nmol/L) was measured using a time-resolved amplified cryptate emission assay on EDTA plasma samples (TRACE BRAHMS MR-proADM Kryptor, BRAHMS AG, Hennigsdorf, Germany). D-dimer values were obtained by an ACL TOP 700 instrument (Instrumentation Laboratory Company, Werfen, Bedford, MA, USA).

2.3. Statistical Analysis

The primary endpoint was the overall in-hospital mortality; the secondary endpoints were the need of noninvasive mechanical ventilation (NIMV) and invasive mechanical ventilation (IMV).

Continuous variables were expressed as mean (standard deviation) or median (interquartile ranges), according to data distribution, and were compared using the Student's *t*-test or the Mann–Whitney U test, when appropriate; categorical variables were expressed as counts and percentages and compared using the Chi-square or Fisher's exact tests, as appropriate. Associations between candidate variables and endpoints were assessed using both univariate and multi-variate Cox regression analyses, and hazard ratios were calculated. We have evaluated survivors compared with non-survivors and patients who needed ventilation (both invasive and non-invasive) compared with patients without ventilation.

The discriminatory power of the analyzed variables for predicting mortality was tested by means of a receiver operating characteristic (ROC) curve analysis with the area under the ROC curve (AUC) determination.

For the regression analysis, variables were dichotomized according to cut-off values derived during the data analysis for this study, using the Youden index arising from the ROC curve analysis.

For each biomarker, sensitivity, specificity, negative and positive predictive values (NPV, PPV), negative and positive likelihood ratio (LR−, LR+) and odds ratio with CI 95% were also reported for mortality, IMV and NIMV.

Kaplan–Meier curves were created to estimate the overall survival and compared using the log-rank test.

All analyses were performed with SPSS software. Tests were considered statistically significant if they yielded two-tailed *p*-values <0.05. For the multivariate analysis, we used variables resulting as statistically significant in the univariate analysis.

3. Results

The demographic and clinical characteristics of the study population are summarized in Table 1. The patient population had a mean age of 63 ± 14.7 years. Hypertension (40.8%), cardiovascular diseases (17.1%) and diabetes (13.1%) represented the most frequent comorbidities (Table 1)

Table 1. Demographic and clinical parameters.

	Overall	Survivors	Non-Surviv	*p* Value	No-IMV	IMV	*p* Value	No-NIMV	NIMV	*p* Value
	N 321	N 224	N 97		N 234	N 87		N 177	N 57	
Age										
Years, mean (SD)	63.3 (14.7)	59.6 (14.6)	71.9 (11.2)	<0.001	61.4 (15.8)	68.6 (9.7)	<0.001	59.6 (16.2)	67 (12.9)	0.002
Sex										
Male, N (%)	215 (67.0)	145 (64.7)	70 (72.2)	0.193	146 (62.4)	69 (79.3)	0.004	107 (60.4)	39 (68.4)	0.280
Female, N (%)	106 (33.0)	79 (35.3)	27 (27.8)		88 (37.6)	18 (20.7)		70 (39.6)	18 (31.6)	
Comorbidities										
Hypertension, N (%)	131 (40.8)	70 (31.3)	61 (62.9)	<0.001	81 (34.6)	50 (57.5)	<0.001	51 (28.8)	30 (52.6)	0.001
Diabetes, N (%)	42 (13.1)	19 (8.5)	23 (23.7)	<0.001	21 (9.0)	21 (24.1)	<0.001	13 (7.3)	8 (14.0)	0.124
Respiratory disease, N (%)	28 (8.7)	14 (6.3)	14 (14.4)	0.017	16 (6.8)	12 (13.8)	0.050	13 (7.3)	3 (5.3)	0.588
Malignancy, N (%)	19 (5.9)	10 (4.5)	9 (9.3)	0.093	11 (4.7)	8 (9.2)	0.129	7 (4.0)	4 (7.0)	0.342
Cardiovasc. disease, N (%)	55 (17.1)	27 (12.1)	28 (28.9)	<0.001	37 (15.8)	18 (20.7)	0.303	26 (14.7)	11 (19.3)	0.407
Renal disease, N (%)	51 (15.9)	13 (5.8)	38 (39.2)	<0.001	17 (7.3)	34 (39.1)	<0.001	8 (4.5)	9 (15.8)	0.004
Obesity, N (%)	15 (4.7)	8 (3.6)	7 (7.2)	0.155	7 (3.0)	8 (9.2)	0.019	7 (4.0)	0 (0)	0.127

Values expressed in percentages (%) indicate the proportion of patients within each group for each variable. Data are presented as mean (standard deviation, SD) where specified. The chi-square ($\chi 2$) test was used to determine significance between the groups for categorical variables, Student's t test for the variable of age. IMV, Invasive Mechanical Ventilation; NIMV, Non Invasive Mechanical Ventilation.

Among the comorbidities reported, obesity did not show significant differences between survivors and non-survivors, whereas malignancy showed a statistical level close to the significance. All the other comorbidities showed a significant difference between the two groups considered. Evaluating the secondary outcomes, cardiovascular disease and malignancy did not show significant differences between IMV and no-IMV, whereas all the other comorbidities showed a significant difference. For the last group of patients, only hypertension and renal disease showed significant differences between NIMV and no-NIMV patients.

All the biomarkers evaluated just after triage showed increased values in non-survivors as compared to survivors as well as in IMV and NIMV compared to no-IMV and no-NIMV, reaching always a statistically significant level (Table 2)

Table 2. Biomarkers values at triage admission.

	Overall	Survivors	Non Survivors	*p* Value	NO IMV	IMV	*p* Value	NO NIMV	NIMV	*p* Value
	N 321	N 224	N 97		N 234	N 87		N 177	N 57	
MR-proADM nmol/L										
Median	0.90	0.75	1.46	<0.001	0.79	1.42	<0.001	0.72	0.99	0.001
(Q1–Q3)	(0.63–0.33)	(0.57–1.0)	(1.14–2.37)		(0.58–1.05)	(1.11–2.14)		(0.55–0.95)	(0.80–1.30)	
CRP mg/L										
Median	61	45.90	134	<0.001	47.5	134	<0.001	35	90	<0.001
(Q1–Q3)	(24–125)	(14–86)	(72–207)		(12.0–93.0)	(68–211)		(10–75)	(48–151)	
PCT ng/mL										
Median	0.08	0.06	0.18	<0.001	0.06	0.19	<0.001	0.05	0.09	0.001
(Q1–Q3)	(0.04–0.20) N 290	(0.03–0.13) N 196	(0.10–0.40) N 94		(0.03–0.13) N 205	(0.10–0.60) N 85		(0.03–0.10) N 150	(0.06–0.20) N 55	
D-dimer ng/mL										
Median	753	647	1295	<0.001	669	1212	<0.001	603	829	0.009
(Q1–Q3)	(446–1437) N 315	(411–1063) N 219	(700–2365) N 96		(417–1148) N 229	(658–2102) N 86		(408–999) N 172	(508–1666) N 57	
LDH UI/L										
Median	349	323	456	<0.001	323	494	<0.001	303	395	<0.001
(Q1–Q3)	(268–487) N 315	(249–432) N 218	(323–597) N 97		(244–427) N 228	(343–616) N 87		(233–413) N 171	(295–500) N 57	

Data are presented as median [first quartile (Q1)-third quartile (Q3)]. The Mann Whitney U test was used to determine significance among biomarker concentrations. MR-pro ADM, mid-regional proadrenomedullin; CRP, C-reactive protein; PCT, procalcitonin; LDH, lactate dehydrogenase. IMV, Invasive Mechanical Ventilation; NIMV, Non Invasive Mechanical Ventilation.

Table 3 shows the results of the univariate Cox regression analysis performed to investigate the possible predictive role of clinical and demographic characteristics in patients with suspected COVID-19 infections. In addition, obesity does not seem to predict 45 days mortality, whereas malignancy showed a statistical level close to the significance in patients evaluated at the triage in the Emergency Department. All the other clinical features have shown significant odds ratio value to predict mortality in this group of patients.

Table 3. Univariate Cox Regression Analysis for biomarkers and clinical characteristics for the primary (survivors) and the secondary (IMV, NIMV) outcomes. Univariate Cox Regression Analysis for the prediction of 45-day mortality and 28-day IMV/NIMV.

	Cut-off	Overall (N)	Non Surv (N)	*p* Value	HR (95% CI)	Overall (N)	IMV (N)	*p* Value	HR (95% CI)	Overall (N)	NIMV (N)	*p* Value	HR (95% CI)
Age		320	96	<0.001	1.06 (1.04–1.07)	321	87	<0.001	1.03 (1.01–1.04)	234	57	0.003	1.03 (1.01–1.04)
Gender		320	96	0.182	1.35 (0.87–2.11)	321	87	0.007	2.03 (1.21–3.41)	234	57	0.301	1.34 (0.77–2.35)
Hypertension		320	96	<0.001	2.87 (1.90–4.34)	321	87	<0.001	2.16 (1.41–3.31)	231	57	0.002	2.28 (1.35–3.80)
Diabetes		320	96	<0.001	2.43 (1.51–3.92)	321	87	<0.001	2.41 (1.48–3.95)	234	57	0.137	1.76 (0.84–3.73)
Respiratory disease		320	96	0.017	2.00 (1.13–3.52)	321	87	0.038	1.91 (1.04–3.51)	234	57	0.669	0.77 (0.24–2.49)
Malignancy		320	96	0.057	1.95 (0.98–3.87)	321	87	0.115	1.80 (0.87–3.72)	234	57	0.298	1.72 (0.62–4.74)
Cardiovasc. disease		320	96	<0.001	2.47 (1.60–3.84)	321	87	0.307	1.31 (0.78–2.20)	234	57	0.385	1.34 (0.70–2.59)
Renal disease		320	96	<0.001	5.44 (3.59–8.25)	321	87	<0.001	4.85 (3.14–7.50)	234	57	0.008	2.92 (1.43–5.97)
Obesity		320	96	0.200	1.65 (0.77–3.57)	321	87	0.007	2.74 (1.32–5.67)	234	57	0.353	[illegible].05 ([illegible]0.0–29.60)

Table 3. *Cont.*

	Cut-off	Overall (N)	Non Surv (N)	*p* Value	HR (95% CI)	Overall (N)	IMV (N)	*p* Value	HR (95% CI)	Overall (N)	NIMV (N)	*p* Value	HR (95% CI)
MR-proADM (nmol/L)	1.105	320	96	<0.001	9.10 (5.64–14.70)	321	87	<0.001	7.22 (4.41–11.83)	234	57	<0.001	4.20 (2.20–8.00)
CRP (mg/L)	95.5	320	96	<0.001	6.28 (4.03–9.78)	321	87	<0.001	4.79 (3.05–7.52)	234	57	<0.001	4.20 (2.40–7.50)
PCT (ng/mL)	0.095	289	93	0.001	4.62 (2.86–7.45)	290	85	<0.001	5.07 (3.01–8.54)	205	55	<0.001	3.10 (1.70–5.80)
D-dimer (ng/mL)	985	314	95	<0.001	4.18 (2.72–6.43)	315	86	<0.001	3.22 (2.08–5.01)	229	57	0.002	2.30 (1.40–4.00)
LDH (UI/L)	439.5	314	96	<0.001	3.52 (2.35–5.27)	315	87	<0.001	4.47 (2.90–6.91)	228	57	<0.001	2.80 (1.60–4.80)

HR, hazard ratio; CI, confidence interval. MR-proADM, mid-regional proadrenomedullin; CRP, C-reactive protein; PCT, procalcitonin; LDH, lactate dehydrogenase. Biomarkers cut-off values derived from ROC (Receiver Operating Characteristic) curves using the Youden index. IMV: Invasive Mechanical Ventilation; NIVM: Non Invasive Mechanical Ventilation. Cut-off changes in IMV: D-dimer 981.5 ng/mL, LDH 437.5 U/L; cut-off changes in NIMV: MR-proADM 0.785 nmol/L, CRP 59.5 mg/mL, LDH 340.5 UI/L.

Concerning the possible role in predicting need of IMV within 28 days in these patients, all the clinical features reached statistical significance except for cardiovascular disease and malignancy, whereas only hypertension and renal diseases showed a significant odds ratio value for NIMV within 28 days. All the biomarkers analyzed showed a significant role in predicting mortality, need of IMV and NIMV in patients admitted at the ED with COVID-19 infection as analyzed by the univariate Cox regression analysis (Table 3).

A multivariate analysis performed pooling together both clinical and laboratory variables, also considering the whole observation period (mortality at 45 days, IMV and NIMV at 28 days), showed that all biomarkers, except PCT, might be considered a valuable predictive tool in the mortality risk stratification at admission to the Emergency Department (Table 4).

Table 4. Multivariate Cox Regression Analysis pooling together biomarkers and clinical characteristics for the primary (survivors) and the secondary (IMV, NIMV) outcomes. Multivariate Cox Regression Analysis for the prediction of 45-day mortality and 28-day IMV/NIMV.

	Overall (N)	Non Surviv (N)	*p* Value	HR (95% CI)	Overall (N)	IMV (N)	*p* Value	HR (95% CI)	Overall (N)	NIMV (N)	*p* Value	HR (95% CI)
Age	284	93	0.083	1.02 (0.99–1.04)	285	85	0.386	0.99 (0.97–1.01)	200	55	0.952	0.99 (0.97–1.02)
Gender					285	85	0.095	1.63 (0.92–2.89)				
Hypertension	284	93	0.970	1.01 (0.63–1.61)	285	85	0.930	1.02 (0.64–1.64)	200	55	0.450	1.30 (0.70–2.30)
Diabetes	284	93	0.880	1.04 (0.61–1.80)	285	85	0.292	1.34 (0.78–2.31)				
Respiratory disease	284	93	0.047	1.86 (1.01–3.41)	285	85	0.248	1.49 (0.76–2.94)				
Malignancy	284	93	0.038	2.28 (1.05–4.95)								
Cardiovascular disease	284	93	0.042	1.78 (1.02–3.10)								
Renal disease	284	93	0.039	1.64 (1.02–2.62)	285	85	0.019	1.82 (1.10–3.00)	200	55	0.745	1.10 (0.50–2.40)
Obesity					285	85	0.259	1.62 (0.70–3.75)				
MR-pro ADM (nmol/L)	284	93	<0.001	2.99 (1.70–5.28)	285	85	0.001	2.83 (1.49–5.36)	200	55	0.071	2.00 (0.90–4.30)
CRP (mg/L)	284	93	<0.001	2.85 (1.73–4.69)	285	85	0.106	1.54 (0.91–2.60)	200	55	0.036	2.00 (1.0–3.70)
PCT (ng/mL)	284	93	0.602	1.17 (0.65–2.10)	285	85	0.288	1.41 (0.75–2.65)	200	55	0.075	1.80 (0.90–3.60)
D-dimer (ng/mL)	284	93	0.024	1.80 (1.08–2.99)	285	85	0.085	1.56 (0.94–2.59)	200	55	0.169	1.50 (0.80–2.80)
LDH (UI/L)	284	93	0.047	1.70 (1.01–2.84)	285	85	0.002	2.18 (1.33–3.57)	200	55	0.078	1.70 (0.90–3.10)

Age, Hpertension, Diabetes, Respiratory disease, Malignancy, Cardiovascular disease and Renal disease were used as adjusting variables within the Multivariate Cox Regression Analysis for the prediction of 45-day mortality. Age, Gender, Hpertension, Diabetes, Respiratory disease and Renal disease were used as adjusting variables within the Multivariate Cox Regression Analysis for the prediction of 28-day IMV. Age, Hypertension and Renal disease were used as adjusting variables within the Multivariate Cox Regression Analysis for the prediction of 28-day NIMV. MR-proADM, mid-regional proadrenomedullin; CRP, C-reactive protein; PCT, procalcitonin; LDH, lactate dehydrogenase.

In fact, patients with MR-proADM level higher than the cut-off value of 1105 nmol/L show an increase of mortality of almost three times (OR 2.97, IC 1.7–5.28); also, CRP levels were independently associated with a higher risk of in-hospital mortality in COVID-19 patients (OR 2.85, IC 1.73–4.69). Similarly, at ED admission, an increase of MR-proADM level was independently associated with an almost three times (OR 2.83, IC 1.49–5.36) higher risk of IMV need as well as for LDH, which showed a smaller but still significant risk (OR 2.18, IC 1.33–3.57). Only CRP showed a significant predictive value (OR 2, IC 1–3.7) for the need of NIMV.

Looking at ROC curve analysis, the prognostic ability of MR-proADM assessed at ED admissions has been shown by the good discrimination performance both for in-hospital mortality (AUC 0.85) and for prediction of IMV (AUC 0.81); it seems to be less effective, as predictive factor for NIMV prediction (AUC 0.71), with the optimal cut-off of 1105 as obtained with the Youden index. The risk stratification role of MR-proADM seems to be more powerful as compared to the other biomarkers as demonstrated by ROC curves and AUC, which resulted as significantly greater for the primary endpoint, i.e., in-hospital mortality, except for CRP (Table 5 and Figure 1).

Table 5. Prognostic accuracy of biomarkers for different outcomes.

	Outcomes	AUC (95% CI)	Cut-off	*p* Value	Sensitivity (95% CI)	Specificity (95% CI)	PPV (95% CI)	NPV (95% CI)	LR+ (95% CI)	LR- (95% CI)	OR (95% CI)
MR-proADM nmol/L	Mortality	**0.848** (0.80–0.90)	1.105		0.77 (0.67–0.85)	0.80 (0.73–0.85)	0.65 (0.58–0.71)	0.87 (0.83–0.91)	3.75 (2.80–5.10)	0.29 (0.20–0.40)	12.76 (7.05–23.08)
	IMV	**0.807** (0.75–0.86)	1.105		0.75 (0.65–0.84)	0.77 (0.70–0.82)	0.58 (0.51–0.64)	0.88 (0.83–0.91)	3.20 (2.43–4.23)	0.32 (0.22–0.47)	9.92 (5.50–17.92)
	NIMV	**0.707** (0.63–0.78)	0.785		0.80 (0.67–0.90)	0.55 (0.46–0.63)	0.40 (0.35–0.45)	0.88 (0.81–0.93)	1.76 (1.41–2.19)	0.37 (0.21–0.64)	4.79 (2.29–10.0)
CRP mg/L	Mortality	**0.785** (0.73–0.84)	95.5	0.090	0.71 (0.61–0.80)	0.78 (0.72–0.84)	0.62 (0.55–0.68)	0.85 (0.80–0.88)	3.24 (2.40–4.40)	0.37 (0.30–0.50)	8.80 (5.01–15.50)
	IMV	**0.759** (0.70–0.82)	95.5	0.242	0.67 (0.56–0.77)	0.74 (0.67–0.80)	0.52 (0.45–0.59)	0.84 (0.79–0.88)	2.58 (1.95–3.40)	0.45 (0.33–0.61)	5.79 (3.34–10.06)
	NIMV	**0.709** (0.63–0.79)	59.5	0.970	0.69 (0.55–0.81)	0.67 (0.59–0.75)	0.44 (0.37–0.51)	0.85 (0.79–0.90)	2.09 (1.56–2.79)	0.46 (0.31–0.70)	4.52 (2.32–8.81)
PCT ng/mL	Mortality	**0.759** (0.70–0.82)	0.095	0.021	0.77 (0.67–0.85)	0.67 (0.60–0.73)	0.53 (0.47–0.59)	0.85 (0.80–0.89)	2.29 (1.80–2.90)	0.35 (0.20–0.50)	6.50 (3.70–11.42)
	IMV	**0.769** (0.71–0.83)	0.095	0.354	0.79 (0.69–0.87)	0.66 (0.59–0.72)	0.49 (0.44–0.55)	0.88 (0.83–0.92)	2.28 (1.83–2.85)	0.32 (0.21–0.49)	7.07 (3.89–12.83)
	NIMV	**0.657** (0.57–0.74)	0.055	0.380	0.76 (0.63–0.87)	0.55 (0.46–0.63)	0.39 (0.34–0.45)	0.86 (0.79–0.91)	1.68 (1.33–2.11)	0.43 (0.26–0.71)	3.87 (1.92–7.81)
D-dimer ng/mL	Mortality	**0.705** (0.64–0.77)	985.5	0.001	0.67 (0.57–0.76)	0.73 (0.66–0.79)	0.55 (0.48–0.61)	0.82 (0.77–0.86)	2.46 (1.90–3.20)	0.45 (0.30–0.60)	5.43 (3.18–9.28)
	IMV	**0.666** (0.60–0.74)	981.5	0.002	0.65 (0.54–0.75)	0.70 (0.63–0.76)	0.47 (0.41–0.54)	0.82 (0.77–0.86)	2.12 (1.63–2.76)	0.51 (0.38–0.69)	4.18 (2.44–7.15)
	NIMV	**0.610** (0.53–0.70)	787.5	0.110	0.60 (0.46–0.73)	0.66 (0.57–0.73)	0.40 (0.33–0.47)	0.81 (0.75–0.86)	1.74 (1.27–2.38)	0.61 (0.43–0.86)	2.85 (1.50–5.40)
LDH UI/L	Mortality	**0.687** (0.62–0.76)	439.5	0.0001	0.55 (0.45–0.66)	0.80 (0.73–0.85)	0.57 (0.49–0.65)	0.78 (0.74–0.82)	2.71 (1.90–3.80)	0.56 (0.40–0.70)	4.83 (2.82–8.26)
	IMV	**0.736** (0.67–0.80)	437.5	0.101	0.61 (0.50–0.72)	0.80 (0.73–0.85)	0.56 (0.48–0.64)	0.83 (0.79–0.87)	2.98 (2.16–4.11)	0.49 (0.37–0.64)	6.30 (3.61–11.00)
	NIMV	**0.649** (0.56–0.73)	340.5	0.320	0.67 (0.53–0.79)	0.61 (0.52–0.69)	0.39 (0.33–0.46)	0.83 (0.77–0.88)	1.71 (1.30–2.25)	0.54 (0.36–0.81)	3.17 (1.65–6.11)

Area Under Curve (AUC) analysis for 45-day mortality prediction and for 28-day IMV or NIMV prediction of study population. P value: differences between area of each biomarker vs MR-pro-ADM. Cut-off derived from ROC (Receiver Operating Characteristic) curves using the Youden index. CI, Confidence Interval; PPV, Positive Predictive Value; NPV, Negative Predictive Value; LR+, Positive Likelihood Ratio; LR-, Negative Likelihood Ratio; OR, Odds Ratio; IMV, Invasive Mechanical Ventilation; NIMV, Non Invasive Mechanical Ventilation. MR-proADM, mid-regional proadrenomedullin; CRP, C-reactive protein; PCT, procalcitonin; LDH, lactate dehydrogenase.

In particular, MR-proADM showed the better PPV (65%) and especially NPV (87%) in predicting mortality, as well as for IMV and NIMV regarding NPV (both 88%) as compared to the other biomarkers, with only CRP showing similar values (Table 5).

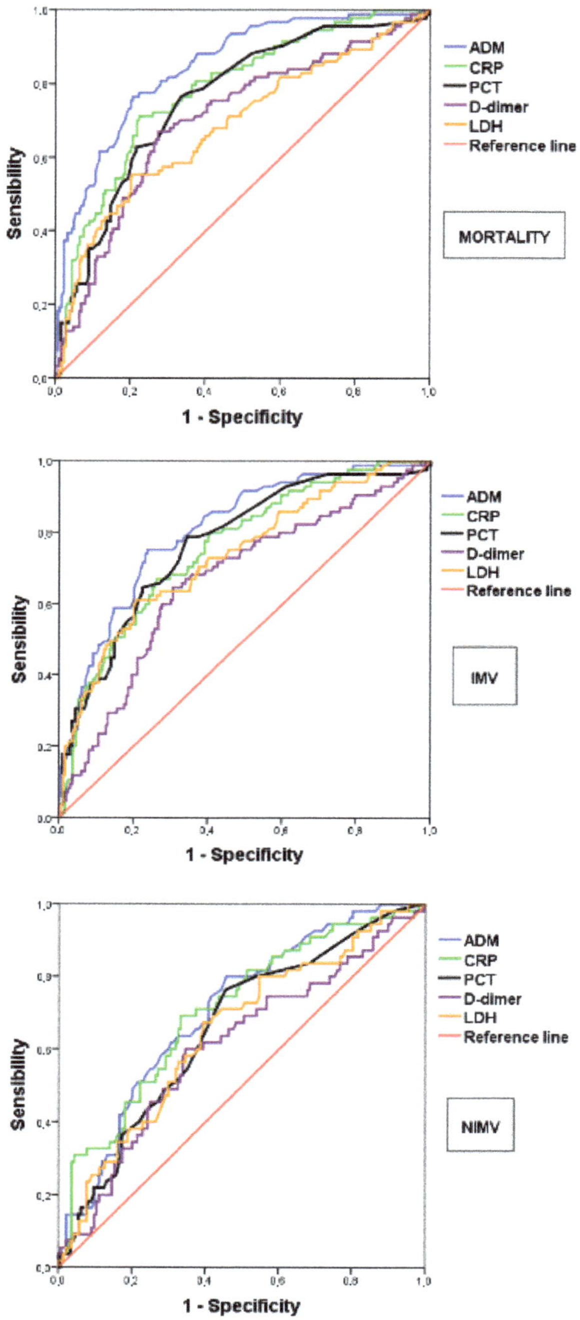

Figure 1. Association of candidate biomarkers with mortality and mechanical ventilation. AU-ROC, Area Under the Receiver Operating Characteristic curve. MR-proADM, mid-regional pro-drenomedullin; CRP, C-reactive protein; PCT, procalcitonin; LDH, lactate dehydrogenase; IMV, Invasive Mechanical Ventilation; NIMV, Non-Invasive Mechanical Ventilation.

The good discrimination performance of MR-proADM for the primary and secondary endpoints is also shown by the survival curves (Figure 2). In fact, a higher survival rate and a reduced mechanical ventilation risk were evident for patients with values less than 1.105 nmol/L for mortality and IMV and 0.785 for NIMV at admission to the ED.

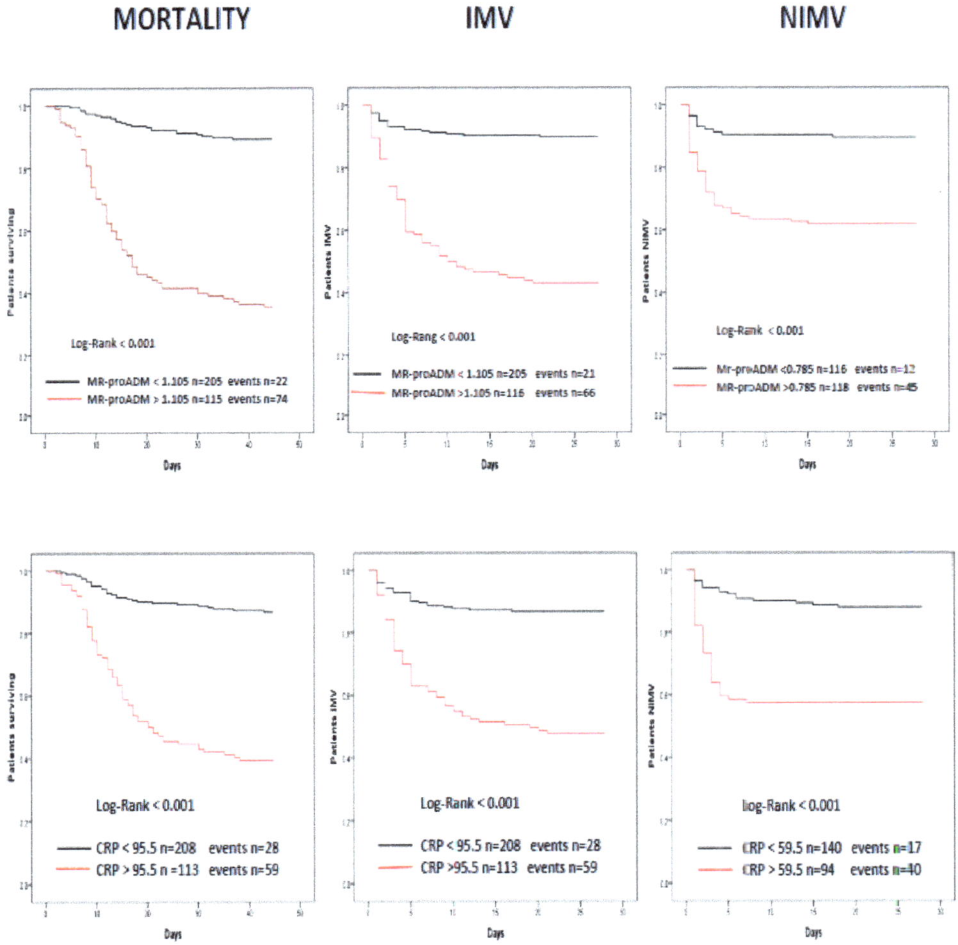

Figure 2. Kaplan–Meier survival curves. Stratification of patients with mid-regional proadrenomedullin (MR-proADM) levels greater or less than 1.105 nmol/L and C-reactive protein (CRP) levels greater or less than 95.5 mg/L at admission in the Emergency Department. IMV, Invasive Mechanical Ventilation; NIMV, Non-Invasive Mechanical Ventilation.

Similar results have been found for CRP (Figure 2), but with less differences when compared to MR-proADM, as shown by the narrower forks. Conversely, the performance was lower for PCT, D-dimer and LDH, as shown in Figure 3, where is evident a poor discrimination power, as shown by the narrower forks.

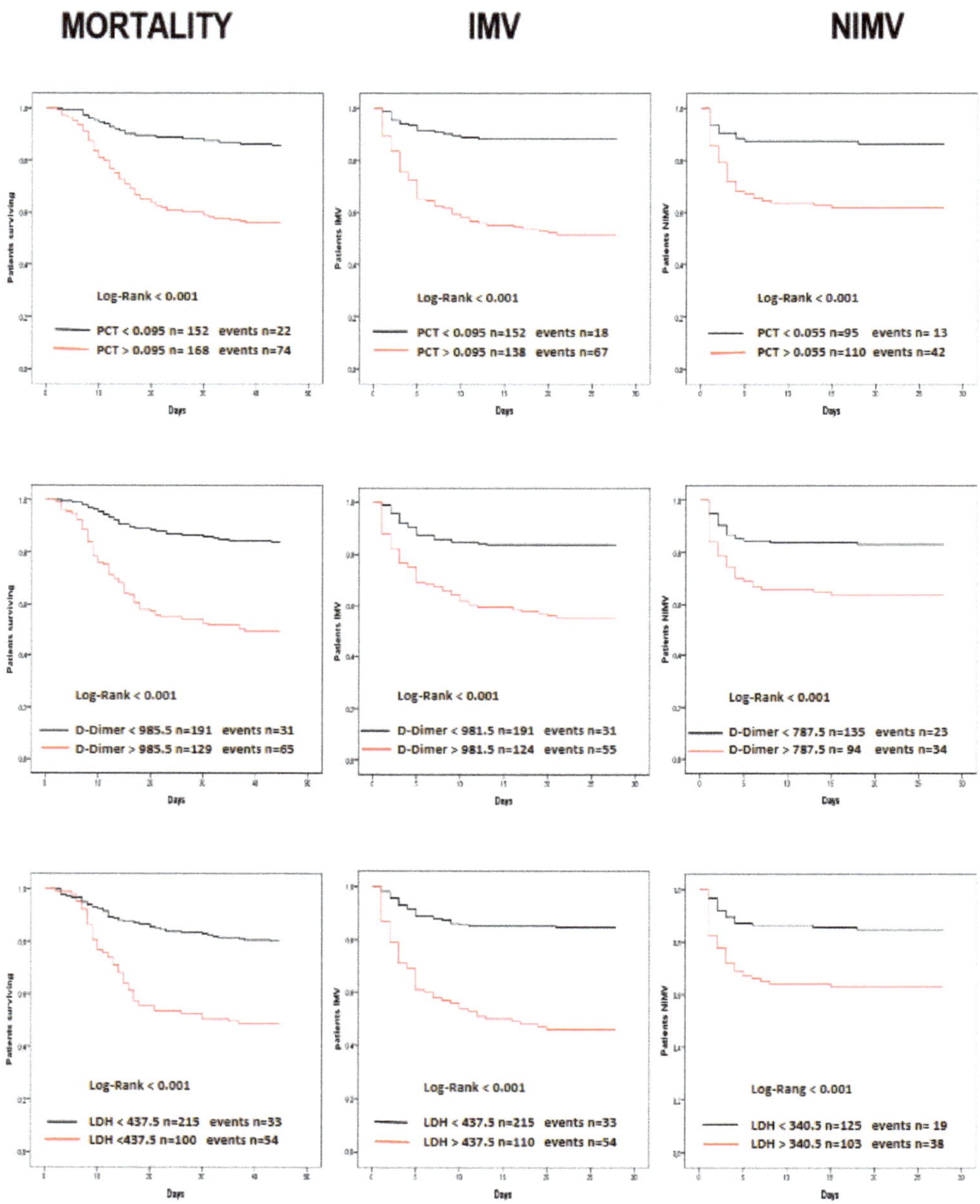

Figure 3. Kaplan–Meier survival curves. Risk stratification of patients with procalcitonin (PCT) levels greater or less than 0.095 ng/mL, D-dimer levels greater or less than 985.5 ng/mL and lactate dehydrogenase (LDH) greater or less than 439.5 U/L at admission in the Emergency Department. IMV, invasive mechanical ventilation; NIMV, non-invasive mechanical ventilation.

4. Discussion

The severe acute respiratory syndrome coronavirus 2 (SARS-CoV-2) infection has heavily affected the worldwide population in the last two years. Although most patients infected by SARS-CoV-2 had only a mild illness, about 5% of them suffered severe lung injury or even multiorgan dysfunction [22], requiring admission in the intensive care unit (ICU).

Consequently, Emergency Departments have seen a dramatic increase in their workload, triggering the need to optimize resources and the decision to hospitalize only seriously ill patients, to face the more adequate care level.

The utilization of biomarkers at the admission to the ED to quickly stratify risks for patients with pneumonia and other diseases has been largely reported [12,23,24]. MR-proADM has already been shown to be effective to stratify the risk in the Emergency Department for patients affected by community-acquired pneumonia (CAP) [11,25].

To our knowledge, this is the first study focused on the ability of new, as MR-proADM, and traditional biomarkers in the global risk stratification of patients with COVID-19 infections at the ED admission.

Previous studies performed with a smaller number of patients have reported that MR-pro ADM can play a role in predicting outcome in already hospitalized patients affected by COVID-19-related pneumonia [19,20].

Accordingly, we have recently shown in ICU critical patients that that MR-proADM seems to represent the most powerful biomarker for predicting death, especially when the outcome can happen earlier, within one week, thus representing a good predictor for disposition of patients from ED to ICU [26]. In addition, our recent data show that also in patients at the triage entry in the Emergency Department, MR-proADM is able to stratify the risk in terms of mortality [21].

In line with these previous studies, the median admission levels of all biomarkers checked in our studies showed significant higher values for all the endpoints considered, i.e., non-survivors vs survivors, IMV vs non-IMV and NIMV vs non-NIMV. This result suggests that these biomarkers might play a predictive role in the early risk stratification of patients with COVID-19 infections. In fact, all the biomarkers considered showed a significant predictive value for the endpoints considered when analyzed with a univariate analysis.

Considering the possible confounding effect of the demographic and clinical features of patients, a multivariate analysis was performed pooling together both the clinical characteristics and the biomarkers assessed in the study. MR-proADM showed the best predictive value for the primary endpoints and for the need of IMV, whereas it did not show significant predictive role for the need of NIMV.

In particular, it is notable that MR-proADM showed the best negative predictive value for all the endpoints considered, thus giving a relevant support to the emergency physician in the eventual decision of the patient rule-out or rule-in. Consequently, MR-proADM could determine an adequate clinical setting of patients affected by COVID-19 being able to predict also the possibility of ventilation need. This relevant information might be helpful for the emergency physician facilitating the decision-making process, thus optimizing the hospital resources. The great power of MR-proADM in the mortality risk stratification of COVID-19 patients has been further confirmed by the analysis of the ROC curves, which showed a significant greater AUC as compared to the other biomarkers analyzed.

Similarly, the survival curves showed a primary role of MR-proADM as a predictive factor in patients affected by COVID-19 in the Emergency Department. This is demonstrated by the wider fork of MR-proADM as compared to the other biomarkers, highlighting a better discrimination power for all the outcomes considered.

Our novel data are in line with previous studies in which the predictive role of MR-proADM has been evaluated in hospitalized COVID-19 patients [19,20].

Some differences among the present and previous studies need to be highlighted. First, the clinical setting. In fact, in previous studies, only critical patients admitted in the hospital wards have been enrolled. Our study, instead, has been primarily focused on patients admitted to the ED and therefore, with different degrees of impairment, ranging from asymptomatic to critical conditions. In these populations, the goal is to predict the trajectory of the illness at the onset of the symptoms and, very often, this is not easy to do.

The second important aspect of our study was the number of patients enrolled, which was considerably greater as compared to previous studies.

Similar conclusions have been reported by previous studies which have analyzed the predictive role of biomarkers in patients affected by CAP, where MR-proADM was shown to be particularly effective in stratifying the risk in patients affected by bacterial pneumonia [9,11,12]. On the other hand, PCT was particularly effective as a diagnostic tool for tailoring antibiotic therapy but with a minor impact on the predictive power [9,27,28].

Although PCR has been confirmed to be a non-specific marker of acute inflammation as it is usually influenced by many other factors [9,28,29], its predictive power, as confirmed in our study, makes it particularly useful in a context such as the Emergency Department, even for COVID-19 patients.

PCT, in line with previous studies [21,26], does not show a sufficient predictive value because, even if the ROC analysis shows a valid result, the multivariate analysis does not show a significant predictive value. This could be due to the different statistical analysis performed. In fact, in previous studies [21,26], PCT showed sufficient discriminatory power for patients who died with COVID-19 after 28 days, presumably due to complications such as bacterial superinfections. In contrast, MR-proADM was particularly effective in predicting death in patients who died rapidly, within a week, because of COVID-19.

Our novel data strengthen the role of biomarkers as a useful tool in the early risk stratification of patients presenting to the ED, especially in the age of COVID-19. We can only speculate that a diagnostic and predictive power could be enhanced by combining score and biomarkers in more complex biomarkers panels [30–32]. For these reasons, it is important to know how biomarkers behave in response to defined diseases such as SARS-CoV-2.

A limitation of our study is that patients were recruited in only one single hospital. Therefore, it would be desirable to extend the study to multiple centers to increase the number of enrolled patients, to confirm our results. A further limitation is represented by the retrospective design of the study.

5. Conclusions

This study, which to our knowledge is the first to evaluate the behavior of the MR-proADM compared to traditional biomarkers in COVID-19 patients at admission to the ED, shows that all the biomarkers utilized can help the physician in the decision-making process. Among them, MR-proADM and CRP seem to represent the most powerful biomarkers for predicting mortality and the need of ventilation in patients admitted to the Emergency Department. Therefore, the assessment of the MR-proADM level strengthens the predictive power of CRP. This is particularly useful for helping the emergency physician in the rule-in or rule-out of COVID-19 patients. Furthermore, the present study extends the previous results giving a support to the emergency physician in deciding the adequate clinical setting according to the possible need of ventilation, thus contributing to optimizing hospital resources.

However, biomarkers might oversimplify the interpretation of important variables, and consequently, they must be considered as a valid help but not as replacing clinician judgment and/or the right consideration of validated severity scores.

Author Contributions: Conceptualization, J.M.L. and M.M. (Marilena Minieri); methodology, M.S.L.; validation, M.S.L., A.G., M.P. and A.T.; formal analysis, V.N.D.L. and M.M. (Massimo Maurici); investigation, M.S.L., F.L., D.J.B.; data curation, V.N.D.L. and M.M. (Massimo Maurici); writing—original draft preparation, M.M. (Marilena Minieri); writing—review and editing, J.M.L. and M.M. (Marilena Minieri); visualization, L.C., C.P., S.L. (Susanna Longo), V.M., M.C., C.D., S.L. (Stefania Levantesi) and M.G.; supervision, S.B.; funding acquisition, M.M. (Marilena Minieri). All authors have read and agreed to the published version of the manuscript.

Funding: The APC will be funded by the Company B.R.A.H.M.S. AG ThermoFisher.

Institutional Review Board Statement: The study was conducted in accordance with the Declaration of Helsinki and approved by the Ethics Committee of Tor Vergata University Hospital (protocol code 87/20, date of approval: July 2020).

Informed Consent Statement: Informed consent was waived due to the emergency condition of patients.

Data Availability Statement: Data supporting reported results are stored in databases created from LIS of the Tor Vergata University Hospital.

Acknowledgments: The Authors wish to thank B.R.A.H.M.S, AG ThermoFisher Company for its support with MR-proADM assay reagents, and Loreta D'Amico and Annarita Cococcia for their valuable technical contribution to the study.

Conflicts of Interest: The authors declare no conflict of interest.

References

1. Zhao, D.; Yao, F.; Wang, L.; Zheng, L.; Gao, Y.; Ye, J.; Guo, F.; Zhao, H.; Gao, R. A Comparative Study on the Clinical Features of Coronavirus 2019 (COVID-19) Pneumonia with other Pneumonias. *Clin. Infect. Dis.* **2020**, *71*, 756–761. [CrossRef] [PubMed]
2. Xie, X.; Zhong, Z.; Zhao, W.; Zheng, C.; Wang, F.; Liu, J. Chest CT for Typical Coronavirus Disease 2019 (COVID-19) Pneumonia: Relationship to Negative RT-PCR Testing. *Radiology* **2020**, *296*, E41–E45. [CrossRef]
3. Xie, J.; Tong, Z.; Guan, X.; Du, B.; Qiu, H.; Slutsky, A.S. Critical Care Crisis and Some Recommendations during the COVID-19 Epidemic in China. *Intensive Care Med.* **2020**, *46*, 837–840. [CrossRef]
4. Melbye, H.; Stocks, N. Point of Care Testing for C-Reactive Protein—A New Path for Australian GPs? *Aust. Fam. Physician* **2006**, *35*, 513–517. [PubMed]
5. Jin, M.; Khan, A.I. Procalcitonin: Uses in the Clinical Laboratory for the Diagnosis of Sepsis. *Lab. Med.* **2010**, *41*, 173–177. [CrossRef]
6. Hedlund, J.; Hansson, L.-O. Procalcitonin and C-Reactive Protein Levels in Community-Acquired Pneumonia: Correlation with Etiology and Prognosis. *Infection* **2000**, *28*, 68–73. [CrossRef] [PubMed]
7. Masiá, M.; Gutiérrez, F.; Shum, C.; Padilla, S.; Navarro, J.C.; Flores, E.; Hernández, I. Usefulness of Procalcitonin Levels in Community-Acquired Pneumonia According to the Patients Outcome Research Team Pneumonia Severity Index. *Chest* **2005**, *128*, 2223–2229. [CrossRef] [PubMed]
8. Krüger, S.; Ewig, S.; Papassotiriou, J.; Kunde, J.; Marre, R.; von Baum, H.; Suttor, N.; Welte, T. Inflammatory Parameters Predict Etiologic Patterns but Do Not Allow for Individual Prediction of Etiology in Patients with CAP—Results from the German Competence Network CAPNETZ. The CAPNETZ study group. *Respir. Res.* **2009**, *10*, 65. [CrossRef] [PubMed]
9. Christ-Crain, M.; Morgenthaler, N.G.; Stolz, D.; Müller, C.; Bingisser, R.; Harbarth, S.; Tamm, M.; Struck, J.; Bergmann, A.; Müller, B. Pro-Adrenomedullin to Predict Severity and Outcome in Community-Acquired Pneumonia [ISRCTN04176397]. *Crit. Care* **2006**, *10*, R96. [CrossRef] [PubMed]
10. Sanchez, F.V.; Mendez, B.V.; Gutierrez, J.R.; Austria, R.B.d; Quiñones, J.R.; Martínez, L.P.; Alemán, I.V.; García, A.E. Initial Levels of Mr-Proadrenomedullin: A Predictor of Severity in Patients with Influenza a Virus Pneumonia. *ICMx J.* **2015**, *3*, A832. [CrossRef]
11. Legramante, J.M.; Mastropasqua, M.; Susi, B.; Porzio, O.; Mazza, M.; Miranda Agrippino, G.; D'Agostini, C.; Brandi, A.; Giovagnoli, G.; Di Lecce, V.N.; et al. Prognostic Performance of MR-pro-Adrenomedullin in Patients with Community Acquired Pneumonia in the Emergency Department Compared to Clinical Severity Scores PSI and CURB. *PLoS ONE* **2017**, *12*, e0187702. [CrossRef] [PubMed]
12. Bello, S.; Lasierra, A.B.; Mincholé, E.; Fandos, S.; Ruiz, M.A.; Vera, E.; de Pablo, F.; Ferrer, M.; Menendez, R.; Torres, A. Prognostic Power of Proadrenomedullin in Community-Acquired Pneumonia Is Independent of Aetiology. *Eur. Respir. J.* **2012**, *39*, 1144–1155. [CrossRef] [PubMed]
13. Temmesfeld-Wollbrück, B.; Hocke, A.; Suttorp, N.; Hippenstiel, S. Adrenomedullin and Endothelial Barrier Function. *Thromb. Haemost.* **2007**, *98*, 944–951. [CrossRef]
14. Hupf, J.; Mustroph, J.; Hanses, F.; Evert, K.; Maier, L.S.; Jungbauer, C.G. RNA-Expression of Adrenomedullin Is Increased in Patients with Severe COVID-19. *Crit. Care* **2020**, *24*, 527. [CrossRef]
15. Li, H.; Liu, L.; Zhang, D.; Xu, J.; Dai, H.; Tang, N.; Su, X.; Cao, B. SARS-CoV-2 and Viral Sepsis: Observations and Hypotheses. *Lancet* **2020**, *395*, 1517–1520. [CrossRef]
16. Huang, C.; Wang, Y.; Li, X.; Ren, L.; Zhao, J.; Hu, Y.; Zhang, L.; Fan, G.; Xu, J.; Gu, X.; et al. Clinical Features of Patients Infected with 2019 Novel Coronavirus in Wuhan, China. *Lancet* **2020**, *395*, 497–506. [CrossRef]
17. Liu, J.; Li, S.; Liu, J.; Liang, B.; Wang, X.; Wang, H.; Li, W.; Tong, Q.; Yi, J.; Zhao, L.; et al. Longitudinal Characteristics of Lymphocyte Responses and Cytokine Profiles in the Peripheral Blood of SARS-CoV-2 Infected Patients. *EBioMedicine* **2020**, *55*, 102763. [CrossRef]
18. Iwasaki, A.; Pillai, P.S. Innate Immunity to Influenza Virus Infection. *Nat. Rev. Immunol.* **2014**, *14*, 315–328. [CrossRef]

19. Gregoriano, C.; Koch, D.; Kutz, A.; Haubitz, S.; Conen, A.; Bernasconi, L.; Hammerer-Lercher, A.; Saeed, K.; Mueller, B.; Schuetz, P. The Vasoactive Peptide MR-pro-Adrenomedullin in COVID-19 Patients: An Observational Study. *Clin. Chem. Lab. Med.* **2021**, *59*, 995–1004. [CrossRef]
20. Spoto, S.; Agrò, F.E.; Sambuco, F.; Travaglino, F.; Valeriani, E.; Fogolari, M.; Mangiacapra, F.; Costantino, S.; Ciccozzi, M.; Angeletti, S. High Value of Mid-regional Proadrenomedullin in COVID-19: A Marker of Widespread Endothelial Damage, Disease Severity, and Mortality. *J. Med. Virol.* **2021**, *93*, 2820–2827. [CrossRef] [PubMed]
21. Minieri, M.; Di Lecce, V.N.; Lia, M.S.; Maurici, M.; Bernardini, S.; Legramante, J.M. Role of MR-ProADM in the Risk Stratification of COVID-19 Patients Assessed at the Triage of the Emergency Department. *Crit. Care* **2021**, *25*, 407. [CrossRef]
22. Guan, W.; Ni, Z.; Hu, Y.; Liang, W.; Ou, C.; He, J.; Liu, L.; Shan, H.; Lei, C.; Hui, D.S.C.; et al. Clinical Characteristics of Coronavirus Disease 2019 in China. *N. Engl. J. Med.* **2020**, *382*, 1708–1720. [CrossRef]
23. Saeed, K.; Wilson, D.C.; Bloos, F.; Schuetz, P.; van der Does, Y.; Melander, O.; Hausfater, P.; Legramante, J.M.; Claessens, Y.-E.; Amin, D.; et al. The Early Identification of Disease Progression in Patients with Suspected Infection Presenting to the Emergency Department: A Multi-Centre Derivation and Validation Study. *Crit. Care* **2019**, *23*, 40. [CrossRef] [PubMed]
24. Cardellini, M.; Rizza, S.; Casagrande, V.; Cardolini, I.; Ballanti, M.; Davato, F.; Porzio, O.; Canale, M.P.; Legramante, J.M.; Mavilio, M.; et al. Soluble ST2 Is a Biomarker for Cardiovascular Mortality Related to Abnormal Glucose Metabolism in High-Risk Subjects. *Acta Diabetol.* **2019**, *56*, 273–280. [CrossRef] [PubMed]
25. Saeed, K.; Legramante, J.M.; Angeletti, S.; Curcio, F.; Miguens, I.; Poole, S.; Tascini, C.; Sozio, E.; Del Castillo, J.G. Mid-Regional pro-Adrenomedullin as a Supplementary Tool to Clinical Parameters in Cases of Suspicion of Infection in the Emergency Department. *Expert Rev. Mol. Diagn.* **2021**, *21*, 397–404. [CrossRef]
26. Leonardis, F.; Minieri, M.; Lia, M.S.; Formica, V.; Dauri, M.; Colella, D.F.; Natoli, S.; Paganelli, C.; Terrinoni, A.; Sarmati, L.; et al. Early Predictive Value of MR-proADM in Critically Ill Patients with Covid-19: An Observational Study in the Emergency Department. *J. Emerg. Med. Care* **2021**, *4*, 102.
27. Christ-Crain, M.; Jaccard-Stolz, D.; Bingisser, R.; Gencay, M.M.; Huber, P.R.; Tamm, M.; Müller, B. Effect of Procalcitonin-Guided Treatment on Antibiotic Use and Outcome in Lower Respiratory Tract Infections: Cluster-Randomised, Single-Blinded Intervention Trial. *Lancet* **2004**, *363*, 600–607. [CrossRef]
28. España, P.P.; Capelastegui, A.; Mar, C.; Bilbao, A.; Quintana, J.M.; Diez, R.; Esteban, C.; Bereciartua, E.; Unanue, U.; Uranga, A. Performance of Pro-Adrenomedullin for Identifying Adverse Outcomes in Community-Acquired Pneumonia. *J. Infect.* **2015**, *70*, 457–466. [CrossRef] [PubMed]
29. Luna, C.M. C-Reactive Protein in Pneumonia. *Chest* **2004**, *125*, 1192–1195. [CrossRef]
30. Asai, N.; Watanabe, H.; Shiota, A.; Kato, H.; Sakanashi, D.; Hagihara, M.; Koizumi, Y.; Yamagishi, Y.; Suematsu, H.; Mikamo, H. Efficacy and Accuracy of QSOFA and SOFA Scores as Prognostic Tools for Community-Acquired and Healthcare-Associated Pneumonia. *Int. J. Infect. Dis.* **2019**, *84*, 89–96. [CrossRef]
31. Spoto, S.; Legramante, J.M.; Minieri, M.; Fogolari, M.; Terrinoni, A.; Valeriani, E.; Sebastiano, C.; Bernardini, S.; Ciccozzi, M.; Angeletti, P.S. How Biomarkers Can Improve Pneumonia Diagnosis and Prognosis: Procalcitonin and Mid-Regional-pro-Adrenomedullin. *Biomark. Med.* **2020**, *14*, 549–562. [CrossRef]
32. Formica, V.; Minieri, M.; Bernardini, S.; Ciotti, M.; D'Agostini, C.; Roselli, M.; Andreoni, M.; Morelli, C.; Parisi, G.; Federici, M.; et al. Complete Blood Count Might Help to Identify Subjects with High Probability of Testing Positive to SARS-CoV-2. *Clin. Med.* **2020**, *20*, e114–e119. [CrossRef]

diagnostics

Article

Application of Procalcitonin for the Rapid Diagnosis of *Clostridioides difficile* Infection in Patients with Inflammatory Bowel Disease

Shuhua Xie †, Peisong Chen †, Dong Wang, Xiaobing Jiang, Zhongwen Wu, Kang Liao, Min Liu, Shihong Zhang * and Yili Chen *

Department of Laboratory Medicine, The First Affiliated Hospital, Sun Yat-sen University, Guangzhou 510080, China

* Correspondence: zhshih@mail.sysu.edu.cn (S.Z.); chenyli3@mail.sysu.edu.cn (Y.C.); Tel.: +86-020-87332200-8461 (S.Z.); Fax: +86-020-87332200 (S.Z.)

† These authors contributed equally to this work.

Abstract: Background: The incidence of *Clostridioides difficile* infection (CDI) has increased in recent years in patients with inflammatory bowel disease (IBD). *C. difficile* is a toxin-producing bacterium, and CDI results in the worsening of underlying IBD, increasing the risk of IBD treatment failure, surgery, and hospitalization. Because the symptoms of CDI overlap with those of IBD, it is challenging to make a differential diagnosis. Therefore, early, rapid, and reliable diagnostic tools that can identify CDI in IBD patients would be valuable to clinicians. **Methods**: This study retrospectively collected 135 patients with IBD. Among them, 44 patients were diagnosed with CDI, and 42 patients were diagnosed with viral or fungal infections. A total of 49 patients without infections were defined as the control group. The diagnostic values of procalcitonin (PCT), C-reactive protein (CRP), and white blood cell (WBC) count in the peripheral blood were examined. **Results**: In this study, PCT levels were significantly higher in patients with CDI than in non-CDI patients (including patients with viral/fungal infections and the control group; $p < 0.001$ and $p < 0.05$, respectively). CRP levels were significantly higher in patients with CDI than in non-CDI patients ($p < 0.05$). The area under the curve (AUC) of PCT and WBC count were compared using DeLong's test: the AUCs of PCT vs. CRP for the detection of the IBD–CDI group and the control group was 0.826 [95% confidence interval (CI) 0.743–0.909] vs. 0.663 [95% confidence interval (CI) 0.551–0.774] ($p < 0.05$), respectively. WBC count was inferior as a diagnostic tool for CDI. The sensitivity was 59.09% (95% CI: 43.2% to 73.7%), the specificity was 89.80% (95% CI: 77.8% to 96.6%), and the positive likelihood ratio LR (+) was 5.79 for PCT for the diagnosis of CDI. **Conclusions**: The present study demonstrates the superiority of PCT over CRP and WBC count for the rapid diagnosis of CDI in IBD patients.

Keywords: *Clostridioides difficile*; inflammatory bowel disease; PCT; CRP; WBC

Citation: Xie, S.; Chen, P.; Wang, D.; Jiang, X.; Wu, Z.; Liao, K.; Liu, M.; Zhang, S.; Chen, Y. Application of Procalcitonin for the Rapid Diagnosis of *Clostridioides difficile* Infection in Patients with Inflammatory Bowel Disease. *Diagnostics* **2022**, *12*, 3108. https://doi.org/10.3390/diagnostics12123108

Academic Editor: Michael Nagler

Received: 18 October 2022
Accepted: 6 December 2022
Published: 9 December 2022

1. Background

Inflammatory bowel disease (IBD), including Crohn's disease (CD) and ulcerative colitis (UC), is a group of chronic disorders of the gastrointestinal tract that is characterized by inflammation of an unknown etiology, and the recurrence rate is high. The clinical characteristics are diverse and include serious complications, such as bleeding, perforation, and abscess formation [1,2].

The resident microorganisms in the human gastrointestinal tract play an important role in maintaining intestinal equilibrium. There are several mechanisms, including the degradation of xenobiotic substances, synthesis of beneficial metabolites and vitamins, immune system regulation, and colonization resistance against invading pathogenic microorganisms [3]. Chronic diarrhea caused by IBD leads to dysbacteriosis, decreasing bacterial diversity in the long term and reducing colonization resistance against *Clostridioides difficile* [4]. Changes in the community structure of the gut microbiota and markedly decreased microbial diversity have been identified in patients with CDI [5]. The long-term use of immunosuppressive medications for the treatment of IBD, such as immunomodulators and glucocorticoids, particularly when these agents are used in combination, result in an increased risk of opportunistic infections. Individuals with IBD have a nearly five-fold-higher risk of *C. difficile* infection than individuals without IBD [6]. In contrast, infections with bacteria other than *C. difficile* are uncommon in IBD patients experiencing a disease flare [7].

C. difficile, a Gram-positive spore-forming anaerobe, produces exotoxins that cause a spectrum of diseases. Serious infections are occasionally complicated by toxic megacolon, perforation, sepsis, and death [8]. It is reported that CDI complicates the course of IBD, leading to longer hospital stays, an increased rate of colectomy, and increased mortality [9]. CDI and IBD have an obvious overlap with the clinical manifestations of colitis. Diarrhea is the most prominent symptom, but bleeding is more likely in IBD; other common symptoms include abdominal discomfort and fever. Therefore, it is difficult to clinically distinguish an acute IBD attack and acute CDI. In all IBD patients, these symptoms or signs may indicate a colitis attack, such as increased blood in the stool. Accordingly, these patients should also be tested for the presence of toxigenic *C. difficile* in the stool. This presents challenges in diagnosis and treatment.

Based on the above evidence, the 2013 ACG guidelines made the following conditional recommendation: in IBD patients with severe colitis, the simultaneous initiation of empirical therapy directed against both CDI and IBD attacks may be needed while awaiting the results of *C. difficile* testing. The escalation of immunosuppression medications should be avoided if there is a possibility of untreated CDI [10]. Therefore, a fast and reliable CDI diagnostic tool is valuable for clinicians.

Nucleic acid amplification tests (NAATs) have a sensitivity greater than 95% for the diagnosis of CDI [11,12]. The positive predictive value is dependent on the presence of diarrhea [13,14]. *C. difficile* can colonize without producing significant amounts of toxin, resulting in a symptomatic infection and making a positive PCR result clinically irrelevant [15,16]. An ideal test to diagnose CDI in a situation in which symptoms are not reliable would be a highly sensitive assay to measure the level of the toxin [17]. Unfortunately, most enzyme immunoassays (EIAs) for toxins have low sensitivities, sometimes even in the 50% to 70% range. Glutamate dehydrogenase (GDH) is produced by all *C. difficile* strains, regardless of whether they are toxigenic, and it cannot be used to identify toxigenic strains of *C. difficile* [18]. Currently, using PCR/NAATs alone leads to overdiagnoses, and the use of EIAs for toxins alone may lead to underdiagnoses [19]. Therefore, reliable, specific, and sensitive markers are needed for CDI in IBD patients.

At present, routine inflammation markers, including procalcitonin levels, C-reactive protein levels, and white blood cell count, are widely used for the diagnosis of infectious diseases. C-reactive protein (CRP) is one of the most important proteins involved in acute inflammation. CRP has a short half-life, so CRP levels increase and decrease rapidly during acute inflammation [20]. The detection of CRP is inexpensive and rapid. Procalcitonin (PCT) is a prohormone of calcitonin, and serum levels of PCT are mainly increased in patients with bacterial infections. PCT has been found to reach a high level within 8–24 h after the onset of a bacterial infection [21]. As an inflammatory marker, PCT has obvious advantages because it is stable, not affected by exogenous bacteria in vitro, and the test is relatively easy to carry out with a fast turnaround time of 2 h [5].

This is a retrospective study. We aimed to determine the efficacy of noninvasive and rapid routine inflammation marker tests in monitoring IBD patients infected with *C. difficile.* Our research aimed to provide auxiliary evidence of CDI infection in IBD patients and carry out antibiotic treatment as early as possible to improve prognosis. Thus, in the present study, we evaluated the diagnostic value of PCT, CRP, and WBC count for CDI in IBD patients.

2. Methods

2.1. Patient Eligibility and Classification

In this study, a total of 445 patients diagnosed with inflammatory bowel disease (IBD) were collected from the medical records of the First Affiliated Hospital of Sun Yat-sen University register from 2016 to 2020. *C. difficile* should be detected in IBD patients with a high risk of CDI, including the following: (1) all active IBD inpatients; (2) IBD patients with diarrhea in remission or recent exposure to risk factors (such as contact with CDI patients, gastrointestinal surgery, tube feeding, bowel preparation, etc.); (3) IBD patients with severe colitis, without bacteriological evidence, and requiring empirical CDI treatment; and (4) patients with low immunity, diabetes, renal failure, malnutrition, etc. An endoscopy was performed in all individuals. IBD lacks the gold standard for diagnosis and needs to be comprehensively analyzed in combination with clinical manifestations, laboratory examinations, endoscopies, imaging examinations, and histopathological examinations. The Crohn's disease activity index (CDAI) is used to evaluate the severity of disease activity. The improved 'Truelove' and 'Witts' disease severity classification criteria were used to grade the active stage of UC [22]. Patients diagnosed with inflammatory bowel disease (IBD) participated in the study, and 135 patients were included. A total of 300 subjects were excluded from the analysis due to incomplete blood tests (n = 270), the presence of other positive bacteria cultures in the stool (n = 10), and other sites of bacterial infection (n = 20). Among all of the included 135 patients, 44 patients were diagnosed with CDI based on NAATs for *Clostridioides difficile*. Forty-two patients were diagnosed with other opportunistic infections, including viral infections and fungal infections. Viral infection diagnosis was based on positive viral IgM antibodies and positive viral PCR results in blood samples, including Epstein–Barr virus (EBV), cytomegalovirus (CMV), herpes simplex virus type 1 (HSV-1), and herpes simplex virus type 2 (HSV-2); meanwhile, fungal infection diagnosis was based on positive stool culture results, mostly of Candida infection. Forty-nine IBD patients without infections were enrolled in the control group (Figure 1). All methods were carried out in accordance with the declaration of Helsinki. Written informed consent was obtained from all subjects. This study was approved by the Clinical Research and Ethics Committee of the First Affiliated Hospital of Sun Yat-sen University.

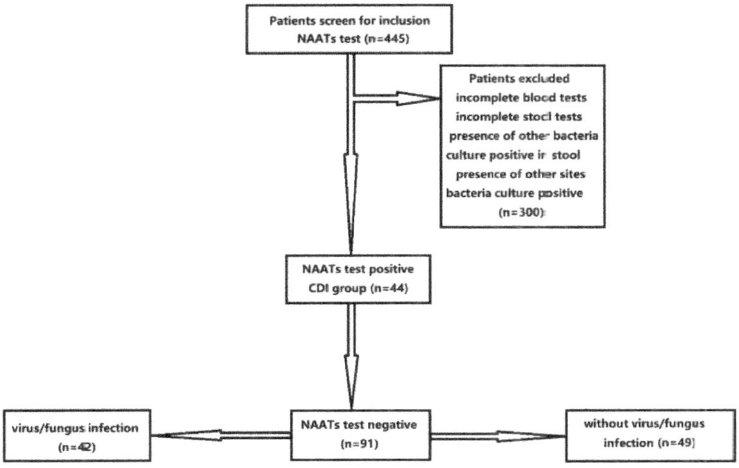

Figure 1. Study algorithm. IBD: inflammatory bowel disease; CDI: Clostridioides difficile infection.

A six-component ATLAS score, which is useful in stratifying the severity of CDI patients [23], was measured on the day of entry into the study (within 48 h of a positive NAAT for *Clostridioides difficile*). The components of ATLAS score were as follows: age (in years); treatment with systemic antibiotics (which occurred on one or more days of

CDI therapy); temperature (in degrees Celsius); total leukocyte count; serum albumin; and serum creatinine as a measure of renal function.

2.2. Detection of Clostridioides Difficile

CDI was diagnosed based on positive nucleic acid amplification tests (NAATs) for the *Clostridioides difficile* tcdA, tcdB, cdtA, and cdtB genes (GeneXpert C. difficile Assay, Cepheid, Sunnyvale, CA, USA).

2.3. Laboratory Measurement of Inflammatory Markers

Serum PCT levels were determined by a quantitative electrochemiluminescence immunometric assay using a COBAS E601 analyzer (e601 module of the Cobas 6000 system, Roche, Switzerland) with a level of <0.05 ng/mL considered normal. Plasma CRP levels were measured by immunoturbidimetry using a Mindray CRP-M100 analyzer (Mindray CRP-M100, Shanghai, China) a Dade Behring BN II (Marburg, Germany). The reference value provided by the manufacturer was < 10 mg/L. Whole WBC levels were measured by a Mindray BC-6800 analyzer (Mindray BC-6800, Shenzhen, China) and SysmexXN-9000 analyzer (SysmexXN-9000, Kobe, Japan) with a level of $<10 \times 10^9$/L considered normal.

2.4. Statistical Analysis

Statistical analyses and graphic presentations were carried out using the software program GraphPad Prism (Version 5.0, GraphPad Software Inc., San Diego, CA, USA), MedCalc (Version 20.1116, Ostend, Belgium) and IBM SPSS software (Version 25.0, SPSS China). Patient characteristics among the groups were analyzed using Pearson's chi-square test, one-way ANOVA, and Tukey's multiple comparison test. Categorical variables were expressed in proportions. Infection marker variables were expressed as the mean ± SEM. DeLong's test was used for the pairwise comparison of AUC and ROC curves. AUCs (with 95% confidence intervals) were calculated to assess the diagnostic values of the tests; AUCs > 0.70 were considered clinically relevant. Youden's index, sensitivity, specificity, and the LR (+) were calculated with MedCalc. In all analyses, statistical tests were two-sided, and p values of less than 0.05 were considered statistically significant.

3. Results

3.1. Patient Characteristics

Of these 135 IBD patients, 83 patients were diagnosed with Crohn's disease and 52 patients were diagnosed with ulcerative colitis. Age and *Clostridioides difficile* infection incidences were different in CD and UC patients ($p < 0.001$ and $p < 0.05$, respectively), while sex and colectomy rate were similar in CD and UC patients ($p > 0.05$). Crohn's disease patients had a younger median age but a higher CDI incidence than ulcerative colitis patients. The clinical characteristics of CD and UC patients are shown in Table 1.

Table 1. Clinical characteristics of CD and UC patients.

Characteristic	Crohn's Disease (n = 83)	Ulcerative Colitis (n = 52)	p-Value
Age, y, median (IQR)	27.0 (15–61)	49.5 (18–70)	<0.0001 ***
Male, n (%)	50 (60.2)	28 (53.8)	0.391
CDI incidence, n (%)	32 (38.6)	12 (23.1)	0.014 *
Colectomy, n (%)	18 (21.7)	6 (11.5)	0.056

* $p < 0.05$; *** $p < 0.001$.

3.2. Evaluation of Diagnostic Value

According to the clinical infection of *Clostridioides difficile*, virus and fungus, the IBD patients were divided into three groups. The sex ratio, clinical characteristics, and outcomes are shown in Table 2. Sex was different in the distribution of the three groups ($p < 0.005$).

There was no difference in endoscopic activity and colectomy among the three groups ($p > 0.05$).

Table 2. IBD patients' characteristics in different infection groups.

Chi-Square Test	Male, n (%) *p*-Value	Colectomy, n (%) *p*-Value	Active Period, n (%) *p*-Value
IBD+CDI vs. IBD+ virus/fungus	0.031 *	0.675	0.675
IBD+CDI vs. IBD	0.127	0.131	0.520
IBD+ virus/fungus vs. IBD	0.476	0.359	0.290

* $p < 0.05$.

The level of inflammatory biomarkers in the different groups in the study are shown in Tables 3 and 4 and Figure 2. The mean of PCT was significantly higher in IBD–CDI patients than in non-CDI patients, as well as in IBD patients with viral or fungal infections (both $p < 0.001$). The mean of CRP was significantly higher in IBD–CDI patients than in IBD patients ($p < 0.05$), and the mean of WBC count was not significantly different among the three groups ($p > 0.05$).

Table 3. Patients' inflammatory biomarker levels of IBD patients in different groups.

Characteristic	IBD with CDI (n = 44)	IBD with Viral/Fungus Infections (n = 42)	IBD Control Group (n = 49)	One-Way ANOVA *p*-Value
PCT (ng/mL) Mean ± SEM	0.178 ± 0.029	0.081 ± 0.007	0.066 ± 0.007	<0.0001 ***
CRP (mg/L) Mean ± SEM	59.70 ± 7.01	40.14 ± 4.91	37.35 ± 5.71	0.0177 *
WBC count ($\times 10^9$/L) Mean ± SEM	8.48 ± 0.57	8.93 ± 0.71	7.38 ± 0.39	0.130

* $p < 0.05$; *** $p < 0.001$.

Table 4. Comparison of inflammatory biomarkers between groups of IBD patients.

Tukey's Multiple Comparison Test	PCT *p*-Value	CRP *p*-Value	WBC *p*-Value
IBD+CDI vs. IBD+ virus/fungus	0.821	0.066	0.840
IBD+CDI vs. IBD	<0.001 *	0.022 *	0.345
IBD+ virus/fungus vs. IBD	<0.001 *	0.942	0.126

* $p < 0.05$.

The receiver operating characteristic (ROC) curves corresponding to each infection biomarker in the IBD–CDI group and the IBD control group for the prediction of CDI in IBD patients are shown in Figure 3. Based on the AUC of the ROC curve, PCT had an AUC of 0.826 (95% CI: 0.743–0.909, $p = 0.000$). CRP had an AUC of 0.663 (95% CI: 0.551–0.774, $p = 0.007$). PCT was more accurate in the diagnosis of CDI or being noninfectious than CRP (DeLong's test, $p = 0.0106$).

The receiver operating characteristic (ROC) curves corresponding to each infection biomarker in the IBD–CDI group and the viral/fungus group for the prediction of CDI infections in IBD patients are shown in Figure 4. Based on the AUC of the ROC curve, PCT had an AUC of 0.716 (95% CI: 0.609–0.824, $p = 0.001$). CRP had an AUC of 0.615 (95% CI 0.496–0.735, $p = 0.067$). PCT was more accurate in the diagnosis of CDI or viral/fungus infections than CRP (DeLong's test, $p = 0.170$).

Youden's index, sensitivity, specificity, and the LR (+) were calculated with MedCalc software. Youden's index associated criterion value >0.09, the sensitivity%, specificity%, and positive likelihood ratio (LR+) predictive values of the best PCT cutoff value for the diagnosis of *Clostridioides difficile* infection are shown in Table 5. The sensitivity% was

59.09% (95% CI: 43.2% to 73.7%), the specificity% was 89.80% (95% CI: 77.8% to 96.6%), and the LR (+) was 5.79.

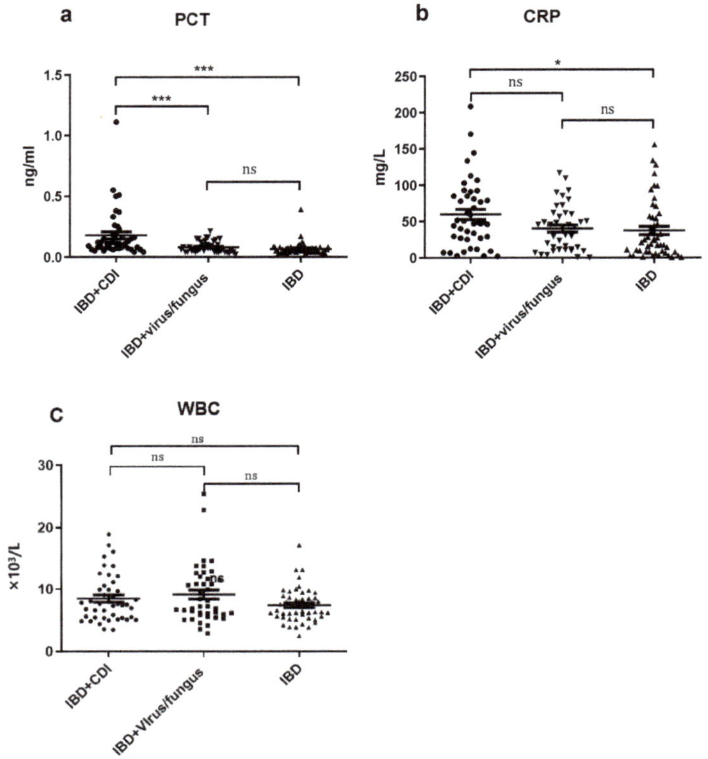

Figure 2. Scatter plots of PCT (**a**), CRP (**b**), and WBC count (**c**) for patients with inflammatory bowel disease with or without *Clostridioides difficile* infection or with viral/fungal infections. (**a**,**b**) In cases of PCT, the mean was obviously higher in IBD–CDI patients than in patients without CDI or with viral/fungal infections. In cases of CRP, the mean was obviously higher in IBD–CDI patients than in patients without CDI infections. (**c**) There was no significant difference in WBC count among patients with or without *Clostridioides difficile* infection or with viral/fungal infections. * $p < 0.05$, *** $p < 0.001$, ns: no significance.

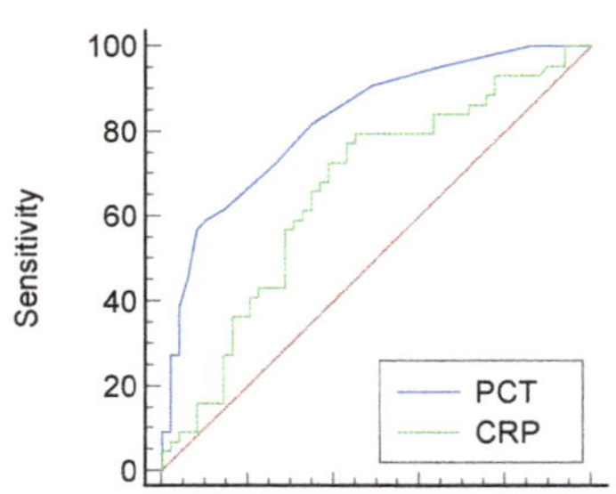

Figure 3. Prediction of *Clostridioides difficile* infections by ROC curve analysis (CDI vs. IBD control group). PCT with an AUC of 0.826 (95% CI: 0.743–0.909, $p = 0.000$). CRP with an AUC of 0.663 (95% CI: 0.551–0.774, $p = 0.007$). DeLong's test, $p = 0.0106$.

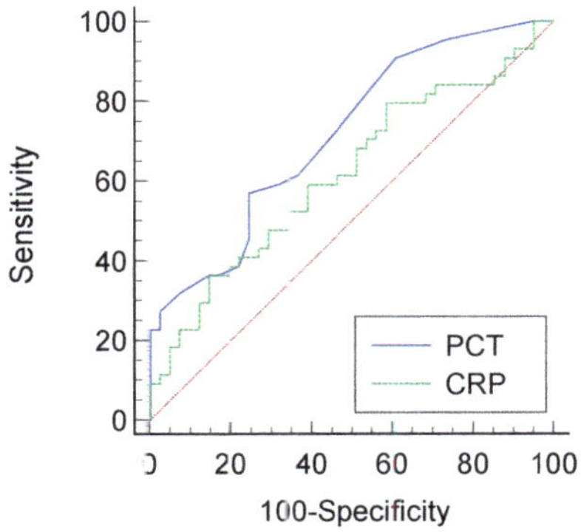

Figure 4. Prediction of *Clostridioides difficile* infections by ROC curve analysis (CDI vs. viral/fungus group). PCT with an AUC of 0.716 (95% CI: 0.609–0.824, p = 0.001). CRP with an AUC of 0.615 (95% CI: 0.496–0.735, p = 0.067). DeLong's test, p = 0.170.

Table 5. Diagnostic performance indices of PCT in *Clostridioides difficile* infection of IBD.

Sensitivity%	95%CI	Specificity%	95%CI	LR−
59.09%	43.2–73.7%	89.80%	77.8–96.6%	5.79

The ATLAS scoring system was used to stratify the severity of CDI patients in this study, which divided CDI patients into seven groups (scores of 0 to 6). PCT and the rate of colectomy were calculated for each stratification, which are shown in Table 6.

Table 6. ATLAS score, PCT, and rate of colectomy of *Clostridioides difficile* infection patients.

ATLAS Score	Number of Patients	PCT Mean ± SEM (ng/mL)	Rate of Colectomy n (%)
0	11	0.1255 ± 0.0285	2 (16.7%)
1	17	0.1888 ± 0.0630	3 (17.6%)
2	6	0.1783 ± 0.0715	1 (16.7%)
3	6	0.2317 ± 0.0766	2 (33.3%)
4	2	0.1000 ± 0.0600	1 (50%)
5	1	0.3800	0
6	1	0.1800	1 (100%)

4. Discussion

IBD patients are at increased risk of developing CDI with worse clinical consequences than the rest of the population. European articles have reported that IBD patients with CDI have an elevated 1.2- to 3-fold risk of gastrointestinal surgery or emergent colectomy [22]. IBD patients are at high risk of CDI recurrence [23]. Crohn's disease patients who were younger at diagnosis are much more likely to obtain CDI than ulcerative colitis patients. The striking part of the epidemiology of CDI is an increase in individuals who are younger [6,24].

A positive stool test for *C. difficile* or its toxins does not absolutely diagnose CDI. The sensitivity of stool culture is limited; it is time-consuming and does not distinguish infection from colonization. NAAT detection methods are more sensitive and can quickly obtain results. NAATs are considered superior to other methods, but it is difficult to accurately distinguish the colonization and infection of *C. difficile*, which may lead to excessive diagnoses and treatments of CDI. The two-step method, namely GDH and toxin A/B, is recommended in the bacterial-infection-diagnosis guidance document [25]. In the three-step method, GDH EIAs or NAATs are used for the initial screening, and A/B EIA tests are used when the initial screening is positive. If the A/B EIAs test is negative, cultures or NAATs are used to confirm the identification [26]. However, these tests are time-consuming, multi-step, and costly.

C-reactive protein is one of the most important proteins in acute inflammation. It is maintained at a low level (less than 1 mg/L) in peripheral blood secreted by hepatocytes in healthy people, but it increases when acute inflammation occurs and is induced by interleukin-6 (IL-6), tumor necrosis factor-α (TNF-α), and interleukin-1β (IL-1β). The CRP value sharply increases, even reaching up to three- to four-hundred times its usual value. CRP remains at a level of 10–40 mg/L, indicating virus infection or chronic inflammation [10]. In this study, we determined that IBD patients suffered from chronic inflammation with a mean CRP level of nearly 37 mg/L and IBD patients with viral or fungal CRP levels remained at a mean of nearly 40 mg/L, while IBD patients suffered from *C. difficile* infection. The mean CRP level increased up to nearly 60 mg/L, nearly 1.5 times higher than that of non-CDI patients.

Procalcitonin is a prehormone of calcitonin. When bacterial infections occur, the circulation levels of PCT increase, mainly due to the presence of bacterial endotoxins and exotoxins, as well as inflammatory cytokines, such as TNF, IL-2, and IL-6. When viral infections occur, interferon-gamma (IFN-γ) is released. PCT production is inhibited by IFN-γ [27]. High levels are found in severe bacterial infections and low levels are found in nonspecific inflammatory diseases and viral infections; therefore, PCT has been considered a promising marker for the diagnosis of bacterial infections [28]. PCT may be used to support clinical decisions in different types of infections, marking the beginning or end of antibiotic therapy [29].

In our study on IBD patients, the purpose of monitoring PCT was to stop immunosuppressive drug treatment as soon as possible and relatively reduce the duration of immunosuppressive treatment. In this present study, we found that mean PCT levels in serum increased nearly three-fold in IBD–CDI patients compared to non-CDI patients with IBD, considering PCT requires more intense stimuli to increase compared with CRP [30]. PCT is characterized by a relatively high specificity but low sensitivity in abdominal local infection diseases, which differentiates bacterial infection from other systemic inflammation or viral infections [28]. In this study, PCT showed a low sensitivity but high specificity for CDI diagnosis in IBD patients as well. The PCT values are relatively high in moderate and severe CDI, suggesting that PCT could be more useful in severe CDI diagnoses, and that more careful differential diagnoses are needed in mild cases. However, more cases need to be collected in the future to clarify the correlation between PCT and CDI severity. Briefly, PCT provides a reference for the clinical withdrawal of immunosuppressants and the initiation of antibiotic therapy in IBD patients. As a marker of inflammation, PCT is stable, easy to operate, specific, and fast.

We found that CRP and WBC count were not ideal diagnostic tools for identifying *C. difficile* infections or other viral and fungal infections in IBD patients. Although CRP levels increased in the CDI of IBD patients, based on the pairwise comparison of AUC and ROC curves by DeLong's test, PCT was more accurate in the diagnosis of CDI than CRP.

Several limitations should be considered in the present study. First, given that it was a retrospective study with a small number of patients, the effect of confounding variables cannot be ruled out. Second, we did not have a long-term follow-up of mortality and recurrence rate in patients with CDI. Third, there were only a few cases detected by EIA in these retrospective studies, so the EIA results were not included in the study.

5. Conclusions

Overall, the present study demonstrates the superiority of PCT over CRP and WBC count in differentiating CDI in IBD patients. PCT can be used as a biomarker for helping with the rapid diagnosis of CDI in IBD patients. Whether PCT levels can help predict the clinical outcome and progress of IBD patients' needs further research to fully evaluate.

Author Contributions: S.X. was responsible for designing the experiments, interpreting the data, and writing the manuscript. P.C. participated in drawing the tables and helping to analyze the data. D.W., X.J., K.L., Z.W. and M.L. participated in the data collection and literature searching. Y.C. and S.Z. were fully responsible for designing the research and provided final approval for the version to be published. All authors have read and agreed to the published version of the manuscript.

Funding: This research received no external funding.

Institutional Review Board Statement: This report was approved by the Clinical Research and Ethics Committee of the First Affiliated Hospital of Sun Yat-sen University.

Informed Consent Statement: Written informed consent was obtained from each participant.

Data Availability Statement: All data generated or analyzed during this study are included in this published article.

Conflicts of Interest: The authors declare no conflict of interest.

Abbreviation

PCT	procalcitonin
CRP	C-reactive protein
WBC	white blood cell
CDI	*Clostridioides difficile* infection
IBD	inflammatory bowel disease
ROC	receiver operating characteristic
CD	Crohn's disease
UC	ulcerative colitis
NAAT	nucleic acid amplification test
PCR	polymerase chain reaction
EIA	enzyme immunoassay
GDH	glutamate dehydrogenase
EBV	Epstein–Barr
CMV	cytomegalovirus
HSV-1	herpes simplex virus type 1
HSV-2	herpes simplex virus type 2
LR	positive likelihood ratio
IL-6	interleukin-6
TNF-α	tumor necrosis factor-α
IL-1β	interleukin-1β
IFN-γ	interferon-gamma

References

1. Danese, S.; Fiocchi, C. Ulcerative colitis. *N. Engl. J. Med.* **2011**, *365*, 1713–1725. [CrossRef] [PubMed]
2. Cosnes, J.; Cattan, S.; Blain, A.; Beaugerie, L.; Carbonnel, F.; Parc, R.; Gendre, J.-P. Long-Term Evolution of Disease Behavior of Crohn's Disease. *Inflamm. Bowel Dis.* **2002**, *8*, 244–250. [CrossRef] [PubMed]
3. Walker, A.W.; Lawley, T.D. Therapeutic modulation of intestinal dysbiosis. *Pharmacol. Res.* **2013**, *69*, 75–86. [CrossRef] [PubMed]
4. Purchiaroni, F.; Tortora, A.; Gabrielli, M.; Bertucci. F.; Gigante, G.; Ianiro, G.; Ojetti, V.; Scarpellini, E.; Gasbarrini, A. The role of intestinal microbiota and the immune system. *Eur. Rev. Med. Pharmacol. Sci.* **2013**, *17*, 323–333.
5. Enguix, A.; Rey, C.; Concha, A.; Medina, A.; Coto. D.; Diéguez, M.A. Comparison of procalcitonin with C-reactive protein and serum amyloid for the early diagnosis of bacterial sepsis in critically ill neonates and children. *Intensiv. Care Med.* **2000**, *27*, 211–215. [CrossRef]
6. Singh, H.; Nugent, Z.; Yu, B.N.; Lix, L.M.; Targownik, L.E.; Bernstein, C.N. Higher Incidence of Clostridium difficile Infection among individuals with inflammatory bowel disease. *Gastroenterology* **2017**, *153*, 430–438. [CrossRef]
7. Hanada, Y.; Khanna, S.; Loftus, E.V.; Raffals, L.E.; Pardi, D.S. Non–Clostridium difficile Bacterial Infections Are Rare in Patients With Flares of Inflammatory Bowel Disease. *Clin. Gastroenterol. Hepatol.* **2018**, *16*, 528–533. [CrossRef]
8. Surawicz, C.M.; Brandt, L.J.; Binion, D.G.; Ananthakrishnan, A.N.; Curry, S.R.; Gilligan, P.H.; McFarland, L.V.; Mellow, M.; Zuckerbraun, B.S. Guidelines for diagnosis, treatment, and prevention of Clostridium difficile infections. *Am. J. Gastroenterol.* **2013**, *108*, 478–498. [CrossRef]
9. Berg, A.M.; Kelly, C.P.; Farraye, F. Clostridium difficile Infection in the Inflammatory Bowel Disease Patient. *Inflamm. Bowel Dis.* **2013**, *19*, 194–204. [CrossRef]

10. Khanna, S.; Shin, A.; Kelly, C.P. Management of Clostridium difficile Infection in Inflammatory Bowel Disease: Expert Review from the Clinical Practice Updates Committee of the AGA Institute. *Clin. Gastroenterol. Hepatol.* **2017**, *15*, 166–174. [CrossRef]
11. Cohen, S.H.; Gerding, D.N.; Johnson, S.; Kelly, C.P.; Loo, V.G.; McDonald, L.C.; Pépin, J.L.; Wilcox, M.H. Clinical Practice Guidelines for Clostridium difficile Infection in Adults: 2010 Update by the Society for Healthcare Epidemiology of America (SHEA) and the Infectious Diseases Society of America (IDSA). *Infect. Control. Hosp. Epidemiol.* **2010**, *31*, 431–455. [CrossRef] [PubMed]
12. Debast, S.; Bauer, M.; Kuijper, E. European Society of Clinical Microbiology and Infectious Diseases: Update of the Treatment Guidance Document for Clostridium difficile Infection. *Clin. Microbiol. Infect.* **2014**, *20* (Suppl. 2), 1–26. [CrossRef] [PubMed]
13. Planche, T.D.; A Davies, K.; Coen, P.G.; Finney, J.M.; Monahan, I.M.; A Morris, K.; O'Connor, L.; Oakley, S.J.; Pope, C.F.; Wren, M.W.; et al. Differences in outcome according to Clostridium difficile testing method: A prospective multicentre diagnostic validation study of C difficile infection. *Lancet Infect. Dis.* **2013**, *13*, 936–945. [CrossRef] [PubMed]
14. Polage, C.R.; Gyorke, C.E.; Kennedy, M.A.; Leslie, J.L.; Chin, D.L.; Wang, S.; Nguyen, H.H.; Huang, B.; Tang, Y.-W.; Lee, L.W.; et al. Overdiagnosis of *Clostridium difficile* Infection in the Molecular Test Era. *JAMA Intern. Med.* **2015**, *175*, 1792–1801. [CrossRef] [PubMed]
15. Kyne, L.; Warny, M.; Qamar, A.; Kelly, C.P. Asymptomatic carriage of Clostridium difficile and serum levels of IgG antibody against toxin A. *N. Engl. J. Med.* **2000**, *342*, 390–397. [CrossRef]
16. Furuya-Kanamori, L.; Marquess, J.; Yakob, L.; Riley, T.V.; Paterson, D.L.; Foster, N.F.; Huber, C.A.; Clements, A.C.A. Asymptomatic Clostridium difficile colonization: Epidemiology and clinical implications. *BMC Infect. Dis.* **2015**, *15*, 516. [CrossRef]
17. Gupta, A.; Khanna, S. Repeat *Clostridium difficile* Testing. *JAMA* **2016**, *316*, 2422–2423. [CrossRef]
18. Tang, Y.M.; Stone, C.D. Clostridium difficile infection in inflammatory bowel disease: Challenges in diagnosis and treatment. *Clin. J. Gastroenterol.* **2017**, *10*, 112–123. [CrossRef]
19. Gupta, A.; Cifu, A.S.; Khanna, S. Diagnosis and Treatment of *Clostridium difficile* Infection. *JAMA* **2018**, *320*, 1031–1032. [CrossRef]
20. Yi, F.; Wu, J. Biomarkers of inflammatory bowel disease. *Dis. Markers* **2014**, *2014*, 710915.
21. van der Galiën, H.T.; Loeffen, E.A.H.; Miedema, K.G.E.; Tissing, W.J.E. Predictive value of PCT and IL-6 for bacterial infection in children with cancer and febrile neutropenia. *Support. Care Cancer* **2018**, *26*, 3819–3826. [CrossRef] [PubMed]
22. Inflammatory Bowel Disease Group; Chinese Society of Gastroenterology; Chinese Medical Association. Chinese consensus on diagnosis and treatment in inflammatory bowel disease (2018, Beijing). *J. Dig. Dis.* **2021**, *22*, 298–317. [CrossRef] [PubMed]
23. A Miller, M.; Louie, T.; Mullane, K.; Weiss, K.; Lentnek, A.; Golan, Y.; Kean, Y.; Sears, P. Derivation and validation of a simple clinical bedside score (ATLAS) for Clostridium difficile infection which predicts response to therapy. *BMC Infect. Dis.* **2013**, *13*, 148. [CrossRef] [PubMed]
24. Jen, M.H.; Saxena, S.; Bottle, A.; Aylin, P.; Pollok, R.C.G. Increased health burden associated with Clostridium difficile diarrhea in patients with inflammatory bowel disease. *Aliment. Pharmacol. Ther.* **2011**, *33*, 1322–1331. [CrossRef] [PubMed]
25. Vitikainen, K.; Haapamäki, J.; Färkkilä, M.; Anttila, V.-J.; Arkkila, P. *Clostridium difficile* infection in patients with inflammatory bowel disease: A case control study. *Scand. J. Gastroenterol.* **2018**, *53*, 947–951. [CrossRef] [PubMed]
26. Crobach, M.; Planche, T.; Eckert, C.; Barbut, F.; Terveer, E.; Dekkers, O.; Wilcox, M.; Kuijper, E. European Society of Clinical Microbiology and Infectious Diseases: Update of the diagnostic guidance document for Clostridium difficile infection. *Clin. Microbiol. Infect.* **2016**, *22*, S63–S81. [CrossRef] [PubMed]
27. Fujii, Y.; Yashiro, M.; Yamada, M.; Kikkawa, T.; Nosaka, N.; Saito, Y.; Tsukahara, K.; Ikeda, M.; Morishima, T.; Tsukahara, H. Serum Procalcitonin Levels in Acute Encephalopathy with Biphasic Seizures and Late Reduced Diffusion. *J. Dis. Markers* **2018**, *2018*, 2380179. [CrossRef]
28. Sager, R.; Kutz, A.; Mueller, B.; Schuetz, P. Procalcitonin-guided diagnosis and antibiotic stewardship revisited. *BMC Med.* **2017**, *15*, 15. [CrossRef]
29. Müller, B.; Becker, K.L.; Schächinger, H.; Rickenbacher, P.R.; Huber, P.R.; Zimmerli, W.; Ritz, R. Calcitonin precursors are reliable markers of sepsis in a medical intensive care unit. *Crit. Care Med.* **2000**, *28*, 977–983. [CrossRef]
30. Muller, F.; Christ-Crain, M.; Bregenzer, T.; Krause, M.; Zimmerli, W.; Mueller, B.; Schuetz, P. Procalcitonin levels predict bacteremia in patients with community-acquired pneumonia: A prospective cohort trial. *Chest* **2010**, *138*, 121–129. [CrossRef]

diagnostics MDPI

Article

Clinical Value of Neutrophil CD64 Index, PCT, and CRP in Acute Pancreatitis Complicated with Abdominal Infection

Biao Wang [1,†], Rongzhu Tang [2,†], Shaohong Wu [3], Ming Liu [4], Fariha Kanwal [5], Muhammad Fayyaz ur Rehman [6,*], Fang Wu [7,*] and Jianping Zhu [3,*]

1 Department of Gastrointestinal Surgery, Renmin Hospital, Hubei University of Medicine, No. 39, Chaoyang Middle Road, Shiyan 442000, China
2 Department of Gastroenterology, Seventh People's Hospital of Shanghai University of Traditional Chinese Medicine, No. 358, Datong Road, Pudong New District, Shanghai 200137, China
3 Department of Critical Care Medicine, Shanghai General Hospital, Shanghai Jiao Tong University School of Medicine, Shanghai 200025, China
4 Department of Emergency, Shanghai General Hospital, Shanghai Jiao Tong University School of Medicine, Shanghai 200025, China
5 Med-X Research Institute, School of Biomedical Engineering, Shanghai Jiao Tong University, Shanghai 200240, China
6 Institute of Chemistry, University of Sargodha, Sargodha 40100, Pakistan
7 Department of Gynecology, Obstetrics and Gynae Hospital, Fudan University, Shanghai 200437, China
* Correspondence: fayyaz9@gmail.com (M.F.u.R.); 13916055069@163.com (F.W.); jianping.zhu@shgh.cn (J.Z.)
† These authors contributed equally to this work.

Abstract: Objective: To study the clinical diagnostic value of neutrophil CD64 index, PCT, and CRP in patients with acute pancreatitis with abdominal infection. Methods: A number of patients with acute pancreatitis (n = 234) participated in the study. According to the infection and health conditions, they were further divided into the non-infection group (n = 122), infection group (n = 78), and sepsis group (n = 34), and 40 healthy subjects were selected in the control group (n = 40). Expression levels of infection indexes, such as CD64 index, PCT, and CRP, were detected and compared. ROC curves were drawn to compare the efficacy of each index in the diagnosis of acute pancreatitis with abdominal infection and sepsis. The study was retrospectively registered under the China Clinical Trial Registry as a trial number ChiCTR2100054308. Results: All indexes were significantly higher in three clinical groups than the healthy control group ($p < 0.05$). The CD64 index, CD64 positive rate, and PCT in the infected group were significantly higher than those in the uninfected group (ALL $p < 0.05$). The PCT of patients infected with Gram-negative bacteria was significantly higher than that of Gram-positive bacteria-infected patients ($p < 0.05$). CD64 index had the best diagnostic efficiency for acute pancreatitis infection, with 82.14% sensitivity, 88.51% specificity, and 0.707 Youden indexes. The CD64 Youden index (0.780) for sepsis diagnosis was the highest, while the AUC of PCT was the highest (0.897). Conclusion: CD64 index combined with PCT has good sensitivity and specificity in diagnosing acute pancreatitis infection and sepsis.

Keywords: neutrophil; CD64 index; acute pancreatitis; abdominal infection

Citation: Wang, B.; Tang, R.; Wu, S.; Liu, M.; Kanwal, F.; Rehman, M.F.u.; Wu, F.; Zhu, J. Clinical Value of Neutrophil CD64 Index, PCT, and CRP in Acute Pancreatitis Complicated with Abdominal Infection. *Diagnostics* **2022**, *12*, 2409. https://doi.org/10.3390/diagnostics12102409

Academic Editors: Yuli Huang, Yong Yuan and Feisong Chen

Received: 22 August 2022
Accepted: 23 September 2022
Published: 5 October 2022

1. Introduction

Acute pancreatitis is a common gastrointestinal disease caused by pancreas inflammation due to systemic inflammatory response that may lead to organ or system impairment [1]. It is also characterized by local inflammatory reactions of the pancreas in common clinically acute and severe cases with or without functional changes of other organs. The disease is self-limiting (one week duration) in patients with mild acute pancreatitis [2]. Moderately severe or severe acute pancreatitis (SAP) has the characteristics of rapid progress, dangerous condition, and high clinical mortality [3]. According to revised Atlanta classification (RAC), acute pancreatitis can be either interstitial edematous pancreatitis (IEP) or acute necrotizing pancreatitis (ANP) [4–6]. Approximately 20–30% of patients with acute pancreatitis will have necrotizing pancreatitis [7]. According to statistics, the overall mortality of patients with acute pancreatitis is about 5–10%, and that of patients with SAP is 20–30% [8]. While 20% patients of AP may develop SAP, there are two peaks in the course of acute pancreatitis; first stage is systemic inflammatory response syndrome with subsequent organ failure and second is infectious complication stage with related organ failure [9,10]. Without intensive care, SAP may progress into critical acute pancreatitis (CAP) with mortality rate of 34% [11–13]. Predicting the risk of complications and carrying out comprehensive treatment in the early stage of acute pancreatitis is the core means to improve the prognosis of patients.

In recent studies, multiple scoring systems have been applied to the severity stratification and early prognosis prediction of patients with acute pancreatitis, but these scores have limitations [14]. Exploring blood markers to predict the condition and complication risk of patients with acute pancreatitis has become an important research direction in recent years. Abdominal infection is an important complication in patients with acute pancreatitis, which can lead to systemic inflammatory response syndrome, multiple organ failure, and septic shock. It is one of the important causes of death in patients with acute pancreatitis [15], but the clinical symptoms and signs of abdominal infection are not specific. However, it is difficult to obtain positive results in the early stage of the disease using imaging diagnosis and etiological examination of peritoneal puncture fluid, thus representing a significant clinical problem.

Although many studies used early infection markers to predict and diagnose abdominal infection in patients with acute pancreatitis, there is no single marker that can meet the clinical needs [16]. It has become a consensus that combining multiple markers can improve the diagnostic effect. The cluster of differentiation 64 (CD64) antigen is considered as an important marker of bacterial infections, systemic inflammation and mortality [17]. It is a transmembrane glycoprotein functioning as a high-affinity IgG receptor (FcγRI) [18]. Cytokines regulate its expression and play a bridge role between humoral immunity and cellular immunity. CD64 is mainly distributed on the surface of monocytes, macrophages, and dendritic cells [19]. Normally, CD64 expression is very low in neutrophils, but the expression is significantly induced in peripheral blood neutrophils during bacterial infection and helps in neutrophil phagocytosis and sterilization [19,20]. Many studies show that CD64 has high specificity in bacterial infection, which is helpful in early diagnosis of bacterial infection and the degree of infection [21,22].

Procalcitonin is another biomarker normally produced in thyroid C cells and enters into blood after its conversion to calcitonin when induced by glucocorticoids, glucagon, gastrin, or β-adrenergic signaling [23]. In infection patients, it is produced by non-thyroid tissues (e.g., adipocytes) upon induction from IL-6 and TNF-α and enter bloods without being converted into calcitonin [24]. Previously, PCT has been reported to be the most sensitive laboratory test for the detection of pancreatitis where low levels of PCT appear to be strong negative predictors of infected necrosis [25]. C-reactive protein (CRP) is a conventional biomarker of systemic inflammatory response (SIRS) and bacterial infections. Based on this current situation, the value of neutrophil CD64 index, PCT, and CRP in predicting the incidence of acute pancreatitis complicated with abdominal infection was analyzed to provide evidence for clinical diagnosis of infectious diseases.

2. Materials and Methods

2.1. Study Design

A total of two hundred and thirty-four patients with acute pancreatitis and met the inclusion criteria of this study were selected from two class III (class A) hospitals from September 2019 to March 2021. According to the infection, they were divided into the No Infection group (n = 122), Infection group (n = 78), and Sepsis group (n = 34). At the same time, 40 healthy persons who underwent physical examination in our hospital in the same period were selected as the Control group (Figure 1). There was no significant difference between baseline data (all $p > 0.05$), which were comparable (Table 1). All patients signed informed consent and voluntarily participated in the study. The study protocol was approved by the hospital ethics committee (Scheme 1). The study was retrospectively registered under the China Clinical Trial Registry as a trial number ChiCTR2100054308.

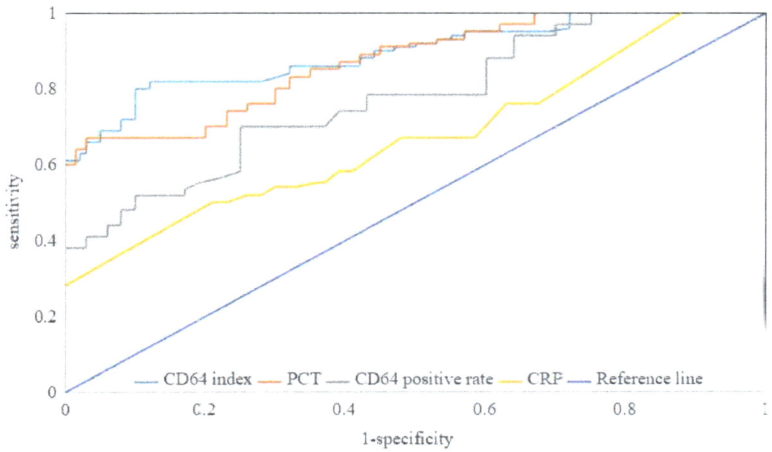

Figure 1. ROC curve of infection indicators for the diagnosis of acute pancreatitis with abdominal infection.

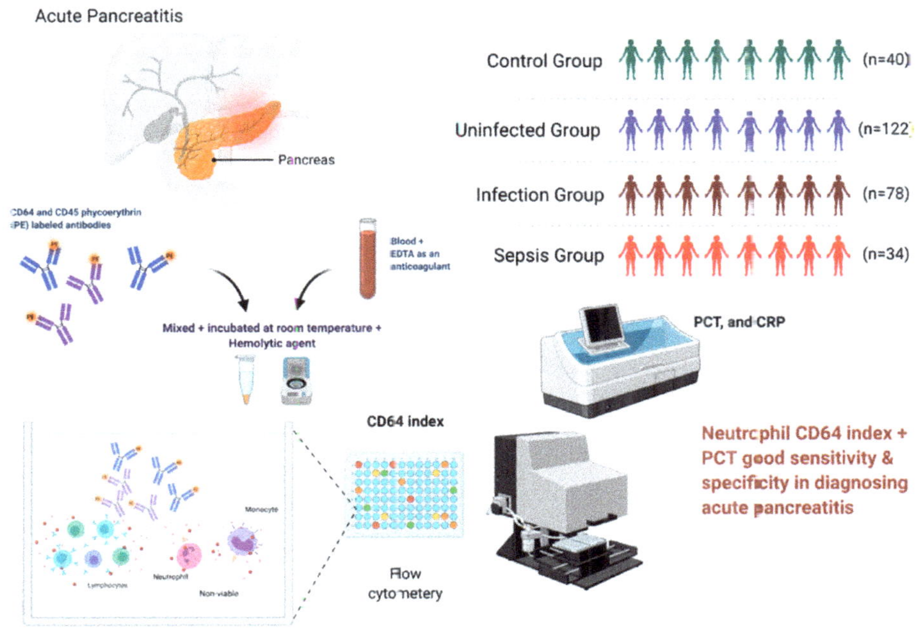

Scheme 1. Study design.

Table 1. Comparison of basic data of patients in each group.

Group	*n*	Sex [*n* (%)]		Age (Years)	Etiology [*n* (%)]		
		Male	Female	(x ± s)	Biliary	Hyperlipidemia	Other Reasons
Control group	40	27 (67.50)	13 (32.50)	45.21 ± 8.24			
Uninfected group	122	74 (60.66)	48 (39.34)	44.49 ± 9.39	66 (54.10)	38 (31.15)	18 (14.75)
Infection group	78	46 (58.97)	32 (41.03)	43.61 ± 9.47	50 (64.10)	20 (25.64)	8 (10.26)
Sepsis group	34	22 (64.71)	12 (35.29)	44.21 ± 8.26	22 (64.71)	6 (17.65)	6 (17.65)
Statistical test		X^2 = 1.001		F = 0.592		X^2 = 4.089	
p-value		0.801		0.555		0.394	

2.1.1. Inclusion Criteria

All patients met the diagnostic criteria of acute pancreatitis in the initial treatment guidelines for acute pancreatitis of the American Gastroenterology Association [26], with complete clinical data and age >18 years. Blood samples were obtained from all patients within 3 days of admission. We selected patients strictly according to the inclusion criteria, which can basically ensure the homogeneity of all included patients.

2.1.2. Exclusion Criteria

The study exclusions include patients with malignant tumors, liver and kidney dysfunction, hematological diseases, cardiovascular and cerebrovascular accidents, immune deficiency diseases or autoimmune diseases; patients diagnosed with an infection in other parts or systemic infection except for pancreas and abdominal cavity at admission; patients with diabetes and acute pancreatitis history. There were patients who used glucocorticoids and immunomodulatory drugs three months before enrollment. Pregnant or nursing mother were also excluded.

2.1.3. Diagnostic Criteria for Abdominal Infection

The management guidelines for abdominal infection [27] updated by the World Society of Emergency Surgery (WSES) in 2019 is used as the diagnostic criteria for abdominal infection: patients generally have symptoms and signs, such as abdominal pain, rebound pain, fever, cessation of anal exhaust and defecation, and Balthazar CT grade of acute pancreatitis is grade D or E, and positive blood culture or ultrasound-guided peritoneal puncture fluid pathogen culture.

2.2. Flow Cytometric Detection of CD64 Index

The detection and calculation formula of CD64 was analyzed by FACS Calibur flow cytometer (Becton Dickinson, NY, USA), supporting reagents, and software CellQuest (Becton Dickinson, New York, NY, USA). 2 mL of whole blood sample (with EDTA as an anticoagulant) was collected from patients with suspected infection and analyzed within two hours. Samples were mixed upside down eight times and added to the special flow tube before processing. CD64 and CD45 phycoerythrin (PE) labeled antibodies (5 μL each) were added to 50 μL whole blood, mixed well, and incubated at room temperature in the dark for 40 min. A hemolytic agent (500 μL) was added to each tube, mixed well, and incubated for 10 min. The samples were centrifuged for 5 min at 2000× *g*, and the supernatant was discarded. Phosphate buffer (PBS) (mL) was added to the pellet and mixed well before another centrifugation round using the same conditions. After discarding the supernatant, sample pellets were resuspended in 300 μL PBS and analyzed by flow cytometry. CD64 expression on the surface of neutrophils was measured as geometric mean fluorescence intensity (MFI) on neutrophils, namely polymorphonuclear leukocytes (PMN), lymphocytes (LYM), and monocytes (MO). CD64 index was calculated using Leuko64 QuantiCALC Software by the following ratio

CD64 index = (MFI_PMN CD64/MFI_LYM CD64)/(MFI_MO CD64/MFI_PMN CD64)

CD64 positive rate = 100% × Number of CD64 positive PMN cells/total number of PMN cells.

2.3. PCT and CRP Index

PCT and CRP were detected from non-anticoagulated peripheral blood by Cobas-e411 Electrochemiluminescence Automatic Immunoanalyzer (Roche, Munich, Germany) and AU5800 Automatic Biochemical Analyzer (Beckman Kurt, Dallas, TX, USA) and their supporting reagents, respectively.

2.4. Statistical Analysis

Utilizing SPSS 20.0 software (SPSS Inc., Chicago, IL, USA), the comparison between multiple groups of data adopts one-way ANOVA, the comparison between two groups is subject to Student–Newman–Keuls (SNK) test, and the comparison between two groups of measurement data adopts an independent sample t-test. Comparison row between person data χ^2 test, Wilcoxon rank-sum test was used to compare grade data. The receiver operator characteristics (ROC) curve was drawn to evaluate the efficacy of various infection indexes in diagnosing acute pancreatitis with abdominal infection and sepsis. $p < 0.05$ was statistically significant. The Youden index [28] measure for the ROC curve was used to measure the overall value of an infection index over the whole region of the ROC curve. Youden index for a diagnostic marker and calculated optimal cut-off point is equivalent to maximizing the sum of sensitivity and specificity for all the possible values of the cut-off point [29].

3. Results

3.1. Comparison of Infection Indexes

When infection indexes were compared, there was a significant difference among the four groups (all $p < 0.05$). The indexes of patients with acute pancreatitis in the three groups were significantly higher than those in the healthy control group (all $p < 0.05$); The CD64 index, CD64 positive rate, and PCT in the infected group were significantly higher than those in the uninfected group (all $p < 0.05$), but there was no significant difference in CRP between the two groups ($p > 0.05$); All indexes in the sepsis group were significantly higher than those in infection group (all $p < 0.05$), as shown in Table 2.

Table 2. Comparison of infection indexes in 4 groups.

Group	*n*	CD64 Positive Rate	CD64 Index	CRP (mg/L)	PCT (μg/L)
Control group	40	9.79 ± 3.56	0.76 ± 0.21	2.29 ± 0.81	0.12 ± 0.07
Uninfected group	122	23.41 ± 7.17 [a]	1.21 ± 0.35 [a]	24.43 ± 4.61 [a]	1.14 ± 0.78 [a]
Infection group	78	43.91 ± 15.82 [ab]	3.34 ± 0.94 [ab]	25.51 ± 4.97 [a]	2.53 ± 1.25 [ab]
Sepsis group	34	67.71 ± 9.64 [abc]	5.06 ± 0.36 [abc]	29.74 ± 4.64 [abc]	5.35 ± 1.75 [abc]
F-value		177.321	304.560	328.296	131.643
p-value		<0.001	<0.001	<0.001	<0.001

Note: Compared with the control group, [a] $p < 0.05$; Compared with the uninfected group, [b] $p < 0.05$; Compared with the infected group, [c] $p < 0.05$.

A comparison of infection indexes among different bacterial infection groups shows that PCT of patients infected with Gram-negative bacteria was significantly higher than that of patients infected with Gram-positive bacterial infections ($p < 0.05$), but there was no significant difference in CD64 index, CD64 positive rate, and CRP indexes between the two groups (all $p > 0.05$), as shown in Table 3.

Table 3. Comparison of infection indexes between different bacterial infection groups.

Bacteria Types	n	CD64 Ositive Rate	CD64 Index	CRP (mg/L)	PCT (μg/L)
Gram-negative bacteria	89	51.91 ± 12.78	4.02 ± 0.81	27.36 ± 5.19	3.91 ±1.11
Gram-positive bacteria	23	49.93 ± 10.55	3.75 ± 0.76	25.954 ± 4.74	2.64 ± 0.75
F-value		0.621	1.259	1.031	4.802
p-value		0.550	0.239	0.329	<0.001

3.2. Evaluation of Diagnostic Efficacy of Acute Pancreatitis with Abdominal Infection

The uninfected acute pancreatitis group is negative, and the infection and sepsis groups of acute pancreatitis are positive. The ROC curve of each index was obtained for diagnosing acute pancreatitis with abdominal infection, as shown in Figure 1. The sensitivity of CD64 index (82.15%), Youden index (0.707), and area under the curve (AUC, 0.892) were greater than other indexes, while the diagnostic specificity of PCT (95.11%) was the highest (Table 4).

Table 4. Diagnostic efficacy of infection indicators for acute pancreatitis complicated with abdominal infection.

Infection Index	AUC *	Sensitivity (%)	Specificity (%)	Youden Index
CD64 positive rate	0.669	71.44	75.42	0.469
CD64 index	0.892	82.15	88.51	0.707
CRP	0.622	58.94	81.96	0.409
PCT	0.867	67.85	95.11	0.630

* AUC; area under the curve.

3.3. Evaluation of Diagnostic Efficacy of Acute Pancreatitis with Sepsis

In this case, the acute pancreatitis infection group is negative, and the sepsis group is positive. The ROC curve of each index for the diagnosis of burn sepsis is drawn (Figure 2). Among them, the CD64 index and Youden index (0.780) are the largest, while the AUC of PCT (0.897) is the highest (Table 5).

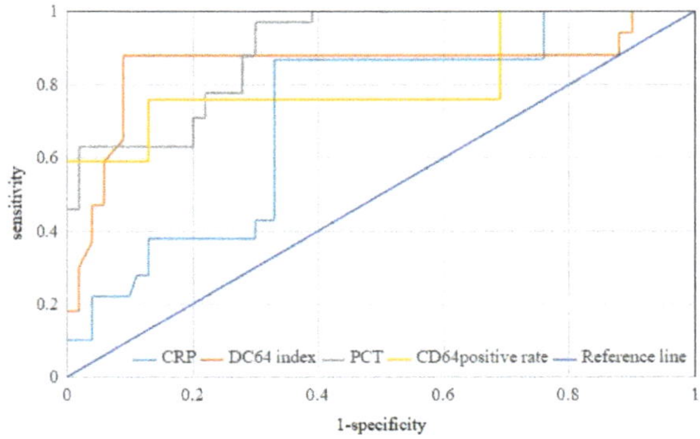

Figure 2. ROC curves of infection indicators for the diagnosis of acute pancreatitis and sepsis.

Table 5. Diagnostic efficacy of infection indicators for acute pancreatitis complicated with sepsis.

Infection Index	AUC	Sensitivity (%)	Specificity (%)	Youden Index
CD64 positive rate	0.815	76.48	87.19	0.637
CD64 index	0.847	88.26	89.76	0.780
CRP	0.627	88.25	66.69	0.549
PCT	0.897	70.61	94.87	0.655

3.4. Combined Diagnosis

The above CD64 index and PCT with good diagnostic effects are used for combined diagnosis, and the positive value is greater than the optimal critical diagnostic value. The sensitivity and Youden index are higher than the single index (Table 6).

Table 6. CD64 index and PCT combined diagnostic efficiency.

Diagnosis of Content	Sensitivity (%)	Specificity (%)	Youden Index
Acute pancreatitis with abdominal infection	89.29 (100/112)	88.52 (108/122)	0.778
Acute pancreatitis with sepsis	94.12 (32/34)	84.62 (66/78)	0.787

4. Discussion

Acute pancreatitis is one of the common emergencies of the digestive system, including mild-acute and severe-acute pancreatitis. Later is prone to septic shock and organ dysfunction. Symptoms and signs or imaging examinations are ineffective in diagnosing acute pancreatitis complicated with abdominal infection. In the clinic, a large number of patients often delay treatment and cause a bad prognosis. So, the mortality of secondary infection is as high as 36–50%. Some patients with acute pancreatitis can suffer from sequelae of pancreatitis [30–32].

Bacteriological culture results and drug sensitivity tests are the gold standards for diagnosing bacterial infection and guiding antibiotic treatment. Still, their positive rate is low, and the isolation time of pathogens is long, which cannot provide the basis for early diagnosis in time. Therefore, finding a rapid and accurate method to evaluate a bacterial infection is particularly important.

In recent years, the application of flow cytometry to detect the expression of CD64 on the surface of peripheral blood neutrophils has attracted extensive attention in the diagnosis of bacterial infection. CD64 mRNA in neutrophils began to express 1~3 h after infection, and CD64 on the cell surface could be detected up-regulated 3~6 h after infection, which has the ability to detect infection early [33]. Rogina et al. also showed that the differentiation of CD64 index and CD64 positive rate in each group of acute pancreatitis was higher. The CD64 index, CD64 positive rate, and PCT in the infected group were significantly higher than those in the uninfected group (all $p < 0.05$), but there was no significant difference in CRP between the two groups ($p > 0.05$). It may be that CRP primarily reflects the acute stress state in vivo, and the basic CRP in patients with acute pancreatitis is higher. The increase was not obvious during infection. Previously, CRP shows low specificity and limited correlation with the disease activity in comparison to other infection indexes [34].

Some studies pointed out that Gram-negative bacteria infection in patients with acute pancreatitis with abdominal infection was significantly higher than that of Gram-positive bacteria [35]. Therefore, this study compared the differences of various indexes between the two types of bacterial infections. The results showed that the PCT of Gram-negative bacterial infections was significantly higher than that of Gram-positive bacterial infections ($p < 0.05$), but there was no significant difference in CD64 index, CD64 positive rate, and CRP between the two groups (all $p > 0.05$). PCT is an important biomarker commonly used to predict bacterial infections in clinics. It can also guide the diagnosis and treatment of infectious diseases. Its sensitivity and specificity are higher than traditional markers such as WBC and CRP. CRP test is controversial in term of limitation sensitivity and specificity [36]. PCT may be better than CD64 index in the early diagnosis of bacterial infection [37] but CD64 index is still preferable and a method of choice in diagnosing severe and complicated bacterial infection in term of sensitivity and specificity [38,39]. When the body is infected or invaded by endotoxins, neutrophil CD64 can increase 4~6 h after stimulation and 0~24 h after sepsis. This study confirmed by the bacterial culture that the peripheral blood CD64 index of acute pancreatitis (AP) patients can be significantly increased within 24 h after concurrent bacterial infection, which is expected to become a new biomarker for the early diagnosis of sepsis.

Ye Z et al. reported that the sensitivity of CD64 on the surface of neutrophils for the diagnosis of bacterial infection is ≥90%, and the specificity can reach 90~100%, which is significantly better than PCT, CRP, and other indicators. However, there are many ways to express the test results, including CD64 MFI, CD64 positive rate, and CD64 index, and the calculation formula of CD64 index is not unified. The calculation formula of the CD64 index for diagnosing infection obtained by various laboratories is quite different from the cut-off value, and the comparability is poor [40].

Because CD64 is expressed in various cells; in monocytes, these are produced in large amounts in physiological and infectious states, while lymphocytes have low expression of CD64 [41]. Considering this expression pattern, this study used (MFI_PMN CD64/MFI_LYM CD64)/(MFI_MO CD64/MFI_PMN CD64) to express the CD64 index, which not only quantified the expression level of CD64, but could also reduce the operation error to make the CD64 index more objective. The results showed that the efficacy of CD64 index in the diagnosis of acute pancreatitis with abdominal infection was better than other indexes, but the specificity was lower than PCT. The Youden index (0.781) was the largest in the diagnosis of sepsis, while the AUC of PCT (0.897) was the highest. The sensitivity of the CD64 index was higher than that of PCT, but the specificity was lower than that of PCT. Previously, the sensitivity and specificity of neutrophil CD64 was found to be more than 80% [42].

When the CD64 index was combined with PCT, the sensitivity and Youden index increased, but the specificity decreased. It is suggested that combined diagnosis is helpful to improve the diagnostic efficiency of acute pancreatitis with abdominal infection and sepsis.

5. Conclusions

In conclusion, neutrophil CD64 index combined with PCT has good sensitivity and specificity in diagnosing acute pancreatitis with abdominal infection and sepsis and has a good prospect of clinical application.

Author Contributions: Data curation, R.T. and M.L.; formal analysis, B.W., R.T., M.L. and F.W.; funding acquisition, S.W.; investigation, B.W., R.T. and S.W.; methodology, B.W. and R.T.; project administration, F.W. and J.Z.; resources, M.L., F.K. and F.W.; software, F.K. and M.F.u.R.; supervision, J.Z.; validation, B.W.; writing—original draft, S.W., M.L., F.K., M.F.u.R., F.W. and J.Z.; writing—review & editing, S.W., M.L. and M.F.u.R. All authors have read and agreed to the published version of the manuscript.

Funding: This research received no external funding.

Institutional Review Board Statement: The study was retrospectively registered under the China Clinical Trial Registry as a trial number ChiCTR2100054308. This study is a part of project approved by Shiyan People's Hospital Medical Ethics Committee on 26-08-2013. This study was executed on 01-12-2021.

Informed Consent Statement: Informed consent was obtained from all subjects involved in the study.

Conflicts of Interest: The authors declare no conflict of interest.

References

1. Johnson, C.D.; Besselink, M.G.; Carter, R. Acute pancreatitis. *Bmj Br. Med. J.* **2014**, *349*, g4859. [CrossRef] [PubMed]
2. Boxhoorn, L.; Voermans, R.P.; Bouwense, S.A.; Bruno, M.J.; Verdonk, R.C.; Boermeester, M.A.; van Santvoort, H.C.; Besselink, M.G. Acute pancreatitis. *Lancet* **2020**, *396*, 726–734. [CrossRef]
3. De Waele, E.; Malbrain, M.L.; Spapen, H.D. How to deal with severe acute pancreatitis in the critically ill. *Curr. Opin. Crit. Care* **2019**, *25*, 150–156. [CrossRef] [PubMed]
4. Banks, P.A.; Bollen, T.L.; Dervenis, C.; Gooszen, H.G.; Johnson, C.D.; Sarr, M.G.; Tsiotos, G.G.; Vege, S.S. Classification of acute pancreatitis—2012: Revision of the Atlanta classification and definitions by international consensus. *Gut* **2013**, *62*, 102. [CrossRef]
5. Foster, B.R.; Jensen, K.K.; Bakis, G.; Shaaban, A.M.; Coakley, F.V. Revised Atlanta Classification for Acute Pancreatitis: A Pictorial Essay. *RadioGraphics* **2016**, *36*, 675–687. [CrossRef]

6. Maheshwari, R.; Subramanian, R.M. Severe acute pancreatitis and necrotizing pancreatitis. *Critical care clinics* **2016**, *32*, 279–290. [CrossRef]
7. Boxhoorn, L.; van Dijk, S.M.; van Grinsven, J.; Verdonk, R.C.; Boermeester, M.A.; Bollen, T.L.; Bouwense, S.A.; Bruno, M.J.; Cappendijk, V.C.; Dejong, C.H. Immediate versus Postponed Intervention for Infected Necrotizing Pancreatitis. *N. Engl. J. Med.* **2021**, *385*, 1372–1381. [CrossRef]
8. Gliem, N.; Ammer-Herrmenau, C.; Ellenrieder, V.; Neesse, A. Management of severe acute pancreatitis: An update. *Digestion* **2021**, *102*, 503–507. [CrossRef]
9. Morató, O.; Poves, I.; Ilzarbe, L.; Radosevic, A.; Vázquez-Sánchez, A.; Sánchez-Parrilla, J.; Burdío, F.; Grande, L. Minimally invasive surgery in the era of step-up approach for treatment of severe acute pancreatitis. *Int. J. Surg.* **2018**, *51*, 164–169. [CrossRef]
10. Brown, L.A.; Hore, T.A.; Phillips, A.R.; Windsor, J.A.; Petrov, M.S. A systematic review of the extra-pancreatic infectious complications in acute pancreatitis. *Pancreatology* **2014**, *14*, 436–443. [CrossRef]
11. Windsor, J.; De-Madaria, E. Critical acute pancreatitis: A category with clinical relevance. *Dig. Liver Dis.* **2021**, *53*, 1588–1589. [CrossRef] [PubMed]
12. Wu, D.; Lu, B.; Xue, H.-d.; Yang, H.; Qian, J.-m.; Lee, P.; Windsor, J.A. Validation of Modified Determinant-Based Classification of severity for acute pancreatitis in a tertiary teaching hospital. *Pancreatology* **2019**, *19*, 217–223. [CrossRef] [PubMed]
13. Wu, D.; Huang, Y.; Xiao, J.; Qin, G.; Liu, H.; Peng, J. Risk Factors for Mortality Among Critical Acute Pancreatitis Patients with Carbapenem-Resistant Organism Infections and Drug Resistance of Causative Pathogens. *Infect. Dis. Ther.* **2022**, *11*, 1089–1101. [CrossRef]
14. Wan, Z.; Shen, B.; Cen, D.; Yu, H.; Cai, X. Minimally Invasive Treatment for Severe Acute Pancreatitis With Superior Mesenteric Vein and Common Bile Duct Stenosis: A Case Report and Review of the Literature. *Pancreas* **2019**, *48*, e61. [CrossRef]
15. Shu, W.; Wan, J.; Chen, J.; He, W.; Zhu, Y.; Lu, N.; Xia, L. Elevated arterial lactate level as an independent risk factor for pancreatic infection in moderately severe acute pancreatitis. *Pancreatology* **2019**, *19*, 653–657. [CrossRef]
16. Huang, H.; Chen, W.; Tang, G.; Liang, Z.; Qin, M.; Qin, M.; Tang, Y.; Qin, H.; Chang, R. Optimal timing of contrast-enhanced computed tomography in an evaluation of severe acute pancreatitis-associated complications. *Exp. Ther. Med.* **2019**, *18*, 1029–1038. [PubMed]
17. Prashant, A.; Vishwanath, P.; Kulkarni, P.; Sathya Narayana, P.; Gowdara, V.; Nataraj, S.M.; Nagaraj, R. Comparative assessment of cytokines and other inflammatory markers for the early diagnosis of neonatal sepsis–a case control study. *PLoS ONE* **2013**, *8*, e68426. [CrossRef]
18. Van Vugt, M.J.; Heijnen, I.; Capel, P.; Park, S.Y.; Ra, C.; Saito, T.; Verbeek, J.S.; van De Winkel, J. FcR gamma-chain is essential for both surface expression and function of human Fc gamma RI (CD64) in vivo. *Blood* **1996**, *87*, 3593–3599. [CrossRef]
19. Qian, L.; Li, S.B.; Zhou, Y.; Teng, S.J.; Guo, J.J. Determination of CD64 for the diagnosis of bacterial chronic prostatitis. *Am. J. Reprod. Immunol.* **2015**, *74*, 309–312. [CrossRef]
20. Qureshi, S.; Lewis, S.; Gant, V.; Treacher, D.; Davis, B.; Brown, K. Increased distribution and expression of CD64 on blood polymorphonuclear cells from patients with the systemic inflammatory response syndrome (SIRS). *Clin. Exp. Immunol.* **2001**, *125*, 258–265. [CrossRef]
21. Pandey, G.; Singh, H.; Chaturvedi, S.; Hatti, M.; Kumar, A.; Mishra, R.; Mishra, P.; Krishna, V.; Bhadauria, A.; Mohindra, S. Utility of neutrophil CD64 in distinguishing bacterial infection from inflammation in severe alcoholic hepatitis fulfilling SIRS criteria. *Sci. Rep.* **2021**, *11*, 19726. [CrossRef] [PubMed]
22. Danikas, D.; Karakantza, M.; Theodorou, G.; Sakellaropoulos, G.; Gogos, C. Prognostic value of phagocytic activity of neutrophils and monocytes in sepsis. Correlation to CD64 and CD14 antigen expression. *Clin. Exp. Immunol.* **2008**, *154*, 87–97. [CrossRef] [PubMed]
23. Vijayan, A.L.; Ravindran, S.; Saikant, R.; Lakshmi, S.; Kartik, R. Procalcitonin: A promising diagnostic marker for sepsis and antibiotic therapy. *J. Intensive Care* **2017**, *5*, 51. [CrossRef]
24. Yuzbasioglu, Y.; Duymaz, H.; Tanrikulu, C.S.; Halhalli, H.C.; Koc, M.O.; Tandoğan, M.; Coskun, F. Role of procalcitonin in evaluation of the severity of acute cholecystitis. *Eurasian J. Med.* **2016**, *48*, 162. [CrossRef]
25. Silva-Vaz, P.; Abrantes, A.M.; Castelo-Branco, M.; Gouveia, A.; Botelho, M.F.; Tralhão, J.G. Multifactorial Scores and Biomarkers of Prognosis of Acute Pancreatitis: Applications to Research and Practice. *Int. J. Mol. Sci.* **2020**, *21*, 338. [CrossRef] [PubMed]
26. Crockett, S.D.; Wani, S.; Gardner, T.B.; Falck-Ytter, Y.; Barkun, A.N.; Crockett, S.; Feuerstein, J.; Flamm, S.; Gellad, Z.; Gerson, L. American Gastroenterological Association Institute guideline on initial management of acute pancreatitis. *Gastroenterology* **2018**, *154*, 1096–1101. [CrossRef]
27. Leppäniemi, A.; Tolonen, M.; Tarasconi, A.; Segovia-Lohse, H.; Gamberini, E.; Kirkpatrick, A.W.; Ball, C.G.; Parry, N.; Sartelli, M.; Wolbrink, D. 2019 WSES guidelines for the management of severe acute pancreatitis. *World J. Emerg. Surg.* **2019**, *14*, 1–20. [CrossRef] [PubMed]
28. Youden, W.J. Index for rating diagnostic tests. *Cancer* **1950**, *3*, 32–35. [CrossRef]
29. Li, C.; Chen, J.; Qin, G. Partial Youden index and its inferences. *J. Biopharm. Stat.* **2019**, *29*, 385–399. [CrossRef]
30. Hart, P.A.; Bradley, D.; Conwell, D.L.; Dungan, K.; Krishna, S.G.; Wyne, K.; Bellin, M.D.; Yadav, D.; Andersen, D.K.; Serrano, J. Diabetes following acute pancreatitis. *Lancet Gastroenterol. Hepatol.* **2021**, *6*, 668–675. [CrossRef]
31. Li, H.; Yang, Z.; Tian, F. Risk factors associated with intolerance to enteral nutrition in moderately severe acute pancreatitis: A retrospective study of 568 patients. *Saudi J. Gastroenterol. Off. J. Saudi Gastroenterol. Assoc.* **2019**, *25*, 362.

32. Kitagawa, S.; Sawai, K. Hypertriglyceridemia-induced acute pancreatitis with normal pancreatic enzymes. *Am. J. Med.* **2018**, *131*, e299–e300. [CrossRef] [PubMed]
33. Rogina, P.; Stubljar, D.; Lejko-Zupanc, T.; Osredkar, J.; Skvarc, M. Expression of CD64 on neutrophils (CD64 index): Diagnostic accuracy of CD64 index to predict sepsis in critically ill patients. *Clin. Chem. Lab. Med.* **2015**, *53*, e89–e91. [CrossRef]
34. Desai, D.; Faubion, W.; Sandborn, W. biological activity markers in inflammatory bowel disease. *Alimentary pharmacology & therapeutics* **2007**, *25*, 247–255.
35. Zhu, C.; Zhang, S.; Zhong, H.; Gu, Z.; Kang, Y.; Pan, C.; Xu, Z.; Chen, E.; Yu, Y.; Wang, Q. Intra-abdominal infection in acute pancreatitis in eastern China: Microbiological features and a prediction model. *Ann. Transl. Med.* **2021**, *9*, 477. [CrossRef]
36. Stirling, A.D.; Moran, N.R.; Kelly, M.E.; Ridgway, P.F.; Conlon, K.C. The predictive value of C-reactive protein (CRP) in acute pancreatitis–is interval change in CRP an additional indicator of severity? *HPB* **2017**, *19*, 874–880. [CrossRef]
37. Gregoriano, C.; Heilmann, E.; Molitor, A.; Schuetz, P. Role of procalcitonin use in the management of sepsis. *J. Thorac. Dis.* **2020**, *12*, S5. [CrossRef]
38. Cong, S.; Ma, T.; Di, X.; Tian, C.; Zhao, M.; Wang, K. Diagnostic value of neutrophil CD64, procalcitonin, and interleukin-6 in sepsis: A meta-analysis. *BMC Infect. Dis.* **2021**, *21*, 384. [CrossRef]
39. Hoffmann, J.J. Neutrophil CD64: A diagnostic marker for infection and sepsis. *Clin. Chem. Lab. Med.* **2009**, *47*, 903–916. [CrossRef]
40. Ye, Z.; Zou, H.; Liu, S.; Mei, C.; Chang, X.; Hu, Z.; Yang, H.; Wu, Y. Diagnostic performance of neutrophil CD64 index in patients with sepsis in the intensive care unit. *J. Int. Med Res.* **2019**, *47*, 4304–4311. [CrossRef]
41. Shi, J.; Tang, J.; Chen, D. Meta-analysis of diagnostic accuracy of neutrophil CD64 for neonatal sepsis. *Ital. J. Pediatrics* **2016**, *42*, 57. [CrossRef] [PubMed]
42. Thiriet, C.; Mahjoub, K.; Courte, G.; Labroca, P.; Cravoisy, A.; Lemarie, J.; Conrad, M.; Nace, L.; Bollaert, P.-E.; Gibot, S. Automated measurement of neutrophil CD64 expression for diagnosing sepsis in critically ill patients. *Minerva Anestesiol.* **2019**, *85*, 943–950. [CrossRef] [PubMed]

diagnostics

Review

Fast Track Diagnostic Tools for Clinical Management of Sepsis: Paradigm Shift from Conventional to Advanced Methods

Ena Gupta [1,†], Juhi Saxena [2,†], Sanni Kumar [1], Umang Sharma [2], Saundarya Rastogi [1], Vijay Kumar Srivastava [1], Sanket Kaushik [1] and Anupam Jyoti [2,*]

1 Amity Institute of Biotechnology, Amity University Rajasthan, Amity Education Valley, Kant Kalwar, NH-11C, Jaipur Delhi Highway, Jaipur 303002, India
2 Department of Biotechnology, University Institute of Biotechnology, Chandigarh University, Sahibzada Ajit Singh Nagar 140413, India
* Correspondence: anupamjyoti03@gmail.com
† These authors contributed equally to this work.

Abstract: Sepsis is one of the deadliest disorders in the new century due to specific limitations in early and differential diagnosis. Moreover, antimicrobial resistance (AMR) is becoming the dominant threat to human health globally. The only way to encounter the spread and emergence of AMR is through the active detection and identification of the pathogen along with the quantification of resistance. For better management of such disease, there is an essential requirement to approach many suitable diagnostic techniques for the proper administration of antibiotics and elimination of these infectious diseases. The current method employed for the diagnosis of sepsis relies on the conventional culture of blood suspected infection. However, this method is more time consuming and generates results that are false negative in the case of antibiotic pretreated samples as well as slow-growing microbes. In comparison to the conventional method, modern methods are capable of analyzing blood samples, obtaining accurate results from the suspicious patient of sepsis, and giving all the necessary information to identify the pathogens as well as AMR in a short period. The present review is intended to highlight the culture shift from conventional to modern and advanced technologies including their limitations for the proper and prompt diagnosing of bloodstream infections and AMR detection.

Keywords: antimicrobial resistance; sepsis; early diagnosis; conventional methods; modern methods; advanced methods

Citation: Gupta, E.; Saxena, J.; Kumar, S.; Sharma, U.; Rastogi, S.; Srivastava, V.K.; Kaushik, S.; Jyoti, A. Fast Track Diagnostic Tools for Clinical Management of Sepsis: Paradigm Shift from Conventional to Advanced Methods. *Diagnostics* **2023**, *13*, 277. https://doi.org/10.3390/diagnostics13020277

Academic Editors: Yuli Huang, Yong Yuan and Peisong Chen

Received: 30 October 2022
Revised: 24 December 2022
Accepted: 9 January 2023
Published: 11 January 2023

1. Introduction

Sepsis is one of the most ambiguous disorders in medicine due to its early onset. Previously, sepsis was thought to be the process through which flesh decomposes, swamps acquire a putrid odour, and wounds deteriorate [1]. Later, it was renamed as systemic infection, which is commonly referred to as "blood poisoning" and was acknowledged as a result of pathogenic organisms growing within the circulation and evading the host's immune system. As a result, it was established that the pathogen, not the host, is the perpetrator in the pathophysiology of sepsis [2]. Pathogens interact in the host immune system during infection, triggering a downstream inflammatory cascade including cytokines and other mediators, eventually producing immunosuppression, which leads to various types of organ failure and subsequent clinical degeneration [3].

Originally, sepsis was thought to be the result of internal organs rotting or decaying, with sepsis being defined as a result of the host's systemic inflammatory response syndrome (SIRS) to infection, severe sepsis (sepsis associated with organ dysfunction, hyperfusion, or hypotension), and septic shock (sepsis induced hypotension that persists despite adequate fluid resuscitation) [4]. However, due to a lack of effective antimicrobials and supportive treatment, the definition of sepsis has evolved with time, preventing patients with sepsis from living long enough to be analyzed or they acquire a sequel of organ failure. As a result, the American College of Chest Physicians (ACCP) and the Society of Critical Care Medicine (SCCM) announced SIRS and published new agreements as a definition of sepsis and associated medical criteria (Sepsis-3) with deadly organ dysfunction caused by a dysregulated host response to an infection, which can be accepted as a shift in complete Sequential Organ Failure Assessment (SOFA) score points $\geq$2 subsequent toward the disease [3–5]. The guidelines of the SOFA score were presumed to be zero in the patients having no infection or organ failure whereas a SOFA score $\geq$2 mirrors a general fatality danger of approx. 10% in a common emergency population with suspected infection [6]. This novel and advanced definition focus on the power of the non-homeostatic response of the host to infection, the potential fatality of the infection that is impressively more than a direct disease, and the requirement for dire acknowledgement. As depicted later, even an unpredictable level of organ dysfunction when it was suspected earlier is related to an in-emergency mortality rate of nearly 10%.

According to the World Health Organization (WHO), in hospitals, sepsis is not just the costliest condition to treat as an additional main source of death, but certain reports have assessed a great many cases at a greater expense and death rates influencing more than 30 million individuals worldwide, thus steadily prompting 6 million deaths with rates of mortality between 20% and 50% [7]. The weight of sepsis is probably most noteworthy in low-income countries [8]. A reported 31.5 million people have are diagnosed with sepsis annually around the world, of which, 19.4 million individuals suffer from severe sepsis while 5.3 million individuals experience death, 3 million cases are estimated in neonates worldwide annually and there are 1.2 million cases in children, with a death rate of 11–19% [9], as well as 75,000 annual deaths in females due to puerperal sepsis globally [5]. Additionally, as per the ongoing safety information of Centers for Disease Control and Prevention (CDC) reports, sepsis inpatient admissions stay high for septic shock, roughly 60%; for severe sepsis, around 36%; for sepsis credited to a particular creature, roughly 31%; and for unknown sepsis, roughly 27% [10]. The latest in-hospital mortality estimates for sepsis patients has decreased from 28% to 18% [11]. Patients with severe sepsis admitted to the ICU increased from 7.2% to 11.1% and hospital mortality in severe sepsis decreased from 35% to 18%, according to a recent study of prospectively collected data from more than 90% of all intensive care unit (ICU) hospitalizations, confirming these trends in both incidence and mortality. Finally, the best epidemiologic evidence shows that severe sepsis is becoming more common while simultaneously becoming less fatal [12].

Because of its narrow window, early and systemic diagnosis of sepsis is a crucial task which may deviate towards shock, organ failure or even death. Delay in every hour in providing the correct treatment results in a decline in the survival rate of sepsis patients by 7.6% [13]. To combat this, non-specific broad-spectrum antibiotics are administered immediately in suspected cases which results not only in poor patient outcomes but also in the development of multi-drug resistance (MDR) [14,15]. Hence, the diagnostic method should be rapid and applicable to detect pathogens along with drug resistance (preferably within 3–5 h of patient admission) [16,17]. It should be capable enough to diagnose polymicrobial infections along with unknown and emerging pathogens. Furthermore, the results of the diagnosis should be able to provide appropriate decisions and antibiotic stewardship within a key time window of hours in order to limit morbidity and death [14]. The procedures that are sensitive, specific and quick for identifying the pathogen are therefore the major operational instruments for critical care units [18,19].

The current review discusses the present status of conventional, modern and advanced methods along with their advantages and disadvantages in the deep specialized, translational and application-related scope (Figure 1). The authors searched PubMed and Google Scholar applying the following keywords: sepsis, early diagnosis, conventional as well as modern methods and cited the papers from the year 2018 to 2022. Many case reports and series, as well as retrospective and prospective investigations, systematic reviews and meta-analyses, clinical recommendations and other narrative reviews, have been included by the authors. The search for literature has been limited to research in English exclusively. Over 100 publications were disclosed in the initial literature search. The authors evaluated all the relevant publications and assessed which research to include, with their conventional and modern techniques, including WHO and CDC clinical guidelines, in the review focused on an early identification of the condition.

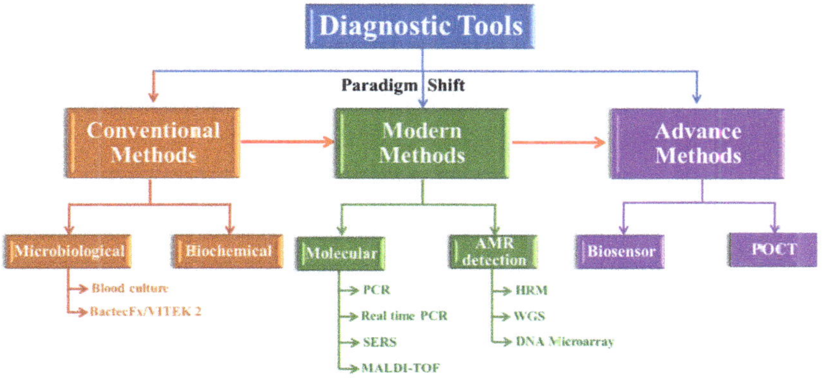

Figure 1. Schematic overview of the diagnostic methods. Conventional methods use the selected cultures and medium in traditional approaches and detect basic aspects of bacterial identification. Molecular detection uses molecular-based (DNA and RNA) approaches to identify resistance genes, as well as mutations in and expression of these genes, or their genomic signature.

2. Conventional Methods for Management of Sepsis

2.1. Microbiological Methods

The detection of pathogens using these techniques is based on the apparent growth of microorganisms on a suitable culture media (solid agar and broth). There are two different methods and they have been discussed in detail in the section below.

2.1.1. Identification through Blood Culture/Gram Staining

For detection as well as identification of causative pathogens, sampling of blood/urine/lavage from patients, their routine culture followed by the Gram staining test remains the gold standard method [20–22]. To enhance the diagnosis, a minimum requirement of blood samples collected aseptically is 0.5–1 mL. Preferably, blood is drawn for blood cultures from two distinct venipunctures sites. One is the central venous catheters, which allows blood to be obtained concurrently from a peripheral and a vascular catheter, allows for faster detection of peripheral bacteremia vs. catheter-related bloodstream infections and appropriate dosing for clinical treatment [23]. Since some microbes can only be identified at the collection site and not in the blood, continuous monitoring of the collection sites is necessary when a positive culture is detected, which facilitates the further processing for pathogen identification in proper sepsis assessment [24].

These are the confirmatory tests to detect the potentiality of microbes in the given sample [25]. Comparatively, gram staining analysis is rapid (<15 min), economical, and provides information about the categorization of the infectious microbe as either Gram-positive, Gram-negative, Gram variable or Gram indeterminate [26,27].

2.1.2. Identification through Bactec Fx/VITEK 2

This automated system is built on sensors that detect any change in pressure within the blood culture bottle or track the CO_2 emitted by actively metabolizing species. Gram-negative microbes take 14 to 24 h in blood culture bottles to detect microbial growth, while Gram-positive bacteria take 24 to 48 h. It is now feasible to eliminate enough pathogenic load for direct identification in Bactec FX without first growing on an agar plate until a blood culture bottle has been marked positive and is suitable for traditional diagnosis [28,29]. Blood culture bottles are prepared, checked and incubated for 5 days at 35 °C with shaking agitation every 10 min in the instrument. By spreading 0.1 mL of consecutive 10-fold dilutions on a blood agar plate, the quantitative plate count technique is used to determine microbial fixation. They are then tested after overnight incubation at 350 °C on plates containing the colonies [30].

The following are the three main issues with using this culture-based method: Firstly, the conventional microbiological tests require 5 days to detect and identify the pathogens involved in sepsis. As a result, the procedure does not provide results promptly, which may be a significant source of anxiety for patients not having the infection. Secondly, the blood culture/Gram stains analysis has a lower sensitivity. Only 30% to 60% of the overall positive result of sensitivity has been estimated despite using it in the proper analytical manner, standardized procedures and accurate collection of amounts of blood sample [28–31]. These recommended false-negative results which range from 40% to 70% might be owing to a scarcity of particular microbial species that flourish in laboratory culture medium, or distantly similar microbial strains [32,33]. Another reason for the false-negative result is self-medication and administration of antibiotics to the patients before the sampling for diagnosis [34]. Thirdly, there is the possibility of false positives due to non-adherence with the sterile condition during sample processing. As a consequence, patients receive antibiotics for those bacteria of which they are not infected. This misuse has led to the prolonged exposure of antibiotics, resulting in allergic reactions, toxicity, development of MDR, a long stay of patients in hospitals and hence increased medical costs [34–36]. As a result, certain points should be proposed for laboratory experts, concerned bodies and other stakeholders to improve the laboratory service by applying and using refined and emerging technology to classify and assess sensitivity to drugs for various microorganisms. However, there is a need for more intensive interventional trials to assess the effect of these technologies on sepsis treatment in clinical practice.

2.2. Biochemical Test

Biochemical tests including the mannitol test, citrate tests, triple sugar iron (TSI) test, indole test, methyl red test and enzymatic tests such as oxidase tests, urease tests and coagulase tests are used to distinguish pathogenic organisms that depend on the various diverging biochemical processes of different bacteria. For the identification of extracellular and intracellular bacterial enzymes, biochemical tests are used. Hence, they are distinguished based on biochemical activities [37].

3. Modern Methods for Management of Sepsis

There is a paradigm shift from conventional culture and biochemical-based detection of pathogens to modern techniques including molecular as well as emerging methods which rely on detection at the strain level in much less time (20 min to 3 h). System biology has been thrust into the spotlight in molecular research due to technological advances and the knowledge provided by the human genome project, which has accelerated multiple methods that may not only produce an expanded understanding of complicated sepsis pathophysiology but can also articulate undetermined methodologies [38,39]. Reviewed here are the modern advancements in early detection, including inventive high-throughput approaches.

3.1. Implementing Molecular Detection for the Identification of Pathogens

Molecular detection using PCR depends upon the amplification of the pathogen's target nucleic acid region using gene-specific primers or probes. For the identification of different pathogens, a variety of molecular targets are used [40–43].

3.1.1. PCR

In 1983, Kary Mullis [44] devised the standard PCR, which allows the detection of a single bacterial pathogen by identifying a specific target DNA sequence [45]. A sensitive examination is portrayed here to classify the microorganisms in the entire bloodstream by the PCR as demonstrated in Figure 2. A particular primer–probe set is intended to reproducibly detect bacteria of purified DNA from whole blood. This assay framework was demonstrated to be comprehensive for all strains, equally from all bacteria. This unique PCR-based test was created to assist amplification from a range of human, bacterial and yeast genomic DNAs due to its inefficiency. A broad sample preparation methodology was designed that was applicable for the DNA purification from various bacteria in whole blood. With the help of this method, it was feasible to distinguish every particular bacterial DNA from whole blood samples inoculated with a minimum of 4 CFU/mL. Co-purified human blood DNA did not influence the sensitivity of detection by PCR [46]. PCR can identify even a single copy of a target DNA sequence under ideal conditions in a given sample. Therefore, prior multiplication enrichment of the microbe is not needed, all things considered with basic DNA probe tests. Thus, PCR-based diagnostic tests have made a significant advanced improvement for infectious agents subsequently [47].

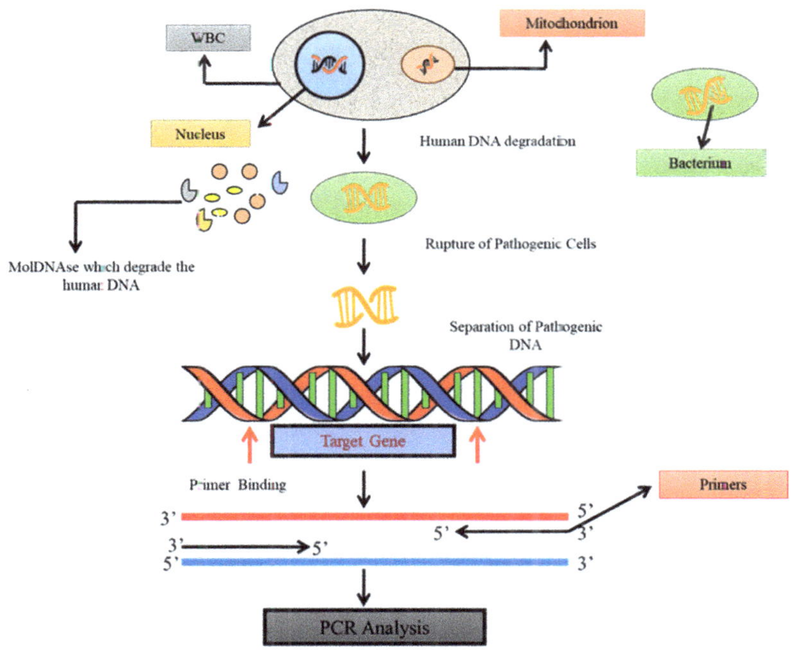

Figure 2. The principle of testing pathogenic DNA by standard PCR in blood samples. Schematic diagram showing the amplification of a segment of DNA, using PCR where the target gene of bacterial DNA is synthesized into two new strands of DNA.

The technology has been updated to expand the usage of traditional PCR. The use of the primers pair in parallel reactions with simultaneous amplification for different target DNA sequences, known as multiplex-PCR, is the first absolute shift. As a result, several DNA sequences replicated in the same processor might be amplified [48]. Another change is nested PCR, which uses two sets of primers with a preferred target for amplification of an internal DNA sequence. The first reaction is carried out using the first set of primers, and the results are subsequently subjected to a second amplification with various primer sets [49].

3.1.2. Real-Time PCR

Despite several advances, high-throughput science research laboratories can redirect standard PCR methods, including nested PCR techniques, due to the immense exposure of excess contamination with amplified products [50]. Traditional PCR procedures rely on automated detection of fluorescence from PCR amplicons, while real-time PCR systems are faster, less sensitive to contamination and require less labour [51,52]. Real-time PCR techniques also allow for infinite and comparable calibration of the desired sequence, which may help to increase the gap in critical microbial potential [53,54]. In addition, compared to traditional PCR, real-time PCR has several novel advantages, including simplicity, quantitative capacity and speed [55].

Fluorescence-based real-time PCR is based on the detection of the fluorescent signal generated during DNA amplification. After a specific number of cycles, the real-time assays synthesize a quantity of target DNA. When the amplification of a PCR product is controlled, it is initially observed during cycling by determining the cycle number at which the reporter dye's discharge energy dominates the background noise. As a result, the threshold cycle is named after this cycle number (Ct). This Ct value is determined during the PCR exponential phase and is inversely proportional to the target's copy number. As a result, the difference in fluorescence signal is observed before the incident when the beginning of the copy number of the target DNA is higher, and the Ct value is lower [56].

SYBR Green, TaqMan, molecular beacons and scorpions are the four types of fluorescent DNA probes currently available for real-time PCR product detection. All of these probes emit a fluorescent signal that can be used to detect the PCR products mentioned in Table 1 [57]. The ability to quantify the process is a key function of RT-PCR [58]. An internal calibrant is put to each well of the assay plates as a reference in each experiment, and PCR is used to examine each well for quantification. The peak heights in the mass spectrum for amplicons determine the proportionate ratios of calibrant to microorganisms [59,60]. The concentration of the pathogen can be thoroughly evaluated using the known initial concentration of calibrant. Having a significant effect on pathogen detection in diagnostic microbiology, real-time PCR is primarily focused on the pathogen genotype [61]. The role of microbes in various human diseases is examined in these essays, which is the most critical use of real-time technology in microbiology. Due to the strength of complexity the technology has positively evaluated the various microbes and their nucleic acid targets present in a single sample, also it has allowed the differentiation of various forms of microbial genotypes in a single reaction tube [62].

Table 1. Comparison of four different types of fluorescent DNA probes currently used in real-time PCR for diagnosing polymicrobial infections.

Fluorescent Molecule	Working	Volume of the Reaction Mixture	Cost-Effective	Sensitivity	Specificity	Sample to Result Time	Detection Limit	Refs.
SYBR Green	Intercalates between the DNA bases to bind to ds-DNA molecules.	10–20 μL	★★★	★	★	2–3 h	60 pg DNA	[63,64]
TaqMan Probes	Taq polymerase performs 5–3 exonuclease activity during hybridization of fluorophore-based detection, cleaving a dual-labeled probe to the corresponding target sequence.	5–10 μL	★	★★	★★★	1–2 h	0.3 pg DNA	[65,66]

Table 1. *Cont.*

Fluorescent Molecule	Working	Volume of the Reaction Mixture	Cost-Effective	Sensitivity	Specificity	Sample to Result Time	Detection Limit	Refs.
Molecular Beacon Probes	The hairpin ring formed by the DNA sequences on the probes ends is designed to be complimentary to one another. The intervening loop portion of the probe is intended to complement the target DNA sequence of interest.	20–25 µL	★	★★	★★★	5–9 h	3–5 pg DNA	[67–69]
Scorpion Probes	On the 5′ and 3′ sides of the probe, complementary stem sequences hold a distinct probe sequence in a hairpin loop shape. After the primer is extended during PCR amplification, the identical probe sequence will attach to its compliment inside the same strand of DNA.	20–25 µL	★	★★	★★★	3–5 h	10 pg DNA	[70–72]

(★ More stars suggest more specificity and higher sensitivity).

The significance of RT-PCR in microbial load detection is that these techniques are extremely useful since they actively reveal the spectrum of increasing infection, host–pathogen interaction and antimicrobial medication effectiveness. It also enables the administration of antibiotics promptly. Real-time assays are useful for distinguishing serotypes within a particular microbial population [73], diagnosing pathogens in clinical samples [74,75] and bacteria (viruses, bacteria, fungi, protozoa or toxins produced by them that cause diseases) used as biological warfare agents [76]. Despite these, molecular techniques including PCR do not provide any information about AMR among pathogens [77].

3.1.3. Surface-Enhanced Raman Spectroscopy (SERS)

Surface-enhanced Raman spectroscopy (SERS) is emerging as an important technology for the identification which amplifies the Raman dispersing of the objective particles on a superficial layer of metal made of graphene or other different materials [78–82]. This technique has the ability to facilitate the label-free nucleic acid identification [83]. Surface plasmons are generated by applying an excitation frequency that is in phase with the particle's plasmon assimilation profile, resulting in a solid electromagnetic field on the metal surface. The emission of the Raman dispersed light is radiated in every direction of the particle is gathered through a microscope and consequently identified. In addition, the Raman dispersed light is then coordinated against a reference profile of microorganisms to be recognized [84–87]. Consequently, SERS can efficiently recognize the occupancy of microbial cells on the outer surface, thus yielding a data-rich spectrum that can be used for microorganism identification [88–94]. In addition to the pathogen identification, SERS has also been utilized for antibiotic susceptibilities in urosepsis [91]. The primary reason the SERS procedure has not been set up as a routine scientific strategy is that it does not withstand its high sensitivity and specificity, which are the significant disadvantage of SERS, and restricted ability in investigating polymicrobial tests which are because of the lower reproducibility of the SERS signal [95]. Thus, SERS combined with different methods should be incorporated which outlines the wide uses of this incredible method.

3.1.4. MALDI-TOF

Matrix-assisted laser desorption ionization-time of flight mass spectroscopy (MALDI-TOF-MS) is another newly discovered procedure that is now being used in clinical research to detect bacterial species. This technique guarantees fast detection of causative bacterial microorganisms demonstrated to be viable in the positive blood cultures that can rapidly recognize bacterial development, hence accelerating the general process of the antimicrobial resistance report [96]. MALDI is a protein identification ionization technique in which the analyte crystallizes in a strong lattice matrix crystal that absorbs laser light, allowing it to ionize and desorb from the matrix. The ionized atoms are isolated and dependent on their mass to charge (m/z) ratio from the entire microscopic organism's test when it flies through a vacuum tube produces an m/z profiles of the apparent multitude of proteins in the sample [97]. Mass spectroscopy has been utilized for the inoculation of different species, along with the microbial suspension. Because of the low microbial concentration, MS cannot conduct direct examinations on human blood samples, which is a major disadvantage. Because of the low reproducibility and changeability in preliminary methodology and the matrix composition, blood culture is typically needed to improve the microscopic organisms to a detachable level [98]. Another drawback of MS is its restricted capacity in arranging various microbes from the polymicrobial samples [99–101] as the spectral profiles formed by this technique are more complicated, making it difficult to deconvolute the composite spectra gathered at the same time from different microbial species in the poly-microbial samples.

3.2. *Broad-Spectrum Genomic Detection of AMR*

The currently employed technique to detect AMR is time taking (3 to 5 days) and hampers the clinical management of sepsis results in a poor outcome. This protocol is discussed herein Figure 3 for some of the methods. With the advent of genome sequencing, AMR can be detected for both targeted (high resolution melting) and non-targeted (sequencing) pathogens.

3.2.1. High Resolution Melting Analysis Technology

High resolution melt (HRM) analysis depends on the detection of differences in melting temperature (Tm) due to the presence of a mutation in a previously amplified target that produces melt curve profiles specific to pathogens. The size and the sequence of the PCR amplicon is the major reason on which the melting curve profile depends [102]. It is so sensitive that even a single point mutation resulting in a Tm shift can be detected [103]. Therefore, it allows molecular detection of resistant genes and hereditary mutations rapidly with a higher output of post-PCR examination which allows the researchers to identify and classify the new hereditary mutations and variations along with single nucleotide polymorphisms without sequencing (gene scanning) or before sequencing in a population [104]. Several antibiotic resistance marker genes conferring to a bacterium have been usually detected (Table 2). A study reported real-time PCR-based rapid identification of Escherichia coli, Staphylococcus aureus, Enterococcus faecalis and Proteus mirabilis using 16S rRNA gene-specific primers. Furthermore, HRM and machine learning algorithm approaches were used to determine the antimicrobial susceptibility test within 6.5 h [105].

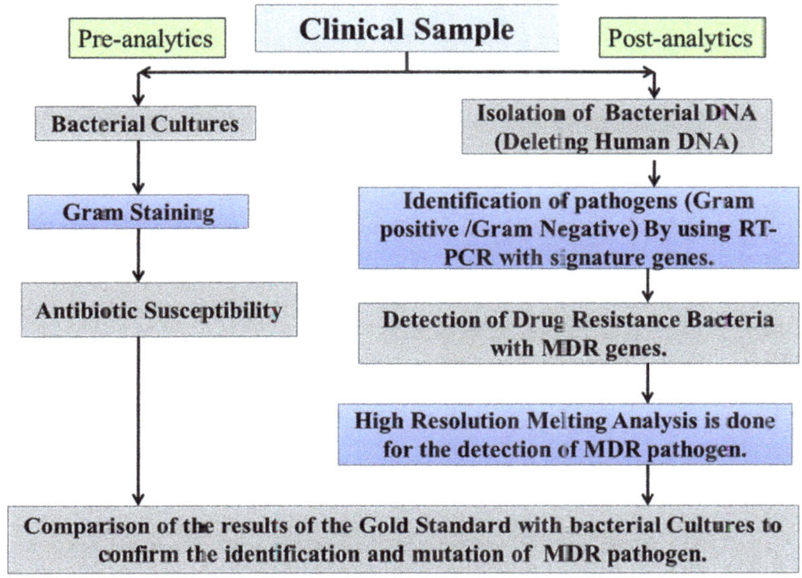

Figure 3. Molecular antimicrobial detecting testing assays. The ideas presented here are relevant to novel antimicrobial testing methods based on nucleic acid amplification. Arrows follow the direction from a clinical specimen to a diagnostic result. A process for the molecular test (right-sided) has been demonstrated, which includes quick sample preparation, amplification and amplicon characterization.

Table 2. List of antibiotics classes, resistance marker genes and bacteria used in real-time PCR-HRM.

Antibiotics Class	Genes	Bacteria	Reference
β-lactamases and Cephalosporins	*ctxM*	*Klebsiella pneumonia*	[106]
	Tem	*K. pneumonia*, *Pseudomonas aeruginosa* *Escherichia coli*, *Acinetobacter baumannii*	[106] [107] [104] [108,109]
	Shv	*K. pneumonia*	[106]
Fluoroquinolones	*parC*	*Streptococcus* sp.	[109]
	gyrA	*Streptococcus* sp.	[109]
Vancomycin	*vanA*	*Staphylococcus aureus*	[110]
	vanB	*Enterococcus faecalis*	[111]
	vanC	*E. faecalis*	[111]
Methicillin	*mecA*	*S.aureus*	[112]
	mecC	*S.aureus*	[113]

3.2.2. Sequencing

In the diagnostic field, the aim of the utilization of genomic rather than gene-based techniques for both bacterial species detection and AMR detection is growing. There is a need for more effective and rapid AMR preventive measures driving the shift to whole-genome sequencing (WGS). Bacterial and AMR gene identification using automated bioinformatics examination methods are easy to perform after an organism has been isolated by culture [114]. WGS allows all genes involved in resistance to be tracked, allowing all genomic data of resistant factors to be present in a bacterial cell to be analyzed. Next-generation sequencing (NGS), which has revolutionized the biological sciences, is another emerging tool. NGS makes large-scale whole-genome sequencing (WGS) affordable and realistic for the average researcher with its super high throughput, versatility and speed. It also allows scientists to sequence the entire human genome in a single experiment, allowing them to research biological processes at a level never before possible. Furthermore, in the age of complex genomic science, which necessitates a deeper understanding of details outside the boundaries of conventional DNA technology, it has filled the gap and become a routine research method to resolve those issues [115]. NGS combined with meta-genomic approaches that essentially include genome sequencing of infectious biological samples such as blood, urine and lavage without culturing them and provide the diverse profile of all species including those that are targeted and untargeted present in the sample. This approach has revolutionized the identification of all including new resistance genes in a single specimen [116,117]. The sequencing-based metagenomic approach has analyzed the kinetics of gut microbiota before, during and after antibiotic treatment [118]. Furthermore, with the help of machine learning, it is now possible to predict the recolonization of micro-biota post-antibiotic treatment, hence will be helpful in individual patient-specific antibiotic treatment regimens. Expensive and more labour-intensive factors while handling a significant number of samples are the major drawbacks in the wide application of these genome-based methodologies despite their promising applications in pathogen detection [119].

3.2.3. DNA Microarray

Microarray innovation has been utilized for longer than 10 years for the identification of microorganisms which have altogether added to our comprehension of pathogenic mechanisms, microbe reactions to ecological improvements and host–microorganism associations with the immediate effect on diagnostic microbiology [120]. Microarrays have been updated as useful methods for bacterial detection and identification due to their strong parallelism in screening for the expression of a wide variety of genes after specific gene amplification by either a broad-range or a multiplex-PCR before microarray analysis [121] Microarrays employ surface-immobilized DNA and RNA probes to collect and categorize DNA/RNA of microorganisms via sequence-specific complementary hybridization, decreasing sample and reagent consumption and costs while permitting precise segregation down to the species or strain level. A study conducted by Ballarini et al. utilized an oligonucleotide-based microarray (BactoChip) for culture-independent detection, quantification as well as differentiation from 21 different bacterial genera among clinical isolates [122]. Additionally, the Verigene stage from Luminex Corporation can distinguish the Gram-positive board of nine species of bacteria and three genes of AMR against methicillin and vancomycin and five species and six AMR genes for carbapenemase and expanded range beta-lactamases [123–125]. Being independent of culture, microarray-based detection is rapid, hence gaining importance in clinics in combination with antimicrobial stewardship [126,127]. Despite these, not a single microarray platform has been commercialized so far to effectively recognize all microorganisms in polymicrobial diseases [128].

4. Advanced Methods for Management of Sepsis

Enhancement in the sensitivity and specificity of the current technology with the correct and rapid data analysis in a short time along with identification of drug susceptibility to cumbersome resistance is needed within an hour [129].

4.1. Biosensors

Since biosensors have more acknowledged extension for the accurate diagnosis of sepsis, accordingly, headway in this field has given novel world-view progressed highlights. The best methodology for the use of this innovation is higher sensitivity for the diagnosis and identification giving a helping hand to clinicians all over the world to control the disturbing death rate [130]. Because species–specific probes or antibodies give an electrical signal after attaching to their targets, and that signal intensity correlates to the target species total signal, the electrochemical method is the primary criteria on which diagnostic biosensors frequently operate towards identification. They do not just distinguish microbes with an incredible specificity very quickly to hours; however, they also give the data about drug sensitivity [131]. Most biosensors have the constraints in their degree and insufficient for the expansive range recognition as their sensitivity might not efficient enough to distinguish BSI where the pathogenic concentration can be a simple 1–10 CFU/mL in the blood [132] but they offer a quick strategy for analysis of a particular infection or microorganisms utilizing just little sample volumes [131]. More research work is required on biosensors to minimize the expenses and enhance their sensitivity, as the specificity and pace of current advances are unusual and may be useful in critical situations.

4.2. Point of Care Test

The restricted advancement which has been previously occurred for the improvement of diagnostics and therapeutics for septicemia is the absence of progress in the immuno responses from sepsis patients has created a challenging part for the advancement of powerful immune therapy along with the proper organization of the anti-infectious agents which activates the dangerous organ dysfunction. Currently, the treatment technique focuses on the administration of anti-toxins, fluid resuscitation and vasopressors [133]. Several studies have shown that early detection of sepsis events and prompt care improves medical outcomes [134–137]. However, several other studies have also shown that early antibiotic therapy has less significant effect as compared to the control patient cohort, demonstrating the disease's variability and the need for ongoing testing and treatment [138,139]. As a result, optimal point-of-care sensors allow for the rapid compilation of data related to a patient's health status, as well as increased health coverage and enhanced healthcare service efficiency while lowering healthcare costs [140,141] (Figure 4). Furthermore, it offers additional pathogen and host–response information practically anywhere with a short processing period, enabling sepsis treatment in two main streams: first, these devices could accelerate the process where optimal care is initiated late, enhancing results. and second, they can quantify multiple entities such as pathogens, plasma proteins and cell-surface proteins, as described as representative of the host immune response, which, when paired with complex data analytics, may help stratify sepsis even at the hospital bedside. Such data could help to expedite the identification of patients who would benefit from the increased treatment [134]. POCT can also potentially diagnose the assessment of the progression of several proteins' biomarkers (IL-6, IL-10, TNF-a, PCT and CRP) to acute sepsis or septic shock in patients in ICUs, as well as to measure the 28-day risk of all-cause mortality [142] to help with the antibiotic. However, certain POCT parameters, such as temperature regulation and optical signal readout, also necessitate resource-intensive instruments. So far, the research community of POCT has not been informed of the crucial need for polymicrobial infection diagnostics research in this sector, which has been substantially insufficient [143].

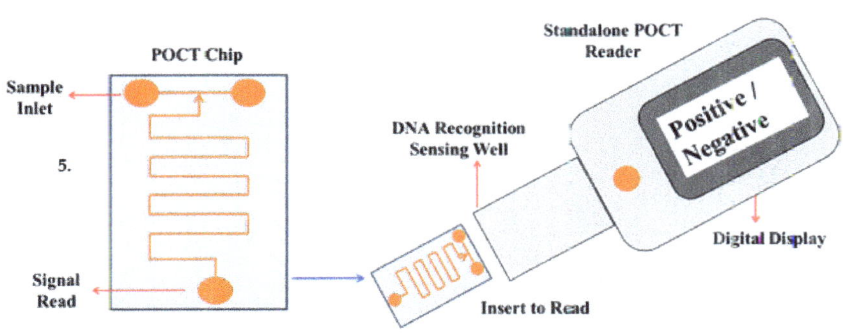

Figure 4. Diagrammatic representation of an ideal POC sensor is shown here which gives the critical information regarding sepsis including both pathogen and host–response information that provides bimolecular identification of the pathogen and quantification of host–response biomarkers.

4.3. CRISPR-Cas9

Advances in biology have been fueled by the technology used to create and modify DNA. The insertion of site-specific changes into the genomes of cells and animals, on the other hand, is still a mystery [144]. CRISPR-Cas9 is a ground-breaking genetic engineering tool that has revolutionized the use of biological RNA-programmable CRISPR-Cas9 in laboratories all over the world [145]. CRISPR/Cas9 is a revolutionary method for editing genomes that is gradually finding applications in numerous domains of biomedical research, including sepsis, as a new way to investigate and cure diseases. CRISPR-Cas9 technology is based on the CRISPR-Cas system, which gives viruses and plasmids in bacteria adaptive immunity. Cas9 is a CRISPR-associated endonuclease that uses the RNA duplex leader sequence tracrRNA:crRNA to base pair with a DNA target sequence, allowing Cas9 to function and induce site-specific double-strand breaks in DNA [146]. The single guide RNA (sgRNA) is designed from the double tracrRNA:crRNA with two key features, i.e., a sequence on the fifth side that determines the DNA target site through base pairing, while a double-stranded RNA structure on the third side connects to Cas9 [147]. Such discovery results in an uncomplicated two-component approach [148] that uses alterations in the Cas9 sgRNA program's leader sequence to target any DNA sequence of interest [149].

The RNA-guided Cas9 in CRISPR/Cas9 technology for targeted genome editing causes a blunt-ended double-stranded break (DSB) three base pairs upstream of the proto adjacent motif (PAM) domain. To repair DSBs, nonhomologous end joining (NHEJ)- or homology-directed repair (HDR)-mediated disruption or mutation of the genome is utilized where double-strand breaks caused by Cas9 can be repaired in one of two ways. Endogenous DNA repair machinery processes DSB ends before rejoining them via the error-prone NHEJ pathway, which may result in random mutations at the junction point. To take advantage of the HDR pathway, which allows for high fidelity and precise editing, a repair template, such as a plasmid or single-stranded oligodeoxynucleotide (ssODN), can also be given [150]. HDR can also be induced by single-stranded DNA nicks. The unique DNA cleavage mechanism, numerous target recognition capabilities and the presence of many varieties of the CRISPR-Cas system have permitted tremendous improvements in the utilization of this inexpensive and simple-to-operate system-using technology. Genomic loci can be precisely identified, altered, modified, regulated and tagged in a variety of cells and organisms [151,152].

This breakthrough gene editing approach allows for the simultaneous genetic modification of many genes in cells, paving the way for a new class of diagnostics and treatments. This technique is perhaps applied to alter the genomes of contagious microbes as well as host cell genes implicated in the pathogenesis of many illnesses, allowing for better treatment [153]. Furthermore, the potential of the immune system to resist infections can be strengthened using CRISPR/Cas9 technology. Despite advancements in general healthcare, the sepsis high death rate and its repercussions remain unacceptably high. The absence of appropriate therapies is the major cause of this high morbidity and death. Because existing clinical investigations failed to establish wide combination targets, developing creative and evolutionary treatment options to enhance clinical outcomes in sepsis is a primary objective in the prognosis of patients with sepsis [154].

To some extent, sepsis might be considered a hereditary disease, and gene therapy could be a promising new treatment option. In this review work [155], we present a brief application of CRISPR/Cas9 gene editing technology in sepsis research. To begin, here are some guidelines for gene screening: TNF stimulates inflammatory responses while also causing vascular endothelial cell death in sepsis to protect the host from infection. The researchers employed a genome-wide CRISPR/Cas9 knockdown screen to look for potential targets in the TNF signaling system, which controls mild inflammation and apoptosis during sepsis. The benefit of genome-scale screening using CRISPR/Cas9 in sepsis is clearly demonstrated in this work [156]. Second, researchers discovered that mitochondrial DNA (mtDNA) stimulated TLR9 signaling in septic patients, which was probably associated with early inflammation and associated with higher mortality rate in severely ill patients [24].

TLR9 signaling may be triggered by mtDNA, resulting in systemic inflammation, which includes inflammatory mediator overproduction and leukocyte activation [157]. To treat people infected with mutant mtDNA, CRISPR-Cas9 technology can be utilized to delete specific sequences of mutated DNA [158]. The Ccl2 gene product regulates the trafficking of inflammatory monocytes/macrophages, basophils and T lymphocytes in response to inflammatory signals such as TNF- and IL-1 [159]. They modified mouse iPSCs using CRISPR/Cas9 and introduced anti-inflammatory chemicals into the Ccl2 gene. According to the researchers, CRISPR/Cas9 might be used to develop a directed cell-based anticytokine vaccine, which could lead to new pathophysiologic insights and treatments for inflammatory diseases [160].

The therapy options for this condition are currently limited, and employing CRISPR/Cas9 technology to better understand the molecular pathophysiology of sepsis might aid in the development of efficient treatment techniques [161]. Despite the fact that CRISPR/Cas9 revolutionized genome editing, it is far from complete and has several limitations [162]. CRISPR is indeed an RNA-based DNA recognition system that uses an RNA-based guide molecule to detect regions in DNA that have certain molecular features. One of the limitations is that the original Cas9 can only cleave a small portion of the genome if it hits on a genomic region with three "NGG" nucleotide base pairs (N can be any nucleotide). Another drawback is that the Cas9 enzyme only cleaves DNA; approximately 2% of the genome encodes proteins directly from DNA, while the rest, 98%, is the regulatory gene sequence [163]. CRISPR/Cas9 technology may be the simplest and most successful way to perform sepsis-related research; however, due to technical instability and targeting gene constraints, CRISPR/Cas9 research for sepsis therapy requires substantial progress [164].

5. Challenges for a New Improved Method

The significant issues confining the fruitful execution of new strategies are the requirement for considerable information, the time it takes to obtain a result, industrial competition and the complexity of the evolving improved methods. In addition, since an experienced individual is needed, these assays can seem difficult to conduct in ordinary microbiology laboratories [165]. Furthermore, the current situation brings benefit to the clinical microbiology lab by allowing for the cost-effective transportation of samples to the lab as well as to the organization of computer-based data results. Nonetheless, the transition time is lengthened by the distance between the laboratory and the time when a sample is collected at the bedside from different hospitals. As a result, laboratories should be prescribed such techniques that prioritize time from sample processing to lab work schedule to patient care, as well as infection prevention steps, for the rapid diagnosis of sepsis. Prior sequencing data for the specific target gene of interest is required for molecular-based approaches. One of the significant drawbacks of molecular processes is that they can only be used to manipulate the identified genes. As a result, accessing the unknown sequence of the gene using some other procedure is an undesirable pitfall of the sequence-based problems that have already been identified [166,167]. When analyzing the data, it is important to remember that the presence of suspected bacteria or DNAemia (the detection of circulating pathogenic DNA responsible for a specific disease) does not always suggest the existence of microbes. Because of the detection of environmental DNA infecting the blood sample or carryover contamination, it might potentially increase the chance of false-positive outcomes. DNAemia is an infection-related condition that can be caused by false septicemia [168,169] or by circulating DNA that persists after many days of successful anti-infectious therapy [170].

Another significant drawback is that they cannot provide any detail on antimicrobial resistance or the detection of the pathogens that may be used to diagnose sepsis. Rapidity in detecting pathogens may allow an excellent and more advanced calibration of this efficient procedure, resulting in greater commercial savings. However, the major issue is the shortcoming of a distinct susceptibility spectrum, particularly with the emergence of multidrug resistant microbes, which could restrict the clinical usage of these assays.

Therefore, according to the current situation, it is very important to invest less energy on limited and steady enhancements to existing innovation, yet rather to target for generous improvements so novel approaches with prevalent execution attributes can be affirmed and promoted at the earliest opportunity [171].

6. Conclusions and Future Direction

Despite understanding the awareness of the common molecular methods for calibrating gene expression, it is critical to learn about the various choices available in all facets of this science. In comparison to traditional PCR, real-time PCR is much more complex and has a major impact on the final performance. As a consequence, molecular methods can be one of the most responsive and effective methodologies for studying any gene expression analysis in a proper outline carried out with the proper controls [172]. The innovations require enhancements of both pre-analytic and post-analytic updates at the clinical level bringing the issues focused on AMR to more extensive public consideration is an absolute necessity. This will require advancing a superior comprehension of antimicrobial use for reasonable clarifications to the overall population about the drug determination along with the improvement of multidrug resistance from an overdose of antimicrobial agents [173].

With the introduction of novel advancements such as low-cost, high-capacity liquid-handling systems and nucleic acid extraction, modern approaches and innovations are becoming more fundamental and interesting for routine diagnostics. Recent advancements in technology have made it strong and reliable while making it reasonable to challenge easy identification and genotyping with fast reactions and the measurement of a single DNA target following the related quality confirmation programs. There is also a need to improve and verify the full assessment of any newly designed assay against previously used standardized assays, as well as the accuracy of specifications with their appropriate curves. However, recent technology shows that this is a breakthrough that has now arrived [174]. A few promising new techniques and standards are being created and applied in the clinical arena, as we have sketched out in this work, and such upgraded processes will aid and oversee AMR.

In Table 3, we summarize and compare the performance of the technologies in this review in terms of cost, sensitivity, specificity, turnaround time and multiplexing capability of polymicrobial infection diagnosis. Despite progress in technical and clinical improvements, many unanswered questions remain in this sector, necessitating collaboration with not only clinical collaborators, but also regulatory and funding bodies, the diagnostics industry and public health agencies before new approaches such as healthcare systems, skilled nursing and public health agencies can be implemented [175]. Furthermore, a range of advancements will be required to work together, such as the ailments for clinical trials, the use of biomarkers as a technological breakthrough, and these revolutionary developments arise as future directions to focus on a new ground-breaking concept to the diagnosis of sepsis. To keep working on these present diagnostic goals, the prior focus must be on a single hand that is actionable and robust in a well-planned approach to include these pathogen identification results for the treatment of patients with verified disease who will have the desired therapeutic outcomes. As a result, infection to host responses have been selected as the most interesting new technological technique and a viable alternative to pathogen-based diagnoses in the assessment of significant progress [176]. Thus, prospective studies should focus on supervised analysis for correct patient treatment, an integrated "ideal" approach to the diagnosis of sepsis [177,178], as well as labour preparation in wards and labs, which are key ingredients for fruitful programs.

Table 3. Different infectious polymicrobial diseases: a comparison of molecular diagnostic technologies.

Technology	Cost-Effective	Sensitivity	Specificity	Turnaround Time	Multiplexing Capability	Refs.
Microbiological Methods						
Blood Culture/Gram Staining	★★★	★	★	★★★	★	[179]
BactecFx/VITEK 2	★★	★★	★★	★★★	★★	[180]
Biochemical Methods	★★★	★	★	★★★	★	[181]
Modern Methods Molecular Methods						
Real-Time PCR	★★	★★★	★★	★★★	★★	[182]
SERS	★★	★★	★★	★★	★★	[183]
MALDI-TOF	★	★	★★	★	★	[184]
AMR Detection Methods						
HRM	★★	★★	★★	★★★	★	[185]
Sequencing	★	★★★	★★★	★	★★★	[173]
DNA Microarray	★	★★	★★	★★	★★★	[186]
Advanced Methods						
Biosensors	★	★★	★★	★	★★★	[187]
POCT	★	★★	★★	★	★★★	[186,187]
CRISPR/Cas9	★	★★	★★	★★	★★★	[188]
CRISPR/Cas9						

(★ More stars suggest more cost effective, less sensitive and specificity, more turnaround time and vice-versa according to the technology).

Author Contributions: Conceptualization, E.G., J.S. and A.J.; formal analysis, S.K. (Sanni Kumar), U.S., S.R., V.K.S., S.K. (Sanket Kaushik) and A.J.; original draft preparation, E.G., J.S. and A.J.; supervision, A.J. All authors have read and agreed to the published version of the manuscript.

Funding: Financial support from the Indian Council of Medical Research, New Delhi, India (2019-5517/CMB-BMS) is acknowledged.

Institutional Review Board Statement: Not applicable.

Informed Consent Statement: Not applicable.

Data Availability Statement: Data included in the manuscript are available and can be shared if required.

Conflicts of Interest: Authors declare no conflict of interest.

References

1. Kumar, S.; Payal, N.; Srivastava, V.K.; Kaushik, S.; Saxena, J.; Jyoti, A. Neutrophil extracellular traps and organ dysfunction in sepsis. *Clin. Chim. Acta* **2021**, *523*, 152–162. [CrossRef] [PubMed]
2. Guirgis, F.; Black, L.P.; DeVos, E.L. Updates and controversies in the early management of sepsis and septic shock. *Emerg. Med. Pract.* **2018**, *20*, 1–28.
3. Kumar, S.; Gupta, E.; Srivastava, V.K.; Kaushik, S.; Saxena, J.; Goyal, L.K.; Mehta, S.; Jyoti, A. Nitrosative stress and cytokines are linked with the severity of sepsis and organ dysfunction. *Br. J. Biomed. Sci.* **2019**, *76*, 29–34. [CrossRef] [PubMed]
4. Gyawali, B.; Ramakrishna, K.; Dhamoon, A.S. Sepsis: The evolution in definition, pathophysiology, and management. *SAGE Open Med.* **2019**, *7*, 2050312119835043. [CrossRef] [PubMed]
5. Rudd, K.E.; Johnson, S.C.; Agesa, K.M.; Shackelford, K.A.; Tsoi, D.; Kievlan, D.R.; Colombara, D.V.; Ikuta, K.S.; Kissoon, N.; Finfer, S.; et al. Global, regional, and national sepsis incidence and mortality, 1990–2017: Analysis for the Global Burden of Disease Study. *Lancet* **2020**, *395*, 200–211. [CrossRef] [PubMed]
6. Rudnov, V.A.; Kulabukhov, V.V. Sepsis-3: Updated main definitions, potential problems and next practical steps. *Messenger Anesthesiol. Resusc.* **2018**, *13*, 4–11. [CrossRef]

7. Masoudifar, M.; Gouya, M.M.; Pezeshki, Z.; Eshrati, B.; Afhami, S.; Farzami, M.R.; Seifi, A. Health care-associated infections, including device-associated infections, and antimicrobial resistance in Iran: The national update for 2018. *J. Prev. Med. Hyg.* **2021**, *62*, E943. [CrossRef] [PubMed]
8. Fleischmann-Struzek, C.; Mellhammar, L.; Rose, N.; Cassini, A.; Rudd, K.E.; Schlattmann, P.; Allegranzi, B.; Reinhart, K. Incidence and mortality of hospital-and ICU-treated sepsis: Results from an updated and expanded systematic review and meta-analysis. *Crit. Care Med.* **2020**, *46*, 1552–1562. [CrossRef]
9. Makic, M.B.F.; Bridges, E. CE: Managing sepsis and septic shock: Current guidelines and definitions. *Am. J. Nurs.* **2018**, *118*, 34–39. [CrossRef]
10. Buchman, T.G.; Simpson, S.Q.; Sciarretta, K.L.; Finne, K.P.; Sowers, N.; Collier, M.; Chavan, S.; Oke, I.; Pennini, M.E.; Santhosh, A.; et al. Sepsis among medicare beneficiaries: 1. The burdens of sepsis, 2012–2018. *Crit. Care Med.* **2020**, *48*, 276. [CrossRef]
11. Chávez-Vivas, M.; Cristo-Martínez, D.; Tascón, A.J. Epidemiological characteristics of patients diagnosed with sepsis and septic shock in a hospital in Cali, Colombia. *Acta Med. Costarric.* **2018**, *60*, 150–156.
12. Fleischmann-Struzek, C.; Mikolajetz, A.; Schwarzkopf, D.; Cohen, J.; Hartog, C.S.; Pletz, M.; Gastmeier, P.; Reinhart, K. Challenges in assessing the burden of sepsis and understanding the inequalities of sepsis outcomes between National Health Systems: Secular trends in sepsis and infection incidence and mortality in Germany. *Intensive Care Med.* **2018**, *44*, 1826–1835. [CrossRef] [PubMed]
13. Abe, T.; Ogura, H.; Kushimoto, S.; Shiraishi, A.; Sugiyama, T.; Deshpande, G.A.; Uchida, M.; Nagata, I.; Saitoh, D.; Fujishima, S.; et al. Variations in infection sites and mortality rates among patients in intensive care units with severe sepsis and septic shock in Japan. *J. Intensive Care* **2019**, *7*, 28. [CrossRef] [PubMed]
14. Gajdács, M. The concept of an ideal antibiotic: Implications for drug design. *Molecules* **2019**, *24*, 892. [CrossRef]
15. van Belkum, A.; Burnham, C.A.D.; Rossen, J.W.; Mallard, F.; Rochas, O.; Dunne, W.M. Innovative and rapid antimicrobial susceptibility testing systems. *Nat. Rev. Microbiol.* **2020**, *18*, 299–311. [CrossRef]
16. Kethireddy, S.; Bilgili, B.; Sees, A.; Kirchner, H.L.; Ofoma, U.R.; Light, R.B.; Mirzanejad, Y.; Maki, D.; Kumar, A.; Layon, A.J.; et al. Culture-negative septic shock compared with culture-positive septic shock: A retrospective cohort study. *Crit. Care Med.* **2018**, *46*, 506–512. [CrossRef]
17. Falcone, M.; Bassetti, M.; Tiseo, G.; Giordano, C.; Nencini, E.; Russo, A.; Graziano, E.; Tagliaferri, E.; Leonildi, A.; Barnini, S.; et al. Time to appropriate antibiotic therapy is a predictor of outcome in patients with bloodstream infection caused by KPC-producing Klebsiella pneumoniae. *Crit. Care* **2020**, *24*, 24. [CrossRef]
18. Nath, P.; Kabir, A.; Khoubafarin Doust, S.; Kreais, Z.J.; Ray, A. Detection of bacterial and viral pathogens using photonic point-of-care devices. *Diagnostics* **2020**, *10*, 841. [CrossRef]
19. Choi, J.A.; Bae, S.M.; Kim, J.W.; Lee, K.J. Development of a Two Triplex Real-Time Polymerase Chain Reaction for Rapid Detection of Six Carbapenemase Genes in Enterobacteriaceae. *Osong Public Health Res. Perspect.* **2020**, *11*, 53. [CrossRef]
20. Mejia-Chew, C.; O'Halloran, J.A.; Olsen, M.A.; Stwalley, D.; Kronen, R.; Lin, C.; Salazar, A.S.; Larson, L.; Hsueh, K.; Powderly, W.G.; et al. Effect of infectious disease consultation on mortality and treatment of patients with candida bloodstream infections: A retrospective, cohort study. *Lancet Infect. Dis.* **2019**, *19*, 1336–1344. [CrossRef]
21. Edmiston, C.E.; Garcia, R.; Barnden, M.; DeBaun, B.; Johnson, H.B. Rapid diagnostics for bloodstream infections: A primer for infection preventionists. *Am. J. Infect. Control* **2018**, *46*, 1060–1068. [CrossRef] [PubMed]
22. Sato, H.; Nakao, A.; Sato, K.; Otomo, Y.; Niijima, S.; Shimizu, T. Comparison of time to positivity of pediatric blood cultures obtained within the first year of life and in later years. *J. Infect. Chemother.* **2020**, *26*, 813–817. [CrossRef] [PubMed]
23. Salinas, M.; López-Garrigós, M.; Flores, E.; Leiva-Salinas, C. Current Practice and Regional Variability in Recommendations for Patient Preparation for Laboratory Testing in Primary Care. *Lab. Med.* **2020**, *51*, e32–e37. [CrossRef] [PubMed]
24. Zelellw, D.A.; Dessie, G.; Worku Mengesha, E.; Balew Shiferaw, M.; Mela Merhaba, M.; Emishaw, S. A Systemic Review and Meta-analysis of the Leading Pathogens Causing Neonatal Sepsis in Developing Countries. *BioMed Res. Int.* **2021**, *2021*, 6626983. [CrossRef] [PubMed]
25. Özenci, V.; Strålin, K. Clinical implementation of molecular methods in detection of microorganisms from blood with a special focus on PCR electrospray ionization mass spectrometry. *Expert Rev. Mol. Diagn.* **2019**, *19*, 389–395. [CrossRef]
26. Quirino, A.; Marascio, N.; Peronace, C.; Gallo, L.; Barreca, G.S.; Giancotti, A.; Lamberti, A.G.; Colosimo, M.; Minchella, P.; Trecarichi, E.M.; et al. Direct antimicrobial susceptibility testing (AST) from positive blood cultures using Microscan system for early detection of bacterial resistance phenotypes. *Diagn. Microbiol. Infect. Dis.* **2021**, *101*, 115485. [CrossRef]
27. Butler-Laporte, G.; Yansouni, C.P.; Paquette, K.; Lawandi, A.; Stabler, S.N.; Akhter, M.; Davidson, A.C.; Gavric, M.; Jinah, R.; Saeed, Z.; et al. September. Real-word time-to-positivity of two widely used commercial blood culture systems in patients with severe manifestations of sepsis: An analysis of the FABLED study. *Open Forum Infect. Dis.* **2020**, *7*, ofaa371. [CrossRef]
28. Rule, R.; Paruk, F.; Becker, P.; Neuhoff, M.; Chausse, J.; Said, M. Diagnostic accuracy of the BioFire FilmArray blood culture identification panel when used in critically ill patients with sepsis. *J. Microbiol. Methods* **2021**, *189*, 106303. [CrossRef]
29. Rodrigues, C.; Siciliano, R.F.; Charbel, C.E.; de Carvalho Sarahyba da Silva, L.; Baiardo Redaelli, M.; de Paula Rosa Passetti, A.P.; Franco, M.R.G.; Rossi, F.; Zeigler, R.; De Backer, D.; et al. The effect of a rapid molecular blood test on the use of antibiotics for nosocomial sepsis: A randomized clinical trial. *J. Intensive Care* **2019**, *7*, 37. [CrossRef]

30. Lin, J.F.; Ge, M.C.; Liu, T.P.; Chang, S.C.; Lu, J.J. A simple method for rapid microbial identification from positive monomicrobial blood culture bottles through matrix-assisted laser desorption ionization time-of-flight mass spectrometry. *J. Microbiol. Immunol. Infect.* **2018**, *51*, 659–665. [CrossRef]
31. Wu, Y.; Yao, Y.M.; Ke, H.L.; Ying, L.; Wu, Y.; Zhao, G.J.; Lu, Z.Q. Mdivi-1 protects CD4+ T cells against apoptosis via balancing mitochondrial fusion-fission and preventing the induction of endoplasmic reticulum stress in sepsis. *Mediators Inflamm.* **2019**, *2019*, 7329131. [CrossRef]
32. Ransom, E.M.; Alipour, Z.; Wallace, M.A.; Burnham, C.A.D. Evaluation of optimal blood culture incubation time to maximize clinically relevant results from a contemporary blood culture instrument and media system. *J. Clin. Microbiol.* **2021**, *59*, e02459-20. [CrossRef] [PubMed]
33. Du Plessis, A. Influence of Blood Culture Results on Antimicrobial Prescribing in a Private Hospital in North West, South Africa. Ph.D. Thesis, North-West University, Potchefstroom, South Africa, 2020.
34. Chou, W.K.; Vaikunthan, M.; Schröder, H.V.; Link, A.J.; Kim, H.; Brynildsen, M.P. Synergy screening identifies a compound that selectively enhances the antibacterial activity of nitric oxide. *Front. Bioeng. Biotechnol.* **2020**, *8*, 1001. [CrossRef] [PubMed]
35. Kaczor, A.; Witek, K.; Podlewska, S.; Sinou, V.; Czekajewska, J.; Żesławska, E.; Doroz-Płonka, A.; Lubelska, A.; Latacz, G.; Nitek, W.; et al. Molecular insights into an antibiotic enhancer action of new morpholine-containing 5-arylideneimidazolones in the fight against MDR bacteria. *Int. J. Mol. Sci.* **2021**, *22*, 2062. [CrossRef]
36. Bakhit, M. Antibiotic Resistance: Patient-Clinician Communication and Decision-Making about Antibiotic Use in Primary Care. Ph. D. Thesis, Bond University, Gold Coast, Australia, 2018.
37. Mehta, Y.; Paul, R.; Rabbani, R.; Acharya, S.P.; Withanaarachchi, U.K. Sepsis Management in Southeast Asia: A Review and Clinical Experience. *J. Clin. Med.* **2022**, *11*, 3635. [CrossRef]
38. Nguyen, M.; Brettin, T.; Long, S.; Musser, J.M.; Olsen, R.J.; Olson, R.; Shukla, M.; Stevens, R.L.; Xia, F.; Yoo, H.; et al. Developing an in silico minimum inhibitory concentration panel test for Klebsiella pneumoniae. *Sci. Rep.* **2018**, *8*, 421. [CrossRef]
39. Kharb, S. Biochemical Tests in Clinical Medicine. In *Mind Maps in Clinical Chemistry (Part I)*; Bentham Science Publishers: Sharjah, United Arab Emirates, 2021; p. 15.
40. Tavassoly, I.; Goldfarb, J.; Iyengar, R. Systems biology primer: The basic methods and approaches. *Essays Biochem.* **2018**, *62*, 487–500. [CrossRef]
41. Pilecky, M.; Schildberger, A.; Orth-Höller, D.; Weber, V. Pathogen enrichment from human whole blood for the diagnosis of bloodstream infection: Prospects and limitations. *Diagn. Microbiol. Infect. Dis.* **2019**, *94*, 7–14. [CrossRef] [PubMed]
42. Cheung, S.W.; Bi, W. Novel applications of array comparative genomic hybridization in molecular diagnostics. *Expert Rev. Mol. Diagn.* **2018**, *18*, 531–542. [CrossRef]
43. Iregbu, K.; Dramowski, A.; Milton, R.; Nsutebu, E.; Howie, S.R.; Chakraborty, M.; Lavoie, P.M.; Costelloe, C.E.; Ghazal, P. Global health systems' data science approach for precision diagnosis of sepsis in early life. *Lancet Infect. Dis.* **2021**, *22*, e143–e152. [CrossRef]
44. Philips, C.A.; Ahamed, R.; Rajesh, S.; George, T.; Mohanan, M.; Augustine, P. Update on diagnosis and management of sepsis in cirrhosis: Current advances. *World J. Hepatol.* **2020**, *2*, 451. [CrossRef] [PubMed]
45. Sune, D.; Rydberg, H.; Augustinsson, Å.N.; Serrander, L.; Jungeström, M.B. Optimization of 16S rRNA gene analysis for use in the diagnostic clinical microbiology service. *J. Microbiol. Methods* **2020**, *170*, 105854. [CrossRef] [PubMed]
46. Llerena, J.P.; Araujo, P.; Mazzafera, P. Optimization of RT-PCR reactions in studies with genes of lignin biosynthetic route in *Saccharum spontaneum*. *An. Acad. Bras. Cienc.* **2018**, *90*, 509–519. [CrossRef] [PubMed]
47. Sreejith, K.R.; Ooi, C.H.; Jin, J.; Dao, D.V.; Nguyen, N.T. Digital polymerase chain reaction technology–recent advances and future perspectives. *Lab. Chip.* **2018**, *18*, 3717–3732. [CrossRef]
48. Manzano, M. Labelled and unlabelled probes for pathogen detection with molecular biology methods and biosensors. *Methods Microbiol.* **2021**, *48*, 79–225. [CrossRef]
49. Ferguson, J.; Duran, J.; Killinen, W.; Wagner, J.; Kulesza, C.; Chatterley, C.; Li, Y. A Field-Deployable and Low-Cost PCR (FLC-PCR) Thermocycler for the Rapid Detection of Environmental *E. coli*. In Proceedings of the 2020 42nd Annual International Conference of the IEEE Engineering in Medicine & Biology Society (EMBC), Montreal, QC, Canada, 20–24 July 2020; IEEE: Piscataway, NJ, USA, 2020; pp. 2209–2212.
50. Dayarathne, M.C.; Mridha, A.U.; Wang, Y. Diagnosis of Fungal Plant Pathogens Using Conventional and Molecular Approaches. In *Diagnostics of Plant Diseases*; IntechOpen: London, UK, 2020.
51. Paul, R.; Ostermann, E.; Wei, Q. Advances in point-of-care nucleic acid extraction technologies for rapid diagnosis of human and plant diseases. *Biosens. Bioelectron.* **2020**, *169*, 112592. [CrossRef]
52. Kumar, S.S.; Ghosh, A.R. Assessment of bacterial viability: A comprehensive review on recent advances and challenges. *Microbiology* **2019**, *165*, 593–610. [CrossRef]
53. Jiang, X.W.; Huang, T.S.; Xie, L.; Chen, S.Z.; Wang, S.D.; Huang, Z.W.; Li, X.Y.; Ling, W.P. Development of a diagnostic assay by three-tube multiplex real-time PCR for simultaneous detection of nine microorganisms causing acute respiratory infections. *Sci. Rep.* **2022**, *12*, 13306. [CrossRef]
54. Mota, F.A.; Pereira, S.A.; Araújo, A.R.; Passos, M.L.; Saraiva, M.L.M. Biomarkers in the diagnosis of wounds infection: An analytical perspective. *Trends Anal. Chem.* **2021**, *143*, 116405. [CrossRef]

55. Saha, O.; Islam, M.R.; Rahman, M.S.; Hoque, M.N.; Hossain, M.A.; Sultana, M. Genome-wide diversity and differentiation of two novel multidrug-resistant populations of *Pasteurella multocida* type B: 2 from fowl cholera. *bioRxiv* **2020**. [CrossRef]
56. Gunsolus, I.L.; Sweeney, T.E.; Liesenfeld, O.; Ledeboer, N.A. Diagnosing and managing sepsis by probing the host response to infection: Advances, opportunities, and challenges. *J. Clin. Microbiol.* **2019**, *57*, e00425-19. [CrossRef] [PubMed]
57. Mishra, D.; Satpathy, G.; Chawla, R.; Venkatesh, P.; Ahmed, N.H.; Panda, S.K. Utility of broad-range 16S rRNA PCR assay versus conventional methods for laboratory diagnosis of bacterial endophthalmitis in a tertiary care hospital. *Br. J. Ophthalmol.* **2019**, *103*, 152–156. [CrossRef] [PubMed]
58. Verbakel, J.Y.; Matheeussen, V.; Loens, K.; Kuijstermans, M.; Goossens, H.; Ieven, M.; Butler, C.C. Performance and ease of use of a molecular point-of-care test for influenza A/B and RSV in patients presenting to primary care. *Eur. J. Clin. Microbiol. Infect. Dis.* **2020**, *39*, 1453–1460. [CrossRef] [PubMed]
59. Reta, D.H.; Tessema, T.S.; Ashenef, A.S.; Desta, A.F.; Labisso, W.L.; Gizaw, S.T.; Abay, S.M.; Melka, D.S.; Reta, F.A. Molecular and immunological diagnostic techniques of medical viruses. *Int. J. Microbiol.* **2020**, *2020*, 8832728. [CrossRef]
60. Nik Zuraina, N.M.N.; Mohamad, S.; Hasan, H.; Goni, M.D.; Suraiya, S. Diagnostic performance of an in-house multiplex PCR assay and the retrospective surveillance of bacterial respiratory pathogens at a teaching hospital, Kelantan, Malaysia. *Pathog. Glob. Health* **2022**, 1–13. [CrossRef]
61. Davidson, K.R.; Ha, D.M.; Schwarz, M.I.; Chan, E.D. Bronchoalveolar lavage as a diagnostic procedure: A review of known cellular and molecular findings in various lung diseases. *J. Thorac. Dis.* **2020**, *12*, 4991. [CrossRef]
62. Cao, X.; Zhao, L.; Zhang, J.; Chen, X.; Shi, L.; Fang, X.; Xie, H.; Chang, Y.; Wang, L. Detection of viable but nonculturable Vibrio parahaemolyticus in shrimp samples using improved real-time PCR and real-time LAMP methods. *Food Control* **2019**, *103*, 145–152. [CrossRef]
63. Yang, W.; Zhang, J.; Ma, R. The prediction of infectious diseases: A bibliometric analysis. *Int. J. Environ. Res. Public Health* **2020**, *7*, 6218. [CrossRef]
64. Pan, Z.; Lu, J.; Wang, N.; He, W.T.; Zhang, L.; Zhao, W.; Su, S. Development of a Taq Man-probe-based multiplex real-time PCR for the simultaneous detection of emerging and reemerging swine coronaviruses. *Virulence* **2020**, *11*, 707–718. [CrossRef]
65. Kadri, K. Polymerase chain reaction (PCR): Principle and applications. In *Synthetic Biology-New Interdisciplinary Science*; IntechOpen: London, UK, 2019. [CrossRef]
66. Pumford, E.A.; Lu, J.; Spaczai, I.; Prasetyo, M.E.; Zheng, E.M.; Zhang, H.; Kamei, D.T. Developments in integrating nucleic acid isothermal amplification and detection systems for point-of-care diagnostics. *Biosens. Bioelectron.* **2020**, *170*, 112674. [CrossRef]
67. Miotto, B.A.; Hora, A.S.D.; Taniwaki, S.A.; Brandão, P.E.; Heinemann, M.B.; Hagiwara, M.K. Development and validation of a modified TaqMan based real-time PCR assay targeting the lipl32 gene for detection of pathogenic Leptospira in canine urine samples. *Braz. J. Microbiol.* **2018**, *49*, 584–590. [CrossRef] [PubMed]
68. Barkallah, M.; Elleuch, J.; Smith, K.F.; Chaari, S.; Neila, I.B.; Fendri, I.; Michaud, P.; Abdelkafi, S. Development and application of a real-time PCR assay for the sensitive detection of diarrheic toxin producer *Prorocentrum lima*. *J. Microbiol. Methods.* **2020**, *178*, 106081. [CrossRef] [PubMed]
69. Marras, S.A.; Tyagi, S.; Antson, D.O.; Kramer, F.R. Color-coded molecular beacons for multiplex PCR screening assays. *PLoS ONE* **2019**, *14*, 0213906. [CrossRef] [PubMed]
70. Inchingolo, R.; Pierandrei, C.; Montemurro, G.; Smargiassi, A.; Lohmeyer, F.M.; Rizzi, A. Antimicrobial resistance in common respiratory pathogens of chronic bronchiectasis patients: A literature review. *Antibiotics* **2021**, *10*, 326. [CrossRef] [PubMed]
71. Candel, F.J.; Sá, M.B.; Belda, S.; Bou, G.; Del Pozo, J.L.; Estrada, O.; Ferrer, R.; del Castillo, J.G.; Julian-Jimenez, A.; Martin-Loeches, I.; et al. Current aspects in sepsis approach. Turning things around. *Span. J. Psychol.* **2018**, *31*, 298.
72. Pashchenko, O.; Shelby, T.; Banerjee, T.; Santra, S. A comparison of optical, electrochemical, magnetic, and colorimetric point-of-care biosensors for infectious disease diagnosis. *ACS Infect. Dis.* **2018**, *4*, 1162–1178. [CrossRef]
73. Han, H.; Sohn, B.; Choi, J.; Jeon, S. Recent advances in magnetic nanoparticle-based microfluidic devices for the pretreatment of pathogenic bacteria. *Biomed. Eng. Lett.* **2021**, *11*, 297–307. [CrossRef]
74. Schmitz, J.E.; Stratton, C.W.; Persing, D.H.; Tang, Y.W. Forty Years of Molecular Diagnostics for Infectious Diseases. *J. Clin. Microbiol.* **2022**, *60*, e02446-21. [CrossRef]
75. Bronder, T.S.; Jessing, M.P.; Poghossian, A.; Keusgen, M.; Schöning, M.J. Detection of PCR-amplified tuberculosis DNA fragments with polyelectrolyte-modified field-effect sensors. *Anal. Chem.* **2018**, *90*, 7747–7753. [CrossRef]
76. Wang, J.; Yang, J.; Gao, S.; Liu, A.; Rashid, M.; Li, Y.; Liu, Z.; Liu, J.; Liu, G.; Luo, J.; et al. Rapid detection and differentiation of *Theileria annulata*, *T. orientalis* and *T. sinensis* using high-resolution melting analysis. *Ticks Tick Borne Dis.* **2020**, *11*, 101312. [CrossRef]
77. Kurbakov, K.A.; Konorov, E.A.; Minaev, M.Y.; Kuznetsova, O.A. Multiplex real-time PCR with HRM for detection of *Lactobacillus sakei* and *Lactobacillus curvatus* in Food Samples. *Food Technol. Biotechnol.* **2019**, *57*, 97–104. [CrossRef] [PubMed]
78. Pohanka, M. Current trends in the biosensors for biological warfare agents assay. *Materials* **2019**, *12*, 2303. [CrossRef] [PubMed]
79. Zheng, W.; Jiang, L.; Lei, Q.; Yang, J.; Gao, X.; Wang, W.; Zhang, Y.; Kong, T.; Chen, Q.; Li, G. Development and validation of quantitative real-time pcr for the detection of residual CHO host cell DNA and optimization of sample pretreatment method in biopharmaceutical products. *Biol. Proced. Online* **2019**, *21*, 17. [CrossRef] [PubMed]

80. Chen, X.; Tang, M.; Liu, Y.; Huang, J.; Liu, Z.; Tian, H.; Zheng, Y.; de la Chapelle, M.L.; Zhang, Y.; Fu, W. Surface-enhanced Raman scattering method for the identification of methicillin-resistant Staphylococcus aureus using positively charged silver nanoparticles. *Mikrochim. Acta* **2019**, *186*, 1–8. [CrossRef]
81. Jeong, K.; Stanwix, P.L.; May, E.F.; Aman, Z.M. Surface-Enhanced Raman Scattering Imaging of Cetylpyridinium Chloride Adsorption to a Solid Surface. *Anal. Chem.* **2022**, *94*, 14169–14176. [CrossRef] [PubMed]
82. Xu, G.; Guo, N.; Zhang, Q.; Wang, T.; Song, P.; Xia, L. An ultrasensitive surface-enhanced Raman scattering sensor for the detection of hydrazine via the Schiff base reaction. *J. Hazard. Mater.* **2022**, *424*, 127303. [CrossRef]
83. Ge, M.; Li, P.; Zhou, G.; Chen, S.; Han, W.; Qin, F.; Nie, Y.; Wang, Y.; Qin, M.; Huang, G.; et al. General surface-enhanced Raman spectroscopy method for actively capturing target molecules in small gaps. *J. Am. Chem. Soc.* **2021**, *143*, 7769–7776. [CrossRef] [PubMed]
84. Sun, Y.; Chen, X.; Zheng, Y.; Song, Y.; Zhang, H.; Zhang, S. Surface-enhanced Raman scattering trace-detection platform based on continuous-rolling-assisted evaporation on superhydrophobic surfaces. *ACS Appl. Nano Mater.* **2020**, *3*, 4767–4776. [CrossRef]
85. Pérez-Jiménez, A.I.; Lyu, D.; Lu, Z.; Liu, G.; Ren, B. Surface-enhanced Raman spectroscopy: Benefits, trade-offs and future developments. *Chem. Sci.* **2020**, *11*, 4563–4577. [CrossRef]
86. Shvalya, V.; Filipič, G.; Zavašnik, J.; Abdulhalim, I.; Cvelbar, U. Surface-enhanced Raman spectroscopy for chemical and biological sensing using nanoplasmonics: The relevance of interparticle spacing and surface morphology. *Appl. Phys. Rev.* **2020**, *7*, 031307. [CrossRef]
87. Zong, C.; Xu, M.; Xu, L.J.; Wei, T.; Ma, X.; Zheng, X.S.; Hu, R.; Ren, B. Surface-enhanced Raman spectroscopy for bioanalysis: Reliability and challenges. *Chem. Rev.* **2018**, *118*, 4946–4980. [CrossRef] [PubMed]
88. Li, P.; Long, F.; Chen, W.; Chen, J.; Chu, P.K.; Wang, H. Fundamentals and applications of surface-enhanced Raman spectroscopy–based biosensors. *Curr. Opin. Biomed. Eng.* **2020**, *13*, 51–59. [CrossRef]
89. Sun, J.; Gong, L.; Wang, W.; Gong, Z.; Wang, D.; Fan, M. Surface-enhanced Raman spectroscopy for on-site analysis: A review of recent developments. *Luminescence* **2020**, *35*, 808–820. [CrossRef] [PubMed]
90. Liu, S.; Hu, Q.; Li, C.; Zhang, F.; Gu, H.; Wang, X.; Li, S.; Xue, L.; Madl, T.; Zhang, Y.; et al. Wide-range, rapid, and specific identification of pathogenic bacteria by Surface-Enhanced Raman Spectroscopy. *ACS Sens.* **2021**, *6*, 2911–2919. [CrossRef]
91. Pyrak, E.; Krajczewski, J.; Kowalik, A.; Kudelski, A.; Jaworska, A. Surface enhanced Raman spectroscopy for DNA biosensors—How far are we? *Molecules* **2019**, *24*, 4423. [CrossRef]
92. Kim, J.; Jang, Y.; Kim, N.J.; Kim, H.; Yi, G.C.; Shin, Y.; Kim, M.H.; Yoon, S. Study of chemical enhancement mechanism in non-plasmonic surface enhanced Raman spectroscopy (SERS). *Front. Chem.* **2019**, *7*, 582. [CrossRef]
93. Han, Y.Y.; Lin, Y.C.; Cheng, W.C.; Lin, Y.T.; Teng, L.J.; Wang, J.K.; Wang, Y.L. Rapid antibiotic susceptibility testing of bacteria from patients' blood via assaying bacterial metabolic response with surface-enhanced Raman spectroscopy. *Sci. Rep.* **2020**, *10*, 12538 [CrossRef]
94. Wang, K.; Li, S.; Petersen, M.; Wang, S.; Lu, X. Detection and characterization of antibiotic-resistant bacteria using surface-enhanced Raman spectroscopy. *Nanomaterials* **2018**, *8*, 762. [CrossRef]
95. Tahir, M.A.; Dina, N.E.; Cheng, H.; Valev, V.K.; Zhang, L. Surface-enhanced Raman spectroscopy for bioanalysis and diagnosis. *Nanoscale* **2021**, *13*, 11593–11634. [CrossRef]
96. Dizaji, A.N.; Ozek, N.S.; Aysin, F.; Calis, A.; Yilmaz, A.; Yilmaz, M. Combining vancomycin-modified gold nanorod arrays and colloidal nanoparticles as a sandwich model for the discrimination of Gram-positive bacteria and their detection via surface-enhanced Raman spectroscopy (SERS). *Analyst* **2021**, *146*, 3642–3653. [CrossRef]
97. Ahmad, W.; Wang, J.; Li, H.; Jiao, T.; Chen, Q. Trends in the bacterial recognition patterns used in surface enhanced Raman spectroscopy. *Trends Anal. Chem.* **2021**, *142*, 116310. [CrossRef]
98. Kumar, M.; Shergill, S.P.S.; Tandel, K.; Sahai, K.; Gupta, R.M. Direct antimicrobial susceptibility testing from positive blood culture bottles in laboratories lacking automated antimicrobial susceptibility testing systems. *Med. J. Armed Forces India* **2019**, *75*, 450–457. [CrossRef]
99. Wang, Y.; Jin, Y.; Bai, Y.; Song, Z.; Chu, W.; Zhao, M.; Hao, Y.; Lu, Z. Rapid method for direct identification of positive blood cultures by MALDI-TOF MS. *Exp. Ther. Med.* **2020**, *20*, 235. [CrossRef]
100. Dai, Y.; Xu, X.; Yan, X.; Li, D.; Cao, W.; Tang, L.; Hu, M.; Jiang, C. Evaluation of a rapid and simplified protocol for direct identification of microorganisms from positive blood cultures by using Matrix Assisted Laser Desorption Ionization Time-Of-Flight Mass Spectrometry (MALDI-TOF MS). *Front. Cell. Infect. Microbiol.* **2021**, *11*, 632679. [CrossRef]
101. Kayin, M.; Mert, B.; Aydemir, S.; Özenci, V. Comparison of rapid BACpro® II, Sepsityper® kit and in-house preparation methods for direct identification of bacteria from blood cultures by MALDI-TOF MS with and without Sepsityper® module analysis. *Eur. J. Clin. Microbiol. Infect. Dis.* **2019**, *38*, 2133–2143. [CrossRef]
102. Homolová, R.; Bogdanová, K.; Bardoň, J.; Kolář, M. Direct identification of bacteria in blood cultures by MALDI-TOF MS. *Clin. Microbiol. Infect.* **2020**, *26*, 45–50.
103. Tsuchida, S.; Nakayama, T. MALDI-Based Mass Spectrometry in Clinical Testing: Focus on Bacterial Identification. *Appl. Sci.* **2022**, *12*, 2814. [CrossRef]
104. Perini, M.; Batisti Biffignandi, G.; Di Carlo, D.; Pasala, A.R.; Piazza, A.; Panelli, S.; Zuccotti, G.V. Comandatore, F. MeltingPlot, a user-friendly online tool for epidemiological investigation using High Resolution Melting data. *BMC Bioinform* **2021**, *22*, 76. [CrossRef]

105. Ahmed, M.O.; Baptiste, K.E. Vancomycin-resistant enterococci: A review of antimicrobial resistance mechanisms and perspectives of human and animal health. *Microb. Drug Resist.* **2018**, *24*, 590–606. [CrossRef]
106. Li, Y.; Xiu, L.; Wang, L.; Zhang, L.; Wang, F.; Peng, J. Rapid Detection of Antimicrobial Resistance in Mycoplasma genitalium by High-Resolution Melting Analysis with Unlabeled Probes. *Microbiol. Spectr.* **2022**, *10*, e01014-22. [CrossRef]
107. Dehshiri, M.; Khoramrooz, S.S.; Zoladl, M.; Khosravani, S.A.; Parhizgari, N.; Motazedian, M.H.; Jahedi, S.; Sharifi, A. The frequency of Klebsiella pneumonia encoding genes for CTX-M, TEM-1 and SHV-1 extended-spectrum beta lactamases enzymes isolated from urinary tract infection. *Ann. Clin. Microbiol. Antimicrob.* **2018**, *17*, 4. [CrossRef]
108. Shalmashi, H.; Farajnia, S.; Sadeghi, M.; Tanoumand, A.; Veissi, K.; Hamishekar, H.; Gotaslou, R. Detection of ESBLs types blaCTX-M, blaSHV and blaTEM resistance genes among clinical isolates of Pseudomonas aeruginosa. *Gene Rep.* **2022**, *28*, 101637. [CrossRef]
109. Zarabadi-Pour, M.; Peymani, A.; Habibollah-Pourzereshki, N.; Sarookhani, M.R.; Karami, A.A.; Javadi, A. Detection of Extended-Spectrum ß-Lactamases among Acinetobacter Baumannii Isolated from Hospitals of Qazvin, Iran. *Ethiop. J. Health Sci.* **2021**, *31*, 229–236. [CrossRef] [PubMed]
110. Abdar, M.H.; Taheri-Kalani, M.; Taheri, K.; Emadi, B.; Hasanzadeh, A.; Sedighi, A.; Pirouzi, S.; Sedighi, M. Prevalence of extended-spectrum beta-lactamase genes in Acinetobacter baumannii strains isolated from nosocomial infections in Tehran, Iran. *GMS Hyg. Infect. Control* **2019**, *14*, 318. [CrossRef]
111. Jordt, H.; Stalder, T.; Kosterlitz, O.; Ponciano, J.M.; Top, E.M.; Kerr, B. Coevolution of host–plasmid pairs facilitates the emergence of novel multidrug resistance. *Nat. Ecol. Evol.* **2020**, *4*, 863–869. [CrossRef]
112. Maharjan, M.; Sah, A.K.; Pyakurel, S.; Thapa, S.; Maharjan, S.; Adhikari, N.; Rijal, K.R.; Ghimire, P.; Thapa Shrestha, U. Molecular Confirmation of Vancomycin-Resistant Staphylococcus aureus with vanA Gene from a Hospital in Kathmandu. *Int. J. Microbiol.* **2021**, *2021*, 3847347. [CrossRef] [PubMed]
113. Fisher, J.F.; Mobashery, S. β-Lactams against the Fortress of the Gram-Positive Staphylococcus aureus Bacterium. *Chem. Rev.* **2020**, *121*, 3412–3463. [CrossRef] [PubMed]
114. Leonard, H.; Colodner, R.; Halachmi, S.; Segal, E. Recent advances in the race to design a rapid diagnostic test for antimicrobial resistance. *ACS Sens.* **2018**, *3*, 2202–2217. [CrossRef]
115. Schürch, A.C.; Arredondo-Alonso, S.; Willems, R.J.L.; Goering, R.V. Whole genome sequencing options for bacterial strain typing and epidemiologic analysis based on single nucleotide polymorphism versus gene-by-gene–based approaches. *Clin. Microbiol. Infect.* **2018**, *24*, 350–354. [CrossRef]
116. Fujii, H.; Kakiuchi, S.; Tsuji, M.; Nishimura, H.; Yoshikawa, T.; Yamada, S.; Omura, N.; Inagaki, T.; Shibamura, M.; Harada, S.; et al. Application of next-generation sequencing to detect acyclovir-resistant herpes simplex virus type 1 variants at low frequency in thymidine kinase gene of the isolates recovered from patients with hematopoietic stem cell transplantation. *J. Virol. Methods* **2018**, *251*, 123–128. [CrossRef]
117. Yan, Q.; Wi, Y.M.; Thoendel, M.J.; Raval, Y.S.; Greenwood-Quaintance, K.E.; Abdel, M.P.; Jeraldo, P.R.; Chia, N.; Patel, R. Evaluation of the CosmosID bioinformatics platform for prosthetic joint-associated sonicate fluid shotgun metagenomic data analysis. *J. Clin. Microbiol.* **2019**, *57*, e01182-18. [CrossRef] [PubMed]
118. Friães, A.; Mamede, R.; Ferreira, M.; Melo-Cristino, J.; Ramirez, M. Annotated Whole-Genome Multilocus Sequence Typing Schema for Scalable High-Resolution Typing of *Streptococcus pyogenes*. *J. Clin. Microbiol.* **2022**, *60*, e00315-22. [CrossRef] [PubMed]
119. Palleja, A.; Mikkelsen, K.H.; Forslund, S.K.; Kashani, A.; Allin, K.H.; Nielsen, T.; Hansen, T.H.; Liang, S.; Feng, Q.; Zhang, C.; et al. Recovery of gut microbiota of healthy adults following antibiotic exposure. *Nat. Microbiol.* **2018**, *3*, 1255–1265. [CrossRef] [PubMed]
120. Rahman, S.F.; Olm, M.R.; Morowitz, M.J.; Banfield, J.F. Machine learning leveraging genomes from metagenomes identifies influential antibiotic resistance genes in the infant gut microbiome. *MSystems* **2018**, *3*, e00123-17. [CrossRef] [PubMed]
121. Sturaro, L.L.; Gonoi, T.; Busso-Lopes, A.F.; Tararam, C.A.; Levy, C.E.; Lyra, L.; Trabasso, P.; Schreiber, A.Z.; Kamei, K.; Moretti, M.L. Visible DNA microarray system as an adjunctive molecular test in identification of pathogenic fungi directly from a blood culture bottle. *J. Clin. Microbiol.* **2018**, *56*, e01908-17. [CrossRef]
122. Dhanjal, D.S.; Chopra, C.; Chopra, R.S. Metagenomic DNA sequencing: Technological advances and applications. In *Metagenomics: Techniques, Applications, Challenges and Opportunities*; Springer: Berlin/Heidelberg, Germany, 2020; pp. 37–53. [CrossRef]
123. Schaack, D.; Siegler, B.H.; Tamulyte, S.; Weigand, M.A.; Uhle, F. The immunosuppressive face of sepsis early on intensive care unit—A large-scale microarray meta-analysis. *PloS ONE* **2018**, *13*, 0198555. [CrossRef] [PubMed]
124. Carroll, K.C.; Reid, J.L.; Thornberg, A.; Whitfield, N.N.; Trainor, D.; Lewis, S.; Wakefield, T.; Davis, T.E.; Church, K.G.; Samuel, L.; et al. Clinical performance of the novel GenMark Dx ePlex blood culture ID Gram-positive panel. *J. Clin. Microbiol.* **2020**, *58*, e01730-19. [CrossRef]
125. Li, Y.; Zhang, F.; Cong, Y.; Zhao, Y. Identification of potential genes and miRNAs associated with sepsis based on microarray analysis. *Mol. Med. Rep.* **2018**, *17*, 6227–6234. [CrossRef]
126. Kuchibiro, T.; Hirano, A.; Ogasawara, S.; Nakamura, T. The microcolony detection method (MCD), a simple and rapid screening test for antimicrobial resistance bacteria on positive blood cultures. *Heliyon* **2020**, *6*, 05494. [CrossRef]
127. She, R.C.; Bender, J.M. Advances in rapid molecular blood culture diagnostics: Healthcare impact, laboratory implications, and multiplex technologies. *J. Appl. Lab. Med.* **2019**, *3*, 617–630. [CrossRef]

128. Huang, T.D.; Melnik, E.; Bogaerts, P.; Evrard, S.; Glupczynski, Y. Evaluation of the ePlex blood culture identification panels for detection of pathogens in bloodstream infections. *J. Clin. Microbiol.* **2019**, *57*, e01597-18. [CrossRef] [PubMed]
129. Żukowska, M.E. Advanced methods of bacteriological identification in a clinical microbiology laboratory. *J. Pre Clin. Clin. Res.* **2021**, *15*, 68–72. [CrossRef]
130. Fournier, C.; Aires-de-Sousa, M.; Nordmann, P.; Poirel, L. Occurrence of CTX-M-15-and MCR-1-producing Enterobacterales in pigs in Portugal: Evidence of direct links with antibiotic selective pressure. *Int. J. Antimicrob. Agents* **2020**, *55*, 105802. [CrossRef]
131. Kumar, S.; Tripathy, S.; Jyoti, A.; Singh, S.G. Recent advances in biosensors for diagnosis and detection of sepsis: A comprehensive review. *Biosens. Bioelectron.* **2019**, *124*, 205–215. [CrossRef]
132. Kundu, S.; Tabassum, S.; Kumar, R. A perspective on sepsis pathogenesis, biomarkers and diagnosis: A concise survey. *Med. Devices Sens.* **2020**, *3*, 10089. [CrossRef]
133. Min, J.; Nothing, M.; Coble, B.; Zheng, H.; Park, J.; Im, H.; Weber, G.F.; Castro, C.M.; Swirski, F.K.; Weissleder, R.; et al. Integrated biosensor for rapid and point-of-care sepsis diagnosis. *ACS Nano* **2018**, *12*, 3378–3384. [CrossRef]
134. Levy, M.M.; Gesten, F.C.; Phillips, G.S.; Terry, K.M.; Seymour, C.W.; Prescott, H.C.; Friedrich, M.; Iwashyna, T.J.; Osborn, T.; Lemeshow, S. Mortality changes associated with mandated public reporting for sepsis. The results of the New York state initiative. *Am. J. Respir. Crit. Care Med.* **2018**, *198*, 1406–1412. [CrossRef]
135. Alam, N.; Oskam, E.; Stassen, P.M.; van Exter, P.; van de Ven, P.M.; Haak, H.R.; Holleman, F.; van Zanten, A.; van Leeuwen-Nguyen, H.; Bon, V.; et al. Prehospital antibiotics in the ambulance for sepsis: A multicentre, open label, randomised trial. *Lancet Respir. Med.* **2018**, *6*, 40–50. [CrossRef]
136. Cheng, M.P.; Stenstrom, R.; Paquette, K.; Stabler, S.N.; Akhter, M.; Davidson, A.C.; Gavric, M.; Lawandi, A.; Jinah, R.; Saeed, Z.; et al. Blood culture results before and after antimicrobial administration in patients with severe manifestations of sepsis: A diagnostic study. *Ann. Intern. Med.* **2019**, *171*, 547–554. [CrossRef]
137. Peltan, I.D.; Mitchell, K.H.; Rudd, K.E.; Mann, B.A.; Carlbom, D.J.; Rea, T.D.; Butler, A.M.; Hough, C.L.; Brown, S.M. Prehospital care and emergency department door-to-antibiotic time in sepsis. *Ann. Am. Thorac. Soc.* **2018**, *15*, 1443–1450. [CrossRef]
138. Rello, J.; Van Engelen, T.S.R.; Alp, E.; Calandra, T.; Cattoir, V.; Kern, W.V.; Netea, M.G.; Nseir, S.; Opal, S.M.; van de Veerdonk, F.L.; et al. Towards precision medicine in sepsis: A position paper from the European Society of Clinical Microbiology and Infectious Diseases. *Clin. Microbiol. Infect.* **2018**, *24*, 1264–1272. [CrossRef] [PubMed]
139. Kumar, V. Targeting macrophage immunometabolism: Dawn in the darkness of sepsis. *Int. Immunopharmacol.* **2018**, *58*, 173–185. [CrossRef] [PubMed]
140. Antonioli, L.; Blandizzi, C.; Fornai, M.; Pacher, P.; Lee, H.T.; Haskó, G. P2X4 receptors, immunity, and sepsis. *Curr. Opin. Pharmacol.* **2019**, *47*, 65–74. [CrossRef] [PubMed]
141. Mirasoli, M.; Bonvicini, F.; Lovecchio, N.; Petrucci, G.; Zangheri, M.; Calabria, D.; Costantini, F.; Roda, A.; Gallinella, G.; Caputo, D.; et al. On-chip LAMP-BART reaction for viral DNA real-time bioluminescence detection. *Sens. Actuators B Chem.* **2018**, *262*, 1024–1033. [CrossRef]
142. Nasseri, B.; Soleimani, N.; Rabiee, N.; Kalbasi, A.; Karimi, M.; Hamblin, M.R. Point-of-care microfluidic devices for pathogen detection. *Biosens. Bioelectron.* **2018**, *117*, 112–128. [CrossRef] [PubMed]
143. Reddy, B.; Hassan, U.; Seymour, C.; Angus, D.C.; Isbell, T.S.; White, K.; Weir, W.; Yeh, L.; Vincent, A.; Bashir, R. Point-of-care sensors for the management of sepsis. *Nat. Biomed. Eng.* **2018**, *2*, 640–648. [CrossRef]
144. Zhang, Y.; Hu, A.; Andini, N.; Yang, S. A 'culture' shift: Application of molecular techniques for diagnosing polymicrobial infections. *Biotechnol. Adv.* **2019**, *37*, 476–490. [CrossRef]
145. Thurtle-Schmidt, D.M.; Lo, T.W. Molecular biology at the cutting edge: A review on CRISPR/CAS9 gene editing for undergraduates. *Biochem. Mol. Biol. Educ.* **2018**, *46*, 195–205. [CrossRef]
146. Butiuc-Keul, A.; Farkas, A.; Carpa, R.; Iordache, D. CRISPR-Cas system: The powerful modulator of accessory genomes in prokaryotes. *Microb. Physiol.* **2022**, *32*, 2–17. [CrossRef]
147. Mohamadi, S.; Bostanabad, S.Z.; Mirnejad, R. CRISPR arrays: A review on its mechanism. *J. Appl. Biotechnol. Rep.* **2020**, *7*, 81–86. [CrossRef]
148. Majumdar, S.; Terns, M.P. CRISPR RNA-guided DNA cleavage by reconstituted Type IA immune effector complexes. *Extremophiles* **2020**, *23*, 19–33. [CrossRef] [PubMed]
149. Gouw, A.M. Challenging the therapy/enhancement distinction in CRISPR gene Editing. In *The Palgrave Handbook of Philosophy and Public Policy*; Palgrave Macmillan: Cham, Switzerland, 2018; pp. 493–508. [CrossRef]
150. Koonin, E.V. CRISPR: A new principle of genome engineering linked to conceptual shifts in evolutionary biology. *Biol. Philos.* **2018**, *34*, 9. [CrossRef] [PubMed]
151. Ewart, D.T.; Peterson, E.J.; Steer, C.J. Gene editing for inflammatory disorders. *Ann. Rheum. Dis.* **2019**, *78*, 6–15. [CrossRef] [PubMed]
152. Knott, G.J.; Doudna, J.A. CRISPR-Cas guides the future of genetic engineering. *Science* **2018**, *361*, 866–869. [CrossRef] [PubMed]
153. Wang, R.; Graham, S.; Gao, L.; Tam, J.; Levesque, M.C. Editing the immune system in vivo in mice using CRISPR/Cas9 ribonucleoprotein (RNP)-mediated gene editing of transplanted hematopoietic stem cells. *Methods* **2021**, *194*, 30–36. [CrossRef]
154. Reyes, M.; Filbin, M.R.; Bhattacharyya, R.P.; Billman, K.; Eisenhaure, T.; Hung, D.T.; Levy, B.D.; Baron, R.M.; Blainey, P.C.; Goldberg, M.B.; et al. An immune-cell signature of bacterial sepsis. *Nat. Med.* **2020**, *26*, 333–340. [CrossRef]

155. Ma, L.; Li, Q.; Cai, S.; Peng, H.; Huyan, T.; Yang, H. The role of NK cells in fighting the virus infection and sepsis. *Int. J. Med. Sci.* **2021**, *18*, 3236. [CrossRef]
156. Wu, M.; Hu, N.; Du, X.; Wei, J. Application of CRISPR/Cas9 technology in sepsis research. *Brief Funct. Genomics* **2020**, *19*, 229–234. [CrossRef]
157. Cai, M.; Li, S.; Shuai, Y.; Li, J.; Tan, J.; Zeng, Q. Genome-wide CRISPR-Cas9 viability screen reveals genes involved in TNF-α-induced apoptosis of human umbilical vein endothelial cells. *J. Cell. Physiol.* **2019**, *234*, 9184–9193. [CrossRef]
158. Grigoriev, E.V.; Salakhov, R.R.; Golubenko, M.V.; Ponasenko, A.V.; Shukevich, D.L.; Matveeva, V.G.; Radivilko, A.S.; Tsepokina, A.V.; Velikanova, E.A.; Kornelyuk, R.S.; et al. Mitochondrial DNA as DAMP in critical conditions. *Bull. Sib. Med.* **2019**, *18*, 134–143. [CrossRef]
159. Panicker, S.; Balijepalli, S.; Zhang, B.; Swamy, S.; Sherman, M.A.; Raghavendran, K.; Suresh, M.V. Role of Toll-like Receptor-9 in Lung Injury. *J. Nat. Sci.* **2019**, *5*, 551.
160. Pustylnikov, S.; Costabile, F.; Beghi, S.; Facciabene, A. Targeting mitochondria in cancer: Current concepts and immunotherapy approaches. *Transl. Res.* **2018**, *202*, 35–51. [CrossRef] [PubMed]
161. Zhong, R.; Miao, L.; Zhang, H.; Tan, L.; Zhao, Y.; Tu, Y.; Prieto, M.A.; Simal-Gandara, J.; Chen, L.; He, C.; et al. Anti-inflammatory activity of flavonols via inhibiting MAPK and NF-κB signaling pathways in RAW264. 7 macrophages. *Curr. Res. Food Sci.* **2022**, *5*, 1176–1184. [CrossRef] [PubMed]
162. Baglaenko, Y.; Macfarlane, D.; Marson, A.; Nigrovic, P.A.; Raychaudhuri, S. Genome editing to define the function of risk loci and variants in rheumatic disease. *Nat. Rev. Rheumatol.* **2021**, *17*, 462–474. [CrossRef]
163. Khan, S.H. Genome-editing technologies: Concept, pros, and cons of various genome-editing techniques and bioethical concerns for clinical application. *Mol. Ther. Nucleic. Acids.* **2019**, *16*, 326–334. [CrossRef]
164. Chew, W.L. Immunity to CRISPR Cas9 and Cas12a therapeutics. *Wiley Interdiscip. Rev. Syst. Biol. Med.* **2018**, *10*, 1408. [CrossRef]
165. Yan, W.X.; Hunnewell, P.; Alfonse, L.E.; Carte, J.M.; Keston-Smith, E.; Sothiselvam, S.; Garrity, A.J.; Chong, S.; Makarova, K.S.; Koonin, E.V.; et al. Functionally diverse type V CRISPR-Cas systems. *Science* **2019**, *363*, 88–91. [CrossRef]
166. Gupta, Y.; Ghrera, A.S. Recent advances in gold nanoparticle-based lateral flow immunoassay for the detection of bacterial infection. *Arch. Microbiol.* **2021**, *203*, 3767–3784. [CrossRef]
167. Costa, S.P.; Carvalho, C.M. Burden of bacterial bloodstream infections and recent advances for diagnosis. *Pathog. Dis.* **2022**, *80*, ftac027. [CrossRef]
168. Di Gaudio, F.; Indelicato, S.; Indelicato, S.; Tricoli, M.R.; Stampone, G.; Bongiorno, D. Improvement of a rapid direct blood culture microbial identification protocol using MALDI-TOF MS and performance comparison with SepsiTyper kit. *J. Microbiol. Methods.* **2018**, *155*, 1–7. [CrossRef]
169. Mirza, F.H.; Baig, F.A.; Syed, S.; Kumar, A.; Shahid, M.A. Role of Presepsin and Comparison with Conventional Markers for Early Diagnosis and Differentiation of Sepsis. *J. Hunan Univ. Nat. Sci.* **2021**, *48*, 72–77.
170. Scerbo, M.H.; Kaplan, H.B.; Dua, A.; Litwin, D.B.; Ambrose, C.G.; Moore, L.J.; Murray, C.C.K.; Wade, C.E.; Holcomb, J.B. Beyond blood culture and Gram stain analysis: A review of molecular techniques for the early detection of bacteremia in surgical patients. *Surg. Infect.* **2016**, *17*, 294–302. [CrossRef] [PubMed]
171. Schenz, J.; Weigand, M.A.; Uhle, F. Molecular and biomarker-based diagnostics in early sepsis: Current challenges and future perspectives. *Expert Rev. Mol. Diagn.* **2019**, *19*, 1069–1078. [CrossRef] [PubMed]
172. Behera, B.; Vishnu, G.A.; Chatterjee, S.; Sreekumar, N.; Nagabhushan, A.; Rajendran, N.; Prathik, B.H.; Pandya, H.J. Emerging technologies for antibiotic susceptibility testing. *Biosens. Bioelectron.* **2019**, *142*, 111552. [CrossRef] [PubMed]
173. Peker, N.; Couto, N.; Sinha, B.; Rossen, J.W. Diagnosis of bloodstream infections from positive blood cultures and directly from blood samples: Recent developments in molecular approaches. *Clin. Microbiol. Infect.* **2018**, *24*, 944–955. [CrossRef] [PubMed]
174. Tabak, Y.P.; Vankeepuram, L.; Ye, G.; Jeffers, K.; Gupta, V.; Murray, P.R. Blood culture turnaround time in US acute care hospitals and implications for laboratory process optimization. *J. Clin. Microbiol.* **2018**, *56*, e00500–e00518. [CrossRef] [PubMed]
175. Green, M.R.; Sambrook, J. Analysis and normalization of real-time polymerase chain reaction (PCR) experimental data. *Cold Spring Harb. Protoc.* **2018**, *2018*, 095000. [CrossRef]
176. Dailey, P.J.; Elbeik, T.; Holodniy, M. Companion and complementary diagnostics for infectious diseases. *Expert Rev. Mol. Diagn.* **2020**, *20*, 619–636. [CrossRef]
177. Briggs, N.; Campbell, S.; Gupta, S. Advances in rapid diagnostics for bloodstream infections. *Diagn. Microbiol. Infect. Dis.* **2021**, *99*, 115219. [CrossRef] [PubMed]
178. Nathwani, D.; Varghese, D.; Stephens, J.; Ansari, W.; Martin, S.; Charbonneau, C. Value of hospital antimicrobial stewardship programs [ASPs]: A systematic review. *Antimicrob. Resist. Infect. Control* **2019**, *8*, 35. [CrossRef]
179. Ulrich, P.S.; Bastian, I.N.; Chen, D.J. Clinical Significance of BD Bactec FX Blood Culture Incubation Beyond 96 Hours (4 Days). *J. Clin. Microbiol.* **2022**, *60*, e00549-22. [CrossRef] [PubMed]
180. Maugeri, G.; Lychko, I.; Sobral, R.; Roque, A.C. Identification and antibiotic-susceptibility profiling of infectious bacterial agents: A review of current and future trends. *Biotechnol. J.* **2019**, *14*, 1700750. [CrossRef] [PubMed]
181. Yang, C.Y.; Lee, C.H.; Hsieh, C.C.; Hong, M.Y.; Chen, M.J.; Lee, C.C. Differential effects of inappropriate empirical antibiotic therapy in adults with community-onset gram-positive and gram-negative aerobe bacteremia. *J. Infect. Chemother.* **2020**, *26*, 222–229. [CrossRef] [PubMed]

182. Wang, W.; Kang, S.; Vikesland, P.J. Surface-enhanced Raman spectroscopy of bacterial metabolites for bacterial growth monitoring and diagnosis of viral infection. *Environ. Sci. Technol.* **2021**, *55*, 9119–9128. [CrossRef]
183. Tsuchida, S.; Umemura, H.; Nakayama, T. Current status of matrix-assisted laser desorption/ionization–time-of-flight mass spectrometry (MALDI-TOF MS) in clinical diagnostic microbiology. *Molecules* **2020**, *25*, 4775. [CrossRef]
184. Tjandra, K.C.; Ram-Mohan, N.; Abe, R.; Hashemi, M.M.; Lee, J.H.; Chin, S.M.; Yang, S. Diagnosis of Bloodstream Infections: An Evolution of Technologies towards Accurate and Rapid Identification and Antibiotic Susceptibility Testing. *Antibiotics* **2022**, *11*, 511. [CrossRef] [PubMed]
185. Brenner, T.; Decker, S.O.; Grumaz, S.; Stevens, P.; Bruckner, T.; Schmoch, T.; Sohn, K. Next-generation sequencing diagnostics of bacteremia in sepsis (Next GeneSiS-Trial): Study protocol of a prospective, observational, noninterventional, multicenter, clinical trial. *Medicine* **2018**, *97*, 9868. [CrossRef]
186. Gopal, A.; Yan, L.; Kashif, S.; Munshi, T.; Roy, V.A.; Voelcker, N.H.; Chen, X. Biosensors and Point-of-Care Devices for Bacterial Detection: Rapid Diagnostics Informing Antibiotic Therapy. *Adv. Healthc. Mater.* **2022**, *11*, 2101546. [CrossRef]
187. Gholizadeh, P.; Köse, Ş.; Dao, S.; Ganbarov, K.; Tanomand, A.; Dal, T.; Kafil, H.S. How CRISPR-Cas system could be used to combat antimicrobial resistance. *Infect. Drug Resist.* **2020**, *13*, 1111. [CrossRef]
188. Legenza, L.; Barnett, S.; Lacy, J.P.; See, C.; Desotell, N.; Eibergen, A.; Piccirillo, J.F.; Rose, W.E. Geographic mapping of *Escherichia coli* susceptibility to develop a novel clinical decision support tool. *Antimicrob. Agents Chemother.* **2019**, *63*, 19. [CrossRef]

diagnostics

Article

Metabolomic Selection in the Progression of Type 2 Diabetes Mellitus: A Genetic Algorithm Approach

Jorge Morgan-Benita [1,†], Ana G. Sánchez-Reyna [1,†], Carlos H. Espino-Salinas [1,†], Juan José Oropeza-Valdez [2], Huizilopoztli Luna-García [1], Carlos E. Galván-Tejada [1], Jorge I. Galván-Tejada [1], Hamurabi Gamboa-Rosales [1], Jose Antonio Enciso-Moreno [2] and José Celaya-Padilla [1,*]

1 Unidad Académica de Ingeniería Eléctrica, Universidad Autónoma de Zacatecas, Jardín Juarez 147, Centro, Zacatecas 98000, Mexico
2 Metabolomics and Proteomics Laboratory, Autonomous University of Zacatecas, Zacatecas 98000, Mexico
* Correspondence: jose.celaya@uaz.edu.mx
† These authors contributed equally to this work.

Citation: Morgan-Benita, J.; Sánchez-Reyna, A.G.; Espino-Salinas, C.H.; Oropeza-Valdez, J.J.; Luna-García, H.; Galván-Tejada, C.E.; Galván-Tejada, J.I.; Gamboa-Rosales, H.; Enciso-Moreno, J.A.; Celaya-Padilla, J. Metabolomic Selection in the Progression of Type 2 Diabetes Mellitus: A Genetic Algorithm Approach. *Diagnostics* **2022**, *12*, 2803. https://doi.org/10.3390/diagnostics12112803

Academic Editors: Yuli Huang, Yong Yuan and Peisong Chen

Received: 15 October 2022
Accepted: 7 November 2022
Published: 15 November 2022

Abstract: According to the World Health Organization (WHO), type 2 diabetes mellitus (T2DM) is a result of the inefficient use of insulin by the body. More than 95% of people with diabetes have T2DM, which is largely due to excess weight and physical inactivity. This study proposes an intelligent feature selection of metabolites related to different stages of diabetes, with the use of genetic algorithms (GA) and the implementation of support vector machines (SVMs), K-Nearest Neighbors (KNNs) and Nearest Centroid (NEARCENT) and with a dataset obtained from the Instituto Mexicano del Seguro Social with the protocol name of the following: "Análisis metabolómico y transcriptómico diferencial en orina y suero de pacientes pre diabéticos, diabéticos y con nefropatía diabética para identificar potenciales biomarcadores pronósticos de daño renal" (differential metabolomic and transcriptomic analyses in the urine and serum of pre-diabetic, diabetic and diabetic nephropathy patients to identify potential prognostic biomarkers of kidney damage). In order to analyze which machine learning (ML) model is the most optimal for classifying patients with some stage of T2DM, the novelty of this work is to provide a genetic algorithm approach that detects significant metabolites in each stage of progression. More than 100 metabolites were identified as significant between all stages; with the data analyzed, the average accuracies obtained in each of the five most-accurate implementations of genetic algorithms were in the range of 0.8214–0.9893 with respect to average accuracy, providing a precise tool to use in detections and backing up a diagnosis constructed entirely with metabolomics. By providing five potential biomarkers for progression, these extremely significant metabolites are as follows: "Cer(d18:1/24:1) i2", "PC(20:3-OH/P-18:1)", "Ganoderic acid C2", "TG(16:0/17:1/18:1)" and "GPEtn(18:0/20:4)".

Keywords: genetic algorithm; machine learning; metabolites; type 2 diabetes

1. Introduction

Diabetes is a chronic and progressive disease that occurs in the pancreas when it is no longer able to make a hormone known as insulin or when the body is unable to use it properly [1]. Adults numbering 537 million (20–79 years) currently live with diabetes in the world, and over 6.7 million deaths in 2021 are reported (approximately one death every 5 s) [2]. Type 2 diabetes mellitus (T2DM) is a progressive condition that is produced by relative insulin deficiencies caused by pancreatic β-cell (cells that synthesize and secrete insulin and amylin) dysfunction and insulin resistance [3]. The International Diabetes Federation (IDF) presented in 2021 that 541 million people in adulthood present a higher risk of developing T2DM [4]. As T2DM progresses, the comorbidities associated with hyperglycemia that induces renal damage directly or via hemodynamic modifications appear, which cause diabetic nephropathy or kidney disease (DN), and these are one of the most common problems developing in 30 to 40% of patients [5]. In recent years, metabolomics has been used as a novel approach for biomarker discovery and, in conjunction with genomics, can potentially provide a systemic understanding of the underlying causes of pathology, highlighting the importance of metabolomic approaches in the clinical sciences and helping provide guidance in clinical interventions. Metabolomics are a powerful and potentially high-throughput approach for biomarkers that can help provide molecular knowledge, identify therapeutic targets and improve the prevention of T2DM and its progression [6]. For a quicker adaptation of biomarker discoveries, portable and wearable technologies are aided by clever data mining, as well as deep learning and artificial intelligence inclusion [7]. Since individuals with decreased functional β-cell mass are at risk of developing T2DM, these individuals must be identified for prevention, but since in vivo detection remains unsuccessful, the use of metabolomics provides readouts of the states of this disease before symptoms appear. Gastrointestinal problems associated with metabolomics and diabetes [8], as well as other comorbidities, can be addressed by identifying novel plasma biomarkers for a loss of functional β-cell mass in the asymptomatic prediabetes stage as a solution to this problem, with non-targeted and targeted metabolomics. In this study performed on mice, Lingzi L. et al. [9] identified 1,5-anhydroglucitol as being associated with the loss of functional β-cell mass and uncovered metabolic similarities between the liver and plasma, providing insights into the systemic effects caused by early declines in β-cells; this deoxyhexose reflects the progressive decline of functional β-cell mass at the asymptomatic prediabetic stage. These findings stated a baseline to be applied in human cohorts so that they can be validated.

Another way to understand the metabolic function in the organs and the development and progression of T2DM is to match two-dimensional metabolic screening in tissue samples from key metabolic tissues such as serum, visceral adipose tissue, liver, pancreatic islets or skeletal muscle of individuals in different states of T2DM. In this way, carnitines are significantly higher in livers, while lysophosphatidylcholines were significantly lower in the muscle and serum of diabetes subjects. Other findings showed that lysophosphatidylcholines are significantly lower in the muscle and the serum of pre-diabetes subjects, and glycodeoxycholic acid was significantly higher in livers [10].

On the other hand, metabolites analyses are potent approaches for unraveling the relations between them and the progression or conditions of a particular disease, and a relation between conditions such as obesity and the progression of diabetes can be determined by major variations in lipid-related metabolites [11]. As there are more than 200,000 different metabolites in the human body, as shown in the human metabolome database 5.0 or HMDB 5.0 [12], data analyses by conventional methods prove to be inefficient and costly. The ML techniques provide a solution for this volume of data by detecting patterns and providing predictions. In order to stage novel results or predictions that are of high quality and usability, the data must be clearly supported by experts in the field and extracted by professional or scientific methods. The dataset also must be presented with the correct protocols for its liberation relative to the experiment or analyses that it was created for.

Peddinti G. et al. [13] implemented ML models based on entire metabolome datasets, and with a combination of glucose, mannose and α-hydroxybutyrate (known biomarkers commonly used as clinical risk factors) introduce predictive biomarkers, such as α-tocopherol, bradykinin hydroxyproline, X-12063 and X-13435, which are other metabolites that showed potential value in making precise predictions on the progression to type 2 diabetes. Moreover, Huang J. et al. [14] proposed a case of potential biomarkers in CKD prediction, and this case involved sphingomyelin C18:1 and phosphatidylcholine diacyl C38:0, which are identified specifically in hyperglycemic individuals.

Metabolites inclusion as part of the ML models brings new possibilities, making detection more robust and accurate. Another biomarker as a lone candidate on non-targeted urinary metabolomics is urine metabolome 3-hydroxy decanoyl-carnitine, which can be used for the identification of individuals with T2DM risks [15]. DN pathogenesis can be diagnosticated early with non-invasive biomarkers such as the base urea cycle, TCA cycle, glycolysis and amino acid metabolism, which includes lactic acid, hippuric acid, allantoin (in urine) and glutamine (in blood) (the latter are suggested as meta-analyses [16]). Valine (or betaine) and 3-(4-methyl-3-pentenyl)thiophene were associated with higher hazards with respect to end-stage kidney diseases [17]. The prognostic biomarkers given by metabolomics have the potential to uncover mechanisms in DN progression. Recent studies present potential target antigens in membranous nephropathy, with a signature of urinary peptides; this adds prognostic information to urinary albumin and implicates circulating inflammatory proteins as potential mediators of DN, demonstrating the importance of kidney bioenergetics as a modifiable factor in acute kidney injury [18].

ML has been widely used in the medical context using clinical data to detect patterns and/or predict different diseases, solving classification problems: using extreme learning machines on malaria parasite detection and classification [19]; using deep learning and image processing in diabetic retinopathy [20]; using Internet of Things (IoT) to provide an intelligent forensic analysis [21]; using FastAI and 1-Cycle Policy in breast cancer metastasis prediction [22]; using various multimodal models such as decision tree, logistic regression or random forest, among others, in Alzheimer's disease progression detection [23]. The ML implementations provide information for the analyst, which can be used to perform a pre-diagnosis if the patient has a particular disease or in identifying significant features; these results as the forms of predictions or classifiers can be ratified by a medical professional, and the professional can give approval to validate or discard this pre-diagnosis.

Most of the studies presented in this work propose diverse techniques to find a relation between control subjects and prediabetes: T2DM or DN. The ones that propose an ML model use metabolomics as a complement for classification; however, there are few studies that entirely use metabolomic data as features for the classification of a disease or provide tools to predict progression. Additional studies are required to replicate and expand upon these findings in independent cohorts, such as this one.

The proposal in this study is to provide a tool for the classification and analysis of the role of metabolomics in four different stages of T2DM. The novelty of this work is to present a genetic algorithm approach that selects the most significant metabolites in each stage of T2DM progression and not an individual classification of a particular stage, as other related work proposed.

The KNN, nearest centroid and support vector machines implemented as proposed ML models inside a genetic algorithm focus on analyzing 80 patients in five different sub-datasets: Control-Prediabetes, Control-T2DM, Prediabetes-T2DM, Control-DN and T2DM-DN. The dataset for this study was acquired from the "Unidad de Investigación Médica Biomédica by the Unidad de Investigación Biomédica located in Zacatecas, México, IMSS" with information on Mexican patients. The database contains anthropometric, clinical, laboratory and metabolomic data on Mexican patients.

This work is divided into five sections. The first is this Introduction. Section 2 describes the data, models and the methodology used to carry out the development of the ensemble model and how it is validated. Section 3 shows the results obtained and a detailed analysis is included using output graphs. Finally, Section 4 shows the discussions, and in Section 5 conclusions and future work.

2. Materials and Methods

The methodology of this study consists of six stages, as shown in Figure 1 and explained as follows: The first stage describes the dataset used (Figure 1A). In the second stage, a preliminary analysis of the data was performed by selecting subjects according to given inclusion criteria (Figure 1B). Subsequently, the data from the dataset were separated into three groups: Control-Prediabetes, Control-Diabetes and Control-DN (Figure 1C). In the fourth stage, feature selection using a genetic algorithm is implemented (Figure 1D). In the fifth stage, machine learning models (support vector machines, k-nearest neighbor and nearest centroid) were developed using the main features of the previous stage (Figure 1E).

Finally, the models were validated by taking into consideration different metrics (accuracy, sensitivity, specificity and AUC) to determine the performance of our models (Figure 1F).

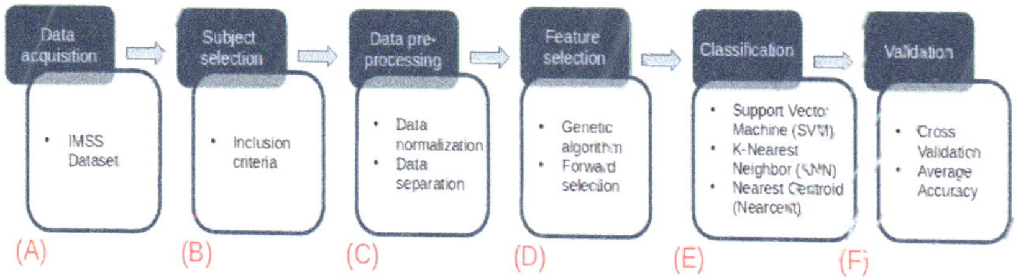

Figure 1. Flowchart of the proposed methodology. The blue squares refer to the data analysis methodology, while the white squares detail the task involved in each step. (**A**) The metabolomics dataset is obtained from the Unidad de Investigación Médica Biomedica. (**B**) The dataset is analyzed and new datasets are created by selecting subjects according to the criteria described in Table 1. (**C**) The data are normalized by probabilistic quotient normalization (PQN) and the dataset's observations are analyzed and separated in three groups: Control-Prediabetes, Control-Diabetes and Control-DN. (**D**) The use of genetic algorithms is implemented to extract the main data features. (**E**) Using the main features for prediabetes, diabetes and diabetic nephropathy detection in patients, several models are generated using support vector machines k-nearest neighbor and nearest centroid. (**F**) The validation of our results is carried out using different metrics (cross-validation by GALGO and average accuracy) to determine which of the models has the best performance.

Table 1. Inclusion criteria.

Inclusion Criteria
1. The age of the patients must be over 18 years
2. There will be no distinction in gender, education, ethnicity, race and marital status.
3. The datasets should contain only the metabolomics of each subject.
4. The dataset should distinguish controls from prediabetes, T2DM and DN.
5. The data of each feature in each subject must be complete.

2.1. Sample

The methodology implemented in order to obtain data of each of the 80 patients and 717 metabolomic data is presented in the following subsections: Sample Preparation; Quality Control (QC) and Quality Assurance (QA); Ultra-performance Liquid Chromatography (UPLC)—Mass Spectrometry Method for Lipid Separation and Data processing.

2.1.1. Sample Preparation

Thawed plasma measuring 100 µL in ice was extracted with 300 µL of precooled isopropanol (LCMS grade, Honeywell, Charlotte, NC USA), vortexed for 1 min and incubated at −20 °C for 1 h for protein precipitation. Subsequently, the extraction mixture was centrifuged at 15,800× *g* for 15 min, and supernatants were collected. For the analysis, each aliquot was transferred into LC vials and diluted to 1:20 with a mixture of isopropanol/acetonitrile/water (2:1:1, *v:v:v*). Sample preparation and analysis orders were randomized to ensure no systematic bias was present.

2.1.2. Quality Control (QC) and Quality Assurance (QA)

The instrument was subjected to maintenance in the chromatography system: the mass analyzer. The sample cone and ion sources were cleaned before the analysis. Calibration and manual tuning were also performed before running samples. A pool of human plasma from every sample in the study served as a technical replica for the entire experiment for QC. The overall variability was established by determining the relative standard deviation (RSD) for all endogenous metabolites present in 100% in the QC process. Experimental samples were randomized across the platform and run with ten QC samples at the start (column equilibration), and one QC sample was acquired for every ten samples.

2.1.3. Ultra-Performance Liquid Chromatography (UPLC)—Mass Spectrometry Method for Lipid Separation

The analysis was performed using an ACQUITY UPLC I-Class (Waters Corp., Milford, MA, USA) coupled to an XEVO-G2 XS quadrupole time-of-flight (ToF) mass spectrometer (Waters, Manchester, NH, USA) with an electrospray ionization source. The samples were analyzed in the positive mode (electrospray ionization). A UPLC CSH C18 column (2.1 × 100 mm, 1.7 µm) with a binary gradient elution of solvents was utilized, as represented in Table 2.

Table 2. Chromatography conditions.

Time (min)	%A	%B	Curve
Initial	60	40	Initial
2	57	43	6
2.1	50	50	1
12	46	54	6
12.1	30	70	1
18	1	99	6
18.1	60	40	6
20	60	40	1
Column temperature 55 °C			

The injection volume was five microliters. Data were acquired using positive electrospray ionization modes with the capillary voltage set to 3.2 kV, the cone voltage to 40 eV and the source temperature to 120 °C. The desolvation gas was nitrogen, with a flow rate of 900 L/h and source temperature of 550 °C. Data were acquired from a range of m/z 50–1200 in the MSE mode in which the collision energy alternated between low energy (6 eV) and high energy (ramped from 10 to 40 eV).

2.1.4. Data Processing

Raw data were processed under default settings as a Unifi Export Package (.uep), which was processed in Progenesis QI (version 2.3, Waters, Milford, MA, USA). Alignment was performed using a retention time range within 0.5–14 min to avoid interference with blank peaks. A peak width of 0.06 s was used. Deconvolution was automatically performed, considering M + H, M + Na and M + NH4 as adducts; manual inspection was performed after, eliminating features with incorrect alignments in chromatograms and neutral masses. An excel file was exported and a signal-to-noise ration was calculated for each sample based on the blank features, and all features with a signal-to-noise ratio (S/N) < 3 in 80% of samples were eliminated. In addition, RSD was calculated by taking QC features as medians, and RSD values > 20% were additionally removed.

2.1.5. Metabolite Identification

Putative annotations were assigned based on accurate mass and fragmentation patterns. Metabolite annotations were determined using the metascope plugin in Progenesis QI, mass measurements were taken with less than 10 ppm in error and fragmentation spectrum matching (when MSMS data were available) was conducted using the HMDB 5.0 [12], LipidMaps [24] and METLIN [25] databases. The top metabolite annotation was selected when the progenesis metascope score was >30.

2.2. IMSS Dataset

The dataset used in this study was provided by the Unidad de Investigación Biomedica located in Zacatecas, Mexico, which is incorporated into the IMSS. All Mexican patients signed an informed consent letter, and the data included in the IMSS dataset meet the R-2017-785-131 dictum approval according to the protocol "Análisis metabolómico y transcriptómico diferencial en orina y suero de pacientes prediabéticos, diabéticos y con nefropatía diabética para identificar potenciales biomarcadores pronósticos de daño renal" (differential metabolomic and transcriptomic analysis in the urine and serum of prediabetic, diabetic and diabetic nephropathy patients to identify potential prognostic biomarkers of renal impairment), which complies with the criteria approved by the National Committee for Scientific Research and Ethics and follows the international ethical standards of the Helsinki convention for research studies in humans. The IMSS dataset has information on metabolomic, anthropometric, clinical and laboratory tests. These assessments can be combined to measure the progression of prediabetes, diabetes and diabetic nephropathy.

2.3. Data Inclusion

The IMSS dataset includes 375 patients and 842 features. From this dataset, a filter was applied to select only patients and features that met the inclusion criteria indicated in Table 1. The resulting filtered dataset (FDS), after applying the inclusion criteria listed in Table 1, contains information corresponding to 80 patients (42 female/38 male), such as age (52.34 ± 10.45). metabolomics and diagnosis (20 patients positive for prediabetes, 20 patients positive for T2DM, 20 patients positive for DN and 20 control patients).

2.4. Data Normalization

The normalization implemented in this study is probabilistic quotient normalization (PQN) and is conducted as follows: For each function, the output mean is calculated over all samples. Then, a reference vector is generated. The median between the resulting reference vector and each sample is calculated, obtaining a vector of related coefficients. Then, each sample is divided by the mean value of the vector of coefficients; this mean value is different for each sample. The purpose of PQN is to account for concentration changes of some metabolite characteristics that affect limited regions of the data. The PQN approach assumes that changes in the concentrations of individual analytes influence only parts of the spectra, while changes in the overall concentration of a sample influence the entire spectrum. In contrast to integral normalization, which assumes that the total integral, covering all signals, is a function of dilution only, PQN instead assumes that the intensity of most signals is a function of dilution only. Therefore, a most likely quotient between the signals of the corresponding spectrum and a reference spectrum is calculated as a normalization factor, which replaces the total integral as a marker of the sample's concentration. This most likely quotient for a specific spectrum can be derived from the distribution of signals from a spectrum divided by the corresponding signal from a reference spectrum [26]:

$$I(i) = \frac{I^{old}(i)}{\sum_k \left(\int_{j_k^l}^{j_k^u} (I(x))^n \mathrm{d}x\right)^{\frac{1}{n}}} \tag{1}$$

where $I^{old}(i)$ and $I(i)$ are the intensities of variable i, which is the spectral feature, wavelength, bin and chemical shift. Before and after normalization, k is an index of the spectral regions used for normalization, j_k^l and j_k^u are the lower and upper borders, respectively, of spectral region k, for which the power n of intensities $I(x)$ is integrated [26].

2.5. Feature Selection

The dataset used to perform this study has 717 different metabolites, and each has a potential significance to become a biomarker or part of it to solve a classification problem; nevertheless, this task could become computationally expensive and complex to process. With genetic algorithms, this complex task can be performed and solved. GALGO [27] is a GA implemented in this study as the R package software used to perform feature selections in 5 sets of this article (Control-Prediabetes, Control-T2DM, Prediabetes-T2DM, Control-DN and T2DM-DN). For this study, GA creates an initial population of chromosomes comprising random sets of metabolites. The fitness of the chromosomes is evaluated by comparing their ability to correctly detect each stage in the progression of T2DM (control → prediabetes → T2DM → DN).

Depending on the obtained fitness score, the chromosome population continues to be replicated and the chromosomes crossover and mutate, as the fittest chromosomes will produce next-generation offspring. The process only stops when it meets the goal criteria (in this study, it is set at 1) or when the bigbangs (iterations) reach the limit proposed (3 times the number of metabolites in this case rounded to 2300 in all GALGO implementations). The GA blast (the implementation of the GALGO model) output is then submitted to a forward selection process to obtain the model that performed best (could be one or more); then, this model or set of features is ready for utilization in an ML model. Forward selection is widely used in genetic algorithms as a complement to presenting the best possible model output of GALGO implementations, such as the following example: Alzheimer's [28], COVID-19 [29] or diabetic retinopathy [30].

GALGO allows the use of different model criteria or parameters. In this study, the k-nearest neighbors (KNN), nearest centroid (NEARCENT) and support vector machines (SVM) were configured, as shown in Table 3.

Table 3. GALGO parameters.

Model	Parameter	Value
KNN	classification.method	knn
	chromosomeSize	5
	maxSolutions	2100
	maxGenerations	200
	goalFitness	1
Nearest Centroid	classification.method	nearcent
	chromosomeSize	5
	maxSolutions	2100
	maxGenerations	200
	goalFitness	1
SVM	classification.method	svm
	svm.kernel	radial
	chromosomeSize	5
	maxSolutions	2100
	maxGenerations	200
	goalFitness	1

2.6. Model Development

Once the main features from different partitioned datasets, described in Section 2.5, have been selected, the ML models are developed. This process is used to ensure that the training and testing results are as accurate as possible, avoiding overfitting or underfitting; the models will be paired with the ones used in GALGO implementations. In this way, the selection will be consistent with the results of the ML implementations, as is presented in Table 4.

2.7. K-Nearest Neighbors

KNN is one of the most fundamental classification methods and is widely used when there is little to no prior knowledge about the distribution of the data. In this study, as there is no certainty about the classification prior to the implementation (only a clear separation of the stages of the T2DM progression), a KNN implementation can provide a clear path to validate this separation and to obtain a discriminant analysis of the data [31]. The Euclidean distance between a test sample and the specified training samples is commonly defined by this model. The Euclidean distance between sample x_i and x_l ($l = 1,2,\ldots,n$) is defined as follows:

$$d(x_i, x_l) = \sqrt{(x_{i1}x_{l1})^2 + (x_{i2}x_{l2})^2 + \cdots + (x_{ip}x_{lp})^2}, \tag{2}$$

where x_i is an input sample with p features $(x_{i1}, x_{i2}, \ldots, x_{ip})$, n is the total number of input samples ($i = 1,2,\ldots,n$) and p is the total number of features ($j = 1,2,\ldots,p$)

Table 4. GALGO models—ML models.

Sub-Dataset	GALGO Model	ML Model
Control-Prediabetes	knn	K-Nearest Neighbours
	nearcent	Nearest Centroid
	svm	Support Vector Machines
Control-T2DM	knn	K-Nearest Neighbours
	nearcent	Nearest Centroid
	svm	Support Vector Machines
Prediabetes-T2DM	knn	K-Nearest Neighbours
	nearcent	Nearest Centroid
	svm	Support Vector Machines
Control-DN	knn	K-Nearest Neighbours
	nearcent	Nearest Centroid
	svm	Support Vector Machines
T2DM-DN	knn	K-Nearest Neighbours
	nearcent	Nearest Centroid
	svm	Support Vector Machines

2.8. Nearest Centroid

The nearest centroid is one of the simplest classifiers, nevertheless, it is capable of classifying data without any feature selection (for example, raw mass spectra [32]). In addition, it is extremely fast and requires low computational power, provides a baseline for the evaluation of feature selection algorithms and allows testing a number of algorithms that were previously inapplicable. NEARCENT and KNN provide similar approaches when there is limited knowledge on the distribution. In this study, they provide validations for the classifications results relative to one another.

$$\vec{\mu}_\ell = \frac{1}{|C_\ell|} \sum_{i \in C_\ell} x_i \tag{3}$$

Given the labeled training samples $(\vec{x}_1, y_1), \ldots, (\vec{x}_n, y_n)$ with class labels $y_i \in \mathbf{Y}$, the per-class centroids $\vec{\mu}_\ell = \frac{1}{|C_\ell|} \sum_{i \in C_\ell} \vec{x}_i$ are computed, where C_ℓ is the set of the indices of samples belonging to class $\ell \in \mathbf{Y}$.

2.9. Support Vector Machines

SVM is included as it is robust and precise for solving binary classification ML problems. This model uses the theory of Structural Risk Minimization to maximize its prediction accuracy and procures avoiding data overfitting [33]. This model can use a wide variety of standard or custom kernels. The radial kernel support vector machine model used in this study fits the closest observations into the new observation, grouping them (similar processes as KNN) based on how much they influence the output of the set classifier for multiple hyperplanes. This kernel has been proven to be one of the most accurate kernals for solving nonlinear separation problems [34].

The radial basis function kernel is defined as follows:

$$K(x, x') = e^{\frac{-||x-x'||^2}{2\sigma^2}}, \tag{4}$$

where x and x' are original observations and new observations, respectively [35].

2.10. Implementation

All models and methodology were implemented in R, which is a well-known open-source software validated by the scientific community, as well as the following packages:

- Genetic algorithms were implemented using "galgo 1.4" [27].
- For support vector machines, "caret" was used [36].

3. Results

The methodology proposed in Figure 1, presents the process followed in this study in six steps: data acquisition, subject selection, data pre-processing, feature selection, classification and validation. A description of the methodology to obtain the sample (Section 2.1) included the following: Sample Preparation (Section 2.1.1); Quality Control (QC) and Quality Assurance (QA) (Section 2.1.2); Ultra-performance Liquid Chromatography (UPLC)—Mass Spectrometry 158 Method for Lipidomic Analysis (Section 2.1.3); data processing (Section 2.1.4); and metabolite identification (Section 2.1.5). The data normalization process is presented in Section 2.4.

The obtained dataset is described in Section 2.2 with the inclusion criteria presented in Table 1. Data inclusion is provided in Section 2.3. After the data inclusion process, the feature selection process begins with the implementation of genetic algorithms in Section 2.5, and 15 runs were required to obtain each combination of the stages, as shown in Table 3. Each set of features obtained in the genetic algorithms integrates a model (Section 2.6), as presented in Table 4. The models are as follows: KNN (Section 2.7), NEARCENT (Section 2.8) and SVM (Section 2.9). Lastly, the implementation in R is presented in Section 2.10.

3.1. Galgo Results

The 15 runs of GALGO, with different datasets combining the samples and comparing them, provided an average accuracy, as presented in Table 5, and a group of features are presented in the next sections.

Table 5. GALGO models—Average Accuracy.

Sub-Dataset	GALGO Model	ML Model	Average Accuracy
Control-Prediabetes	knn	K-Nearest Neighbours	0.8143
	nearcent	Nearest Centroid	0.8321
	svm	**Support Vector Machines**	**0.8464**
Control-T2DM	knn	K-Nearest Neighbours	0.9268
	nearcent	**Nearest Centroid**	**0.9286**
	svm	Support Vector Machines	0.9036
Prediabetes-T2DM	knn	K-Nearest Neighbours	0.7821
	nearcent	Nearest Centroid	0.7982
	svm	**Support Vector Machines**	**0.8214**
Control-DN	**knn**	**K-Nearest Neighbours**	**0.9893**
	nearcent	Nearest Centroid	0.9714
	svm	Support Vector Machines	0.9857
T2DM-DN	knn	K-Nearest Neighbours	0.9054
	nearcent	**Nearest Centroid**	**0.9125**
	svm	Support Vector Machines	0.8804

Bold text represent the model with the highest Average accuracy in each sub-dataset

3.1.1. GALGO Implementation with the Control-Prediabetes Dataset

The features obtained by GALGO with the SVM classification model and the forward selection best model, in this case model 1, are presented in Table 6 and in Figure 2, with an average accuracy of 0.8464, as shown in Table 5. Derived from low-quantity data against the large quantity of features included in the model, these results prove that the resultant metabolites included are extremely significant, as is presented in the gene rank's stability (Figure 3) and fitness (Figure 4).

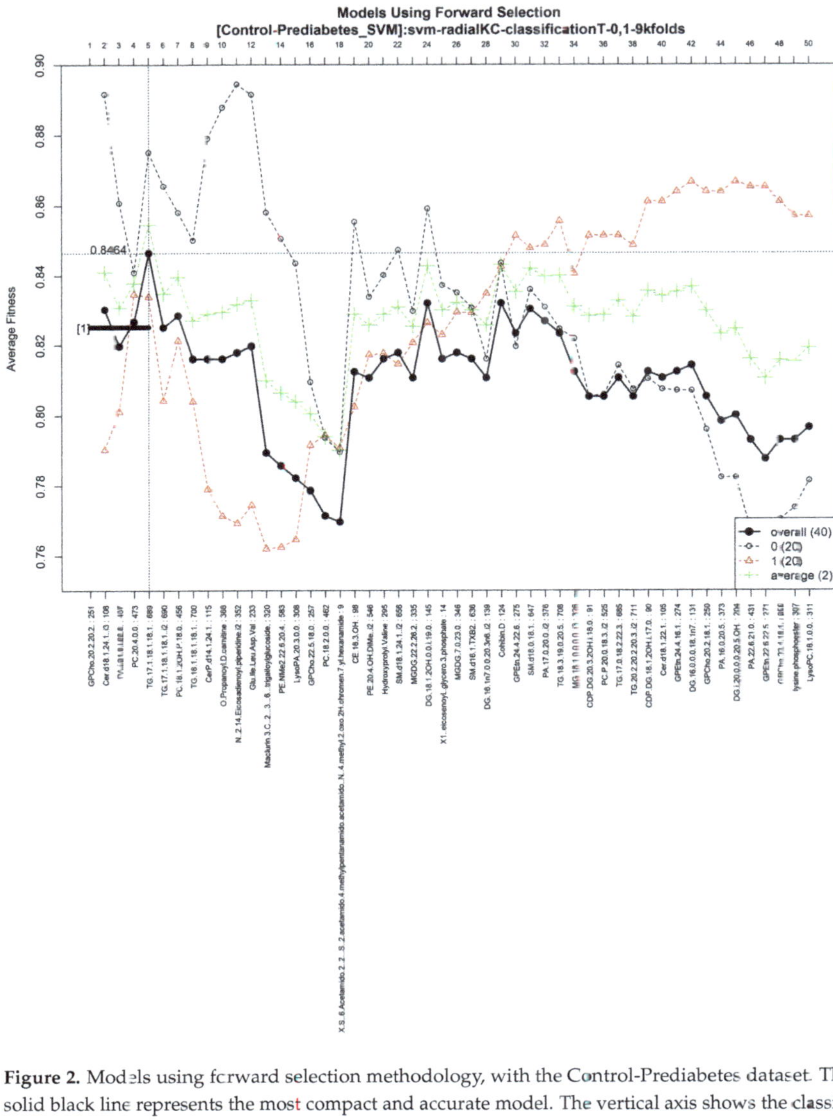

Figure 2. Models using forward selection methodology, with the Control-Prediabetes dataset. The solid black line represents the most compact and accurate model. The vertical axis shows the classification's accuracy. The horizontal axis represents the ordered features, each of which corresponds to a number and a specific feature.

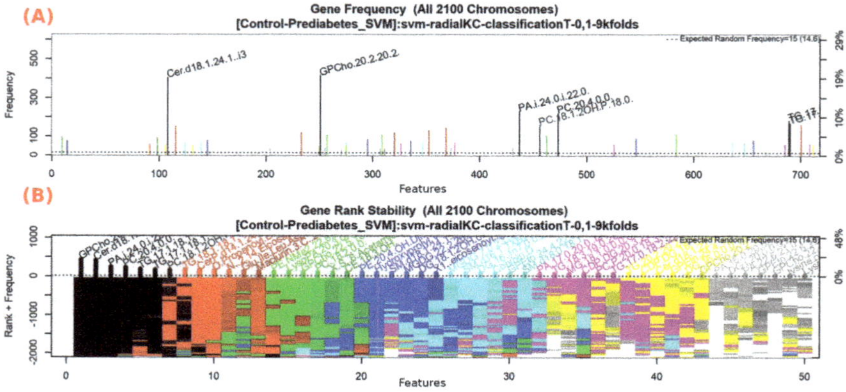

Figure 3. Gene frequency and gene rank stability in the models are ascertained by applying GA with SVM to select the main features from the Control-Prediabetes dataset. (**A**) The gene frequency allows observing the number of times each feature appears in the models. (**B**) The gene rank allows observing the stability and frequency of each feature within the models and orders them by rank.

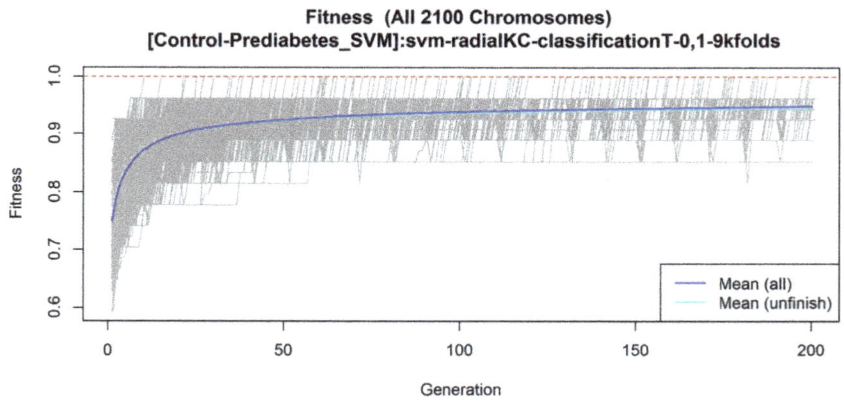

Figure 4. Evolution of maximum fitness scores over generations, using the Control-Prediabetes dataset. The vertical axis represents the fitness score, while the horizontal axis represents a given generation. The solid blue line represents the average fitness across all models. The average unfinished fitness plotted as a solid cyan line represents the average worst case expectation for all failed searches in a given generation. The red dotted line shows the established GA fitness goal.

Table 6. Features obtained by the GALGO SVM method in the Control-Prediabetes dataset.

Result Features
"GPCho(20:2/18:1)", "Cer(d18:1/24:1) i2", "PA(i-19:0/20:3-2OH)", "PC(20:3-OH/P-18:1)", "TG(17:1/17:2/22:5)"

All features in this table were obtained with GALGO with an SVM method implementation, with 2100 bigbangs and 200 generations.

3.1.2. GALGO Implementation with the Control-T2DM Dataset

The features obtained by GALGO with the NEARCENT classification method and the forward selection best model, in this case model 1, are presented in Table 7 and in Figure 5, with an average accuracy of 0.9286 (as shown in Table 5). Derived from low-quantity data against the large quantity of features included in the model, these results prove that the resultant metabolites included are extremely significant, as presented in the gene rank's stability (Figure 6) and in fitness (Figure 7).

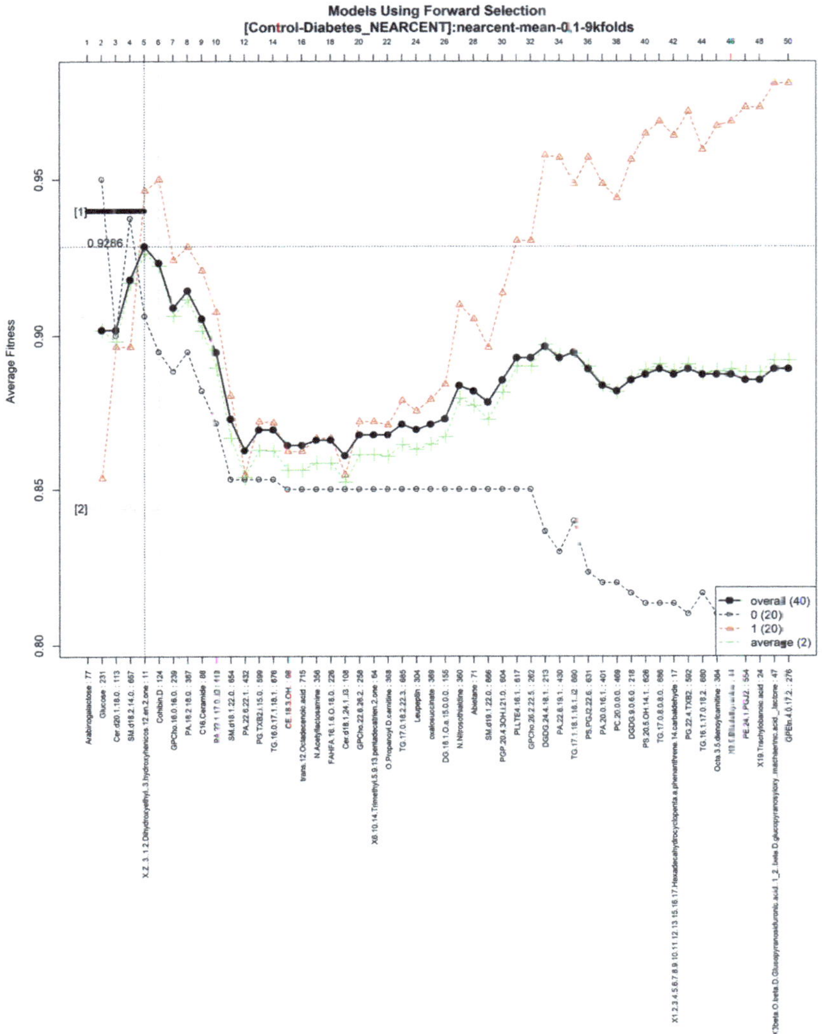

Figure 5. Models using the forward selection methodology, with the Control-T2DM dataset. The solid black line represents the most compact and accurate models. The vertical axis shows the classification accuracy. The horizontal axis represents the ordered features, each of which corresponds to a number.

Table 7. Features obtained by GALGO NEARCENT method in the Control-T2DM dataset.

Result Features
"Androst-16-ene", "Ganoderic acid C2", "Cer(d18:2/20:4-3OH)", "SM(d18:1/24:1) i2", "(Z)-11-Hexadecenal"

All features in this table were obtained with GALGO with a NEARCENT method implementation with 2100 bigbangs and 200 generations.

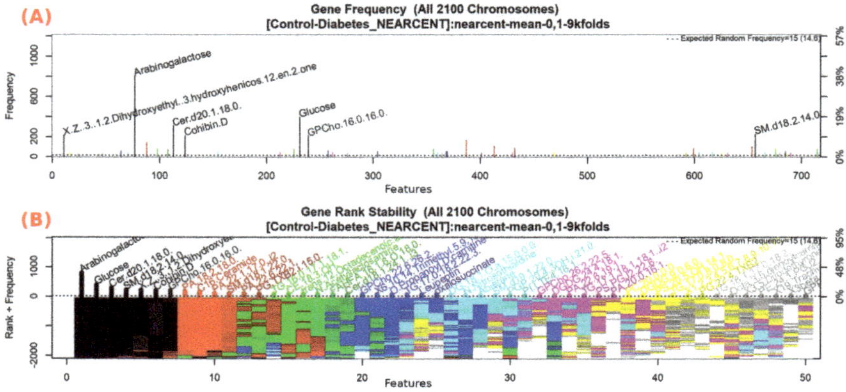

Figure 6. Gene frequency and gene rank stability in the models are ascertained by applying GA with NEARCENT to select the main features from the Control-T2DM dataset. (**A**) Gene frequency allows observing the number of times each feature appears in the models. (**B**) Gene rank allows observing the stability and frequency of each feature within the models and orders them by rank.

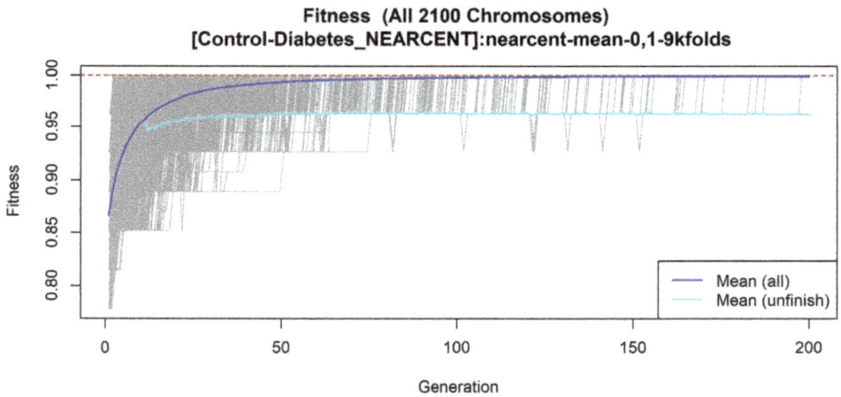

Figure 7. Evolution of maximum fitness scores over generations, using the Control-T2DM dataset. The vertical axis represents the fitness score, while the horizontal axis represents a given generation. The solid blue line represents the average fitness across all models. The average unfinished fitness is plotted as a solid cyan line and represents the average worst case expectation for all failed searches in a given generation. The red dotted line shows the established GA goal fitness.

3.1.3. GALGO Implementation with the Prediabetes-T2DM Dataset

The features obtained by GALGO with the SVM classification method and the forward selection best model, in this case model 4, are presented in Table 8 and in Figure 8, with an average accuracy of 0.8214, as shown in Table 5. Derived from low-quantity data against the large quantity of features included in the model, these results prove that the resultant metabolites included are extremely significant, as presented in the gene rank's stability (Figure 9) and fitness (Figure 10).

3.1.4. GALGO Implementation with the Control-DN Dataset

The features obtained by GALGO with the KNN classification method and the forward selection best model, in this case model 6, are presented in Table 9 and in Figure 11 with an average accuracy of 0.9893, as shown in Table 5. Derived from low-quantity data against the large quantity of features included in the model, these results prove that the resultant metabolites included are extremely significant, as presented in the gene rank's stability (Figure 12) and fitness (Figure 13).

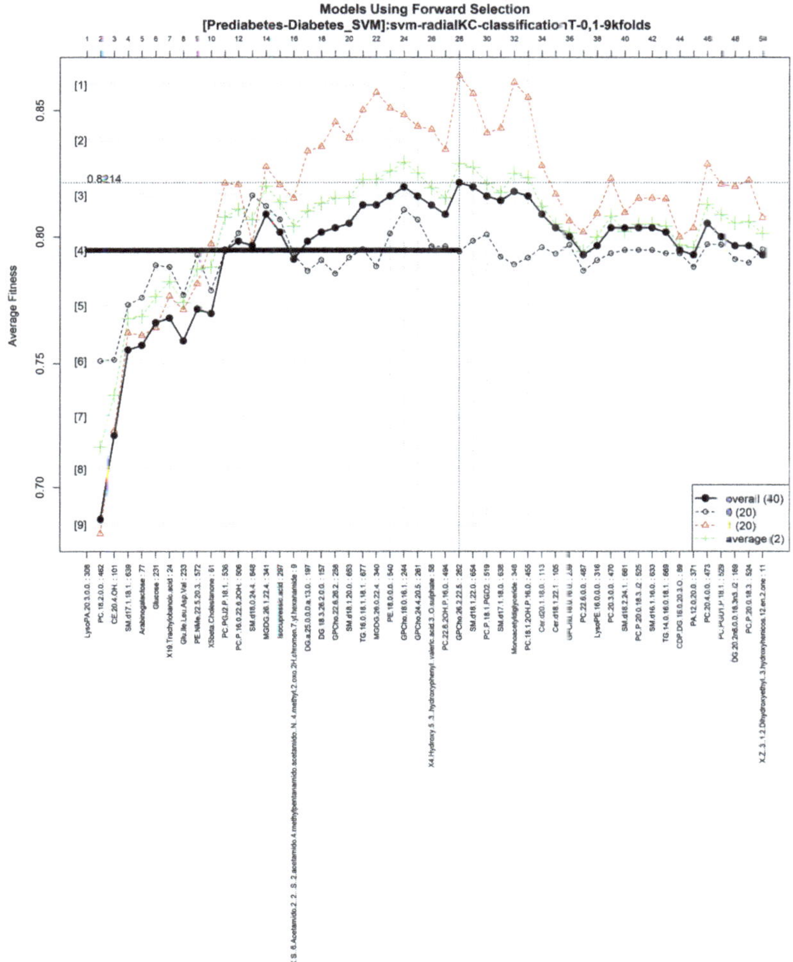

Figure 8. Models using forward selection methodology, with the Prediabetes-T2DM dataset. The solid black line represents the most compact and accurate model. The vertical axis shows the classification accuracy. The horizontal axis represents ordered features, each of which corresponds to a number.

Table 8. Features obtained by the GALGO SVM method in the Prediabetes-T2DM dataset.

Result Features
"lysine phosphoester", "PC(18:1-O/20:1)", "CE(20:4-2OH)", "SM(d17:1/18:0)", "Androst-16-ene", "Ganoderic acid C2", "12-Methylheptadecanoylcarnitine", "Glufosinate", "PE-NMe(22:0/18:1)", "5alpha-Androsta-16-ene-3-ol", "PC(PGJ2/DiMe)", "PC(P-16:0/20:4-3OH)", "SM(d18:0/18:1)", "MGDG(26:0/22:4)", "Isobehenic acid", "(Melle-4)cyclosporin", "DG(a-17:0/0:0/8:0) i2", "DG(18:3/18:1/0:0)", "GPCho(22:5/18:0)", "SM(d18:1/18:1)", "TG(16:0/17:1/18:1)", "MGDG(24:1/18:1)", "PE(15:0/18:4)", "GPCho(16:1/16:1)", "GPCho(24:1/22:6)", "4-Hydroperoxycyclophosphamide", "PC(22:6-2OH/24:0) i2", "GPCho(24:4/20:5)"

All features in this table were obtained with GALGO with an SVM method implementation with 2100 bigbangs and 200 generations.

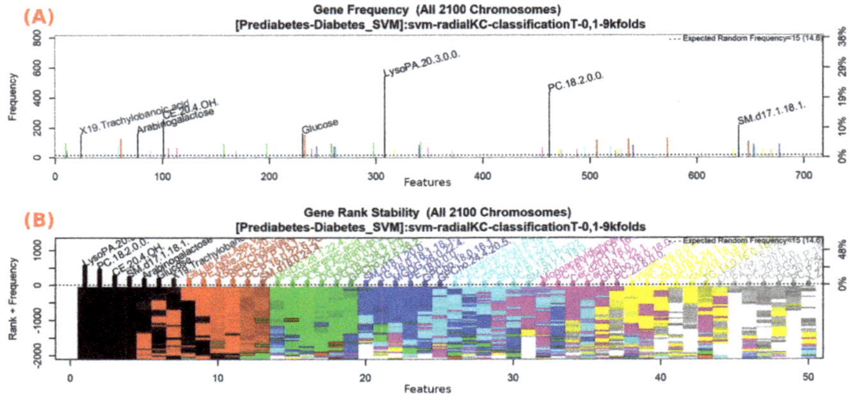

Figure 9. Gene frequency and gene rank stability in the models are ascertained by applying GA with SVM to select the main features from the Prediabetes-T2DM dataset. (**A**) Gene frequency allows observing the number of times each feature appears in the models. (**B**) Gene rank allows observing the stability and frequency of each feature within the models and orders them by rank.

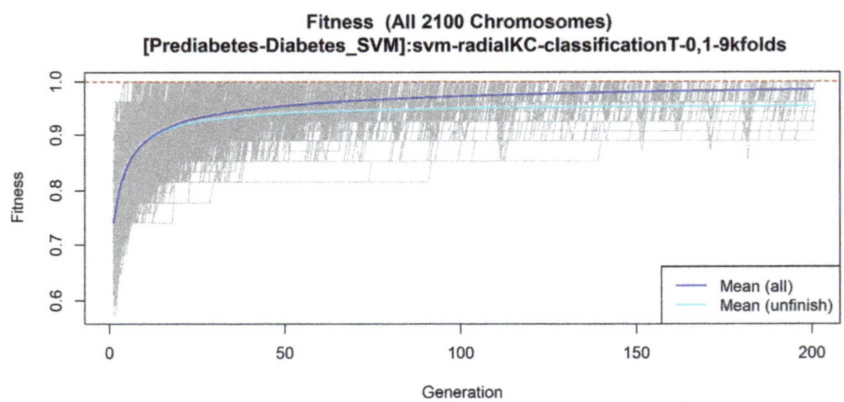

Figure 10. Evolution of maximum fitness scores over generations, using the Prediabetes-T2DM dataset. The vertical axis represents the fitness score, while the horizontal axis represents a given generation. The solid blue line represents the average fitness across all models. The average unfinished fitness plotted as a solid cyan line represents the average worst case expectation for all failed searches in a given generation. The red dotted line shows the established GA goal fitness.

Table 9. Features obtained by the GALGO KNN method in the Control-DN dataset.

Result Features
"5beta-Cholestanone", "GPSer(18:1/11:0)", "GPEtn(18:0/20:4)", "TG(16:0/17:1/18:1)", "1,1-Dimethylbiguanide", "5-Pentahydroxy-5-cucurbiten-11-one 3-[glucosyl-(1->6)-glucoside]", "PG(PGF1alpha/i-16:0)", "TG(16:0/18:1/18:1)", "DG(i-18:0/22:6-OH/0:0)", "Butyl methacrylate", "DG(20:4/16:0/0:0)", "Androst-16-ene", "Threonyltyrosine", "DG(a-25:0/0:0/a-13:0)", "GPCho(20:2/18:0)", "Cholest-8-en-3-ol", "Cer(d18:1/24:1) i2", "9-Decenoylcarnitine", "PA(22:1/17:0)", "PE(TXB2/DiMe)", "(2Z,4E,6Z)-Decatrienoylcarnitine", "delta-24-Cholesterol", "3-methoxy-4-hydroxy-5-all-trans-hexaprenylbenzoate", "N-Methylethanolaminium phosphate"

All features in this table were obtained with GALGO with a KNN method implementation with 2100 bigbangs and 200 generations.

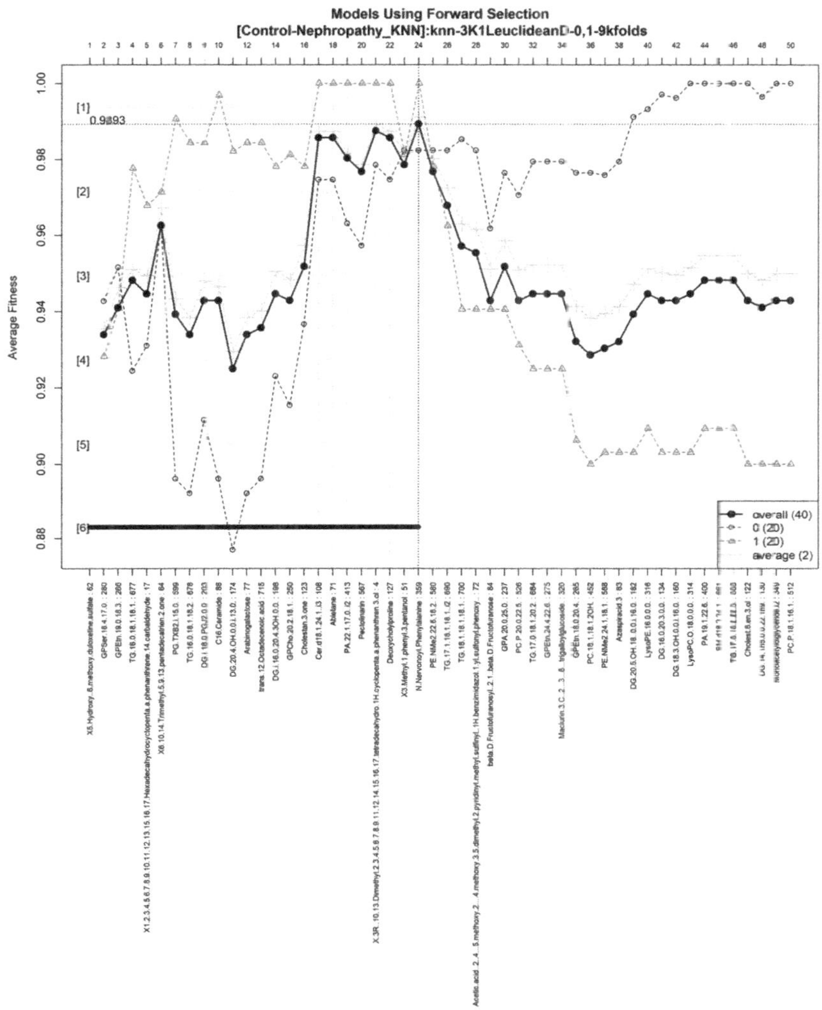

Figure 11. Models using forward selection methodology, with the Control-DN dataset. The solid black line represents the most compact and accurate models. The vertical axis shows the classification accuracy. The horizontal axis represents the ordered features, each of which corresponds to a number.

3.1.5. GALGO Implementation with the T2DM-DN Dataset

The features obtained by GALGO with the NEARCENT classification method and the forward selection best model, in this case model 5, are presented in Table 10 and in Figure 14, with an average accuracy of 0.9125, as shown in Table 5. Derived from low-quantity data against the large quantity of features included in the model, these results prove that the resultant metabolites included are extremely significant, as presented in the gene rank's stability (Figure 15 and fitness (Figure 16).

Comparing Tables 11–15, the metabolite addressed as "Cer(d18:1/24:1) i2" was found in three out of five stages of T2DM, Control-Prediabetes, Control-ND and T2DM-ND. The metabolite "PC(20:3-OH/P-18:1)" was found in Control-Prediabetes and Prediabetes-T2DM. The metabolite "Ganoderic acid C2" was found in Control-T2DM and Prediabetes-

T2DM. The metabolites "TG(16:0/17:1/18:1)" and "GPEtn(18:0/20:4)" were found in Control-ND and T2DM-ND. This findings were obtained by matching the results of the most significant metabolites presented in each table of common metabolites found.

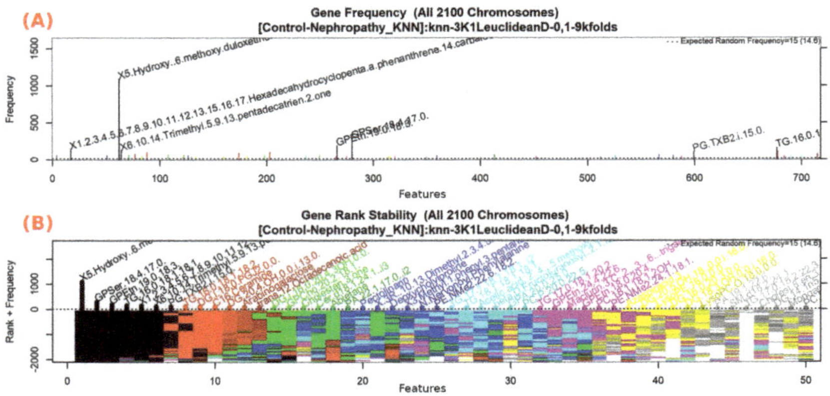

Figure 12. Gene frequency and gene rank stability in models are ascertained by applying GA with KNN to select the main features from the Control-DN dataset. (**A**) Gene frequency allows observing the number of times each feature appears in the models. (**B**) Gene rank allows observing the stability and frequency of each feature within the models and orders them by rank.

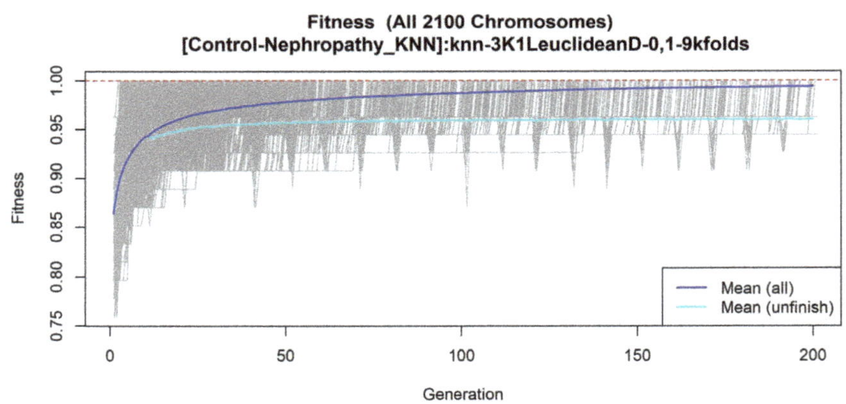

Figure 13. Evolution of maximum fitness scores over generations, using the Control-DN dataset. The vertical axis represents the fitness score, while the horizontal axis represents a given generation. The solid blue line represents the average fitness across all models. The average unfinished fitness plotted as a solid cyan line represents the average worst case expectation for all failed searches in a given generation. The red dotted line shows the established GA goal fitness.

Table 10. Features obtained by GALGO NEARCENT method in the T2DM-DN dataset.

Result Features
"SM(d19:0/20:3-OH)", "TG(16:0/17:1/18:1)", "5-Pentahydroxy-5-cucurbiten-11-one 3-[glucosyl-(1->6)-glucoside]", "PS(PGJ2/22:6)", "TG(16:0/18:1/18:1)", "GPEtn(18:0/20:4)", "Butyl methacrylate", "PC(22:6-2OH/P-16:0)", "Cer(d18:1/24:1) i2", "PC(P-20:0/14:1)", "PA(19:1/16:0)", "CerP(d15:0/2:0)", "PC(22:6-2OH/P-18:0)", "SM(d16:1/18:0)", "MG(20:4/0:0/0:0)", "MG(18:1/0:0/0:0)", "1,1-Dimethylbiguanide"

All features in this table were obtained with GALGO with a NEARCENT method implementation with 2100 bigbangs and 200 generations.

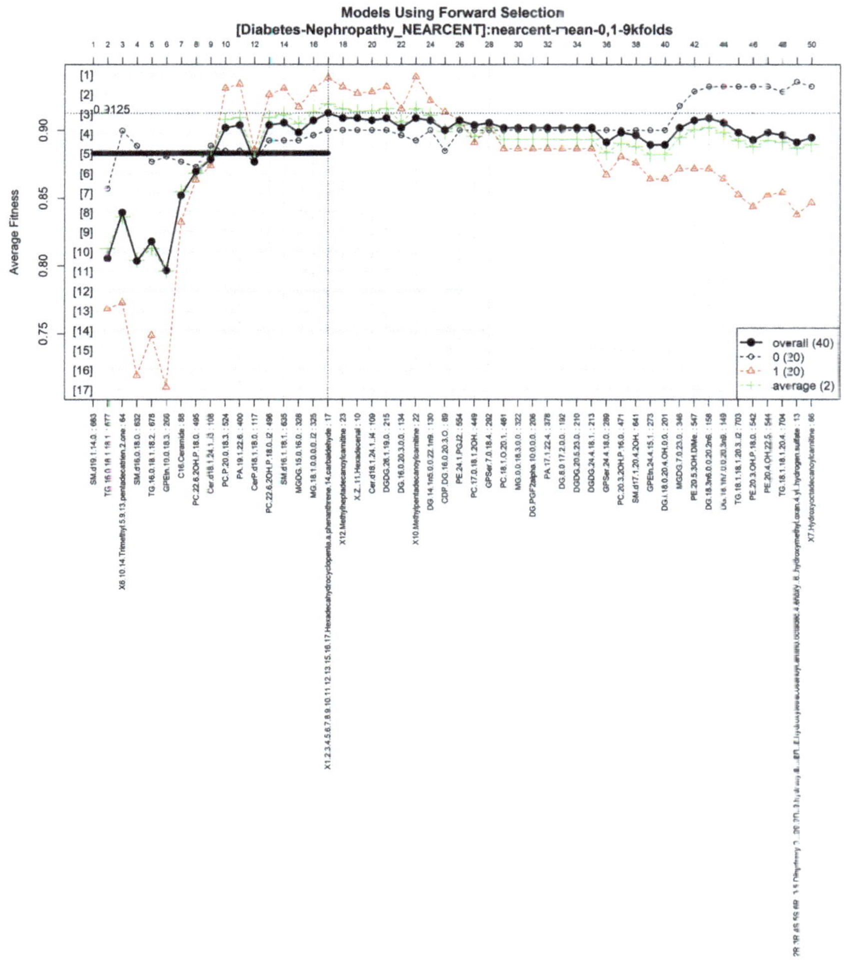

Figure 14. Models using forward selection methodology, with the T2DM-DN dataset. The solid black line represents the most compact and accurate model. The vertical axis shows the classification accuracy. The horizontal axis represents the ordered features, each of which corresponds to a number.

Table 11. Common metabolites found in each model for Control-Prediabetes.

KNN and NEARCENT and SVM	KNN and NEARCENT	KNN and SVM	NEARCENT and SVM
"PC(20:3-OH/P-18:1)" "TG(17:1/17:2/22:5)"	"TG(17:1/18:1/18:1)" "Glufosinate" "PC(18:1-O/20:1)" "DGDG(20:5/14:0)" "TG(18:0/18:1/20:2)" "lysine phosphoester" "TG(17:0/18:1/20:2)"	"GPCho(20:2/18:1)" "Cer(d18:1/24:1) i2"	

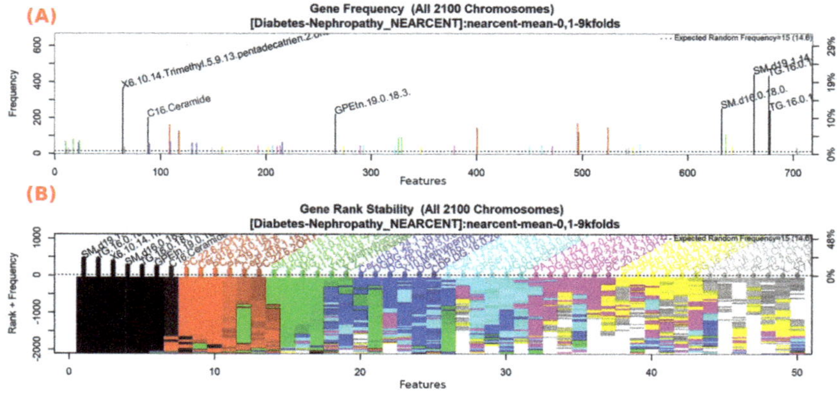

Figure 15. Gene frequency and gene rank stability in the models are ascertained by applying GA with NEARCENT to select the main features from the T2DM-DN dataset. (**A**) Gene frequency allows observing the number of times each feature appears in the models. (**B**) Gene rank allows observing the stability and frequency of each feature within the models and orders them by rank.

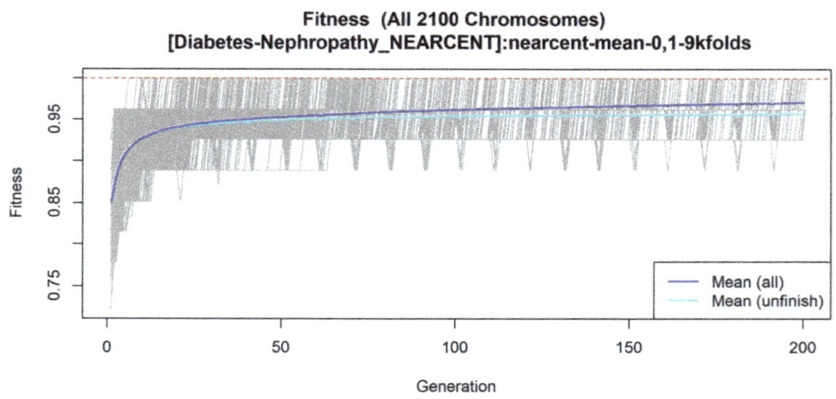

Figure 16. Evolution of maximum fitness scores over generations, using the T2DM-DN dataset. The vertical axis represents the fitness score, while the horizontal axis represents a given generation. The solid blue line represents the average fitness across all models. The average unfinished fitness is plotted as a solid cyan line and represents the average worst case expectation for all failed searches in a given generation. The red dotted line shows the established GA goal fitness.

Table 12. Common metabolites found in each model for Control-Diabetes.

KNN and NEARCENT and SVM	KNN and NEARCENT	KNN and SVM	NEARCENT and SVM
"Androst-16-ene"		"Cholestan-3-one"	"(Z)-11-Hexadecenal"
"Ganoderic acid C2"		"PA(18:1/18:2) i2"	
"Cer(d18:2/20:4-3OH)"		"GPA(26:2/6:0)"	
"SM(d18:1/24:1) i2"		"CE(18:2+=O)"	

Table 13. Common metabolites found in each model for Prediabetes-Diabetes.

KNN and NEARCENT and SVM	KNN and NEARCENT	KNN and SVM	NEARCENT and SVM
	"PC(20:3-OH/P-18:1)"	"Ganoderic acid C2"	

Table 14. Common metabolites found in each model for Control-Nephropathy.

KNN and NEARCENT and SVM	KNN and NEARCENT	KNN and SVM	NEARCENT and SVM
"5beta-Cholestanone" "DG(a-25:0/0:0/a-13:0)" "PG(PGF1alpha/i-16:0)" "DG(20:4/16:0/0:0)" "N-Methylethanolaminium phosphate" "GPEtn(18:0/20:4)" "TG(16:0/17:1/18:1)"		"Cer(d18:1/24:1) i2"	"GPSer(18:1/11:0)" "PC(18:0/18:1-2OH)"

Table 15. Common metabolites found in each model for Diabetes-Nephropathy.

KNN and NEARCENT and SVM	KNN and NEARCENT	KNN and SVM	NEARCENT and SVM
"Butyl.methacrylate" "TG(16:0/17:1/18:1)" "Cer(d18:1/24:1) i2" "TG(16:0/18:1/18:1)" "1,1-Dimethylbiguanide" "PS(PGJ2/22:6)" "5-Pentahydroxy-5-cucurbiten-11-one 3-[glucosyl-(1->6)-glucoside]"	"SM(d19:0/20:3-OH)" "GPEtn(18:0/20:4)" "PC(22:6-2OH/P-16:0)" "CerP(d15:0/2:0)" "PC(22:6-2OH/P-18:0)" "SM(d16:1/18:0)" "PA(19:1/16:0)"		

4. Discussion

The use of ML is proven to be an effective classifier and feature selection tool; however, the metabolomic community has some concerns with the lack of explanation on where does this biomarker's significance come from. There are some methods that unveil these doubts; the statistical validation, for example, has the most widely known validator, which is the AUC or area under the curve [37]. In this proposal, this metric however will not be used; instead, the strict use of average accuracies in the feature selection with genetic algorithms is proposed as it comes from a validator given by the forward selection method provided by GALGO.

The proposed selection and model implementation shows the potential to establish significant metabolites in each stage of the disease described. The progression asseveration can be made as features collide and the model's products of the features and the ML method can be used for fitting classifiers. As the metabolomics obtained are mainly set from serum samples, the families of lipids involved in the features obtained reveal metabolites of potential risk and more specifically propose a base to create a support tool for personalized diagnosis.

To establish a connection in each stage, the five different sets were created within a previous classification and direct transition, so it can be asseverated that there will be a statistical relation between each stage. The proposal provides a selection of features independently of the type of data selected, as in this case of metabolomics, which has been a wide field to cover but is a powerful ally for diagnosing a disease or for identifying progression.

The identification of metabolites as a potential or candidate biomarker of the incidence of DN in hyperglycemic subjects can be performed with intelligent feature selection [14]; nevertheless, in the case of LASSO, the biomarkers obtained can be diverse depending not only on the type of metabolomics in the dataset; since genetic algorithms prove the superiority and diversity of models against LASSO, new approximations can be made. The prediction of the progression from T2DM to DN remains difficult, even with potential biomarkers, with respect to detecting one or another, and novel biomarkers are needed to for detecting the progression of the disease. Nevertheless, ML methods can detect potential biomarkers that could otherwise escape identification using a conventional statistical method. Even identifying a potential biomarker such as the urinary 1-methylpyridin-1-ium

(NMP) [38] with non-targeted metabolites still needs other features as complements in order to be marked as clinically usable; in contrast, a group of significant metabolites can be useful for the detection of each stage of the disease. Limiting the number of features to find strong predictors could highlight differences in pathways leading to accurate predictions on more specific populations and provide novel clues to lead to features of strong significance [39]; however, there is a need to have sufficient observations to make a model that does not overfit, and even with this, it becomes difficult in T2DM's progression detection as there is no certainty as to when this disease will show symptoms or metrics that can be registered, as it can takes decades to develop. In cases of a small number of observations and even with a leave-one-out cross validation that seems to overfit, feature selection can stays stable with good average accuracy values (over 0.82 in this study) when there are plenty of features (717 in this case).

The metabolite's biomarker identification can widely assist in understanding progression and in obtaining a more precise assumption on how a disease works in a specific part of the human body or how it can progress in terms of increasing risks towards developing a comorbidity. For example, in gestational diabetes, a series of metabolites can identify potential risks in developing T2DM by only perceiving the variations and finding patterns with ML and data-mining techniques [40].

In comparison with clinical or anthropometric data, metabolite data are presented as complements of models; however, the metabolites presented in this study can perform better than clinical and anthropometric data [41]. This behavior can be explained by the use of genetic algorithms as selectors and the proper preselection and prediction of the most important metabolites or family-related metabolites with respect to the disease.

The methods that best performed in the GALGO implementations were as follows: In the Control-Prediabetes dataset, SVM had 0.8464 with respect to the average accuracy and five features (to see features, view Table 6); in the Control-T2DM dataset, NEARCENT had 0.9286 with respect to the average accuracy and five features (to see features, view Table 7); in the Prediabetes-T2DM dataset, SVM had 0.8214 with respect to the average accuracy and 28 features (to see features, view Table 8); in the Control-DN dataset, KNN had 0.9893 with respect to the average accuracy and 24 features (to see features, view Table 9); and in T2DM-DN dataset, NEARCENT had 0.9125 with respect to the average accuracy and 17 features (to see features, view Table 10). The comparisons were made with an average accuracy of each GALGO selection in each clasificator method implemented in each dataset (see Table 16). Since the methods that resulted in the best average accuracies were relatively small (the larger set of metabolites selected was 28 of 717), a comparable model can be established with the same quantities and measured as percentage of features selected from the total, with other related works that had less than 10 features as the final outputs [14,38,39].

Table 16. Comparison of feature selection with related work.

Title	Feature Selection Technique	Validation Metric	Result
Machine Learning Approaches Reveal Metabolic Signatures of Incident Chronic Kidney Disease in Individuals With Prediabetes and Type 2 Diabetes [14]	LASSO	AUC	0.857
Potential progression biomarkers of diabetic kidney disease determined using comprehensive machine learning analysis of non-targeted metabolomics [38]	NON	AUC	0.775
Predictive Modeling of Type 1 Diabetes Stages Using Disparate Data Sources [39]	Repeated Optimization for Feature Interpretation	AUC	0.91
Machine Learning–Derived Prenatal Predictive Risk Model to Guide Intervention and Prevent the Progression of Gestational Diabetes Mellitus to Type 2 Diabetes: Prediction Model Development Study [40]	CatBoost tree ensembles	AUC	0.86

Table 16. *Cont.*

Title	Feature Selection Technique	Validation Metric	Result
Data-Driven Machine-Learning Methods for Diabetes Risk Prediction [41]	Pearson Correlation, Gain Ratio, Naive Bayes and Random Forest	AUC	0.942
Interpretable machine learning-derived nomogram model for early detection of diabetic retinopathy in type 2 diabetes mellitus: a widely targeted metabolomics study [42]	Classification and Regression Tree	AUC	0.95
Environmental chemical exposure dynamics and machine learning-based prediction of diabetes mellitus [43]	Lasso	AUC	0.78
This Work in Control-Prediabetes	**Genetic Algorithm with GALGO-svm**	**Accuracy**	**0.8464**
This Work in Control-T2DM	**Genetic Algorithm with GALGO-Nearcent**	**Accuracy**	**0.9286**
This Work in Prediabetes-T2DM	**Genetic Algorithm with GALGO-svm**	**Accuracy**	**0.8214**
This Work in Control-DN	**Genetic Algorithm with GALGO-knn**	**Accuracy**	**0.9893**
This Work in T2DM-DN	**Genetic Algorithm with GALGO-nearcent**	**Accuracy**	**0.9125**

Bold text represent this work models with the highest Average accuracy in each sub-dataset

5. Conclusions

The metabolite selection method presented in this study has the potential to classify each stage of the T2DM with superior average accuracies. These metabolites can provide potential biomarkers for classification. The comparison between each classification method provided two assertions: Even when the methods used different approaches, they have similar results in a few features, and the features repeated in two of three OR three of three methods for each dataset are those that are extremely significant. In this case, those particular metabolites presented in each dataset are significant and provide the tools for classification between stages. These particular metabolites colliding between stages can be considered potential biomarkers for progression, and these metabolites are as follows: "Cer(d18:1/24:1) i2", "PC(20:3-OH/P-18:1)", "Ganoderic acid C2", "TG(16:0/17:1/18:1)" and "GPEtn(18:0/20:4)". Further investigations are required to establish more relations or to corroborate with the observation that these metabolites are extremely significant.

As a disadvantage, the small number of observations presented limitations in this study and could not be used as part of an ML model for more validation as this could present potential overfitting during performances (100% in AUC). More observations with the same metabolites can validate or complement this proposal.

Author Contributions: Conceptualization, J.M.-B., A.G.S.-R. and C.H.E.-S.; Data curation, J.J.O.-V. and J.A.E.-M.; Formal analysis, J.M.-B., A.G.S.-R. and C.H.E.-S.; Funding acquisition, J.C.-P., J.I.G.-T. and H.G.-R.; Investigation, J.M.-B., A.G.S.-R. and C.H.E.-S.; Methodology, J.M.-B., A.G.S.-R., C.H.E.-S., H.L.-G. and J.C.-P.; Project administration, J.M.-B.; Resources, J.C.-P., J.J.O.-V. and J.A.E.-M.; Supervision, C.E.G.-T., H.L.-G., J.C.-P. and J.M.-B.; Validation, J.C.-P.; Visualization, J.M.-B., A.G.S.-R. and C.H.E.-S. Writing-original draft, J.M.-B., A.G.S.-R. and C.H.E.-S. Writing-review & editing, J.M.-B., A.G.S.-R. and C.H.E.-S. All authors have read and agreed to the published version of the manuscript.

Funding: This research received no external funding.

Institutional Review Board Statement: The dataset used in this study was provided by the Unidad de Investigación Biomedica located in Zacatecas, Mexico, which is incorporated into the IMSS. All Mexican patients signed an informed consent letter, and the data included in the IMSS dataset meet the R-2017-785-131 dictum approval according to the protocol "Análisis metabolómico y transcriptómico diferencial en orina y suero de pacientes prediabéticos, diabéticos y con nefropatía diabética para identificar potenciales biomarcadores pronósticos de daño renal" (differential metabolomic and transcriptomic analysis in the urine and serum of prediabetic, diabetic and diabetic nephropathy patients to identify potential prognostic biomarkers of renal impairment), which complies with the criteria approved by the National Committee for Scientific Research and Ethics and follows the international ethical standards of the Helsinki convention for research studies in humans.

Informed Consent Statement: Informed consent was obtained from all subjects involved in the study.

Data Availability Statement: Not applicable.

Conflicts of Interest: The authors declare no conflict of interest. The founder had no role in the design of the study; in the collection, analyses or interpretation of data; in the writing of the manuscript; or in the decision to publish the results.

References

1. What Is Diabetes. Available online: https://www.idf.org/aboutdiabetes/what-is-diabetes.html (accessed on 2 October 2022).
2. IDF Diabetes Atlas. Tenth Edition. Available online: https://diabetesatlas.org/#:%7E:text=Diabetes%20around%20the%20world%20in%202021%3A&text=Over%203%20in%204%20adults,over%20the%20last%2015%20years (accessed on 2 October 2022).
3. Chatterjee, S.; Khunti, K.; Davies, M. Type 2 diabetes. *Lancet* **2017**, *389*, 2239–2251. [CrossRef]
4. International Diabetes Federation. Diabetes Facts & Figures. Available online: https://idf.org/aboutdiabetes/what-is-diabetes/facts-figures.html (accessed on 1 October 2022).
5. Schena, F.; Gesualdo, L. Pathogenetic Mechanisms of Diabetic Nephropathy. *J. Am. Soc. Nephrol.* **2005**, *16*, S30–S33. [CrossRef] [PubMed]
6. Jin, Q.; Ma, R. Metabolomics in Diabetes and Diabetic Complications: Insights from Epidemiological Studies. *Cells* **2021**, *10*, 2832. [CrossRef]
7. Drupad, K.; TrivediKatherine, A.; Royston, G. Metabolomics for the masses: The future of metabolomics in a personalized world. *New Horiz. Transl. Med.* **2017**, *3*, 294–305. [CrossRef]
8. Pagnini, C.; Di Paolo, M.C.; Mariani, B.M.; Urgesi, R.; Pallotta, L.; Vitale, M.A.; Villotti, G.; d'Alba, L.; De Cesare, M.A.; Di Giulio, E.; et al. Mayo Endoscopic Score and Ulcerative Colitis Endoscopic Index Are Equally Effective for Endoscopic Activity Evaluation in Ulcerative Colitis Patients in a Real Life Setting. *Gastroenterol. Insights* **2021**, *12*, 217–224. [CrossRef]
9. Li, L.; Krznar, P.; Erban, A.; Agazzi, A.; Martin-Levilain, J.; Supale, S.; Kopka, J.; Zamboni, N.; Maechlerr, P. Metabolomics Identifies a Biomarker Revealing In Vivo Loss of Functional β-Cell Mass Before Diabetes Onset. *Diabetes* **2019**, *68*, 2272–2286. [CrossRef] [PubMed]
10. Diamanti, K.; Cavalli, M.; Pan, G.; Pereira, M.; Kumar, C.; Skrtic, S.; Grabherr, M.; Risérus, U.; Eriksson, J.; Komorowski, J.; et al. Intra- and inter-individual metabolic profiling highlights carnitine and lysophosphatidylcholine pathways as key molecular defects in type 2 diabetes. *Sci. Rep.* **2019**, *9*, 9653. [CrossRef]
11. Park, S.; Sadanala, K.; Kim, E. A Metabolomic Approach to Understanding the Metabolic Link between Obesity and Diabetes. *Mol. Cells* **2015**, *38*, 587–596. [CrossRef]
12. Wishart, D.; Guo, A.; Oler, E.; Wang, F.; Anjum, A.; Peters, H.; Dizon, R.; Sayeeda, Z.; Tian, S.; Lee, B.; et al. HMDB 5.0: The Human Metabolome Database for 2022. *Nucleic Acids Res.* **2021**, *50*, D622–D631. [CrossRef]
13. Peddinti, G.; Cobb, J.; Yengo, L.; Froguel, P.; Kravić, J.; Balkau, B.; Tuomi, T.; Aittokallio, T.; Groop, L. Early metabolic markers identify potential targets for the prevention of type 2 diabetes. *Diabetologia* **2017**, *60*, 1740–1750. [CrossRef]
14. Huang, J.; Huth, C.; Covic, M.; Troll, M.; Adam, J.; Zukunft, S.; Prehn, C.; Wang, L.; Nano, J.; Scheerer, M.F.; et al. Machine Learning Approaches Reveal Metabolic Signatures of Incident Chronic Kidney Disease in Individuals With Prediabetes and Type 2 Diabetes. *Diabetes* **2020**, *69*, 2756–2765. [CrossRef] [PubMed]
15. Salihovic, S.; Broeckling, C.D.; Ganna, A.; Prenni, J.E.; Sundström, J.; Berne, C.; Lind, L.; Ingelsson, E.; Fall, T.; Ärnlöv, J.; et al. Non-targeted urine metabolomics and associations with prevalent and incident type 2 diabetes. *Sci. Rep.* **2020**, *10*, 16474. [CrossRef] [PubMed]
16. Roointan, A.; Gheisari, Y.; Hudkins, K.L.; Gholaminejad, A. Non-invasive metabolic biomarkers for early diagnosis of diabetic nephropathy: Meta-analysis of profiling metabolomics studies. *Nutr. Metab. Cardiovasc. Dis.* **2021**, *31*, 2253–2272. [CrossRef] [PubMed]
17. Zhang, J.; Fuhrer, T.; Ye, H.; Kwan, B.; Montemayor, D.; Tumova, J.; Darshi, M.; Afshinnia, F.; Scialla, J.J.; Anderson, A.; et al. High-Throughput Metabolomics and Diabetic Kidney Disease Progression: Evidence from the Chronic Renal Insufficiency (CRIC) Study. *Am. J. Nephrol.* **2022**, *53*, 215–225. [CrossRef] [PubMed]
18. Dubin, R.F.; Rhee, E.P. Proteomics and Metabolomics in Kidney Disease, including Insights into Etiology, Treatment, and Prevention. *Clin. J. Am. Soc. Nephrol.* **2019**, *15*, 404–411. [CrossRef] [PubMed]
19. Dutta, A.K.; Mageswari, R.U.; Gayathri, A.; Dallfin Bruxella, J.M.; Ishak, M.K.; Mostafa, S.M.; Hamam, H. Barnacles Mating Optimizer with Deep Transfer Learning Enabled Biomedical Malaria Parasite Detection and Classification. *Comput. Intell. Neurosci.* **2022**, *2022*, 7776319. [CrossRef]
20. Kaushik, H.; Singh, D.; Kaur, M.; Alshazly, H.; Zaguia, A.; Hamam, H. Diabetic Retinopathy Diagnosis From Fundus Images Using Stacked Generalization of Deep Models. *IEEE Access* **2021**, *9*, 108276–108292. [CrossRef]
21. Mazhar, M.S.; Saleem, Y.; Almogren, A.; Arshad, J.; Jaffery, M.H.; Rehman, A.U.; Shafiq, M.; Hamam, H. Forensic Analysis on Internet of Things (IoT) Device Using Machine-to-Machine (M2M) Framework. *Electronics* **2022**, *11*, 1126. [CrossRef]
22. Vulli, A.; Srinivasu, P.N.; Sashank, M.S.K.; Shafi, J.; Choi, J.; Ijaz, M.F. Fine-Tuned DenseNet-169 for Breast Cancer Metastasis Prediction Using FastAI and 1-Cycle Policy. *Sensors* **2022**, *22*, 2988. [CrossRef]

23. El-Sappagh, S.; Saleh, H.; Sahal, R.; Abuhmed, T.; Islam, S.R.; Ali, F.; Amer, E. Alzheimer's disease progression detection model based on an early fusion of cost-effective multimodal data. *Future Gener. Comput. Syst.* **2021**, *115*, 680–699. [CrossRef]
24. LIPID MAPS. Available online: https://www.lipidmaps.org/ (accessed on 8 November 2022).
25. METLIN. Available online: https://metlin.scripps.edu/ (accessed on 8 November 2022).
26. Dieterle, F.; Ross, A.; Schlotterbeck, G.; Senn, H. Probabilistic Quotient Normalization as Robust Method to Account for Dilution of Complex Biological Mixtures. Application in ^{1}H NMR Metabonomics. *Anal. Chem.* **2006**, *78*, 4281–4290. [CrossRef] [PubMed]
27. Trevino, V.; Falciani, F. GALGO: An R package for multivariate variable selection using genetic algorithms. *Bioinformatics* **2006**, *22*, 1154–1156. [CrossRef] [PubMed]
28. Sánchez-Reyna, A.; Celaya-Padilla, J.; Galván-Tejada, C.; Luna-García, H.; Gamboa-Rosales, H.; Ramirez-Morales, A.; Galván-Tejada, J. Multimodal Early Alzheimer's Detection, a Genetic Algorithm Approach with Support Vector Machines. *Healthcare* **2021**, *9*, 971. [CrossRef] [PubMed]
29. Celaya-Padilla, J.M.; Villagrana-Bañuelos, K.E.; Oropeza-Valdez, J.J.; Monárrez-Espino, J.; Castañeda-Delgado, J.E.; Oostdam, A.S.H.-V.; Fernández-Ruiz, J.C.; Ochoa-González, F.; Borrego, J.C.; Enciso-Moreno, J.A.; et al. Kynurenine and Hemoglobin as Sex-Specific Variables in COVID-19 Patients: A Machine Learning and Genetic Algorithms Approach. *Diagnostics* **2021**, *11*, 2197. [CrossRef] [PubMed]
30. Shen, Z.; Wu, Q.; Wang, Z.; Chen, G.; Lin, B. Diabetic Retinopathy Prediction by Ensemble Learning Based on Biochemical and Physical Data. *Sensors* **2021**, *21*, 3663. [CrossRef] [PubMed]
31. Peterson, L. K-nearest neighbor. *Scholarpedia* **2009**, *4*, 1883. [CrossRef]
32. Levner, I. Feature selection and nearest centroid classification for protein mass spectrometry. *BMC Bioinform.* **2005**, *6*, 68. [CrossRef]
33. Karaçalı, B.; Ramanathb, R.; Snyder, W. A comparative analysis of structural risk minimization by support vector machines and nearest neighbor rule. *Pattern Recognit. Lett.* **2004**, *25*, 63–71. [CrossRef]
34. Zhou, B.; Cheema, A.; Ressom, H. SVM-based spectral matching for metabolite identification. In Proceedings of the 2010 Annual International Conference of the IEEE Engineering in Medicine and Biology, Buenos Aires, Argentina, 31 August–4 September 2010; pp. 756–759. [CrossRef]
35. Amari, S.; Wu, S. Improving support vector machine classifiers by modifying kernel functions. *Neural Netw.* **1999**, *12*, 783–789. [CrossRef]
36. Kuhn, M. Building Predictive Models in R Using the caret Package. *J. Stat. Softw. Artic.* **2008**, *28*, 1–26.
37. Barberis, E.; Khoso, S.; Sica, A.; Falasca, M.; Gennari, A.; Dondero, F.; Afantitis, A.; Manfredi, M. Precision Medicine Approaches with Metabolomics and Artificial Intelligence. *Int. J. Mol. Sci.* **2022**, *23*, 11269. [CrossRef] [PubMed]
38. Hirakawa, Y.; Yoshioka, K.; Kojima, K.; Yamashita, Y.; Shibahara, T.; Wada, T.; Nangaku, M.; Inagi, R. Potential progression biomarkers of diabetic kidney disease determined using comprehensive machine learning analysis of non-targeted metabolomics. *Sci. Rep.* **2022**, *12*, 16287. [CrossRef] [PubMed]
39. Frohnert, B.; Webb-Robertson, B.; Bramer, L.; Reehl, S.; Waugh, K.; Steck, A.; Norris, J.; Rewers, M. Predictive Modeling of Type 1 Diabetes Stages Using Disparate Data Sources. *Diabetes* **2019**, *69*, 238–248. [CrossRef] [PubMed]
40. Kumar, M.; Ang, L.; Ho, C.; Soh, S.; Tan, K.; Chan, J.; Godfrey, K.; Chan, S.; Chong, Y.; Eriksson, J.; et al. Machine Learning–Derived Prenatal Predictive Risk Model to Guide Intervention and Prevent the Progression of Gestational Diabetes Mellitus to Type 2 Diabetes: Prediction Model Development Study. *JMIR Diabetes* **2022**, *7*, e32366. [CrossRef]
41. Dritsas, E.; Trigka, M. Data-Driven Machine-Learning Methods for Diabetes Risk Prediction. *Sensors* **2022**, *22*, 5304. [CrossRef]
42. Li, J.; Guo, C.; Wang, T.; Xu, Y.; Peng, F.; Zhao, S.; Li, H.; Jin, D.; Xia, Z.; Che, M.; et al. Interpretable machine learning-derived nomogram model for early detection of diabetic retinopathy in type 2 diabetes mellitus: A widely targeted metabolomics study. *Nutr. Diabetes* **2022**, *12*, 36. [CrossRef]
43. Wei, H.; Sun, J.; Shan, W.; Xiao, W.; Wang, B.; Ma, X.; Hu, W.; Wang, X.; Xia, Y. Environmental chemical exposure dynamics and machine learning-based prediction of diabetes mellitus. *Sci. Total Environ.* **2022**, *806*, 150674. [CrossRef]

diagnostics

MDPI

Article

Postpartum Assessment of the Correlation between Serum Hormone Levels of Estradiol, Progesterone, Prolactin and ß-HCG and Blood Pressure Measurements in Pre-Eclampsia Patients

Mariz Kasoha [1,*], Zoltan Takacs [1], Jacob Dumé [1], Sebastian Findeklee [1], Christoph Gerlinger [1], Romina-Marina Sima [2], Liana Ples [2], Erich-Franz Solomayer [1] and Bashar Haj Hamoud [1]

[1] Department of Gynecology, Obstetrics and Reproductive Medicine, University Medical School of Saarland, 66421 Homburg, Germany; zoltan.takacs@uks.eu (Z.T.); jacob.dume@uks.eu (J.D.); sebastian.findeklee@uks.eu (S.F.); c.gerlinger@gmx.de (C.G.); erich.solomayer@uks.eu (E.-F.S.); bashar.hajhamoud@uks.eu (B.H.H.)

[2] Department of Obstetrics and Gynecology, 'Carol Davila' University of Medicine and Pharmacy, 020021 Bucharest, Romania; romina.sima@yahoo.es (R.-M.S.); liana.ples@umfcd.ro (L.P.)

* Correspondence: mariz.kasoha@uks.eu; Tel.: +49-(0)-6841-16-28199; Fax: +49-(0)-684-16-28110

Citation: Kasoha, M.; Takacs, Z.; Dumé, J.; Findeklee, S.; Gerlinger, C.; Sima, R.-M.; Ples, L.; Solomayer, E.-F.; Haj Hamoud, B. Postpartum Assessment of the Correlation between Serum Hormone Levels of Estradiol, Progesterone, Prolactin and ß-HCG and Blood Pressure Measurements in Pre-Eclampsia Patients. *Diagnostics* **2022**, *12*, 1700. https://doi.org/10.3390/diagnostics12071700

Academic Editors: Yuli Huang, Yong Yuan and Peisong Chen

Received: 19 May 2022
Accepted: 28 June 2022
Published: 12 July 2022

Abstract: Background: Preeclampsia is a pregnancy-related hypertensive disease. Aberrant hormone levels have been implicated in blood pressure disorders. This study investigated the association of postpartum maternal serum hormone levels of estradiol, progesterone, prolactin, and ß-HCG with poorer PE-related complications including arterial hypertension. **Methods:** Thirty patient women with preeclampsia, and twenty women with uncomplicated pregnancy were included in this study. Serum levels of estradiol, progesterone, prolactin, and ß-HCG were determined immediately after delivery, and on the first and third postpartum days by means of ECLIA. **Results:** Compared with normal pregnancy cases, preeclampsia cases had higher serum levels of ß-HCG levels on Day-0 (319%), of progesterone on Day-0 (207%) and Day-1 (178%), and of estradiol on Day-1 (187%) and Day-3 (185%). Increased prolactin levels were positively associated with disease severity and estradiol and progesterone levels were decreased in poorer preeclampsia features including disease onset and IUGR diagnosis. No significant correlation between different hormone levels and blood pressure measurements was reported. **Conclusions:** This study is the first that detected postpartum maternal serum hormone levels and their correlation with blood pressure measurements in preeclampsia. We believe that the persistent arterial hypertension in the puerperium in preeclampsia as well as poorer disease specifications are most likely not of hormonal origin. Larger, well-defined prospective studies are recommended.

Keywords: preeclampsia; hypertension; hormones; postpartum; serum

1. Introduction

Preeclampsia (PE) is a pregnancy-related hypertensive disease that is characterised by elevated arterial blood pressure (BP) (≥140/90 mmHg) associated with proteinuria (≥300 mg/dL in 24-h urine) and/or other organ failure, beginning after the 20th week of gestation [1]. Approximately 5% of pregnant women develop PE, which in turn causes almost 40% of early-term deliveries and is a major cause of maternal and fetal mortality in low-income countries [2]. Despite the amount of effort put into understanding PE, its pathophysiology and pathogenesis are still not fully understood. One of the more plausible and widely accepted theories is that trophoblast invasion is hampered early in pregnancy, resulting in impaired spiral artery remodelling in the placenta and restricted blood flow, resulting in a reduced nutrient supply to the foetus and signs of foetal growth restriction (FGR) [3]. Furthermore, it has been demonstrated that incomplete spiral artery remodelling in the uterus contributes to placental ischemia and, as a result, the release of antiangiogenic factors into the maternal circulation, resulting in endothelial damage [4]. PE can manifest in multiple organs, but it always results in arterial hypertension, which can last into the puerperium and increase the risk of future cardiovascular disease [5].

A hormonal genesis of PE-induced hypertension has been studied and obtained data is ambiguous. Female sex hormones including estradiol and progesterone are produced in the ovaries and in the placenta during pregnancy. Their blood levels increase steadily during pregnancy and continue to be high until birth. Estradiol and progesterone are crucial for a successful pregnancy via promoting and regulating different physiologic changes in the mother such as metabolic, cardiovascular, stimulating of mammary gland development, maternal tolerance of the feotus, and initiation of labour [6]. Maliqueo and colleagues showed that estradiol and progesterone act as regulators of uterine blood flow and the production of angiogenic factors in the placental tissue [7]. Therefore, abnormalities in these factors have been suggested as associated with different pregnancy complications including PE. Studies showed that decreased maternal serum estradiol levels appear to be associated with the development of PE [8,9]. In addition, exogenous estradiol supplementation had a BP-lowering effect in PE patients [10,11]. Furthermore, altered levels of progesterone were found to be involved in dysregulated trophoblastic proliferation and aberrant development of placental vasculature in PE patients [12,13].

Prolactin is a polypeptide hormone that has a significant role in the physiology of the breast and its level is imperative for normal lactation capabilities, with significant effects on the menstrual cycle also [14]. It is normally synthesized and secreted from the lactotrophs in the anterior pituitary gland. During early pregnancy, decidua and placenta secrete different types of prolactin-like molecules that can bind to the prolactin receptor, which in turn are involved in regulating pituitary prolactin secretion. This function is controlled by the local decidual prolactin-releasing factors and progesterone levels [15,16]. Apart from its role in maintaining pregnancy and lactation, prolactin has been implicated in immune system regulation, osmoregulation, and angiogenesis [17]. In normal pregnancy, the rise of maternal prolactin reaches 5–10 fold of its level in a non-normal state during the third trimester, and dysregulated maternal prolactin levels have been linked to pregnancy complications such as gestational diabetes and hypertension [18–20].

Beta-Human chorionic gonadotropin (β-HCG), better known as "pregnancy hormone", is a glycoprotein produced by trophoblast cells. The main function of this hormone during pregnancy is to stimulate the production of progesterone and promote uterine and umbilical cord growth as well as placental development [21]. The association between aberrant β-HCG levels in the first trimester and the development of PE has been reported by several studies and β-HCG has been considered a serum marker for PE screening at 8–14 weeks of gestation [22–24].

To date, the only treatment for PE is delivery of the placenta, making this disease one of the leading causes of preterm birth. Therefore, the clinical use of several biochemical markers of placental dysfunction in combination with other relevant ultrasound and clinical manifestations provides the potential to improve pregnancy outcomes by allowing early detection and appropriate decision-making for referral of pregnant women at higher risk to proper care centres [25]. Effective screening and intervention strategies for preterm PE have been seriously considered. However, little is known about the value of maternal parameters in predicting PE complications postpartum [26]. The objective of this study is to investigate the correlation between maternal serum hormone levels of estradiol, progesterone, prolactin, and ß-HCG and their influence on the systolic and diastolic BP measurements immediately after delivery, on the first and the third postpartum days.

2. Materials and Methods

2.1. Study Cases and Performed Measurements

The current study included fifty women. Thirty women with PE as the patient group and twenty women with uncomplicated pregnancy as the control group. PE cases were obtained, respectively, from our blood bank. Control cases were prospectively enrolled in the study during their delivery in our department.

All study cases were older than 18 years and had a gestational age between 23 + 5 and 41 + 6 weeks of gestation. Cases with chronic hypertension, chronic renal or lung disease, inflammatory or infectious disease, nicotine abuse, and gestations complicated by chromosomal or fetal anomalies were excluded.

Diagnosis of PE, determination of PE severity and onset, diagnosis of HELLP syndrome, and intrauterine growth retardation (IUGR) were carried out in accordance with the German national guideline for hypertensive illnesses during pregnancy, as described in our previous study [27]. PE was diagnosed using two criteria: blood pressure of 140/90 mmHg and proteinuria of more than 300 mg/24 h after the 20th week of pregnancy. At least one of the following parameters was required for the diagnosis of severe PE: progressive renal insufficiency (creatinine 0.9 mg/dL), blood pressure of 160/110 mmHg, thrombocytopenia, impaired liver function (elevated transaminases, upper abdominal pain), central nervous system dysfunction abnormalities (severe headache, blurred vision), pulmonary oedema, or IUGR. The presence of pathological Doppler of the umbilical and/or uterine artery and foetal estimated weight less than the 10th percentile indicated the diagnosis of IUGR. Late-onset PE was defined as PE patients with a gestational age $\geq$ 34 weeks. In addition, increased liver enzymes (transaminases > 35 U/L), decreased platelet count (thrombocytes < 100,000/µL), and lactate dehydrogenase [LDH] 2 times bigger than the upper level of the normal range [0–262 U/L]) or hemolysis (haptoglobin $\leq$ 25 mg/dL) indicated a diagnosis of HELLP syndrome.

Aseptic venous blood samples from control subjects were collected using serum gel monovette between 8 a.m. and 10 a.m. on the day of delivery immediately after giving birth and on the first and third day postpartum. Samples were centrifuged and supernatants were transferred in aliquots into Eppendorf tubes that were stored at −80° until they could be analysed. All clinical data were collected by reviewing electronic patients' medical records.

Serum levels of estradiol, progesterone, prolactin, and ß-HCG were tested in the Central Laboratory of our University hospital by means of electrochemiluminescence immunoassay (ECLIA) from Roche® according to the manufacturer's instructions. The technical variability of the hormonal assays ranged from 1.68–2.30% for ß-HCG, 2.07–2.55% for prolactin, 3.12–3.26% for progesterone, and 1.61–1.83% for estradiol. In addition, BP readings were taken regularly during blood sampling days using the BP monitor from the company Erka®. Measurements were performed in the early morning, at noon, in the afternoon, and in the evening and were recorded in the patient registry. The highest BP reading was used in the study.

2.2. Statistical Analysis

Following the intent-to-treat principle, all patients that provided data at baseline were included in the analyses. As appropriate for explorative analyses comparison-wise two-sided significance level α of 5% was used. The statistical analyses were performed using RStudio version 1.2.5042 and R version 4.0.3 running under Ubuntu 20.04 LTS. The R packages gtsummary, and tidyverse were used.

3. Results

3.1. Clinical and Laboratory Parameters of Study Cases

Briefly, control and patient women were age-matched [Mean ± STD (years): 31 ± 7 vs. 32 ± 5, respectively, $p > 0.05$]. PE cases had higher BMI values [Mean ± STD: 36.9 ± 8.7 vs. 32.3 ± 5.7, respectively, $p = 0.030$] as well as higher creatinine, ALT, and AST concentrations. Furthermore, gestation age at delivery and number of gravida and para were significantly lower in PE cases compared with control cases. The existence of one or more pathological signs such as headache, severe edema, vaginal bleeding, epigastric pain, and vision changes was observed in 20% of PE cases (6/30) but not in control cases; however, this difference was not statistically significant. All parameters are displayed in Table 1.

Table 1. Clinical and laboratory parameters of included cases.

Characteristic	Control Cases (N = 20)	PE Cases (N = 30)	P
* Age (Years)	31 ± 7	32 ± 5	NS
* BMI (Kg/m^2)	32.3 ± 5.7	36.9 ± 8.7	0.030
‡ Gestation age at the time of delivery (Weak)	38.7 (34.3–41.9)	34.4 (26.6–40.1)	<0.0001
* Creatinin (mg/dL)	0.6 ± 0.12	0.7 ± 0.16	<0.0001
‡ Alanine aminotransferase (ALT) (U/L)	11.50 (8–28) $^{\varphi}$	20.5 (8–392)	0.003
‡ Aspartate aminotransferase (AST) (U/L)	20 (15–34) $^{\varphi}$	31.5 (7–242)	0.018
† Gravida [N (%)]			
One Gravida	3 (15)	31 (70)	<0.0001
≥2 Gravida	17 (75)	9 (30)	
† Para [N (%)]			
0 Para	5 (25)	25 (83)	<0.0001
≥1 Para	15 (75)	6 (17)	
†,† Symptoms [N (%)]			
No	20 (100)	24 (80)	NS
Yes	0 (0)	6 (20)	

*: Results are shown as [Mean ± STD] and p-value is calculated using Student's t-test. ‡: Results are shown as [Median (Range)] and p-value is calculated using Mann-Whitney-U- Test. †: Results are shown as number (%) and p-value is calculated using Chi-squared Test. †: Existence of one or more of the following signs: headache, severe edema, vaginal bleeding, epigastric pain, and vision changes. $^{\varphi}$: Data were available from 14 control cases. N: Number of cases. NS: Not significant.

3.2. Postpartum Serum Hormone Levels and BP Measurements in Normal and PE Cases

Serum hormone levels of estradiol, progesterone, prolactin, and ß-HCG were detected on the day of delivery immediately after giving birth (Day-0) and on the first day (Day-1) and third day (Day-3) postpartum. Blood samples of PE cases were available from all 30 subjects on Day-0, from 29 subjects on Day-1 and from 26 subjects on Day-3. In addition, BP measurements on Day-0, Day-1, and Day-3 were obtained from medical records of patients of 29, 28, and 26 subjects, respectively. On the other hand, all hormones and BP measurements of the control cases were performed for all 20 subjects on all study days.

As presented in Table 2, systolic and diastolic BP measurements were significantly higher in PE cases compared with control cases during all study days. Furthermore, higher serum hormone levels of ß-HCG, progesterone, and estradiol were reported in PE cases compared with normal cases. These increases were statistically significant in ß-HCG levels on Day-0, in progesterone levels on Day-0 and Day-1, and in estradiol levels on Day-1 and Day-3. Serum prolactin levels showed no significant differences at all.

Table 2. Serum hormone levels of ß-HCG, prolactin, progesterone, and estradiol and systolic- and diastolic BP measurements in control and PE cases on all study days.

Parameter	Control Cases (N = 20)	PE Cases ℷ	*p*
Systolic blood pressure (mm Hg)			
Day-0	123 (105–170)	160 (130–200)	<0.0001
Day-1	125 (100–145)	157 (130–194)	<0.0001
Day-3	120 (80–150)	150 (130–190)	<0.0001
Φ p	NS *	Day-1 vs. Day-0: 0.032 Day-3 vs. Day-0: 0.029 Day-3 vs. Day-1: NS	
Diastolic blood pressure (mm Hg)			
Day-0	80 (60–90)	100 (60–140)	<0.0001
Day-1	80 (50–90)	92 (70–120)	<0.0001
Day-3	80 (65–90)	90 (70–110)	<0.0001
Φ p	NS *	NS *	
ß-HCG (mIU/mL)			
Day-0	8357 (1786–95,328)	26,623 (2675–172,281)	0.005
Day-1	3379 (948–17,942)	4685 (487–43,860)	NS
Day-3	519 (6–2103)	916 (138–6975)	NS
Φ p	<0.0001 *	<0.0001 *	
Prolactin (µIU/mL)			
Day-0	5682 (1868–9277)	6327 (2489–17,660)	NS
Day-1	5687 (369–14,576)	5766 (314–10,661)	NS
Day-3	5310 (193–11,383)	5570 (149–12,275)	NS
Φ p	NS	NS	
Progesterone (ng/mL)			
Day-0	55.8 (8.16–250)	115.5 (16.9–485)	0.046
Day-1	8.3 (3.1–24.3)	14.8 (3.8–103)	0.002
Day-3	1.3 (0.2–3.7)	1.8 (0.8–11.7)	NS
Φ p	<0.0001 *	<0.0001 *	
Estradiol (pg/mL)			
Day-0	2544 (215–55,080)	12,675 (2126–63,086)	NS
Day-1	227 (75–817)	425 (85–10,665)	0.020
Day-3	48 (12–278)	89 (12–681)	0.008
Φ p	<0.0001 *	<0.0001 *	

ℷ: N = 30 at delivery day and Day-1; N = 29 at Day-3. *Φ*: *p*-value within each group. *: Significant differences between values from 3 days. N: Number of cases. NS: Not significant. Results are shown as [Median (Range)] and *p*-value is calculated using Mann-Whitney-*U*-Test.

We also found that serum hormone levels of ß-HCG, progesterone, and estradiol were significantly reduced during study days in each study group separately. Systolic BP measurements in PE cases decreased significantly on Day-1 and Day-3 compared with Day-0 [Median (Range) (mm Hg): Day-0: 160 (130–200), Day-1: 157 (130–194), Day-3: 150 (130–190); $p = 0.032$ and $p = 0.029$ respectively)]. No significant differences in systolic BP measurements in the control group or in diastolic BP measurements in both groups were observed (Table 2).

Then, we calculated the percent ratio of observed differences in serum hormone levels as well as in BP measurements in order to scrutinize the scenario of these differences between study groups and within each group separately during study days. For this purpose, we included 20 control cases and 25 PE cases in which all measurements of all tested parameters were available. Our data showed that the biggest differences in serum hormone levels and in BP measurements between control cases and PE cases were reported on Day-0 (Figure 1A). In addition, we found that both study groups presented a similar schema of reduction percent in ß-HCG, progesterone, and estradiol serum levels during study days with maximal reduction percent between Day-0 and Day-3 (Figure 1B and Figure 1C).

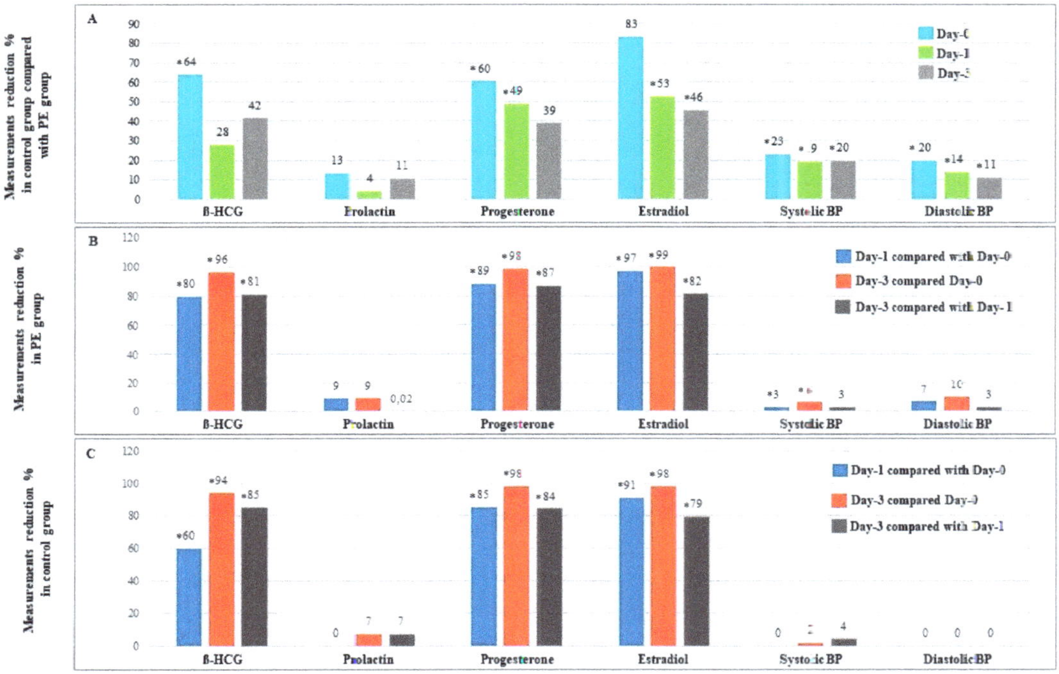

Figure 1. Measurements reduction % in serum hormone levels and BP measurements during study days. (**A**) Reduction of serum hormone levels and BP measurements in 20 normal cases compared with 25 PE cases. (**B**) Reduction of serum hormone levels and BP measurements in 25 PE cases. (**C**) Reduction of serum hormone levels and BP measurements in 20 normal cases with normal pregnancy. * Significant difference ($p < 0.05$ using Mann-Whitney-*U*-Test). Results are presented as percent.

3.3. *Association between Serum Hormone Levels with PE and BP Measurements*

First, we tested the association between PE and parameters that differed significantly between the PE group and control group using logistic regression. We found that PE was negatively associated with gestation weeks and positively associated with creatinine concentrations, serum ß-HCG levels on Day-0, and serum progesterone levels on Day-1 (Table 3). However, only gestation age and creatinine concentrations showed as independent predictors of PE (data are not shown).

Next, as shown in Table 1, systolic and diastolic BP measurements were significantly higher in the PE group compared with the control group on all study days. Logistic regression was used to investigate various parameters that could predict these measurements. Serum ß-HCG levels on Day-0 were found to predict diastolic BP on Day-0 and systolic BP measurements on Day-3. Serum ß-HCG levels on Day-1 and Day-3 were found to predict systolic BP measurements on Day-3 (Table 4). When these observations were adjusted for other variables, they became no more statistically significant. We found that PE diagnosis was an independent predictor of systolic and diastolic BP measurements on all study days (data are not shown).

Table 3. Association between PE and parameters that differ significantly between PE group and control group.

Predictor Variable *	B	*p*	Exp(B)	95% CI for Exp(B)	
				Lower	Upper
BMI $^{\Delta}$	0.085	0.053	1.089	0.999	1.188
Gestation age (Week)	-0.324	0.004	0.723	0.581	0.900
Creatinin (mg/dL)	8.895	0.002	7294.2	25.7	20,741,143.8
ALT (U/L)	0.117	0.038	1.124	1.006	1.256
AST (U/L)	0.067	0.081	1.069	0.992	1.153
ß-HCG-Day-0 (mIU/mL)	0.031	0.030	1.03	1.003	1.061
Progesterone-Day-0 (ng/mL)	0.007	0.074	1.007	0.999	1.014
Progesterone-Day-1 (ng/mL)	0.145	0.018	1.156	1.025	1.304
Estradiol-Day-1 (pg/mL)	0.246	0.061	1.278	0.989	1.635
Estradiol-Day-3 (pg/mL)	0.100	0.109	1.100	0.978	1.250

*: Predictor variables included only parameters that were significantly different between PE group and control group. $^{\Delta}$: Actual values.

Table 4. Simple linear regression of BP measurements with different parameters in all study cases.

	Systolic BP Measurements (mm Hg)			Diastolic BP Measurements (mm Hg)		
Predictor Variable	**Day-0**	**Day-1**	**Day-3**	**Day-0**	**Day-1**	**Day-3**
PE diagnosis						
P	<0.0001	<0.0001	<0.0001	<0.0001	<0.0001	<0.0001
β	37.800	29.7	36.4	21.8	13.4	13.6
BMI						
P	0.017	NS	NS	0.017	NS	NS
β	1.071	—	—	0.710	—	—
Gestation age (Weeks)						
P	<0.0001	<0.0001	<0.0001	0.005	0.024	0.002
β	−2.471	−2.467	−2.841	−1.475	−0.966	−1.060
Creatinin (mg/dL)						
P	0.001	0.003	0.003	0.003	0.018	NS
β	66.9	53.2	61.2	41.6	27.3	17.1
ALT (U/L)						
P	0.023	NS	0.040	NS	NS	NS
β	0.134	—	0.122	—	—	—
AST (U/L)						
P	NS	NS	0.042	NS	NS	NS
β	—	—	0.192	—	—	—
ß-HCG (mIU/mL) *						
Day-0 *P*	NS	NS	0.013	0.028	NS	NS
β	—	—	4.928	0.136	—	—
Day-1 *P*	NT	NS	0.004	NT	NS	NS
β	—	—	4.846	—	—	—
Day-3 *P*	NT	NT	0.014	NT	NT	NS
β	—	—	4.769	—	—	—

*: Only results of β-HCG were shown because the results of all other hormones were not significant. NS: Not significant. NT: Not tested.

3.4. Relevant Diagnosis and Characteristics within the PE Group

Figure 2 illustrates PE features among our patient women cohort. Briefly, severe PE and late-onset PE were reported in 54% and 57% of the patient women, respectively. Seven women were diagnosed with HELLP syndrome (23%) and sixteen women were characterised by IUGR (53%).

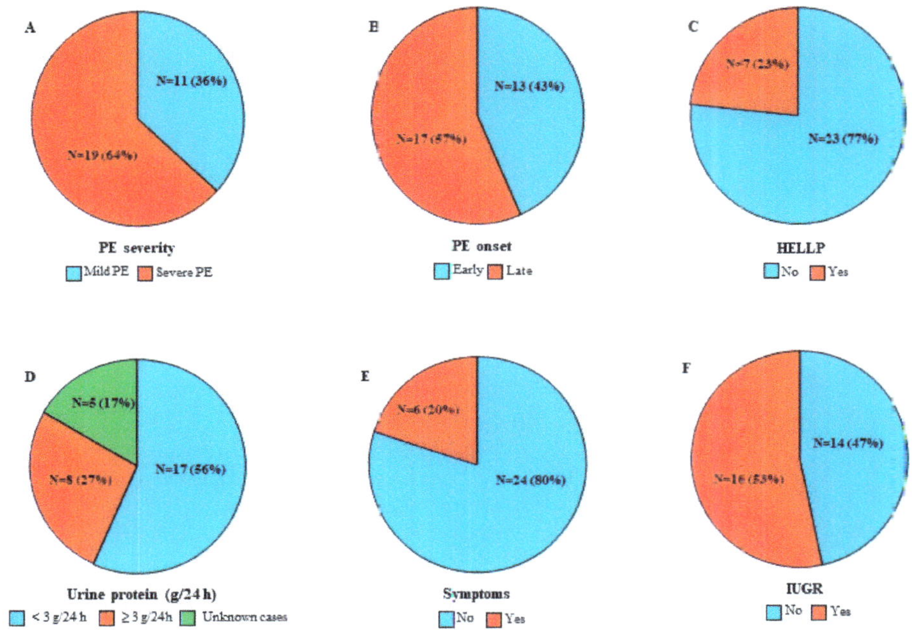

Figure 2. PE features among PE cases (N = 30). (**A**) PE severity. (**B**) PE onset. (**C**) HELLP cases. (**D**) Protein urea concentration. (**E**) Existence of one or more of the following signs: headache, severe edema, vaginal bleeding, epigastric pain, and vision changes. (**F**) IUGR cases. Results are presented as N (number of cases) (%).

Differences in serum hormone levels and BP measurements according to the pathological feature were tested using the Mann-Whitney test. Results showed that severe PE cases (N = 19) had significantly higher serum prolactin levels compared with mild PE cases (N = 11) on Day-0 [Median (Range) (µIU/mL): 7711 (2489–17,660) vs. 4702 (3056–9624), respectively, $p = 0.030$]. However, no significant differences in PB measurements according to disease severity were reported (Data are not shown).

Next, we found that decreased progesterone and estradiol levels on Day-0 as well as increased systolic BP measurements on Day-3 were associated with worse PE features including early disease onset and IUGR diagnosis (Figure 3). Consequently, predictor variables for systolic PB on Day-3 in PE cases were tested using logistic regression. Data showed that IUGR diagnosis, early disease onset, and lower gestation age predicted increased systolic BP measurements on Day-3 [(β-*p*): (12.100–0.041), (−17.537–0.002) and (−1.559–0.021), respectively]. No significant results were reported for serum hormone levels (Data are not shown).

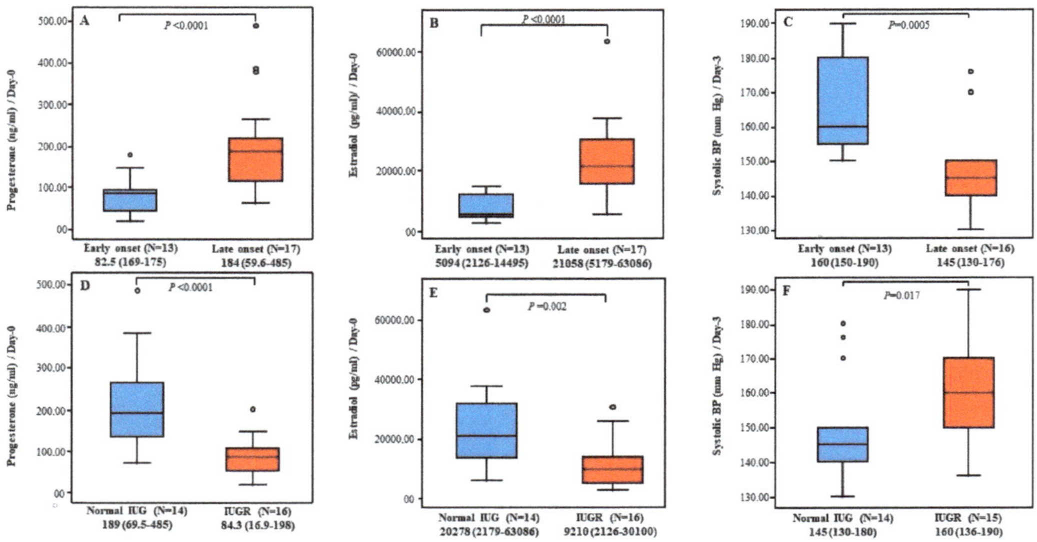

Figure 3. Postpartum serum progesterone and estradiol levels and systolic BP within PE cases. (**A–C**) Serum progesterone and estradiol levels at Day-0 and systolic PB measurements at Day-3 in cases with early PE onset vs. cases with late PE onset respectively. (**D–F**) Serum progesterone and estradiol levels at Day-0 and systolic PB measurements at Day-3 in cases with normal IUG vs. cases with IUGR respectively. Results are presented as Median (Range). *p*-value is tested using Mann-Whitney-*U*-Test. N: number of cases.

4. Discussion

PE has been extensively studied. Large cohort studies yielded a wealth of information on risk factors, symptoms, pathogenicity, clinical and laboratory findings, etc. PE characteristics such as obesity, nulligravida, and elevated liver enzymes and creatinine concentrations were confirmed in this study [28–30]. We also reported shorter gestation age in PE women. This is due to the fact that, to date, pregnancy termination is the only cure for this disorder [31].

PE, as previously stated, is classified as a hypertensive disease of pregnancy. Excessive maternal inflammatory response, caused by disrupted spiral artery remodeling in the placenta, provokes generalized maternal endothelial dysfunction, which contributes to the clinically visible elevations in maternal BP [32]. We found that despite a slight decrease in BP measurements during the puerperium, the significant increase in BP measurements in the PE group compared with the normal group persisted for the first three postpartum days. PE-induced hypertension is responsible for a substantial proportion of maternal mortality and for a wide range of postpartum complications that mandate long-term follow-up and contribute to the overall healthcare spending in developed countries [33]. Effective screening and intervention strategies for high BP in the postpartum period have been seriously considered [1,34,35]. To date, no specific biological marker has been identified that can accurately predict PE postpartum consequences. In a previous study, we assessed the predictive value of soluble fms-like tyrosine kinase 1 (sFlt-1), placental growth factor (PlGF), and their ratio to estimate the short-term postpartum maternal outcome. According to our findings, increased serum PlGF levels were associated with elevated systolic BP measurements. However, none of these markers could predict the general worsening of postpartum PE [26].

It is well known that hormonal changes, such as estradiol [36], progesterone [37], ß-HCG [38], and prolactin [39] are associated with the pathology of PE. In this study, we looked at whether the alteration schema in serum levels of the aforementioned hormones differs between the PE group and normal group over three postpartum days, and what value these biomarkers might have as predictors of PB measurements and poorer disease features. We found that, compared to control cases, PE cases had generally higher serum hormone levels. Only ß-HCG levels on Day-0 were significantly elevated, as were progesterone levels on Day-0 and Day-1, and estradiol levels on Day-1 and Day-3. These elevations were no more significant when adjusted to gestation age and creatinine concentrations. In addition, all serum prolactin level comparison tests were insignificant. As a result, with the current study design, it was not possible to determine the role of the tested hormones in PE pathogenicity. Nevertheless, this was not our study goal. For this purpose, further prospective studies in a large multicentric cohort are warranted.

Next, we found that the alteration schema in serum hormone levels of estradiol, progesterone, and ß-HCG was similar in both study groups, with a gradual decrease over the course of the study (Figure 1B and Figure 1C), which could imply that the early postpartum profile of tested hormones is unaffected by the presence of PE.

Furthermore, our findings revealed that serum levels of tested hormones had no predictive value for PB measurements, with the exception of ß-HCG levels, which predicted systolic on Day-3. However, when adjusted for PE diagnosis and gestational age, this effect was no more significant. In a retrospective case-control study, Chen and colleagues showed that free early pregnancy β-hCG levels in PE groups were significantly lower than in the control group. This study included 680 hypertensive disorders of pregnancy (HDP)-free pregnant women and 222 HDP-affected pregnant women, 90 of whom had PE and 71 of whom had severe PE. However, the risk calculation model developed using a combination of different factors including β-hCG after adjusting for maternal weight and gestation age outperformed the original method of using β-hCG alone [40].

Lan et al. found that, throughout pregnancy, the PE group had lower estradiol levels than the control group, but the two groups had similar progesterone levels [9]. Low levels of 2-methoxyestradiol, an estradiol metabolite, were found to be negatively correlated with systolic peak arterial pressure in patients with early-onset PE. On the contrary, in patients with late-onset PE, a rise in 2-methoxyestradiol serum levels has been reported. It has been proposed that this increase is due to the activation of compensatory mechanisms to keep normal serum estrogen levels. Exogenous administration of estradiol has been shown in animal models as well as in PE patients to normalize blood pressure and other associated symptoms of PE [41].

Prolactin has been linked to the pathogenesis of pregnancy-associated hypertension by modulating the activity of endothelial nitric oxide synthase [42]. Hypertensive pregnant women had higher blood prolactin levels than non-hypertensive pregnant women, but the difference was not statistically significant. Nonetheless, cord blood prolactin levels were significantly higher in babies born to hypertensive mothers than in normal pregnancy mothers [43].

As evidenced by the studies listed below, investigating better predictors of HDP has been a research hot spot in the field of prenatal medicine. As a result, there is no comparable data for our postpartum study. In this study, we assume that postpartum serum levels of the tested hormones could not be implicated in the pathogenesis of PE-induced hypertension.

Our data showed that serum prolactin levels in the PE group on Day-0 were lower by 40% in mild PE cases compared with severe PE cases. A reduction of 20% was also reported by Leaños-Miranda et al. who also found that urinary prolactin levels, as well as the presence of antiangiogenic prolactin fragments in urine, are associated with PE severity at the time of clinical manifestation [44]. Regardless, we found no significant differences in PB measurements based on disease severity, as well as no correlation between prolactin levels and PB measurements, despite the fact that early-onset disease and IUGR diagnosis could have predicted systolic BP measurements in our PE group on Day-3, and that these two subgroups had significantly lower serum estradiol and progesterone levels. While a causal relationship between these two hormone levels and BP measurements could not be established, it is tempting to propose that dysregulated serum prolactin, estradiol, and progesterone levels may be involved in poorer PE features via mechanisms unrelated to hypertension.

The small sample size limits the interpretation of our results. Although we were unable to establish a link between postpartum maternal hormone status and BP measurements, a larger number of cases could significantly refute or affirm our results. Another limitation is the notable difference in gestation age between study groups, which could affect the levels of tested hormones. Serum progesterone and estradiol concentrations in normal pregnant women, for example, increase with advanced pregnancy and reach 100-fold or more at delivery than before pregnancy [45]. The "snapshot" nature of single-point hormone testing is another limitation because hormone production follows a circadian rhythm, making it difficult to determine whether serum levels represent a peak, a valley, or something in between. As a result, larger, well-defined prospective studies are needed to demonstrate the postpartum pattern and actions of these hormones, as well as their role in predicting PE-related arterial hypertension.

Author Contributions: Conceptualization: M.K., J.D., S.F., B.H.H. and E.-F.S., Methodology: M.K. and J.D., Formal Analysis: M.K., J.D. and E.-F.S., Investigation: M.K. and J.D., Resources: M.K., J.D. and Z.T., Data Curation: M.K., J.D. and B.H.H., Statistical analysis: M.K. and C.G., Writing: M.K., Original Draft Preparation: M.K. and J.D., Review and Editing: M.K., J.D., S.F., Z.T., C.G., R.-M.S., L.P. and B.H.H., Supervision: M.K., B.H.H. and E.-F.S. Each author has approved the submitted version and agrees to be personally accountable for the author's own contributions and for ensuring that questions related to the accuracy or integrity of any part of the work, even ones in which the author was not personally involved, are appropriately investigated, resolved, and documented in the literature. All authors have read and agreed to the published version of the manuscript.

Funding: This research received no external funding.

Institutional Review Board Statement: The study was conducted in the Department of Gynaecology, Obstetrics and Reproductive Medicine at Saarland University Medical Centre in accordance with the Declaration of Helsinki, and approved by the Ethics Committee of Saarland (Identification number: 127/21; date of approval: 28 May 2021).

Informed Consent Statement: Informed consent was obtained from all subjects involved in the study to be part of the study and to publish this paper.

Data Availability Statement: The data presented in this study are available on request from the corresponding author. The data are not publicly available due to privacy restrictions.

Conflicts of Interest: The authors declare no conflict of interest.

References

1. Stepan, H.; Kuse-Fohl, S.; Klockenbusch, W.; Rath, W.; Schauf, B.; Walther, T.; Schlembach, D. Diagnosis and Treatment of Hypertensive Pregnancy Disorders. Guideline of DGGG (S1-Level, AWMF Registry No. 015/018, December 2013). *Geburtshilfe Frauenheilkunde* **2015**, *75*, 900–914. [CrossRef]
2. Peres, G.M.; Mariana, M.; Cairrão, E. Pre-Eclampsia and Eclampsia: An Update on the Pharmacological Treatment Applied in Portugal. *J. Cardiovasc. Dev. Dis.* **2018**, *17*, 3. [CrossRef] [PubMed]
3. Ahmed, A.; Rezai, H.; Broadway-Stringer, S. Evidence-Based Revised View of the Pathophysiology of Preeclampsia. *Adv. Exp. Med. Biol.* **2017**, *956*, 355–374. [CrossRef] [PubMed]
4. Phipps, E.; Prasanna, D.; Brima, W.; Jim, B. Preeclampsia: Updates in pathogenesis, definitions, and guidelines. *Clin. J. Am. Soc. Nephrol.* **2016**, *11*, 1102–1113. [CrossRef] [PubMed]
5. Wu, P.; Haththotuwa, R.; Kwok, C.S.; Babu, A.; Kotronias, R.A.; Rushton, C.; Mamas, M.A. Preeclampsia and Future Cardiovascular Health: A Systematic Review and Meta-Analysis. *Circ. Cardiovasc. Qual. Outcomes* **2017**, *10*, 003497. [CrossRef] [PubMed]
6. Napso, T.; Yong, H.E.; Lopez-Tello, J.; Sferruzzi-Perri, A.N. The Role of Placental Hormones in Mediating Maternal Adaptations to Support Pregnancy and Lactation. *Front. Physiol.* **2018**, *17*, 1091. [CrossRef]
7. Maliqueo, M.; Echiburu, B.; Crisosto, N. Sex steroids modulate uterine-placental vasculature: Implications for obstetrics and neonatal outcomes. *Front. Physiol.* **2016**, *7*, 152. [CrossRef]
8. Shen, Z.; Wu, Y.; Chen, X.; Chang, X.; Zhou, Q.; Zhou, J.; Wang, K. Decreased maternal serum 2-methoxyestradiol levels are associated with the development of preeclampsia. *Cell Physiol. Biochem.* **2014**, *34*, 2189–2199. [CrossRef]

9. Lan, K.C.; Lai, Y.J.; Cheng, H.H.; Tsai, N.C.; Su, Y.T.; Tsai, C.C.; Hsu, T.Y. Levels of sex steroid hormones and their receptors in women with preeclampsia. *Reprod. Biol. Endocrinol.* **2020**, *18*, 12. [CrossRef]
10. Babic, G.M.; Markovic, S.D.; Varjacic, M.; Djordjevic, N.Z.; Nikolic, T.; Stojic, I.; Jakovljevic, V. Estradiol decreases blood pressure in association with redox regulation in preeclampsia. *Clin. Exp. Hypertens.* **2018**, *40*, 281–286. [CrossRef]
11. Djordjević, N.Z.; Babić, G.M.; Marković, S.D.; Ognjanović, B.I.; Štajn, A.Š.; Saičić, Z.S. The antioxidative effect of estradiol therapy on erythrocytes in women with preeclampsia. *Reprod. Toxicol.* **2010**, *29*, 231–236. [CrossRef] [PubMed]
12. Kiprono, L.V.; Wallace, K.; Moseley, J.; Martin, J., Jr.; LaMarca, B. Progesterone blunts vascular endothelial cell secretion of endothelin-1 in response to placental ischemia. *Am. J. Obstet. Gynecol.* **2013**, *209*, 44.e1–44.e6. [CrossRef] [PubMed]
13. He, G.; Xu, W.; Chen, Y.; Liu, X.; Xi, M. Abnormal apoptosis of trophoblastic cells is related to the up-regulation of CYP11A gene in placenta of preeclampsia patients. *PLoS ONE* **2013**, *8*, e59609. [CrossRef]
14. Bernard, V.; Young, J.; Binart, N. Prolactin-a pleiotropic factor in health and disease. *Nat. Rev. Endocrinol.* **2019**, *15*, 356–365. [CrossRef]
15. Maslar, I.A.; Ansbacher, R. Effects of progesterone on decidual prolactin production by organ cultures of human endometrium. *Endocrinology* **1986**, *118*, 2102–2108. [CrossRef] [PubMed]
16. Toft, D.J.; Linzer, D.I. Prolactin (PRL)-Like Protein J, a Novel Member of the PRL/Growth Hormone Family, Is Exclusively Expressed in Maternal Decidua. *Endocrinology* **1999**, *140*, 5095–5101. [CrossRef] [PubMed]
17. Freeman, M.E.; Kanyicska, B.; Lerant, A.; Nagy, G. Prolactin: Structure, function, and regulation of secretion. *Physiol. Rev.* **2000**, *80*, 1523–1631. [CrossRef]
18. Ekinci, E.I.; Torkamani, N.; Ramchand, S.K.; Churilov, L.; Sikaris, K.A.; Lu, Z.X.; Houlihan, C.A. Higher maternal serum prolactin levels are associated with reduced glucose tolerance during pregnancy. *J. Diabetes Investig.* **2017**, *8*, 697–700. [CrossRef]
19. Marlettini, M.G.; Cassani, A.; Morselli-Labate, A.M.; Crippa, S.; Contarini, A.; Miniero, R.; Orlandi, C. Maternal and fetal prolactin in pregnancy-induced hypertension. *Arch. Gynecol. Obstet.* **1990**, *247*, 73–81. [CrossRef]
20. Chen, B.L.; Zhang, Z.H.; Liu, N.B.; Huang, K.S. Prolactin in normal pregnancy and severe pregnancy-induced hypertension. *Hunan Yi Ke Da Xue Xue Bao* **2001**, *28*, 67–69.
21. Evans, J. Hyperglycosylated hCG: A Unique Human Implantation and Invasion Factor. *Am. J. Reprod. Immunol.* **2016**, *75*, 333–340. [CrossRef] [PubMed]
22. Liu, H.Q.; Wang, Y.H.; Wang, L.L.; Hao, M. Predictive Value of Free β-Hcg Multiple of the Median for Women With Preeclampsia. *Gynecol. Obstet. Investig.* **2015**, *81*, 137–147. [CrossRef] [PubMed]
23. Zheng, Q.; Deng, Y.; Zhong, S.; Shi, Y. Human Chorionic Gonadotropin, Fetal Sex and Risk of Hypertensive Disorders of Pregnancy: A Nested Case-Control Study. *Pregnancy Hypertens.* **2016**, *6*, 17–21. [CrossRef] [PubMed]
24. Long, W.; Zhou, Q.; Wang, H.; Lu, B.; Chen, Y.; Zhang, B.; Yu, B. Second-Trimester Maternal Serum Screening Biomarkers in the Risk Assessment for Preeclampsia. *Ann. Clin. Lab. Sci.* **2018**, *48*, 308–313.
25. Poon, L.C.; Sahota, D. Screening and Prevention of Preeclampsia. *Matern.-Fetal Med.* **2019**, *1*, 25–30. [CrossRef]
26. Hamza, A.; Gerlinger, C.; Radosa, J.; Solomayer, E.F.; Hagmann, J.; Sester, U.; Kasoha, M. Pilot study: Placental biomarker predictive capability (sFlt-1, PlGF and their ratio) of postpartum maternal outcome. *Arch. Gynecol. Obstet.* **2019**, *299*, 1557–1566. [CrossRef]
27. Kasoha, M.; Takacs, Z.; Fackiner, L.; Gerlinger, C.; Sklavounos, P.; Radosa, J.; Hamza, A. Comparison of maternal serum levels and placental mRNA levels of dickkopf-1 in preeclamptic and normal pregnant women at delivery. *Geburtshilfe Frauenheilkunde* **2021**, *81*, 1247–1255. [CrossRef]
28. Phipps, E.A.; Thadhani, R.; Benzing, T.; Karumanchi, S.A. Pre-eclampsia: Pathogenesis, novel diagnostics and therapies. *Nat. Rev. Nephrol.* **2019**, *15*, 275–289. [CrossRef]
29. Austdal, M.; Tangerås, L.H.; Skråstad, R.B.; Salvesen, K.Å.; Austgulen, R.; Iversen, A.C.; Bathen, T.F. First Trimester Urine and Serum Metabolomics for Prediction of Preeclampsia and Gestational Hypertension: A Prospective Screening Study. *Int. J. Mol. Sci.* **2015**, *16*, 21520–21538. [CrossRef]
30. Sibai, B.M. Diagnosis, controversies, and management of the syndrome of hemolysis, elevated liver enzymes, and low platelet count. *Obs. Gynecol.* **2004**, *103*, 981–991. [CrossRef]
31. Berger, R.; Rath, W.; Abele, H.; Garnier, Y.; Kuon, R.J.; Maul, H. Reducing the Risk of Preterm Birth by Ambulatory Risk Factor Management. *Dtsch. Arztebl. Int.* **2019**, *116*, 858–864. [CrossRef] [PubMed]
32. Tomimatsu, T.; Mimura, K.; Matsuzaki, S.; Endo, M.; Kumasawa, K.; Kimura, T. Preeclampsia: Maternal Systemic Vascular Disorder Caused by Generalized Endothelial Dysfunction Due to Placental Antiangiogenic Factors. *Int. J. Mol. Sci.* **2019**, *20*, 4246. [CrossRef] [PubMed]
33. Granger, J.P.; Alexander, B.T.; Bennett, W.A.; Khalil, R.A. Pathophysiology of pregnancy-induced hypertension. *Am. J. Hypertens.* **2001**, *53*, 178S–185S. [CrossRef]
34. Sibai, B.M. Etiology and management of postpartum hypertension-preeclampsia. *Am. J. Obs. Gynecol.* **2012**, *206*, 470–475. [CrossRef] [PubMed]
35. Brown, M.A.; Magee, L.A.; Kenny, L.C.; Karumanchi, S.A.; McCarthy, F.P.; Saito, S.; Ishaku, S. The hypertensive disorders of pregnancy: ISSHP classification, diagnosis & management recommendations for international practice. *Pregnancy Hypertens.* **2018**, *13*, 291–310. [CrossRef]

36. Cantonwine, D.E.; McElrath, T.F.; Trabert, B.; Xu, X.; Sampson, J.; Roberts, J.M.; Troisi, R. Estrogen metabolism pathways in preeclampsia and normal pregnancy. *Steroids* **2019**, *144*, 8–14. [CrossRef] [PubMed]
37. Di Renzo, G.C.; Giardina, I.; Clerici, G.; Brillo, E.; Gerli, S. Progesterone in normal and pathological pregnancy. *Horm. Mol. Biol. Clin. Investig.* **2016**, *27*, 35–48. [CrossRef]
38. Panwar, M.; Kumari, A.; Arora, R.; Singh, V.; Bansiwal, R. Raised neutrophil lymphocyte ratio and serum beta hCG level in early second trimester of pregnancy as predictors for development and severity of preeclampsia. *Drug Discov. Ther.* **2019**, *13*, 34–37. [CrossRef]
39. Lenke LMartínez de la Escalera, G.; Clapp, C.; Bertsch, T.; Triebel, J.A. Dysregulation of the Prolactin/Vasoinhibin Axis Appears to Contribute to Preeclampsia. *Front. Endocrinol* **2020**, *10*, 893. [CrossRef]
40. Chen, Y.; Xie, Z.; Wang, X.; Xiao, Q.; Lu, X.; Lu, S.; Lv, S. A risk model of prenatal screening markers in first trimester for predicting hypertensive disorders of pregnancy. *EPMA J.* **2020**, *11*, 343–353. [CrossRef]
41. Shu, C.; Han, S.; Xu, P.; Wang, Y.; Cheng, T.; Hu, C.E. strogen and Preeclampsia: Potential of Estrogens as Therapeutic Agents in Preeclampsia. *Drug Des. Devel. Ther.* **2021**, *15*, 2543–2550. [CrossRef] [PubMed]
42. Chang, A.S.; Grant, R.; Tomita, H.; Kim, H.S.; Smithies, O.; Kakoki, M. Prolactin alters blood pressure by modulating the activity of endothelial nitric oxide synthase. *Proc. Natl. Acad. Sci. USA* **2016**, *113*, 12538–12543. [CrossRef] [PubMed]
43. Alawad, Z.M.; Al-Omary, H.L. Maternal and cord blood prolactin level and pregnancy complications. *Pak. J. Med. Sci.* **2019**, *35*, 1122–1127. [CrossRef] [PubMed]
44. Leaños-Miranda, A.; Campos-Galicia, I.; Ramírez-Valenzuela, K.L.; Chinolla-Arellano, Z.L.; Isordia-Salas, I. Circulating angiogenic factors and urinary prolactin as predictors of adverse outcomes in women with preeclampsia. *Hypertension* **2013**, *61*, 1118–1125. [CrossRef] [PubMed]
45. Mendelson, C.R. Minireview: Fetal-maternal hormonal signaling in pregnancy and labor. *Mol. Endocrinol.* **2009**, *23*, 947–954. [CrossRef] [PubMed]

diagnostics

MDPI

Review

Biofluid Biomarkers in the Prognosis of Chronic Subdural Hematoma: A Systematic Scoping Review

Georgios Georgountzos [1,2], Ioannis Gkalonakis [1], Lykourgos Anastasopoulos [3], George Stranjalis [3,4] and Theodosis Kalamatianos [2,3,4,*]

[1] Department of Neurosurgery, General Hospital of Nikaia-Piraeus "Agios Panteleimon", 18454 Athens, Greece
[2] Clinical and Experimental Neuroscience Research Group, Department of Neurosurgery, National and Kapodistrian University of Athens, Evangelismos Hospital, 10676 Athens, Greece
[3] Department of Neurosurgery, National and Kapodistrian University of Athens, Evangelismos Hospital, 10676 Athens, Greece
[4] Hellenic Centre for Neurosurgery Research, "Professor Petros S. Kokkalis", 10675 Athens, Greece
* Correspondence: tkalamatian@med.uoa.gr or theokalamatianos@gmail.com

Abstract: The present systematic scoping review aimed at mapping and analyzing the available literature on biological fluid (biofluid) biomarkers showing promise in the prediction of chronic subdural hematoma (cSDH) recurrence and the prognosis of neurological/functional patient outcome. Twenty-three studies published between 2003 and 2023 investigating a diverse range of biomarkers in hematoma fluid and/or the circulation in 3749 patients were included. Immune cell populations and inflammatory/anti-inflammatory cytokines comprised the most studied category of biomarkers displaying significant findings. A notable time trend in biomarker studies was a recent shift in research focus towards the analysis of circulating biomarkers. Several biomarkers were indicated as independent predictors of cSDH recurrence and/or functional/neurological outcome, including circulating fibrinogen degradation products (FDP), brain natriuretic peptide (BNP-1) and high-density lipoprotein (HDL), as well as blood urea nitrogen (BUN) and the ratios of blood neutrophil to lymphocyte (NLR) or red blood cell distribution width to platelet count (RPR). While studies on cSDH prognostic biomarkers have gained, in recent years, momentum, additional multicenter prospective studies are warranted to confirm and extend their findings. The identification of prognostic biofluid biomarkers in cSDH is an active field of research that may provide future tools, guiding clinical decisions and allowing for the design of treatments based on risk stratification.

Keywords: chronic subdural hematoma; prognosis; recurrence; biomarker

Citation: Georgountzos, G.; Gkalonakis, I.; Anastasopoulos, L.; Stranjalis, G.; Kalamatianos, T. Biofluid Biomarkers in the Prognosis of Chronic Subdural Hematoma: A Systematic Scoping Review. *Diagnostics* **2023**, *13*, 2449. https://doi.org/10.3390/diagnostics13142449

Academic Editors: Yuli Huang, Yong Yuan and Peisong Chen

Received: 18 May 2023
Revised: 3 July 2023
Accepted: 13 July 2023
Published: 22 July 2023

1. Introduction

Chronic subdural hematoma (cSDH) is a common neurosurgical entity that typically affects elderly patients and is increasing in incidence. The presenting symptoms of cSDH are variable but commonly include gait disturbances and mental deterioration as well as limb weakness [1]. The diagnosis of cSDH is based on imaging indicating the presence of a hematoma in the subdural space, often in association with a history of mild or moderate head injury, weeks to months prior to hospitalization [2]. Histologically, cSDH is characterized by an external and an internal membrane, encapsulating the hematoma, both containing fibroblast layers with only the former showing the persistent presence of immature, leaky vessels that have been postulated to drive hematoma expansion [3].

Surgical evacuation remains the standard of care for cSDH with immediate results postoperatively [2]. Nevertheless, the postoperative recurrence of cSDH is not uncommon, ranging between 5 and over 30% in previous studies [2]. cSDH recurrence thus currently represents a significant nosological consideration, the more so since reoperations increase healthcare-associated costs and are a substantial burden on patient morbidity and mortality [1,4]. In this context, previous studies have sought to establish the factors associated with an increased risk of cSDH recurrence and/or which can aid in the prediction of neurological/functional outcomes. Thus far, several clinical factors, e.g., increased patient age, radiological characteristics, e.g., bilateral hematoma, and specific external membrane features, e.g., enhanced presence of angiogenic factors, have been previously indicated as risk factors for cSDH recurrence [5–8]. Additional research has focused on identifying prognostic molecular or cellular factors within the hematoma fluid and/or peripheral blood of cSDH patients. This class of potential prognostic biofluid biomarkers includes molecules and/or cellular populations that are implicated in cSDH pathogenetic and pathophysiological processes such as cerebrospinal fluid leaks, inflammation and fibrinolysis [3,9,10]. While at present no such biomarker has been validated for clinical use, the overarching goal of this research field is to provide valuable tools guiding future clinical decision making in cSDH [11].

To the best of our knowledge, this is the first systematic scoping review that maps the pertinent literature on biofluid biomarkers showing promise in postoperative cSDH prognosis. We included studies reporting differences in biomarker levels between recurrent and non-recurrent patients as well as those indicating associations between levels and patient outcomes.

2. Materials and Methods

The PRISMA checklist (http://prisma-statement.org, accessed on 6 April 2023) for scoping reviews was used to address patients with chronic subdural hematoma (Participants) in terms of biomarkers measured in peripheral blood or the hematoma fluid or both (Intervention) in studies that may additionally recruit neurologically intact patients (Controls), with the aim to determine their association with disease prognosis (Outcome).

A literature search of the MEDLINE database was conducted on 4 April 2023. The search algorithm included keywords such as "chronic subdural", "recurrence", "prognosis" and "markers" and can be found in the Appendix A. The search did not have a language restriction and the search algorithm included free text and Medical Subject Headings (MeSH) terms. Two independent reviewers (I.G. and L.A.) screened the results of the search primarily according to the title and abstract. Any differences in opinion were resolved through discussion and by reaching a consensus with a 3rd independent reviewer (G.G.).

Inclusion criteria were set as follows: (1) study types: observational, cohort studies and randomized control studies (RCTs), (2) English-language manuscripts, (3) human studies which included surgical treatment; (4) studies that involved biofluid markers (e.g., hematoma fluid or circulation cellular/molecular markers) and (5) prognosis relevance. Exclusion criteria were (1) reviews, case reports, letters and hypotheses, (2) animal studies, (3) studies on imaging markers and (4) studies with results of no statistical significance or without statistical analysis of the results. According to these criteria, full text screening was performed by two independent reviewers (I.G. and L.A.), and when a conflict came up, a third reviewer intervened (G.G.).

Data were extracted by two independent reviewers (I.G. and L.A.) and verified by a third (G.G.) reviewer. Data extraction included DOIs, publication year, country, study type, population characteristics (sample size, age, co-morbidities), outcome measures associated with prognosis and conclusions. Reviewers aimed at incorporating studies that address the following questions and provided significant findings from different levels of analysis: (1) biomarkers the levels of which differ between patients with recurrence and those without, (2) biomarkers that have been indicated as independent risk factors for CSDH recurrence and (3) biomarkers associated with prognosis in terms of neurological/functional outcome. Studies were summarized in tables.

3. Results

A Pubmed search identified a total of 1913 articles (Figure 1). After the removal of duplicates, 1506 records remained and were screened based on title and abstract, and 50 reports were further assessed by full text reading. From these, 27 were excluded based on statistically insignificant results ($n = 11$), wrong topic ($n = 2$), not being related to prognosis ($n = 8$), non-biofluid marker ($n = 3$), study design ($n = 2$, one case report and one case series with three patients) and lack of full text availability ($n = 1$). Twenty-three studies were finally included in the review synthesis.

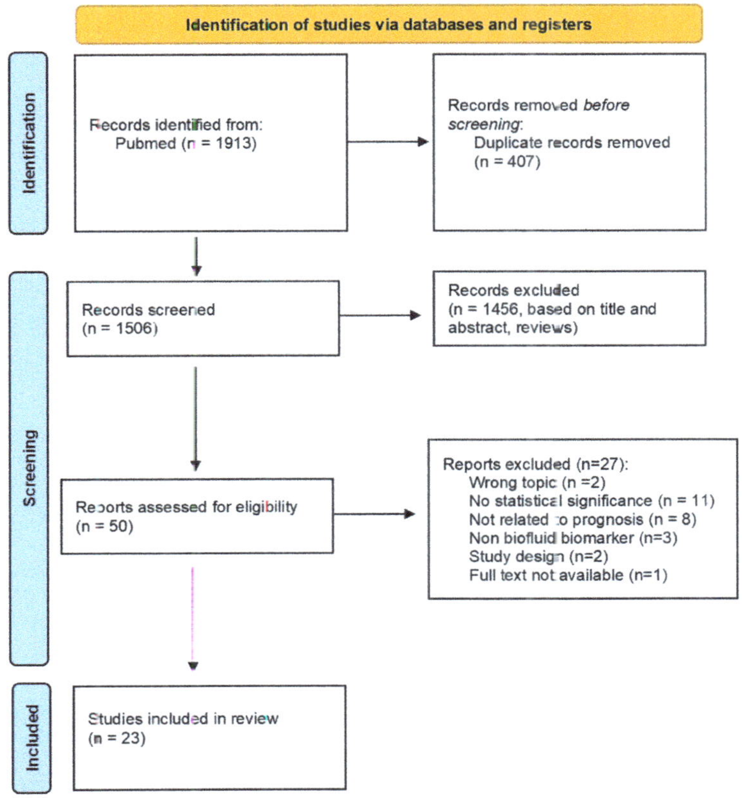

Figure 1. Prisma flow diagram.

3.1. Study and Population Characteristics

Twenty-three studies on a total of 3749 patients were included in the current review (Table 1). Over half of the included studies ($n = 12$) had a prospective design, whilst no RCTs were found. The median (interquartile range, IQR) patient sample size of the included studies was 93 (60–258). The median (IQR) recurrence rate of twenty-one studies (two studies did not provide recurrence data) was 17.5% (5–21).

Studies were published in 17 different journals. The *Journal of Neurosurgery* ranked first in terms of number of published manuscripts ($n = 5$). The most common corresponding author's address was a German one ($n = 8$).

3.2. Biofluids and Biomarker Types

Biomarkers were investigated in peripheral blood and/or hematoma fluid, with one study examining (additionally) hygroma fluid (Table 1).

Investigated biomarkers displaying significant results, which are summarized below, fell within one of the following categories (Table 1): (i) immune- and/or circulatory-system-associated cellular populations and/or their ratios (ten studies), (ii) fibrinolysis-/coagulation-associated molecules, complexes, degradation products or assays (six studies), (iii) cytokines (five studies) and (iv) other/miscellaneous (six studies), such as lipoproteins, ABO (blood) types, CSF markers and markers of cardiovascular/kidney/liver function.

3.3. *Biomarker Findings*

3.3.1. Immune- and/or Circulatory-System-Associated Cellular Populations

Several studies reported significant findings on blood platelets (PLTs) in relation to CSDH recurrence. Thus, a prospective cohort study on 114 patients was the earliest study identified which reported significantly lower preoperative levels in patients with recurrence compared with non-recurrent cases [12]. A similar result was indicated by the univariate (but not multivariable) analysis of a more recent retrospective cohort study [13]. Two additional retrospective studies by a Japanese group incorporating larger patient cohorts indicated preoperative thrombocytopenia as an independent risk factor for CSDH recurrence [14,15].

Blood eosinophil levels were investigated by two studies. An eosinophil-rich count of $\geq 100 \times 10^6$/L preoperatively was highlighted as an independent risk factor for cSDH recurrence in the study of Matsubara et al. [15]. In contrast, a second study on a smaller patient cohort indicated that the preoperative eosinophil count in their non-recurrent CSDH group was significantly higher than that in the recurrent group ($0.14 \pm 0.14 \times 10^9$/L vs. $0.09 \pm 0.08 \times 10^9$/L) and that an eosinophil count < 150×10^6/L was an independent risk factor for recurrence [16].

Three additional studies investigated peripheral blood cellular population ratios that were considered to be inflammatory biomarkers. A prospective cohort study on 61 surgically treated patients indicated that a low admission blood platelet-to-lymphocyte ratio (PLR) was associated with worse outcomes using the Glasgow outcome scale (GOS) and Lagos brain disability examination scale (LABDES) [17]. In a retrospective study on 297 patients, Guresir et al. [18] showed that circulating high red blood cell distribution width to platelet count ratio (RPR) (>0.0568) is an independent predictor of recurrence. In addition, patients with low RPR improved more in terms of Karnofsky performance status at 3 months [18]. The third study, a retrospective cohort study on 160 patients, indicated that a postoperative neutrophil-to-lymphocyte ratio (NLR) ≥ 1 (i.e., NLR that remains high compared to its preoperative value) and the postoperative absolute neutrophil counts were independently associated with recurrence [19].

Lastly, a prospective study on a small cohort (30 patients) measured preoperative and postoperative circulating endothelial progenitor cells (EPCs) and reported that patients with recurrence had a lower mean difference in postoperative and preoperative EPCs in comparison with those without relapse (MV 8.54 vs. 35.52) [20].

3.3.2. Cytokines

A prospective cohort study on 66 patients indicated significantly higher levels of IL-6 in the hematoma fluid of recurrent patients [6]. Another prospective cohort study on 35 patients undergoing surgery indicated that in addition to IL-6, IL-8 hematoma levels were significantly higher in patients with recurrence [21]. Nevertheless, in the retrospective cohort of Pripp and Stanisic [10] that included 93 patients, an association between increased hematoma fluid CXCL8 (IL-8) and reduced risk of recurrence in need of reoperation was indicated. In the same study, a positive association between hematoma fluid CCL5 (also termed regulated on activation, normal T-cell expressed and secreted, RANTES) levels and recurrence as well as a negative association between IL-5, IL-13, IFN-γ and CXCL10 levels and recurrence was also indicated. In the earlier prospective cohort study of these two researchers on 57 patients, an association between increased hematoma fluid (but not serum) of a panel of biomarkers considered an anti-inflammatory index (IL-4, IL-5, IL-10,

IL-13, IL-1 RA) and a reduced risk of recurrence was indicated [22]. In the retrospective cohort pilot study of Puccio et al. [11], higher IL-5 hematoma fluid levels were associated with a better outcome in terms of the extended Glasgow outcome scale (GOS-E) at 3, 6 and 12 months. Moreover, in this study, a lower concentration of hematoma RANTES (<1649.24 pg/mL) was associated with recurrence [11].

3.3.3. Fibrinolysis-/Coagulation-Associated Molecules/Complexes, Products and Assays

Four prospective and two retrospective cohort studies were identified. A prospective cohort study on 60 patients investigated the levels of several biomarkers, including tissue plasminogen activator (TPA), in both hematoma fluid and peripheral blood [23]. The authors indicated that the TPA levels in the hematoma fluid of recurrent cases were significantly higher than those in non-recurrent cases [23]. A second prospective cohort study was on 18 patients with spontaneous cSDH and examined plasma F XIII activity [24]. Patients with recurrence of the spontaneous cSDH after the first operation had significantly lower FXIII activity in comparison to the non-recurrence group (47.5% vs. 78.5%, respectively). Moreover, the authors indicated a cut-off value of 68.5% in association with recurrence [24]. The third prospective study identified was on 61 patients treated surgically for cSDH [17]. This study indicated associations between elevated activated partial thromboplastin time (APTT), prothrombin time (PT) and international normalized ratio (INR) and worse outcome [17]. Reduced levels of serum fibrinogen in cases manifesting recurrence of cSDH compared with non-recurrence cases were indicated in the univariate analysis conducted by Wang et al. [25].

A retrospective cohort study on 90 patients showed that higher preoperative PT was an independent risk factor for recurrence after surgical evacuation [26]. Another retrospective cohort study on 92 patients showed that a serum fibrinogen degradation products (FDPs) value > 5 μg/mL on admission was an independent risk factor for cSDH recurrence within 90 days [27].

3.3.4. Other/Miscellaneous

In the largest prospective cohort study identified (653 patients), blood urea nitrogen (BUN) levels and additional variables (PLTs, fibrinogen, leukocytes, erythrocytes, hemoglobin, creatinine) were analyzed [25]. The authors indicated that elevated postoperative BUN levels (> 6.4 mmol/L) was the only measure independently associated with cSDH recurrence [25]. In a second prospective cohort study on 119 patients, Chihi et al. showed that the preoperative plasma brain natriuretic peptide (BNP-1) level on admission is an independent predictor of poor functional outcome at the 5- to 6-month follow-up [28]. The authors of another prospective study on 75 patients undergoing burr hole hematoma evacuation for subdural hematoma or hygroma measured the levels of the CSF marker beta-trace protein (β-TP) in serum, hematoma or hygroma [9]. The authors of this study reported that when patients with hematoma and hygroma were grouped together, the β-TP levels measured during the first operation were significantly higher in recurrent cases [9].

A large retrospective cohort study on 274 patients showed that preoperative triglycerides levels and serum lipids were positively correlated with recurrence, whilst high-density lipoprotein (HDL) was negatively associated [29]. In this study, a cut-off level of HDL at 37.45 mg/dL was the only variable indicated as an independent risk factor for cSDH recurrence [29]. In the retrospective cohort study of Mainka et al. on 256 cSDH patients, dehydration status on admission (indicated by serum urea-to-creatinine ratios (U/Cr)) was highlighted as another independent predictor for recurrence [30]. Lastly, Hirai et al. retrospectively investigated the relationship between ABO type and cSDH recurrence after burr hole drainage in 320 patients. The authors showed that type A was an independent risk factor for recurrence [14].

Table 1. Biomarker study characteristics and findings.

Author, Year [Ref. Number]	Study Type	Sample Size, Recurrence %	Biofluids Utilized	Biomarkers Analyzed	Significant Biomarker Findings (Recurrence/Outcome)
König et al., 2003 [12]	Prospective Cohort	114, 17.5%	Blood and plasma	PLTs, INR, APTT, fibrinogen, FXIII	↓ PLTs in recurrence
Frati et al., 2004 [21]	Prospective Cohort	35, 14.3%	Serum and hematoma	IL-6, IL-8	↑ Hematoma IL-6 and IL-8 in recurrence
Katano et al., 2006 [23]	Prospective Cohort	60, 8.1%	Serum and hematoma	Plasminogen, antiplasmin, TPA, PIC, HGF, VEGF, bFGF	↑ Hematoma TPA in recurrence
Kristof et al., 2008 [9]	Prospective Cohort	75, 22.7%	Serum, hematoma and hygroma	β-TP	↑ β-TP of hematoma and hygroma in recurrence
Hong et al., 2009 [6]	Prospective Cohort	66, 21.21%	Hematoma	IL-6, VEGF, bFGF	↑ IL-6 in recurrence
Song et al., 2013 [20]	Prospective Cohort	30, 20%	Blood	RBCs, WBCs, PLTs, Hb, PT, APTT, INR, TT, EPCs	↓ EPCs in recurrence
Bosche et al., 2013 [24]	Prospective Cohort	18, 33.3%	Plasma	INR, PTT, PLTs, fibrinogen, FXIII	FXIII activity 68.5% (cut-off) → recurrence (100% sensitivity, 75% specificity)
Pripp et al., 2014 [22]	Prospective Cohort	57, 12.5%	Blood and hematoma	TNF-a, IL-1b, IL-2, IL-2R, IL-6, IL-7, IL-12, IL-15, IL-17, CCL2, CXCL8, CXCL9, CXCL1 and IL-1RA, IL-4, IL-5, IL-10 and IL-13	↑ Hematoma anti-inflammatory activity (IL-4, IL-5, IL-10, IL-1 RA, IL-13) → ↓ recurrence
Pripp et al., 2017 [10]	Retrospective Cohort	93, 17.2%	Hematoma	IL-1-IL-17, TNF-a, IFN-g, CXCL8, GM-CSF, VEGF	↑ CXCL8 (IL8), CXCL10, IL-5, IL-13, IFN-γ → ↓ recurrence ↑ CCL5 (RANTES) → ↑ recurrence
Hori et al., 2018 [27]	Retrospective Cohort	92, NR	Serum	FDP, APTT, PT, D-dimer, PLTs, INR, Hb	FDP > 5 μg/mL = independent recurrence risk factor
Wang et al., 2020 [25]	Prospective Cohort	653, 14.7%	Serum	Glucose, Cr, leukocytes, neutrophils, lymphocytes, erythrocytes, Hg, PLTs, PT, INR, fibrinogen, BUN	↓ Serum leukocytes, neutrophils, PLTs, fibrinogen in recurrence BUN > 6.4 mmol/L = independent recurrence risk factor
Suero-Molina et al., 2020 [13]	Retrospective Cohort	148, 23.6%	Blood	PLTs	↓ PLTs in recurrence

Table 1. *Cont.*

Author, Year [Ref. Number]	Study Type	Sample Size, Recurrence %	Biofluids Utilized	Biomarkers Analyzed	Significant Biomarker Findings (Recurrence/Outcome)
Liu et al., 2021 [29]	Retrospective Cohort	274, 84.67%	Serum	TG, total cholesterol, LDL, HDL	↑ Serum TG in recurrence Low HDL = independent recurrence risk factor (HDL > 37.5 mg/dL, ↓ risk)
Martinez-Perez et al., 2021 [26]	Retrospective Cohort	90, 18.9%	Blood	PLTs, Hb, PT	↑ PT = independent recurrence risk factor
Chihi et al., 2021 [28]	Prospective Cohort	119, 13.3%	Plasma	BNP-1	BNP-1 = independent predictor for functional outcome
Hirai et al., 2021 [14]	Retrospective Cohort	307, 10.6%	Blood	PLTs, ABO type	↓ PLTs = independent recurrence risk factor; blood type A = independent recurrence risk factor (64.9% sensitivity, 58.8% specificity)
De Oliveira et al., 2022 [19]	Retrospective Cohort	160, 22.5%	Blood	Neutrophils, NLR	Absolute neutrophil count = independent recurrence risk factor; NLR $\geq$ 1 = independent recurrence risk factor (77.8% sensitivity, 66.7% specificity); validation on independent sample
Idowu et al., 2022 [17]	Prospective Cohort	61, NR	Blood	INR, PT, APTT, NLR, CRP, ESR, PLR	↑ APTT, ↑ PT, ↑ INR, ↓ PLR → worse outcome
Mainka et al., 2022 [30]	Retrospective Cohort	256, 32%	Serum	U/CR	U/Cr > 80 = independent recurrence risk factor
Chen et al., 2022 [16]	Retrospective Cohort	258, 14.3%	Blood	WBCs, PLTs, basophils, eosinophils	Eosinophils < 0.15×10^9 = independent recurrence risk factor
Guresir et al., 2022 [18]	Retrospective Cohort	297, 12.5%	Blood	CRP, Hb, WBC, PLTs, RDW, RPR	RPR $\geq$ 0.0568 = independent recurrence risk factor (70.3% sensitivity, 56.2% specificity); low RPR → higher Karnofsky at 3 months
Matsubara et al., 2023 [15]	Retrospective Cohort	466, 9.3%	Blood	Eosinophils, PLTs	↓ PLTs and eosinophils $\geq$ 100/μL = independent recurrence risk factors

Table 1. *Cont.*

Author, Year [Ref. Number]	Study Type	Sample Size, Recurrence %	Biofluids Utilized	Biomarkers Analyzed	Significant Biomarker Findings (Recurrence/Outcome)
Puccio et al., 2023 [11]	Prospective Cohort	20, 42% (21% requiring reoperation)	Hematoma	CXCL9, MCP-1, G-CSF, IL-6, VEGF-A, IL-8, PDGF-AB/BB, PDGF-AA, MDC, IL-1α, IP-10, IL-5, RANTES, IL-10, IL-1β, IL-1 Ra, M-CSF	↑ IL-5 → ↑ GOS-E RANTES < 1649.24 pg/mL → ↑ recurrence

Abbreviations. ABO: ABO blood group system, APTT: activated partial thromboplastin clotting time, bFGF: basic fibroblast growth factor, β-TP: beta-trace protein, BNP-1: brain natriuretic peptide-1, BUN: blood urea nitrogen, CBC: complete blood count, CCL: C-C Motif Chemokine Ligand, CRP: C-reactive protein, CXCL: C-X-C Motif Chemokine Ligand, ECL: electrochemiluminescence, ELISA: enzyme-linked immunosorbent assay, EPCs: endothelial progenitor cells, ESR: erythrocyte sedimentation rate, FDP: fibrinogen degradation products, FXIII: factor XIII, GM-CSF: granulocyte-macrophage colony-stimulating factor, GOS-E: Glasgow outcome scale extended, Hb: hemoglobin, HDL: high-density lipoprotein, HGF: hepatocyte growth factor, IFN-g: interferon gamma, IL-: interleukin-, IL-2R: interleukin-2 receptor, IL-1RA: interleukin-1 receptor antagonist protein, INR: international normalized ratio, IP-10: interferon-γ-inducible protein 10, LEIN: latex-enhanced immune nephelometry, LDL: low-density lipoprotein, MCP-1: monocyte chemoattractant protein-1, M-CSF: macrophage colony-stimulating factor, MDC: macrophage-derived chemokine, NLR: neutrophil-to-lymphocyte ratio, NPR: negative predictive value, NR: not reported, PDGF-AA: platelet-derived growth factor-AA, PDGF-AB/BB: platelet-derived growth factor-AB/-BB, PIC: a2-plasmin inhibitor–plasmin complex, PLR: platelet–lymphocyte ratio, PLTs: platelets, PPV: positive predictive value, PT: prothrombin time, RANTES: regulated upon activation, normal T cell expressed and presumably secreted, RBCs: red blood cells, RDW: red blood cell distribution width, RPR: red blood cell distribution width to platelet count ratio, TG: triglycerides, TNF-a: tumor necrosis factor alpha, TPA: tissue plasminogen activator, TT: thrombin time, U/CR: urea-to-creatinine ratio, VEGF: vascular endothelial growth factor, WBCs: white blood cells, ↓: reduced (levels/counts/events of), ↑: increased (levels/counts/events of), → associated with.

4. Discussion

4.1. Biomarker Study Characteristics/Designs and Time Trends

Twenty-three studies published between 2003 and 2023 on a total of 3749 patients were identified in the present review. The median postoperative recurrence rate of 17.5% falls well within the range of previous estimates (5 to 30%) [2]. The notable limitations of the incorporated studies were their retrospective cohort/single-center designs (n =11 studies) and/or small cohort sizes (<100 patients, n = 12 studies).

A noteworthy trend is a shift in research focus towards circulating biomarkers on larger patient cohorts during the last four years (2020–2023: circulation focus in 12 of 13 studies, median patient size: 256 patients) compared with previous studies (2003–2018: circulation focus in 3 of 10 studies, median patient size: 63). While hematoma biomarkers allow for the analysis of local pathophysiological processes [3], the distinct advantages of assessing circulating prognostic biomarkers are the ease of sampling and that they offer the opportunity of tailoring patient treatment according to risk stratification preoperatively. In this context, current guidelines on cSDH treatment incorporate adjuvant drug therapies such as cholesterol-lowering statins and anti-inflammatory corticosteroids that are designed to improve patient neurological function and/or reduce recurrence [2,31]. From an analytical perspective, the protein analysis of biomarkers with significant findings ranged from single enzyme-linked immunosorbent [6,21,23], nephelometric [9] and electrochemiluminescent [28] assays to multiplex immunoassays [10,11,22].

4.2. Biomarkers in Recurrence Prediction

Numerous studies on biomarkers associated with inflammatory processes were identified. Enhanced levels of cytokine biomarkers in hematoma fluid have been highlighted as evidence for a robustly elevated local inflammatory index, as well as for the involvement of inflammatory processes in recurrence. The replication of findings was evident for IL-6 [6,21], with less consistent findings regarding the direction of change (increased or decreased levels) in recurrent cSDH for IL-8 and RANTES [10,11,21]. The notion of increased local anti-inflammatory activity (reflected by the levels of a panel of cytokine biomarkers) in association with reduced cSDH recurrence rates was introduced by the studies of a single group [10,22].

More recent studies assessing circulating cellular population ratios as markers of inflammation support the notion of a systemic inflammatory response in cSDH and its association with recurrence [18,19]. Thus, while Guresir et al. [18] proposed preoperative RPR as a novel independent predictor of cSDH recurrence, De Oliveira et al. [19] highlighted similar value for the postoperative NLR ratio. Of note, evidence for a systemic inflammatory response triggered by cSDH was previously obtained by studies comparing levels of cell inflammatory markers in patients and healthy controls [32,33].

Aside from the aforementioned cellular biomarkers, circulating eosinophil counts were indicated as independent predictors of recurrence by two studies, albeit with contradictory findings. Thus, Chen et al. [16] indicated that lower peripheral eosinophil counts are independently associated with recurrence, while Matsubara et al. [15] reported the opposite. The findings of Chen et al. appear consistent with evidence suggesting a protective role of hematoma-membrane-infiltrating eosinophils against recurrence in a previous histological study [34].

A substantial body of evidence supports a role for local coagulation and fibrinolysis in cSDH development [35–39], including evidence for excessive activation of the former during the early stages of development, succeeded by hyperfibrinolysis [40]. In relation to cSDH recurrence, Katano et al. [23] found higher levels of TPA, a major fibrinolytic mediator, in the hematoma fluid of patients that experienced recurrences, while Wang et al. [25] indicated lower levels of circulating fibrinogen in their recurrent cases. Circulating FDP (but not D-dimer) levels on patient admission as an independent predictor for recurrence was indicated by Hori et al. [27]. The authors of this study hypothesized that the systemic elevation of FDP induces a systemic inflammatory response, which in turn influences expansion of the hematoma and can lead to its recurrence [27]. The potential of coagulation cascade mediators as peripheral biomarkers in the prediction of cSDH recurrence is further supported by studies reporting preoperative thrombocytopenia or increased PT as independent risk factors for cSDH recurrence [15,26,41] and of lower circulating factor XIII in recurrent cases [24].

The formation of immature, leaky vessels in the outer membrane of cSDH upon which damaging forces of inflammation and of abnormal cerebral pulsations are exerted has been postulated as a driving mechanism for cSDH expansion, as well as a risk factor for its recurrence [3,20]. A study indicating significantly lower postoperative levels of peripheral EPCs in patients that experienced cSDH recurrences suggested that this decrease may lead to a reduced capacity for endothelium repair, thus increasing the risk of cSDH recurrence [20]. Following a similar logic, an attenuated protective role on vascular integrity and endothelial function due to decreased peripheral HDL levels was postulated by a second study indicating lower HDL levels as an independent predictor of recurrence [29].

Three additional studies indicated the potential value of peripheral biomarkers as independent predictors of recurrence. The dehydration status of cSDH patients on admission, estimated using the blood U/Cr ratio, was reported as an independent predictor of cSDH recurrence by Mainka et al., 2022 [30]. The authors proposed that the impact of dehydration is due to a reduction in brain volume and/or instigation of inflammatory processes that predispose patients to the occurrence/recurrence of cSDH [30]. Wang et al. investigated BUN levels in cSDH patients based on previous findings indicating the prognostic significance of BUN in neurological disorders such as ischemic stroke [25]. Their findings indicated that an elevated postoperative level of BUN is an independent risk factor for recurrence via mechanisms that remain to be established [25]. Blood type A was also highlighted as an independent risk factor for cSDH recurrence by Hirai et al. [14]. Nevertheless, it is worth pointing out that a second study on another Asian cohort, reporting significantly lower recurrence rates, did not find an association between ABO types and cSDH recurrence [41]. Larger prospective cohort studies that may include additional population samples are warranted to establish the impact of ABO blood types on cSDH recurrence.

4.3. Biomarkers in Neurological/Functional Outcome Prognosis

A smaller number of studies assessed biomarkers in relation to cSDH patient neurological/functional outcome. The pilot study of Puccio et al. indicated strong correlations between higher hematoma IL-5 levels and more favorable GOS-E scores [11]. Interestingly, higher hematoma IL-5 levels were associated with lower rates of cSDH recurrence in two other studies included in the present review [10,22]. Consistent with the notion of a significant association between a lower preoperative peripheral inflammatory index and a favorable functional outcome are the findings of Guresir et al. [18], indicating significantly higher Karnofsky scores 3 months postoperatively in the low compared to the high RPR patient group. Moreover, Idowu et al. [17] indicated significant associations between a low admission PLR, which represents another inflammatory marker, and poor outcome at 3 months using the GOS and LABDES. Notably, in the same study, inflammatory markers, such as c-reactive protein (CRP) and the NLR, did not display associations with the outcome [17].

Lastly, on the basis of previous evidence for significant associations between admission circulating BNP-1 levels and functional outcome in neurological disease, including traumatic brain injury and stroke, Chihi et al. [28] investigated circulating peptide levels in cSDH. Their findings indicated that preoperative BNP-1 is an independent predictor of functional outcome at 5–6 months postoperatively [28].

5. Conclusions

In the present systematic scoping review, we mapped and analyzed the pertinent literature on biofluid biomarkers showing promise in the prediction of cSDH recurrence and for the prognosis of neurological/functional patient outcomes. Twenty-three studies spanning three decades, analyzing a diverse range of biomarkers in hematoma fluid and/or the circulation, were included. Immune cell populations and inflammatory/anti-inflammatory cytokines comprised the most studied category of biomarkers displaying significant findings. A notable time trend in biomarker studies was a recent shift in research focus towards the analysis of circulating biomarkers on larger patient cohorts. Some of the studies on peripheral biomarkers highlighted the notion that aside from local inflammatory processes, cSDH triggers a systemic inflammatory response, and biomarkers associated with the latter are of prognostic value. Several biomarkers were indicated as independent predictors of cSDH recurrence and/or functional outcome, including circulating FDP, BNP-1 and HDL as well as BUN and blood NLR and RPR ratios. In recent years, studies on cSDH prognostic biomarkers have gained momentum. Nevertheless, additional multicenter prospective studies are warranted to confirm and extend their findings. The identification of prognostic biofluid biomarkers in cSDH is an active field of research that may provide future tools guiding clinical decisions and allowing the design of treatments based on risk stratification.

Author Contributions: Conceptualization, T.K. and G.G.; methodology, T.K., G.G. and L.A.; investigation, T.K., G.G., L.A. and I.G.; formal analysis and review process, G.G., L.A., I.G. and T.K.; supervision, T.K. and G.S.; manuscript editing G.G., I.G., G.S. and T.K. All authors have read and agreed to the published version of the manuscript.

Funding: This research received no external funding.

Institutional Review Board Statement: This study is a scoping review for which an ethical review and approval were not required.

Informed Consent Statement: This study is a scoping review for which informed consent was not required.

Data Availability Statement: Not applicable.

Conflicts of Interest: The authors declare no conflict of interest.

Appendix A

The search strategy comprised the following: (("chronic subdural haematoma"[All Fields] OR "hematoma, subdural, chronic"[MeSH Terms] OR ("hematoma"[All Fields] AND "subdural"[All Fields] AND "chronic"[All Fields]) OR "chronic subdural hematoma"[All Fields] OR ("chronic"[All Fields] AND "subdural"[All Fields] AND "hematoma"[All Fields])) AND ("biomarkers"[All Fields] OR "biomarkers"[MeSH Terms] OR "biomarkers"[All Fields] OR "biomarker"[All Fields])), (("chronic subdural haematoma"[All Fields] OR "hematoma, subdural, chronic"[MeSH Terms] OR ("hematoma"[All Fields] AND "subdural"[All Fields] AND "chronic"[All Fields]) OR "chronic subdural hematoma"[All Fields] OR ("chronic"[All Fields] AND "subdural"[All Fields] AND "hematoma"[All Fields])) AND ("recurrance"[All Fields] OR "recurrence"[MeSH Terms] OR "recurrence"[All Fields] OR "recurrences"[All Fields] OR "recurrencies"[All Fields] OR "recurrency"[All Fields] OR "recurrent"[All Fields] OR "recurrently"[All Fields] OR "recurrents"[All Fields])), (("chronic subdural haematoma"[All Fields] OR "hematoma, subdural, chronic"[MeSH Terms] OR ("hematoma"[All Fields] AND "subdural"[All Fields] AND "chronic"[All Fields]) OR "chronic subdural hematoma"[All Fields] OR ("chronic"[All Fields] AND "subdural"[All Fields] AND "hematoma"[All Fields])) AND ("prognosis"[MeSH Terms] OR "prognosis"[All Fields] OR "prognoses"[All Fields])).

References

1. Kolias, A.G.; Chari, A.; Santarius, T.; Hutchinson, P.J. Chronic subdural haematoma: Modern management and emerging therapies. *Nat. Rev. Neurol.* **2014**, *10*, 570–578. [CrossRef]
2. Tamura, R.; Sato, M.; Yoshida, K.; Toda, M. History and current progress of chronic subdural hematoma. *J. Neurol. Sci.* **2021**, *429*, 118066. [CrossRef] [PubMed]
3. Edlmann, E.; Giorgi-Coll, S.; Whitfield, P.C.; Carpenter, K.L.H.; Hutchinson, P.J. Pathophysiology of chronic subdural haematoma: Inflammation, angiogenesis and implications for pharmacotherapy. *J. Neuroinflamm.* **2017**, *14*, 108. [CrossRef] [PubMed]
4. Weigel, R.; Schmiedek, P.; Krauss, J.K. Outcome of contemporary surgery for chronic subdural haematoma: Evidence based review. *J. Neurol. Neurosurg. Psychiatry* **2003**, *74*, 937–943. [CrossRef] [PubMed]
5. Gandhoke, G.S.; Kaif, M.; Choi, L.; Williamson, R.W.; Nakaji, P. Histopathological features of the outer membrane of chronic subdural hematoma and correlation with clinical and radiological features. *J. Clin. Neurosci.* **2013**, *20*, 1398–1401. [CrossRef] [PubMed]
6. Hong, H.J.; Kim, Y.J.; Yi, H.J.; Ko, Y.; Oh, S.J.; Kim, J.M. Role of angiogenic growth factors and inflammatory cytokine on recurrence of chronic subdural hematoma. *Surg. Neurol.* **2009**, *71*, 161–165. [CrossRef]
7. Mishra, R.; Deora, H.; Florez-Perdomo, W.A.; Moscote-Salazar, L.R.; Garcia-Ballestas, E.; Rahman, M.M.; Shrivastava, A.; Raj, S.; Chavda, V.; Montemurro, N.; et al. Clinical and Radiological Characteristics for Recurrence of Chronic Subdural Hematoma: A Systematic Review and Meta-Analysis. *Neurol. Int.* **2022**, *14*, 683–695. [CrossRef]
8. Miah, I.P.; Tank, Y.; Rosendaal, F.R.; Peul, W.C.; Dammers, R.; Lingsma, H.F.; den Hertog, H.M.; Jellema, K.; van der Gaag, N.A. Radiological prognostic factors of chronic subdural hematoma recurrence: A systematic review and meta-analysis. *Neuroradiology* **2021**, *63*, 27–40. [CrossRef] [PubMed]
9. Kristof, R.A.; Grimm, J.M.; Stoffel-Wagner, B. Cerebrospinal fluid leakage into the subdural space: Possible influence on the pathogenesis and recurrence frequency of chronic subdural hematoma and subdural hygroma. *J. Neurosurg.* **2008**, *108*, 275–280. [CrossRef]
10. Pripp, A.H.; Stanišić, M. Association between biomarkers and clinical characteristics in chronic subdural hematoma patients assessed with lasso regression. *PLoS ONE* **2017**, *12*, e0186838. [CrossRef]
11. Puccio, D.J.; Deng, H.; Eagle, S.R.; Okonkwo, D.O.; Nwachuku, E.L. Pilot Biomarker Analysis and Decision Tree Algorithm Modeling of Patients with Chronic Subdural Hematomas. *Neurotrauma Rep.* **2023**, *4*, 184–196. [CrossRef] [PubMed]
12. König, S.A.; Schick, U.; Döhnert, J.; Goldammer, A.; Vitzthum, H.E. Coagulopathy and outcome in patients with chronic subdural haematoma. *Acta Neurol. Scand.* **2003**, *107*, 110–116. [CrossRef] [PubMed]
13. Suero Molina, E.; Borscheid, L.; Freistühler, M.; Zawy Alsofy, S.; Stummer, W.; Schipmann, S. Risk-assessment in chronic subdural hematoma evaluated in 148 patients—A score for predicting recurrence. *Clin. Neurol. Neurosurg.* **2020**, *195*, 106020. [CrossRef] [PubMed]
14. Hirai, S.; Yagi, K.; Hara, K.; Kanda, E.; Matsubara, S.; Uno, M. Postoperative recurrence of chronic subdural hematoma is more frequent in patients with blood type A. *J. Neurosurg.* **2021**, *135*, 1203–1207. [CrossRef]

15. Matsubara, M.; Yagi, K.; Minami, Y.; Kanda, E.; Sunada, Y.; Tao, Y.; Takai, H.; Shikata, E.; Hirai, S.; Matsubara, S.; et al. Preoperative elevated eosinophils in peripheral blood for prediction of postoperative recurrence of chronic subdural hematoma. *J. Neurosurg.* **2023**, *1*, 1–6. [CrossRef]
16. Chen, S.; Shao, L.; Ma, L. Peripheral blood eosinophil and classification of residual hematoma help predict the recurrence of chronic subdural hematoma after initial surgery. *Front. Surg.* **2022**, *9*, 970468. [CrossRef]
17. Idowu, O.E.; Oyeleke, S.O.; Vitowanu, J.M. Impact of inflammatory cell ratio, biomarkers, activated partial thromboplastin time and prothrombin time on chronic subdural haematoma severity and outcome. *Eur. J. Trauma Emerg. Surg.* **2022**, *48*, 1085–1092. [CrossRef]
18. Güresir, Á.; Coch, C.; Heine, A.; Mass, E.; Lampmann, T.; Vatter, H.; Velten, M.; Schmitz, M.T.; Güresir, E.; Wach, J. Red Blood Cell Distribution Width to Platelet Count Ratio Facilitates Preoperative Prediction of Recurrence in Surgically Treated Chronic Subdural Hematoma. *Front. Neurol.* **2022**, *13*, 884231. [CrossRef]
19. de Oliveira, A.J.M.; Solla, D.J.F.; de Oliveira, K.F.; Amaral, B.S.; Andrade, A.F.; Kolias, A.G.; Paiva, W.S. Postoperative neutrophil-to-lymphocyte ratio variation is associated with chronic subdural hematoma recurrence. *Neurol. Sci.* **2022**, *43*, 427–434. [CrossRef]
20. Song, Y.; Wang, Z.; Liu, L.; Wang, D.; Zhang, J. The level of circulating endothelial progenitor cells may be associated with the occurrence and recurrence of chronic subdural hematoma. *Clinics* **2013**, *68*, 1084–1088. [CrossRef]
21. Frati, A.; Salvati, M.; Mainiero, F.; Ippoliti, F.; Rocchi, G.; Raco, A.; Caroli, E.; Cantore, G.; Delfini, R. Inflammation markers and risk factors for recurrence in 35 patients with a posttraumatic chronic subdural hematoma: A prospective study. *J. Neurosurg.* **2004**, *100*, 24–32. [CrossRef]
22. Pripp, A.H.; Stanišić, M. The correlation between pro- and anti-inflammatory cytokines in chronic subdural hematoma patients assessed with factor analysis. *PLoS ONE* **2014**, *9*, 970468. [CrossRef]
23. Katano, H.; Kamiya, K.; Mase, M.; Tanikawa, M.; Yamada, K. Tissue plasminogen activator in chronic subdural hematomas as a predictor of recurrence. *J. Neurosurg.* **2006**, *104*, 79–84. [CrossRef]
24. Bosche, B.; Molcanyi, M.; Noll, T.; Kochanek, M.; Kraus, B.; Rieger, B.; El Majdoub, F.; Dohmen, C.; Löhr, M.; Goldbrunner, R.; et al. Occurrence and recurrence of spontaneous chronic subdural haematoma is associated with a factor XIII deficiency. *Clin. Neurol. Neurosurg.* **2013**, *115*, 13–18. [CrossRef] [PubMed]
25. Wang, N.; Hu, J.; Oppong-Gyebi, A.; Zhu, X.; Li, Y.; Yang, J.; Ruan, L.; Zhuge, Q.; Ye, S. Elevated blood urea nitrogen is associated with recurrence of post-operative chronic subdural hematoma. *BMC Neurol.* **2020**, *20*, 411. [CrossRef] [PubMed]
26. Martinez-Perez, R.; Tsimpas, A.; Rayo, N.; Cepeda, S.; Lagares, A. Role of the patient comorbidity in the recurrence of chronic subdural hematomas. *Neurosurg. Rev.* **2021**, *44*, 971–976. [CrossRef] [PubMed]
27. Hori, Y.S.; Ebisudani, Y.; Aoi, M.; Fukuhara, T. Elevated Serum Fibrinogen Degradation Products on Admission Is a Novel Predictive Factor for Recurrence of Chronic Subdural Hematoma. *World Neurosurg.* **2018**, *118*, e753–e757. [CrossRef] [PubMed]
28. Chihi, M.; Gembruch, O.; Darkwah Oppong, M.; Rauschenbach, L.; Rauscher, S.; Jabbarli, R.; Wrede, K.H.; Sure, U.; Maslehaty, H. Role of brain natriuretic peptide in the prediction of long-term surgical outcome of chronic subdural hematoma. *J. Neurol. Sci.* **2021**, *420*, 117240. [CrossRef] [PubMed]
29. Liu, W.C.; Lin, Q.Q.; Jin, J.; Wang, M.; You, W.D.; Gu, J.; Pan, J.W. An association of low high-density lipoprotein levels with recurrence of chronic subdural hematoma. *Acta Neurochir.* **2021**, *163*, 1061–1068. [CrossRef] [PubMed]
30. Mainka, N.; Borger, V.; Hadjiathanasiou, A.; Hamed, M.; Potthoff, A.L.; Vatter, H.; Schuss, P.; Schneider, M. Dehydration Status at Admission Predicts Recurrence in Patients with Traumatic Chronic Subdural Hematoma. *J. Clin. Med.* **2022**, *11*, 1178. [CrossRef]
31. Zhang, J.; Fei, Z.; Feng, H.; Gao, G.; Hao, J.; Hou, L.; Hu, J.; Huang, Y.; Jiao, B.; Ji, H.; et al. Expert consensus on drug treatment of chronic subdural hematoma. *Chinese Neurosurg. J.* **2021**, *7*, 47. [CrossRef]
32. Zhang, Y.; Yang, Y.; Long, S.; Li, G. Assessment of peripheral blood cell inflammatory markers in patients with chronic subdural hematoma. *Clin. Neurol. Neurosurg.* **2020**, *191*, 105738. [CrossRef]
33. Fan, Y.; Wu, D.; Zhang, X.; Jiang, W.; Nie, M.; Liu, X.; Xiang, T.; Liu, M.; Chen, Y.; Feng, D.; et al. The inflammatory cellular feature in the peripheral blood of chronic subdural hematoma patients. *J. Clin. Lab. Anal.* **2022**, *36*, e24706. [CrossRef]
34. Davidson, B.; Narvacan, K.; Munoz, D.G.; Rotondo, F.; Kovacs, K.; Zhang, S.; Cusimano, M.D. The Crucial Role of Eosinophils in the Life Cycle, Radiographical Architecture, and Risk of Recurrence of Chronic Subdural Hematomas. *Neurotrauma Rep.* **2021**, *2*, 76–83. [CrossRef]
35. Yamashima, T.; Kubota, T.; Yamamoto, S. Eosinophil degranulation in the capsule of chronic subdural hematomas. *J. Neurosurg.* **1985**, *62*, 257–260. [CrossRef]
36. Murakami, H.; Hirose, Y.; Sagoh, M.; Shimizu, K.; Kojima, M.; Gotoh, K.; Mine, Y.; Hayashi, T.; Kawase, T. Why do chronic subdural hematomas continue to grow slowly and not coagulate? Role of thrombomodulin in the mechanism. *J. Neurosurg.* **2002**, *96*, 877–884. [CrossRef]
37. Fujisawa, H.; Ito, H.; Kashiwagi, S.; Nomura, S.; Toyosawa, M. Kallikrein-kinin system in chronic subdural haematomas: Its roles in vascular permeability and regulation of fibrinolysis and coagulation. *J. Neurol. Neurosurg. Psychiatry* **1995**, *59*, 388–394. [CrossRef]
38. Y, K.; M, C.; T, T.; Y, S. Coagulation and fibrinolysis in chronic subdural hematoma. *Neurosurgery* **1989**, *25*, 25. [CrossRef]
39. Ito, H.; Yamamoto, S.; Komai, T.; Mizukoshi, H. Role of local hyperfibrinolysis in the etiology of chronic subdural hematoma. *J. Neurosurg.* **1976**, *45*, 26–31. [CrossRef]

40. Suzuki, M.; Kudo, A.; Kitakami, A.; Doi, M.; Kubo, N.; Kuroda, K.; Ogawa, A. Local hypercoagulative activity precedes hyperfibrinolytic activity in the subdural space during development of chronic subdural haematoma from subdural effusion. *Acta Neurochir.* **1998**, *140*, 261–266. [CrossRef]
41. Ou, Y.; Yu, X.; Wu, L.; Zhang, D.; Liu, W. Recurrence of Chronic Subdural Hematoma Is Independent of ABO Blood Type: A Retrospective Cohort Study. *Front. Neurol.* **2022**, *13*, 833958. [CrossRef]

diagnostics

MDPI

Review

Molecular Biomarkers in Perthes Disease: A Review

Vesna Spasovski [1], Sanja Srzentić Dražilov [1], Gordana Nikčević [1], Zoran Baščarević [2,3], Maja Stojiljković [1], Sonja Pavlović [1] and Duško Spasovski [2,3,*]

[1] Institute of Molecular Genetics and Genetic Engineering, University of Belgrade, Vojvode Stepe 444a, 11010 Belgrade, Serbia
[2] School of Medicine, University of Belgrade, Dr Subotica 8, 11000 Belgrade, Serbia
[3] Institute of Orthopedics "Banjica", Mihajla Avramovića 28, 11000 Belgrade, Serbia
* Correspondence: duskosp@gmail.com; Tel.: +381-64-1455514

Abstract: Background: Perthes disease is a juvenile form of osteonecrosis of the femoral head that affects children under the age of 15. One hundred years after its discovery, some light has been shed on its etiology and the biological factors relevant to its etiology and disease severity. Methods: The aim of this study was to summarize the literature findings on the biological factors relevant to the pathogenesis of Perthes disease, their diagnostic and clinical significance, and their therapeutic potential. A special focus on candidate genes as susceptibility factors and factors relevant to clinical severity was made, where studies reporting clinical or preclinical results were considered as the inclusion criteria. PubMed databases were searched by two independent researchers. Sixty-eight articles were included in this review. Results on the factors relevant to vascular involvement and inflammatory molecules indicated as factors that contribute to impaired bone remodeling have been summarized. Moreover, several candidate genes relevant to an active phase of the disease have been suggested as possible biological therapeutic targets. Conclusions: Delineation of molecular biomarkers that underlie the pathophysiological process of Perthes disease can allow for the provision of earlier and more accurate diagnoses of the disease and more precise follow-ups and treatment in the early phases of the disease.

Keywords: Perthes disease; candidate genes; biomarkers; biological therapy

Citation: Spasovski, V.; Srzentić Dražilov, S.; Nikčević, G.; Baščarević, Z.; Stojiljković, M.; Pavlović, S.; Spasovski, D. Molecular Biomarkers in Perthes Disease: A Review. *Diagnostics* **2023**, *13*, 471. https://doi.org/10.3390/diagnostics13030471

Academic Editors: Yuli Huang, Yong Yuan and Peisong Chen

Received: 18 December 2022
Revised: 6 January 2023
Accepted: 16 January 2023
Published: 27 January 2023

1. Introduction

Perthes disease (LCPD OMIM#150600) is an idiopathic, debilitating structural disorder that affects the growing femoral head of children, mostly boys. The male-to-female ratio is reported to be 3.1:1 [1]. Annual incidence is highly variable between populations and, according to 21 studies from 16 countries, the incidence rate was found to range from 0.2 incidences per 100,000 people to 19.1 incidences per 100,000 people [2].

It is generally accepted that the main hallmark of Perthes disease is the disruption of the blood supply to the femoral capital physis, which can lead to the collapse of the femoral head, loss of sphericity and subsequent degenerative changes in the joint and surrounding tissue. The natural course of the disease goes through four phases (hip synovitis, femoral epiphysis condensation and fragmentation, reossification, and remodeling) that last several years, eventually causing permanent mechanical disturbance of the joint in varying degrees of intensity. Research conducted over a period of more than 100 years included external factors connected with socioeconomic status, such as low birth weight and the mother's smoking habits, and other factors such as obesity and hyperactivity. The elucidation of the etiology of the disease from this research brought many insights, though a definite cause is still unknown [3]. However, the results of more recent research that explored biological and physiological mechanisms have given many new answers, and we are one step closer to a better understanding of the disease's pathogenesis and more accurate treatment for this disabling disease.

Partial- or full-weight bearing restrictions and operative containment methods (femoral and/or pelvic osteotomies) are still considered standard treatment options and are in broad use worldwide [3–5]. During the last decade, a class of new therapeutics based on biomarkers has been introduced and their biological potential has been explored in animal studies. This biological therapy provides a possibility to resolve the disease in its initial phase and prevent subsequent complications, thus diminishing the need for operative procedures.

This review summarizes the results from research on biomarkers that are involved in endothelial activation and dysfunction that could be involved in impaired blood supply to the femoral head in Perthes disease. The findings of studies that examined markers relevant to angiogenesis, inflammation and apoptosis have been introduced. An overview of the familial and sporadic occurrence of collagen mutations is presented. We also discuss the possible usage of new techniques for better diagnostics of the disease and the development of biomarker-based therapeutics that directly target the cause of the disease and could be effective in the early phases of the disease, precluding the femoral head collapse of an affected hip.

2. Materials and Methods

Literature Search Strategy and Selection Criteria

The inclusion criteria for this narrative review encompassed available studies that examined factors influencing the development of Perthes disease. This included hereditary factors that could act as susceptibility factors; factors that could contribute to vascular impairment, including markers of angiogenesis, inflammation and apoptosis; and inflammatory biomarkers relevant for clinical presentation. Studies describing biological therapeutics used in preclinical models of Perthes disease were also included. Studies of any level of evidence published in peer review journals were further analyzed.

Exclusion criteria included non-English written articles, conference presentations and expert opinions. Articles without an accessible abstract were also excluded.

Online databases PubMed and Science Direct were searched by two independent authors. The searches were conducted using the following keywords: "Perthes" or "avascular necrosis of femoral head in children" combined with the terms "biomarkers", "endothelial", "inflammatory", "collagen mutations", "susceptibility", "hereditary", "inflammation", "angiogenesis", "vascular", "apoptosis", "biological therapy" and "genetic factors" in order to gain the most specific results. A time range was set from 1 January 1992. to 1 October 2022. All incongruities were discussed with senior investigators until a consensus was reached.

3. Results and Discussion

3.1. Endothelial Dysfunction in Perthes Disease

Vascular involvement in the etiology of Perthes disease has been proposed since its discovery and studied in epidemiological and experimental studies. The first genetic markers to be studied in Perthes disease were markers of thrombophilia, and their significance is described elsewhere [3,6]. In this review, we highlight endothelial markers that could be relevant to the vascular dysfunction observed in Perthes patients.

The basic building block of blood vessels is the endothelial cells (ECs). They form a permeability barrier, thus playing a primary role in the maintenance of hemostasis and blood fluidity, preventing platelet aggregation and thrombosis. They contribute to the retention of vasomotor tone, play a decisive role in angiogenesis and vasculogenesis, and are involved in immune and inflammatory responses. They also represent a multifunctional paracrine and endocrine organ [7]. For this reason, endothelial cell injury, activation and/or dysfunction play an important role in the pathogenesis of diseases that involve thrombosis, angiogenesis, vascular repair and/or inflammation, which is the case with Perthes disease.

Dysfunction of the vascular system and its life-long consequences were analyzed by Hailer et al. in a study that involved 3141 patients and 15,595 individuals without Perthes disease who were selected from the Swedish Inpatient Register [8]. The study examined Perthes patients' risk for vascular diseases during life and included subjects older than 30 years of age at the follow-up. Patients with Perthes disease were found to have significantly higher risks for blood diseases, including anemias, coagulation defects and hypertension, than the control group, implying the involvement of the vascular system in Perthes disease [8].

Anatomical differences in Perthes patients were examined in a study where the size of blood vessels was measured using the technique of flow-mediated dilation of the brachial artery in response to an ischemic stimulus [9]. A notable reduction in blood velocity and blood flow was detected in children with Perthes disease compared to controls, which was caused by the smaller artery caliber found in affected children [9].

3.1.1. VEGF

Vascular endothelial growth factors (VEGFs) are a family of signal peptides, with VEGF-A as the founding member [10]. VEGF-A is an extracellular signal protein that is produced by many cells with the main function of stimulating the formation of blood vessels during embryonic development and in adults. Angiogenesis is a process that is essential during pregnancy and in tissue growth and repair but also in disease development and progression. Among many other functions, VEGF-A acts as a key regulator of endochondral ossification [11] and, therefore, represents a potential biomarker for Perthes disease (Table 1).

Several studies explored the significance of VEGF for Perthes disease. Experimental studies on piglet and rabbit models of Perthes disease explored its role in cartilage remodeling and ossification [12,13]. Kim et al. investigated the role of VEGF in restoring endochondral ossification in the epiphyseal cartilage after ischemic necrosis in a piglet model of Perthes disease [12]. The results of this study showed ectopic expression of VEGF in the proliferative zone of the epiphyseal cartilage, with VEGF mRNA expression upregulation as early as 24 h after the induction of ischemia. This study proposes a possible role of VEGF in osteoclast recruitment and differentiation and the resorption of the necrotic cartilage and bone and suggests that VEGF is a factor that induces the restoration of endochondral ossification in the epiphyseal cartilage through the coupling of angiogenesis, cartilage remodeling and ossification after ischemic damage [12].

Similar findings came from a study on rabbits which showed increased VEGF mRNA and protein expression in the proliferative zone of epiphyseal cartilage and downregulation of VEGF in the hypertrophic zone of epiphyseal cartilage after necrosis [13].

Two studies explored serum levels of VEGF in the active phase of Perthes disease. In a Turkish study conducted by Sezgin et al., serum VEGF levels were compared between 28 patients and 25 controls. There was no significant difference in serum VEGF levels between the two groups [14]. Nevertheless, a study on Indian children showed that median serum levels of VEGF were significantly lower in the patient group compared to the normal, healthy, control group [15]. The study also examined the level of VEGF at four different stages of the disease but failed to find any significant differences. If future studies confirm the significant role of VEGF, especially if differences between various stages of the disease are established, the serum VEGF level may act as a marker for the follow-up of the disease and a possible therapeutic target for Perthes disease.

3.1.2. eNOS

Among markers that could influence endothelial cell fate in the context of osteonecrosis of the femoral head, the endothelial nitric oxide synthase gene (eNOS) has been much investigated. eNOS is primarily responsible for the generation of NO in the vascular endothelium [16]. It is involved in a variety of physiologic processes, including angiogenesis, thrombosis and bone turnover, making it a promising marker for Perthes disease.

There are three polymorphisms of the eNOS gene that have been identified thus far, including a −786 T > C (rs2070744) polymorphism in the promoter region, an 894 G > T (rs1799983) polymorphism in exon 7 and a variable number of tandem 4a4b repeats (rs61722009) in intron 4 [17,18]. A Chinese study that examined 4a4b polymorphism in intron 4 and 894 G > T polymorphism in exon 7 found that both polymorphisms may be risk factors for Perthes disease [19]. An Iranian study that examined all three polymorphisms of eNOS highlighted the association of 894 G > T and −786 T > C but not the 4a4b polymorphism with susceptibility to Perthes disease [20].

3.1.3. Plasma Exosomes and Microparticles

The number of endothelial progenitor cells in circulation represents a valuable tool for the justification of the state of the endothelium and might also be explored in the context of physiological and pathological events. Additionally, circulating endothelial microparticles (EMPs) and exosomes, which represent extracellular vesicles secreted by endothelial cells, may vary in number and content during normal and pathological conditions, and could have clinical significance. The involvement of endothelial cells in the development of Perthes disease was recently studied by the analysis of plasma extracellular vesicles [21]. The study showed an increased number of EMPs in the circulation of Perthes patients and suggested that the mechanism of action of these circulating EMPs might be endothelial cell apoptosis, endothelial dysfunction and deregulated angiogenesis in Perthes patients [21]. In a study by Huang et al., epigenetic factors that might contribute to Perthes disease were examined in the plasma exosomes of both Perthes patients and controls [22]. They showed the differential expression of several miRNAs involved in endothelial cell dysfunction (miR-3133, miR-4644, miR-4693-3p and miR-4693-5p) and osteoclastogenesis (miR-3133, miR-4693-3p, miR-4693-5p, miR-141-3p and miR-30a) that might contribute to the development of Perthes disease [22].

3.2. *Inflammatory Factors in Perthes Disease*

Inflammation plays a fundamental role in Perthes disease as significant and persistent synovitis is present throughout the course of the disease [23]. Synovitis causes cartilage edema, the deterioration of cartilage mechanical properties, cartilage hypermetabolism and cartilage hypertrophy. It is the primary cause of hip pain in affected children. Synovitis has a negative impact on bone formation during the repair process and is involved in the subsequent ischemic osteonecrosis present in Perthes patients. During the fragmentation phase of the disease, an unsuccessful attempt of the body to repair brittle bone tissue of the collapsing femoral head occurs; instead of forming normal bone matter with high strength and low compressibility, a fibrovascular tissue with fragments of necrotic bone is formed [24]. Yet, only in recent years have researchers gained significant data about the nature of the inflammatory process in Perthes disease.

The process of bone turnover represents a fine balance between bone formation and bone resorption, which is based on the activity of two major cell populations in the bone: osteoclasts (OCs) and osteoblasts (OBs) [25–27]. It is known that IL-6 positively regulates OCs [28,29] and negatively regulates OBs [29,30]. Its elevation could trigger the uncoupling of bone resorption and bone formation, which is characteristic of this stage of the disease and leads to ischemic osteonecrosis of the developing hip.

The significance of genetic markers of inflammation for the development and activity of Perthes disease, specifically IL-6, TLR4, TNF-α and IL-3, were first studied in associative studies [31,32]. Two promoter polymorphisms in IL-6, 174 G > C and −597 G > A, and two SNVs that encode mutations in the ectodomain of TLR4, D299G and T399I, were examined [31]. The significance of TLR4 was not shown, but heterozygous carriers of promoter variants *IL-6* −174 G > C and −597 G > A had a lower chance of developing Perthes disease than carriers of both homozygote genotypes [31]. Next, variants in TNF-α −308 G > A and IL-3 −16 C > T were analyzed [32]. TNF-α has been recognized as a skeletal catabolic agent that stimulates osteoclastogenesis while simultaneously inhibiting osteoblast function [33], and gene variants in IL-3 were shown to be associated with rheumatoid arthritis [34]. No significant differences between the frequencies of analyzed gene variants in TNF-α or IL-3 genes between the patient and control groups were found [32]. An Iranian study from 2019 examined the significance of *IL-6* −174 G > C and −572 G > C genotypes in developing Perthes disease in Iranian children [35]. The study showed that the mutant homozygote genotype CC of IL-6 −174 G > C polymorphism was associated with an increased risk of Perthes disease in the Iranian population.

Several studies examined the expression of inflammatory cytokines in the serum of Perthes patients and animal models of the disease.

In a study by Kamiya et al., 27 inflammatory cytokines and related factors were analyzed in the synovial fluid of 13 Perthes patients and significant overexpression of IL-6 was observed [36]. In addition, this study showed that in the active stage of Perthes disease IL-6 is a key inflammatory cytokine present in the synovial fluid, with strong impact on bone turnover. The detailed mechanism of action of IL-6 was studied by Yamaguchi et al., and the relevance of HIF-1α in the activation of IL-6 was designated [23]. Namely, when osteonecrotic articular chondrocytes were cultured under hypoxic conditions, a significant increase in the expression of *IL-1β* and *TNF-α* genes was observed, which was inhibited by the IL-6 receptor blocker tocilizumab. This work highlighted tocilizumab as a potential therapeutic in the treatment of synovitis.

A murine model of Perthes disease was used to further delineate the role of IL-6 in the process of revascularization and new bone formation following ischemic osteonecrosis [37]. Ischemic osteonecrosis was surgically induced in wild-type and IL-6 knockout (KO) mice. Significant differences were observed in OB and OC numbers and bone formation and mineral apposition rates in the osteonecrotic side of the IL-6 KO mice compared to the wild-type mice.

TLR4

In order to delineate the mechanism of the inflammatory cascade responsible for the disturbed process of bone remodeling and the receptor mechanisms involved in sensing the necrotic bone observed in this phase, the role of macrophages was studied in detail in a piglet model of Perthes disease [38]. The study showed that macrophages that are exposed to soluble factors from the necrotic bone were activated (M1 phenotype) and showed increased secretion of proinflammatory cytokines TNF-α, IL1-β and IL-6. The upregulation of pattern recognition receptor TLR4 in activated M1 macrophages was determined and the relevance of the TLR4 signaling pathway for proliferation, migration and proinflammatory cytokine responses in activated macrophages was established [38]. It was suggested that the process of inflammation, driven by proinflammatory mediators, could be responsible for the distortion of the regenerative process, where instead of new bone formation, the repair process aids the formation of fibrotic tissue. The relevance of the TLR4 signaling pathway in this process was ultimately confirmed by knockdown experiments using the CRSPR-Cas9 approach and the TLR4 inhibitor, TAK242 [38]. TLR4 is a known receptor where both infectious and non-infectious stimuli converge to elicit a pro-inflammatory response. It is involved in the lipopolysaccharide sensing of gram-negative bacteria and binds the endogenous molecules produced as a result of tissue injury [39]. Hence, these results illuminated a new paradigm for therapy that could be based on TLR4 inhibitors. Since the TLR4 signaling pathway regulates OB and OC formation [40], the damage caused by its misregulation could be repaired using such an approach.

3.3. Role of Apoptotic Factors in Uncoupled Bone Metabolism

Several studies explored the mechanism of programmed cell death in the avascular necrosis of the femoral head (ANFH) and reported that the processes leading to the death of femoral head cells involve an increased rate of apoptosis rather than solely bone cell necrosis [32,41–43]. The importance of the inner apoptotic pathway was also observed in the peripheral blood mononuclear cells of Perthes patients, where the upregulation of *BAX* gene expression levels and an increased *BAX*/*BCL2* ratio were first shown by Srzentic et al. [32]. Results from a recent study by Zhu et al. revealed elevated levels of *BAX* mRNA and protein in the cartilage, serum and chondrocytes of Perthes patients compared with a healthy control group. Moreover, this study examined the underlying mechanism of the elevation of *BAX* gene expression and showed that miRNA-214 is a direct regulator of the BAX gene [42]. It showed that the downregulation of miR-214 promotes chondrocyte viability and decreases apoptosis via the downregulation of BAX. Taken together, these results indicated that the expression levels of miR-214 and *BAX* might represent reliable biomarkers and potential therapeutic targets in the treatment of Perthes disease.

3.4. Familial Clustering and Tween Studies of Perthes Disease

Since the first mentions of Perthes disease described among siblings [2,3], familial cases have been occasionally reported, but statistical analysis usually negates the influence of genetic factors on the genesis of Perthes' disease. One of the largest studies that investigated the familial incidence of Perthes disease was conducted among 310 Scottish patients [44]. Of the 310 patients with a family history of the disease (246 male, 64 female; male to female sex ratio 3.8:1), 35 had bilateral disease (11.3%). Five dizygous and one monozygous twin pairs were present among the patients, with only one sibling affected in each of them. This study highlighted low overall proportions of affected relatives, with only a 1.6% frequency of appearance among siblings.

The largest study on twins explored the incidence of Perthes disease among 31 twin pairs of Danish children with at least one affected sibling [45]. Nevertheless, among the 10 pairs of monozygotic, 51 dizygotic and 20 unknown zygosity (UZ) twins described in this study, only 4 concordant pairs were found: 2 of each in the dizygotic and UZ groups. As expected, statistical analysis showed a low probability of a hereditary component of this disorder.

Interestingly, there are families that are prone to joint diseases similar to Perthes disease, such as avascular necrosis or secondary osteoarthritis. A common marker for this type of heritable susceptibility is collagen type II protein.

3.5. COL2A1 Mutations

Collagen type II is a fibrillary structural protein mainly found in hyaline and articular cartilage, but also in other tissues, such as the intervertebral disc (nucleus pulposus), retina (as a part of the clear gel that fills the eyeball (vitreous humor)), sclera and lens of the eye, nose, inner ear, and external ear [46]. As with other types of collagens, collagen type II is a triple helix consisting of three type α1(II) monomers. Fibrils of collagen type II are cross-linked in the viscous proteoglycan matrix with collagen type IX, forming a structure that provides the tissue with strength and compressibility that allows it to resist large deformations in shape, giving joints the ability to absorb shocks [47]. Mutations are dispersed throughout the entire gene, and no mutational hot spots within the *COL2A1* gene have been identified. Several publications reported the presence of missense mutations in the *COL2A1* gene (COL2A1 OMIM#120140) in Perthes patients in sporadic but also in familial cases [48–51]. All mutations found in Perthes patients are missense types of mutations leading to the substitution of the amino acid glycine with another, usually larger, amino acid, predominantly serine (Figure 1). Glycine is a small, nonpolar amino acid, and its substitution with a larger amino acid leads to structural changes in the triple-helical collagen type II structure [47].

The recurrent p.Gly1170Ser mutation (c.3665 G > A) in exon 50 of the *COL2A1* gene was first reported in two Taiwanese families with ANFH [52] and then in Japanese and Chinese families with Perthes disease [48,49]. Miyamoto et al. described three generations containing seven family members affected by Perthes disease, in whom the p Gly1170Ser mutation was detected [49]. Another comprehensive genetic study was conducted in order to trace the same mutation through a five-generation Chinese family affected by three different joint diseases [48]. Of the forty-two family members analyzed in this study, sixteen members were affected by one of three joint diseases: Perthes disease (five affected members), avascular necrosis (AVN) of the femoral head (six affected members) or osteoarthritis (OA) (five affected members). Genetic analysis showed the presence of the c.3665 G > A mutation of the *COL2A1* gene in the heterozygous state in all affected family members, with the exception of one 8-year-old carrier, who was clinically asymptomatic. The authors concluded that this genetic variant was a disease-causing mutation in this family.

In another family study, a similar phenotypic presentation was described as a consequence of another mutation in the *COL2A1* gene, c.1888 G > A, p.Gly630Ser in exon 29 [51]. In order to detect this disease-causing mutation and trace its segregation, forty-five members of a four-generation Chinese family were recruited and analyzed. The study found that six family members were affected by either Perthes disease or ANFH and, in all of their DNA samples, the mutation c.1888 G > A of the *COL2A1* gene was detected. The same mutation was detected in a 7-year-old girl who was asymptomatic. The mutation was absent in samples from healthy family members and controls.

These two family studies are examples of the presence of a disease-causing mutation in the *COL2A1* gene, with a different onset of clinical manifestation. Factors that could influence the closure of the femoral head epiphysis, such as hormonal status, body weight or circulation problems, appear to be critical factors in determining phenotypic expression and defining the age of the onset of disease phenotypes.

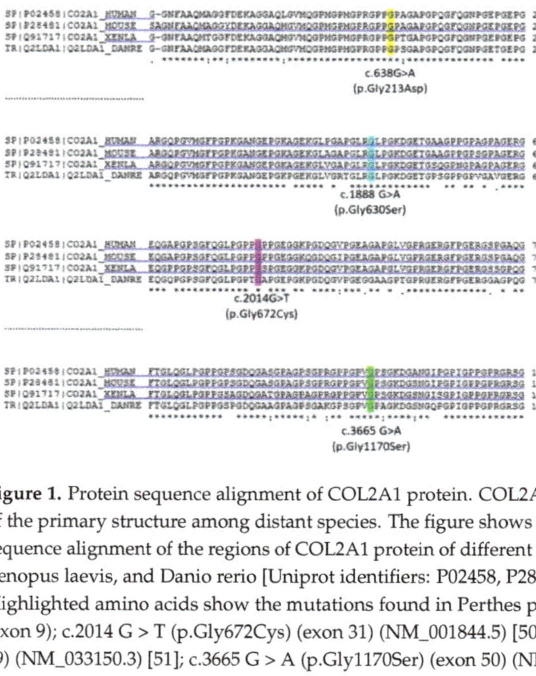

Figure 1. Protein sequence alignment of COL2A1 protein. COL2A1 proteins show high conservation of the primary structure among distant species. The figure shows CLUSTAL OMEGA (1.2.4) multiple sequence alignment of the regions of COL2A1 protein of different organisms: Human, Mus musculus, Xenopus laevis, and Danio rerio [Uniprot identifiers: P02458, P28481, Q91717, Q2LDA, respectively]. Highlighted amino acids show the mutations found in Perthes patients. c.638 G > A (p.Gly213Asp) (exon 9); c.2014 G > T (p.Gly672Cys) (exon 31) (NM_001844.5) [50]; c.1888 G > A (p.Gly630Ser) (exon 29) (NM_033150.3) [51]; c.3665 G > A (p.Gly1170Ser) (exon 50) (NM_033150.3) [48,49]. The alignment of the whole protein sequence is given in Supplementary Material.

COL2A1 Mutations in Patients with Bilateral Presentation

Mutations in the *COL2A1* gene were rarely found in sporadic cases [53]. The only study where mutations in the *COL2A1* gene were found in sporadic cases was the study by Kannu et al., which described two children with bilateral presentations of Perthes disease [50]. Two different mutations were found in probands (c.2014 G > T (p.Gly672Cys) and c.638 G > A (p.Gly213Asp)), but not in their family members. This very interesting finding highlights the need to seek COL2A1 mutations in bilateral cases of Perthes disease, which could be beneficial for differential diagnosis and management of patients.

Table 1. Studies on biomarkers in Perthes disease.

Biomarker	SNV	Experimental Model	Reference
VEGF		piglet model	Kim et al. [12]
VEGF		rabbit model	Li et al [13]
VEGF		human serum	Sezgin et al. [14]
VEGF		human serum	Tiwari et al. [15]
eNOS	894 G > T 4a4b	human peripheral blood	Zhao et al [19]
eNOS	−786 T > C; 894 G > T 4a4b	human peripheral blood	Azarpira et al. [20]
EMPs vesicles		human plasma	Li et al. [21]
exosomes		human plasma	Huang et al. [22]
IL-6 *TLR4*	174 G > C; −597 G > A D299G; T399I	human peripheral blood	Srzentic et al. [31]
TNF-α *IL-3*	−308 G > A −16 C > T	human peripheral blood	Srzentic et al. [32]
IL-6	−174 G > C; −572 G > C	human peripheral blood	Akbarian-Bafghi et al. [35]
IL-6		human synovial fluid	Kamiya et al. [36]
IL-6		piglet model	Yamaguchi et al. [23]
IL-6		murine model	Kuroyanagi et al. [37]
TLR4		piglet model	Adapala et al. [38]
BAX, BCL-2		human peripheral blood	Srzentic et al. [32]
Bax, miR-214		human cartilage, serum and chondrocytes	Zhu et al [42]
COL2A1	c.3665 G > A	human peripheral blood	Su et al. [48]
COL2A1	c.3665 G > A	human peripheral blood	Miyamoto et al. [49]
COL2A1	c.638 G > A c.2014 G > T	human peripheral blood	Kannu et al. [50]
COL2A1	c.1888 G > A	human peripheral blood	Li et al. [51]

3.6. Biological Therapy for Perthes Disease

The accumulation of data on particular pathophysiological changes at various stages of the disease makes biological therapy a promising option for the treatment of patients (Table 2). Bone turnover relies on biogenesis and the activity of OB and CC, and their number and function are tightly regulated. In this respect, the receptor activator of nuclear factor κ-B ligand (RANKL) plays a fundamental role, and the fine-tuning of this process is controlled by its receptors: a transmembrane receptor for RANK and a soluble decoy receptor, osteoprotegerin (OPG) [24].

The mode of action of therapies that can target OC-mediated bone resorption and prevent femoral head deformity is elucidated. The fragmentation (resorptive) stage of Perthes disease lasts for 1–2 years [24] and agents that could attenuate bone resorption and stimulate bone formation, such as bisphosphonates (BP) and RANKL inhibitors, could be beneficial at this stage. Data about the usage of BP (Ibandronate and Zoledronic acid) from a pig model of ischemic necrosis [54], rat model of Perthes disease [55] and rat model of traumatic osteonecrosis [56] have shown that systemically administered BP has a protective role regarding the femoral head. The mode of action of these drugs is the stimulation of OC apoptosis, without affecting the inhibition of OC or new bone formation [57]. Nevertheless, BP have many limitations, especially when used with children. Their long half-life can have inhibitory effects on bone growth, and their systemic administration has limited access to the necrotic bone [57]. An ongoing clinical trial in Australia (ACTRN12610000407099) will compare the results from the intravenous administration of zolendronic acid in 100 children receiving the standard treatment [58].

However, subcutaneous injections of osteoprotegerin (human OPG-Fc) have been shown to have a protective role in the femoral head in a piglet model of Perthes disease [24]. Results indicate that RANKL inhibition produced a decreased total number of OC in repair tissue. Despite the finding that RANKL inhibitors do not bind to bone and that their effect is reversible since they are rapidly cleared from circulation, RANKL inhibition can be considered a novel therapeutic strategy for femoral healing deformity diseases.

3.6.1. Anti-IL-6 Therapy

The influence of IL-6 on bone turnover and the effects of the blockade of the IL-6 receptor have been proven in in vitro and animal models and clinical studies of rheumatoid and juvenile arthritis [59–62]. Since the first confirmation of the role of IL-6 in Perthes disease [31,36], an in vitro study showed that the expression of IL-6 is HIF-1 dependent and is attenuated by the anti-IL-6 monoclonal antibody, tocilizumab [23]. These findings converged in the preclinical study of Ren et al., who used tocilizumab on a piglet model of Perthes disease to explore its influence on hip synovitis and bone healing [63]. Biweekly intravenous injections in doses of 15 to 20 mg/kg were applied in this preclinical setting and, after examination of the experimental and control groups, encouraging results were obtained. Significantly higher bone volume and a lower number of OC per bone surface were shown in the tocilizumab group compared to the control group. Additionally, the number of synovial macrophages and vessels was significantly lower in the tocilizumab group compared to the control group, showing that both the effect on attenuated osteoclastogenesis and the inhibition of angiogenesis had a positive influence on alleviated hip synovitis, supporting bone healing after treatment.

3.6.2. Cellular Therapy as an Option for the Treatment of Perthes Disease

Constant improvements in understanding the biology and course of action of mesenchymal stem cells (MSCs) have encouraged their application and modernized the field of skeletal regenerative medicine. Pediatric cases of bone and joint diseases could potentially have enormous benefits for treatment using stem cells. Severe congenital impairments of bone metabolism, such as osteogenesis imperfecta (OI), have already been successfully treated using MSCs prenatally [64,65] and postnatally [66]. MSCs could be a promising option for diseases where no other treatment options are available and for chronic diseases that require lifelong therapy or where standard therapy gives unsatisfactory results.

Table 2. Biological treatments of Perthes disease.

Therapy	Experimental Model	Reference
Osteoprotegerin	piglet model	Kim et al. [24]
Ibandronate	pig model	Kim et al. [54]
Zoledronic acid	rat model	Little et al. [55]
Zoledronic acid	rat model	Little et al. [56]
Tocilizumab	piglet model	Ren et al. [63]
ADSC	rabbit model	Wang et al. [67]

Usage of MSCs in the pediatric population must be preceded by sufficient and encouraging results from preclinical trials and safety and efficacy clinical studies and adequate results from treatments of similar diseases in adults. A pioneering attempt in the treatment of Perthes disease using stem cells was described in a study by Wang et al., where the authors used a combination of drilling through the growth plate with injections of adipose-derived mesenchymal stem cells (ADSC) and bone morphogenetic protein 2 (BMP-2) on an ischemic model of epiphyseal ischemic necrosis of the femoral head in juvenile rabbits [67]. The treatment showed new bone formation and prevented the collapse of the femoral head epiphysis in an experimental model of Perthes disease.

There are also available results from the treatment of other pediatric joint diseases, such as juvenile idiopathic arthritis (JIA), which is a form of juvenile rheumatoid arthritis. Ten patients who were treated using umbilical cord-derived MSCs were followed for 2 years. No side effects were detected. The patients' symptoms improved and the dose of antirheumatic drugs decreased during the follow-up period with good tolerance [68]. This is an important result, suggesting a new paradigm of treatment for this severe condition.

4. Conclusions

The delineation of molecular biomarkers that underlie the pathophysiological process of Perthes disease allows for the possibility to provide earlier and more accurate diagnoses and more precise follow-ups and treatment in the early phases of the disease. In this respect, techniques that analyze extracellular particles from the patient's plasma could be used as a new tool for early diagnosis and follow-up. Markers involved in endothelial cell dysfunction; markers of angiogenesis, inflammation and apoptosis; and genes involved in bone turnover are possible new targets for biological therapy. In the years to come, high throughput methods of sequencing, such as whole exome sequencing (WES) and whole genome sequencing (WGS), will provide additional data on other genetic loci that could be of interest for the development of Perthes disease.

Supplementary Materials: The following supporting information can be downloaded at: https://www.mdpi.com/article/10.3390/diagnostics13030471/s1.

Author Contributions: Conceptualization, D.S. and S.P.; data curation V.S. and S.S.D.; methodology, V.S. and M.S.; validation, D.S. and G.N.; formal analysis, M.S. and Z.B.; investigation, V.S., S.S.D., M.S. and Z.B.; writing—original draft preparation, V.S. and D.S.; writing—review and editing, V.S., S.S.D., G.N., Z.B., M.S. and S.P.; project administration, M.S. and S.P.; funding acquisition S.P. All authors have read and agreed to the published version of the manuscript.

Funding: This research was funded by Ministry of Education, Science and Technological Development, Republic of Serbia, grant numbers III41004 and 451-03-68/2022-14/200042.

Institutional Review Board Statement: Not applicable.

Informed Consent Statement: Not applicable.

Data Availability Statement: The data presented in this study are publicly available on PubMed (https://pubmed.ncbi.nlm.nih.gov/ (accessed on 1 October 2022)), and in Supplementary Material.

Conflicts of Interest: The authors declare no conflict of interest.

References

1. Johansson, T.; Lindblad, M.; Bladh, M.; Josefsson, A.; Sydsjö, G. Incidence of Perthes' disease in children born between 1973 and 1993. *Acta Orthop.* **2016**, *88*, 96–100. [CrossRef] [PubMed]
2. Perry, D.C.; Machin, D.M.G.; Pope, D.; Bruce, C.E.; Dangerfield, P.; Platt, M.J.; Hall, A.J. Racial and Geographic Factors in the Incidence of Legg-Calvé-Perthes' Disease: A Systematic Review. *Am. J. Epidemiol.* **2012**, *175*, 159–166. [CrossRef] [PubMed]
3. Pavone, V.; Chisari, E.; Vescio, A.; Lizzio, C.; Sessa, G.; Testa, G. Aetiology of Legg-Calvé-Perthes disease: A systematic review. *World J. Orthop.* **2019**, *10*, 145–165. [CrossRef]
4. Braito, M.; Wolf, S.; Dammerer, D.; Giesinger, J.; Wansch, J.; Biedermann, R. Global differences in the treatment of Legg-Calvé-Perthes disease: A comprehensive review. *Arch. Orthop. Trauma Surg.* **2021**, *141*, 1–16. [CrossRef] [PubMed]
5. Kaneko, H.; Kitoh, H.; Mishima, K.; Matsushita, M.; Hattori, T.; Noritake, K.; Ishiguro, N.; Yoshihashi, Y. Comparison of surgical and nonsurgical containment methods for patients with Legg-Calvé-Perthes disease of the onset ages between 6.0 and 8.0 years: Salter osteotomy versus a non-weight-bearing hip flexion-abduction brace. *J. Pediatr. Orthop. B* **2020**, *29*, 542–549. [CrossRef] [PubMed]
6. Asadollahi, S.; Neamatzadeh, H.; Namiranian, N.; Sobhan, M.R. Genetics of Legg-Calvé-Perthes Disease: A Review Study. *J. Pediatr. Rev.* **2021**, *9*, 301–308. [CrossRef]
7. E Sumpio, B.; Riley, J.T.; Dardik, A. Cells in focus: Endothelial cell. *Int. J. Biochem. Cell Biol.* **2002**, *34*, 1508–1512. [CrossRef]

8. Hailer, Y.D.; Montgomery, S.M.; Ekbom, A.; Nilsson, O.S.; Bahmanyar, S. Legg-Calvé-Perthes Disease and Risks for Cardiovascular Diseases and Blood Diseases. *Pediatrics* **2010**, *125*, e1308–e1315. [CrossRef]
9. Perry, D.C.; Green, D.J.; Bruce, C.E.; Pope, D.; Dangerfield, P.; Platt, M.J.; Hall, A.J.; Jones, H. Abnormalities of Vascular Structure and Function in Children with Perthes Disease. *Pediatrics* **2012**, *130*, e126–e131. [CrossRef]
10. Holmes, D.I.; Zachary, I. The vascular endothelial growth factor (VEGF) family: Angiogenic factors in health and disease. *Genome Biol.* **2005**, *6*, 209. [CrossRef]
11. Yang, Y.-Q.; Tan, Y.-Y.; Wong, R.; Wenden, A.; Zhang, L.-K.; Rabie, A.B.M. The role of vascular endothelial growth factor in ossification. *Int. J. Oral Sci.* **2012**, *4*, 64–68. [CrossRef] [PubMed]
12. Kim, H.K.; Bian, H.; Randall, T.; Garces, A.; Gerstenfeld, L.C.; A Einhorn, T. Increased VEGF Expression in the Epiphyseal Cartilage After Ischemic Necrosis of the Capital Femoral Epiphysis. *J. Bone Miner. Res.* **2004**, *19*, 2041–2048. [CrossRef] [PubMed]
13. Li, W.; Fan, Q.; Ma, B.; Ren, M.; Li, H.; Gou, Y.; Qian, J. An animal model of Perthes disease and an experimental research of VEGF expression. *Zhongguo Xiu Fu Chong Jian Wai Ke Za Zhi* **2008**, *22*, 814–819. [PubMed]
14. Sezgin, H.; Gulman, B.; Cirakli, A.; Bedir, A.; Usta, D.; Coskun, S.; Yildiran, A. Effects of circulating endothelial progenitor cells, serum vascular endothelial growth factor and hypogammaglobulinemia in Perthes disease. *Acta Orthop. Traumatol. Turc.* **2014**, *48*, 628–634. [CrossRef] [PubMed]
15. Tiwari, V.; Poudel, R.; Khan, S.; Mehra, S.; Chauhan, S.S.; Raje, A. Is VEGF under-expressed in Indian children with Perthes disease? *Musculoskelet. Surg.* **2018**, *102*, 81–85. [CrossRef]
16. Fish, J.E.; Marsden, P.A. Endothelial nitric oxide synthase: Insight into cell-specific gene regulation in the vascular endothelium. *Cell. Mol. Life Sci.* **2005**, *63*, 144–162. [CrossRef] [PubMed]
17. Chang, J.; Pan, F.; Tang, Q.; Wu, W.; Chen, M.; Lu, C.; Ding, H.; Hu, L.; Chen, D.; Xia, Y.; et al. eNOS gene T786C, G894T and 4a4b polymorphisms and male infertility susceptibility: A me-ta-analysis. *Andrologia* **2017**, *49*, e12646. [CrossRef] [PubMed]
18. Xie, X.; Shi, X.; Xun, X.; Rao, L. Endothelial nitric oxide synthase gene single nucleotide polymorphisms and the risk of hypertension: A meta-analysis involving 63,258 subjects. *Clin. Exp. Hypertens.* **2017**, *39*, 175–182. [CrossRef] [PubMed]
19. Zhao, Y.; Liao, S.; Lu, R.; Dang, H.; Zhao, J.; Ding, X. Endothelial nitric oxide synthase gene polymorphism is associated with Legg-Calvé-Perthes disease. *Exp. Ther. Med.* **2016**, *11*, 1913–1917. [CrossRef]
20. Azarpira, M.R.; Ghilian, M.M.; Sobhan, M.R.; Mahdinezhad-Yazdi, M.; Aghili, K.; Ahrar, H.; Neamatzadeh, H. Association of eNOS 27-bp VNTR, 894G>T and 786T>C polymorphisms with susceptibility to Legg-Calve-Perthes Disease in Iranian children. *J. Orthop.* **2019**, *16*, 137–140. [CrossRef]
21. Li, B.; Huang, Q.; Lin, C.; Lu, R.; Wang, T.; Chen, X.; Liu, Z.; Liu, Y.; Wu, J.; Wu, Y.; et al. Increased circulating CD31+/CD42b-EMPs in Perthes disease and inhibit HUVECs angiogenesis via endothelial dysfunction. *Life Sci.* **2021**, *15*, 118749. [CrossRef] [PubMed]
22. Huang, Q.; Li, B.; Lin, C.; Chen, X.; Wang, T.; Lu, J.; Liu, Y.; Lu, R.; Liao, S.; Ding, X. MicroRNA sequence analysis of plasma exosomes in early Legg-Calvé-Perthes disease. *Cell. Signal.* **2022**, *91*, 110184. [CrossRef] [PubMed]
23. Yamaguchi, R.; Kamiya, N.; Adapala, N.S.; Drissi, H.; Kim, H.K. HIF-1-Dependent IL-6 Activation in Articular Chondrocytes Initiating Synovitis in Femoral Head Ischemic Osteonecrosis. *J. Bone Jt. Surg.* **2016**, *98*, 1122–1131. [CrossRef] [PubMed]
24. Kim, H.K.; Morgan-Bagley, S.; Kostenuik, P. RANKL Inhibition: A Novel Strategy to Decrease Femoral Head Deformity After Ischemic Osteonecrosis. *J. Bone Miner. Res.* **2006**, *21*, 1946–1954. [CrossRef]
25. Weitzmann, M.N.; Pacifici, R. The role of T lymphocytes in bone metabolism. *Immunol. Rev.* **2005**, *208*, 154–168. [CrossRef]
26. Teti, A. Bone Development: Overview of Bone Cells and Signaling. *Curr. Osteoporos. Rep.* **2011**, *9*, 264–273. [CrossRef]
27. Bonewald, L.F. The amazing osteocyte. *J. Bone Miner. Res.* **2011**, *26*, 229–238. [CrossRef]
28. Tamura, T.; Udagawa, N.; Takahashi, N.; Miyaura, C.; Tanaka, S.; Yamada, Y.; Koishihara, Y.; Ohsugi, Y.; Kumaki, K.; Taga, T. Soluble interleukin-6 receptor triggers osteoclast formation by interleukin 6. *Proc. Natl. Acad. Sci. USA* **1993**, *90*, 11924–11928. [CrossRef]
29. De Benedetti, F.; Rucci, N.; Del Fattore, A.; Peruzzi, B.; Paro, R.; Longo, M.; Vivarelli, M.; Muratori, F.; Berni, S.; Ballanti, P.; et al. Impaired skeletal development in interleukin-6-transgenic mice: A model for the impact of chronic inflammation on the growing skeletal system. *Arthritis Rheum.* **2006**, *54*, 3551–3563. [CrossRef]
30. Kaneshiro, S.; Ebina, K.; Shi, K.; Higuchi, C.; Hirao, M.; Okamoto, M.; Koizumi, K.; Morimoto, T.; Yoshikawa, H.; Hashimoto, J. IL-6 negatively regulates osteoblast differentiation through the SHP2/MEK2 and SHP2/Akt2 pathways in vitro. *J. Bone Miner. Metab.* **2014**, *32*, 378–392. [CrossRef] [PubMed]
31. Srzentic, S.; Spasovski, V.; Spasovski, D.; Zivkovic, Z.; Matanovic, D.; Bascarevic, Z.; Terzic-Supic, Z.; Stojiljkovic, M.; Karan-Djurasevic, T.; Stankovic, B.; et al. Association of gene variants in TLR4 and IL-6 genes with Perthes disease. *Srp. Arh. Celok. Lek.* **2014**, *142*, 450–456. [CrossRef] [PubMed]
32. Srzentić, S.; Nikčević, G.; Spasovski, D.; Baščarević, Z.; Živković, Z.; Terzic-Šupić, Z.; Matanović, D.; Djordjević, V.; Pavlović, S.; Spasovski, V. Predictive genetic markers of coagulation, inflammation and apoptosis in Perthes disease—Serbian experience. *Eur. J. Pediatr.* **2015**, *174*, 1085–1092. [CrossRef] [PubMed]
33. Nanes, M.S. Tumor necrosis factor-alpha: Molecular and cellular mechanisms in skeletal pathology. *Gene* **2003**, *4*, 1–15. [CrossRef] [PubMed]

34. Yamada, R.; Tanaka, T.; Unoki, M.; Nagai, T.; Sawada, T.; Ohnishi, Y.; Tsunoda, T.; Yukioka, M.; Maeda, A.; Suzuki, K.; et al. Association between a Single-Nucleotide Polymorphism in the Promoter of the Human Interleukin-3 Gene and Rheumatoid Arthritis in Japanese Patients, and Maximum-Likelihood Estimation of Combinatorial Effect That Two Genetic Loci Have on Susceptibility to the Disease. *Am. J. Hum. Genet.* **2001**, *68*, 674–685. [CrossRef]
35. Akbarian-Bafghi, M.J.; Dastgheib, S.A.; Morovati-Sharifabad, M.; Sobhan, M.R.; Moghimi, M.; Mahdinezhad-Yazdi, M.; Lookzadeh, M.H.; Khajehnoori, S.; Neamatzadeh, H. Association of IL-6-174G > C and -572G > C Polymorphisms with Risk of Legg-Calve-Perthes Disease in Iranian Children. *Fetal Pediatr. Pathol.* **2019**, *22*, 1–8. [CrossRef]
36. Kamiya, N.; Yamaguchi, R.; Adapala, N.S.; Chen, E.; Neal, D.; Jack, O.; Thoveson, A.; Gudmundsson, P.; Brabham, C.; Aruwajoye, O.; et al. Legg-Calvé-Perthes disease produces chronic hip synovitis and elevation of in-terleukin-6 in the synovial fluid. *J. Bone Miner. Res.* **2015**, *30*, 1009–1013. [CrossRef]
37. Kuroyanagi, G.; Adapala, N.S.; Yamaguchi, R.; Kamiya, N.; Deng, Z.; Aruwajoye, O.; Kutschke, M.; Chen, E.; Jo, C.; Ren, Y.; et al. Interleukin-6 deletion stimulates revascularization and new bone formation following ischemic osteonecrosis in a murine model. *Bone* **2018**, *116*, 221–231. [CrossRef]
38. Adapala, N.; Yamaguchi, R.; Phipps, M.; Aruwajoye, O.; Kim, H.K.W. Necrotic Bone Stimulates Proinflammatory Responses in Macrophages through the Activation of Toll-Like Receptor 4. *Am. J. Pathol.* **2016**, *196*, 2987–2999. [CrossRef]
39. Molteni, M.; Gemma, S.; Rossetti, C. The Role of Toll-Like Receptor 4 in Infectious and Noninfectious Inflammation. *Mediat. Inflamm.* **2016**, *2016*, 1–9. [CrossRef]
40. Nakamura, H.; Fukusaki, Y.; Yoshimura, A.; Shiraishi, C.; Kishimoto, M.; Kaneko, T.; Hara, Y. Lack of Toll-like receptor 4 decreases lipopolysaccharide-induced bone resorption in C3H/HeJ mice in vivo. *Oral Microbiol. Immunol.* **2008**, *23*, 190–195. [CrossRef]
41. Calder, J.D.F.; Buttery, L.; Revell, P.A.; Pearse, M.; Polak, J.M. Apoptosis—A significant cause of bone cell death in osteonecrosis of the femoral head. *J. Bone Jt. Surg.* **2004**, *86*, 1209–1213. [CrossRef] [PubMed]
42. Zhu, H.; Qi, X.; Liu, Y.; Liao, W.; Sun, X.; Tang, Y. The role and underlying mechanisms of microRNA-214 in Legg-Calve-Perthes disease. *Mol. Med. Rep.* **2019**, *20*, 685–692. [PubMed]
43. Zhang, W.; Yuan, Z.; Pei, X.; Ma, R. In vivo and in vitro characteristic of HIF-1α and relative genes in ischemic femoral head necrosis. *Int. J. Clin. Exp. Pathol.* **2015**, *8*, 7210–7216.
44. Wynne-Davies, R.; Gormley, J. The aetiology of Perthes' disease. Genetic, epidemiological and growth factors in 310 Edinburgh and Glasgow patients. *J. Bone Jt. Surg.* **1978**, *60*, 6–14. [CrossRef] [PubMed]
45. Metcalfe, D.; Van Dijck, S.; Parsons, N.; Christensen, K.; Perry, D.C. A Twin Study of Perthes Disease. *Pediatrics* **2016**, *137*, e20153542. [CrossRef] [PubMed]
46. Gelse, K.; Pöschl, E.; Aigner, T. Collagens-structure, function, and biosynthesis. *Adv. Drug Deliv. Rev.* **2003**, *28*, 1531–1546. [CrossRef] [PubMed]
47. Liu, F.; Xiong, Z.; Liu, Q.; Hu, J.; Li, W.; Zhang, N. CCL2A1 mutation (c.3508G>A) leads to avascular necrosis of the femoral head in a Chinese family: A case report. *Mol. Med. Rep.* **2018**, *18*, 254–260. [CrossRef]
48. Su, P.; Li, R.; Liu, S.; Zhou, Y.; Wang, X.; Patil, N.; Mow, C.S.; Mason, J.C.; Huang, D.; Wang, Y. Age at onset-dependent presentations of premature hip osteoarthritis, avascular necrosis of the femoral head, or Legg-Calvé-Perthes disease in a single family, consequent upon a p.Gly1170Ser mutation of COL2A1. *Arthritis Rheum.* **2008**, *58*, 1701–1706. [CrossRef]
49. Miyamoto, Y.; Matsuda, T.; Kitoh, H.; Haga, N.; Ohashi, H.; Nishimura, G.; Ikegawa, S. A recurrent mutation in type II collagen gene causes Legg-Calvé-Perthes disease in a Japanese family. *Hum. Genet.* **2007**, *121*, 625–629. [CrossRef]
50. Kannu, P.; Irving, M.; Aftimos, S.; Savarirayan, R. Two Novel COL2A1 Mutations Associated with a Legg-Calvé-Perthes Disease-like Presentation. *Clin. Orthop. Relat. Res.* **2011**, *469*, 1785–1790. [CrossRef]
51. Li, N.; Yu, J.; Cao, X.; Wu, Q.-Y.; Li, W.-W.; Li, T.-F.; Zhang, C.; Cui, Y.-X.; Li, X.-J.; Yin, Z.-M.; et al. A novel p. Gly630Ser mutation of COL2A1 in a Chinese family with presentations of Legg-Calvé-Perthes disease or avascular necrosis of the femoral head. *PLoS ONE* **2014**, *9*, e100505. [CrossRef] [PubMed]
52. Liu, Y.-F.; Chen, W.-M.; Lin, Y.-F.; Yang, R.-C.; Lin, M.-W.; Li, L.-H.; Chang, Y.-H.; Jou, Y.-S.; Lin, P.-Y.; Su, J.-S.; et al. Type II Collagen Gene Variants and Inherited Osteonecrosis of the Femoral Head. *New Engl. J. Med.* **2005**, *352*, 2294–2301. [CrossRef] [PubMed]
53. Kenet, G.; Ezra, E.; Wientroub, S.; Steinberg, D.M.; Rosenberg, N.; Waldman, D.; Hayek, S. Perthes' disease and the search for genetic associations: Collagen mutations, Gaucher's disease and thrombophilia. *J. Bone Joint Surg. Br.* **2008**, *90*, 1507–1511. [CrossRef]
54. Kim, H.; Randall, T.; Bian, H.; Jenkins, J.; Garces, A.; Bauss, F. Ibandronate for prevention of femoral head deformity after ischemic necrosis of the capital femoral epiphysis in immature pigs. *J. Bone Joint Surg. Am.* **2005**, *87*, 550–557. [CrossRef]
55. Little, D.G.; McDonald, M.; Sharpe, I.T.; Peat, R.; Williams, P.; McEvoy, T. Zoledronic acid improves femoral head sphericity in a rat model of perthes disease. *J. Orthop. Res.* **2005**, *23*, 862–868. [CrossRef]
56. Little, D.; Peat, R.A.; Mcevoy, A.; Williams, P.R.; Smith, E.J.; Baldock, P.A. Zoledronic acid treatment results in retention of femoral head structure after traumatic osteonecrosis in young Wistar rats. *J. Bone Miner. Res.* **2003**, *18*, 2016–2022. [CrossRef] [PubMed]
57. Kim, H.K. Pathophysiology and New Strategies for the Treatment of Legg-Calvé-Perthes Disease. *J. Bone Jt. Surg.* **2012**, *94*, 659–669. [CrossRef]

58. Jamil, K.; Zacharin, M.; Foster, B.; Donald, G.; Hassall, T.; Siafarikas, A.; Johnson, M.; Tham, E.; Whitewood, C.; Gebski, V.; et al. Protocol for a randomised control trial of bisphosphonate (zoledronic acid) treatment in childhood femoral head avascular necrosis due to Perthes disease. *BMJ Paediatr. Open* **2017**, *1*, e000084. [CrossRef]
59. Mihara, M.; Ohsugi, Y.; Kishimoto, T. Tocilizumab, a humanized anti-interleukin-6 receptor antibody, for treatment of rheumatoid arthritis. *Open Access Rheumatol. Res. Rev.* **2011**, *3*, 19–29. [CrossRef]
60. Terpos, E.; Fragiadaki, K.; Konsta, M.; Bratengeier, C.; Papatheodorou, A.; Sfikakis, P.P. Early effects of IL-6 receptor inhibition on bone homeostasis: A pilot study in women with rheumatoid arthritis. *Ann. Rheum. Dis.* **2011**, *29*, 921–925.
61. Garnero, P.; Thompson, E.; Woodworth, T.; Smolen, J.S. Rapid and sustained improvement in bone and cartilage turnover markers with the anti-interleukin-6 receptor inhibitor tocilizumab plus methotrexate in rheumatoid arthritis patients with an inadequate response to methotrexate: Results from a substudy of the multicenter double-blind, placebo-controlled trial of tocilizumab in inadequate responders to methotrexate alone. *Arthritis Rheum.* **2010**, *62*, 33–43.
62. Bielak, M.; Husmann, E.; Weyandt, N.; Haas, J.-P.; Hügle, B.; Horneff, G.; Neudorf, U.; Lutz, T.; Lilienthal, E.; Kallinich, T.; et al. IL-6 blockade in systemic juvenile idiopathic arthritis—Achievement of inactive disease and remission (data from the German AID-registry). *Pediatr. Rheumatol.* **2018**, *16*, 22. [CrossRef]
63. Ren, Y.; Deng, Z.; Gokani, V.; Kutschke, M.; Mitchell, T.W.; Aruwajoye, O.; Adapala, N.S.; Kamiya, N.; Abu-Amer, Y.; Kim, H.K. Anti-Interleukin-6 Therapy Decreases Hip Synovitis and Bone Resorption and Increases Bone Formation Following Ischemic Osteonecrosis of the Femoral Head. *J. Bone Miner. Res.* **2020**, *36*, 357–368. [CrossRef] [PubMed]
64. Le Blanc, K.; Götherström, C.; Ringdén, O.; Hassan, M.; McMahon, R.; Horwitz, E.; Anneren, G.; Axelsson, O.; Nunn, J.; Ewald, U.; et al. Fetal mesenchymal stem-cell engraftment in bone after in utero transplantation in a patient with severe osteogenesis imperfecta. *Transplantation* **2015**, *79*, 1607–1614. [CrossRef]
65. Götherström, C.; Westgren, M.; Shaw, S.S.; Åström, E.; Biswas, A.; Byers, P.H.; Mattar, C.N.; Graham, G.E.; Taslimi, J.; Ewald, U.; et al. Pre- and Postnatal Transplantation of Fetal Mesenchymal Stem Cells in Osteogenesis Imperfecta: A Two-Center Experience. *Stem Cells Transl. Med.* **2013**, *3*, 255–264. [CrossRef] [PubMed]
66. Horwitz, E.M.; Prockop, D.J.; Gordon, P.L.; Koo, W.W.K.; Fitzpatrick, L.A.; Neel, M.D.; McCarville, M.E.; Orchard, P.J.; Pyeritz, R.E.; Brenner, M.K. Clinical responses to bone marrow transplantation in children with severe osteogenesis imperfecta. *Blood* **2001**, *97*, 1227–1231. [CrossRef]
67. Wang, Z.-L.; He, R.-Z.; Tu, B.; He, J.-S.; Cao, X.; Xia, H.-S.; Ba, H.-L.; Wu, S.; Peng, C.; Xiong, K. Drilling Combined with Adipose-derived Stem Cells and Bone Morphogenetic Protein-2 to Treat Femoral Head Epiphyseal Necrosis in Juvenile Rabbits. *Curr. Med. Sci.* **2018**, *38*, 277–288. [CrossRef]
68. Wang, L.; Zhang, Y.; Li, H.; Hong, J.; Chen, X.; Li, M.; Bai, W.; Wang, J.; Liu, Y.; Wu, M. Clinical Observation of Employment of Umbilical Cord Derived Mesenchymal Stem Cell for Juvenile Idiopathic Arthritis Therapy. *Stem Cells Int.* **2016**, *2016*, 1–7. [CrossRef] [PubMed]

MDPI
St. Alban-Anlage 66
4052 Basel
Switzerland
www.mdpi.com

Diagnostics Editorial Office
E-mail: diagnostics@mdpi.com
www.mdpi.com/journal/diagnostics

www.ingramcontent.com/pod-product-compliance
Lightning Source LLC
Chambersburg PA
CBHW061621220926
51850CB00005B/560
* 9 7 8 3 7 2 5 8 0 5 8 7 7 *